Y0-AAX-927

Methods in Enzymology

Volume 332

REGULATORS AND EFFECTORS OF SMALL GTPases

Part F

Ras Family I

METHODS IN ENZYMOLOGY

EDITORS-IN-CHIEF

John N. Abelson   Melvin I. Simon

DIVISION OF BIOLOGY
CALIFORNIA INSTITUTE OF TECHNOLOGY
PASADENA, CALIFORNIA

FOUNDING EDITORS

Sidney P. Colowick and Nathan O. Kaplan

Methods in Enzymology

Volume 332

Regulators and Effectors of Small GTPases

Part F
Ras Family I

EDITED BY

W. E. Balch

THE SCRIPPS RESEARCH INSTITUTE
LA JOLLA, CALIFORNIA

Channing J. Der

LINEBERGER COMPREHENSIVE CANCER CENTER
THE UNIVERSITY OF NORTH CAROLINA AT CHAPEL HILL
CHAPEL HILL, NORTH CAROLINA

Alan Hall

UNIVERSITY COLLEGE LONDON, LONDON, ENGLAND

ACADEMIC PRESS
San Diego London Boston New York Sydney Tokyo Toronto

This book is printed on acid-free paper. 

Academic Press
A Harcourt Science and Technology Company
525 B Street, Suite 1900, San Diego, California 92101-4495, USA
http://www.academicpress.com

Academic Press
Harcourt Place, 32 Jamestown Road, London NW1 7BY, UK
http://www.academicpress.com

International Standard Book Number: 0-12-182233-8

PRINTED IN THE UNITED STATES OF AMERICA
01 02 03 04 05 06 07 SB 9 8 7 6 5 4 3 2 1

Table of Contents

Contributors to Volume 332 ix

Preface xiii

Volumes in Series xv

Section I. Protein Expression and Protein–Protein Interactions

1. Mammalian Expression Vectors for Ras Family Proteins: Generation and Use of Expression Constructs to Analyze Ras Family Function	James J. Fiordalisi, Ronald L. Johnson II, Aylin S. Ülkü, Channing J. Der, and Adrienne D. Cox	3
2. Protein Transduction: Delivery of Tat–GTPase Fusion Proteins into Mammalian Cells	Adamina Vocero-Akbani, Meena A. Chellaiah, Keith A. Hruska, and Steven F. Dowdy	36
3. Green Fluorescent Protein-Tagged Ras Proteins for Intracellular Localization	Edwin Choy and Mark Philips	50
4. Targeting Proteins to Membranes, Using Signal Sequences for Lipid Modifications	John T. Stickney, Michelle A. Booden, and Janice E. Buss	64
5. Targeting Proteins to Specific Cellular Compartments to Optimize Physiological Activity	Garabet G. Toby and Erica A. Golemis	77
6. Mapping Protein–Protein Interactions with Alkaline Phosphatase Fusion Proteins	Montarop Yamabhai and Brian K. Kay	88
7. Assays of Human Postprenylation Processing Enzymes	Yun-Jung Choi, Michael Niedbala, Mark Lynch, Marc Symons, Gideon Bollag, and Anne K. North	103
8. *In Vivo* Prenylation Analysis of Ras and Rho Proteins	Paul T. Kirschmeier, David Whyte, Oswald Wilson, W. Robert Bishop, and Jin-Keon Pai	115

9. Ras Interaction with RalGDS Effector Targets — SHINYA KOYAMA AND AKIRA KIKUCHI — 127

10. RAS Interaction with RIN1 Effector Target — YING WANG AND JOHN COLICELLI — 139

11. Ras and Rap1 Interaction with AF-6 Effector Target — BENJAMIN BOETTNER, CHRISTIAN HERRMANN, AND LINDA VAN AELST — 151

Section II. Screening Analyses

12. Analysis of Protein Kinase Specificity by Peptide Libraries and Prediction of *in Vivo* Substrates — ZHOU SONGYANG — 171

13. Peptide Library Screening for Determination of SH2 or Phosphotyrosine-Binding Domain Sequences — ZHOU SONGYANG AND DAN LIU — 183

14. Expression Cloning of Farnesylated Proteins — DOUGLAS A. ANDRES — 195

15. Expression Cloning to Identify Monomeric GTP-Binding Proteins by GTP Overlay — DOUGLAS A. ANDRES — 203

16. Retrovirus cDNA Expression Library Screening for Oncogenes — GWENDOLYN M. MAHON AND IAN P. WHITEHEAD — 211

17. Identification of Ras-Regulated Genes by Representational Difference Analysis — JANIEL M. SHIELDS, CHANNING J. DER, AND SCOTT POWERS — 221

18. Differential Display Analysis of Gene Expression Altered by *ras* Oncogene — HAKRYUL JO, YONG-JIG CHO, HONG ZHANG, AND PENG LIANG — 233

19. cDNA Array Analyses of K-Ras-Induced Gene Transcription — GASTON G. HABETS, MARC KNEPPER, JAINA SUMORTIN, YUN-JUNG CHOI, TAKEHIKO SASAZUKI, SENJI SHIRASAWA, AND GIDEON BOLLAG — 245

20. Ras Signaling Pathway for Analysis of Protein–Protein Interactions — AMI ARONHEIM — 260

21. Isolation of Effector-Selective Ras Mutants by Yeast Two-Hybrid Screening — KIRAN J. KAUR AND MICHAEL A. WHITE — 270

22. Two-Hybrid Dual Bait System to Discriminate Specificity of Protein Interactions in Small GTPases — ILYA G. SEREBRIISKII, OLGA V. MITINA, JONATHAN CHERNOFF, AND ERICA A. GOLEMIS — 277

23. Functional Proteomics Analysis of GTPase Signaling Networks — Gordon Alton, Adrienne D. Cox, L. Gerard Toussaint III, and John K. Westwick — 300

Section III. Analyses of Mitogen-Activated Protein Kinase Cascades

24. Analyzing JNK and p38 Mitogen-Activated Protein Kinase Activity — Alan J. Whitmarsh and Roger J. Davis — 319

25. Phospho-Specific Mitogen-Activated Protein Kinase Antibodies for ERK, JNK, and p38 Activation — Said A. Goueli and Bruce W. Jarvis — 337

26. Immunostaining for Activated Extracellular Signal-Regulated Kinases in Cells and Tissues — Daniel G. Gioeli, Maja Zecevic, and Michael J. Weber — 343

27. Dominant Negative Mutants of Mitogen-Activated Protein Kinase Pathway — M. Jane Arboleda, Derek Eberwein, Barbara Hibner, and John F. Lyons — 353

28. Scaffold Protein Regulation of Mitogen-Activated Protein Kinase Cascade — Andrew D. Catling, Scott T. Eblen, Hans J. Schaeffer, and Michael J. Weber — 368

29. Bacterial Expression of Activated Mitogen-Activated Protein Kinases — Julie L. Wilsbacher and Melanie H. Cobb — 387

30. Steroid Receptor Fusion Proteins for Conditional Activation of Raf–MEK–ERK Signaling Pathway — Martin McMahon — 401

31. Pharmacologic Inhibitors of MKK1 and MKK2 — Natalie G. Ahn, Theresa Stines Nahreini, Nicholas S. Tolwinski, and Katheryn A. Resing — 417

32. Analysis of Pharmacologic Inhibitors of Jun N-Terminal Kinases — Brion W. Murray, Brydon L. Bennett, and Dennis T. Sasaki — 432

Author Index 453

Subject Index 477

Contributors to Volume 332

Article numbers are in parentheses following the names of contributors.
Affiliations listed are current.

Natalie G. Ahn (31), *Department of Chemistry and Biochemistry, Howard Hughes Medical Institute, University of Colorado, Boulder, Colorado 80309*

Gordon Alton (23), *Celgene Corporation Signal Research Division, Department of Informatics and Functional Genomics, San Diego, California 92121*

Douglas A. Andres (14, 15), *Department of Biochemistry, University of Kentucky, Lexington, Kentucky 40536-0084*

M. Jane Arboleda (27), *Onyx Pharmaceuticals, Richmond, California 94806*

Ami Aronheim (20), *Department of Molecular Genetics, The B. Rappaport Faculty of Medicine, Israel Institute of Technology, Haifa 31096, Israel*

Brydon L. Bennett (32), *Signal Pharmaceuticals, Inc., San Diego, California 92121*

W. Robert Bishop (8), *Department of Tumor Biology, Schering Plough Research Institute, Kenilworth, New Jersey 07033*

Benjamin Boettner (11), *Cold Spring Harbor Laboratory, Cold Spring Harbor, New York 11724*

Gideon Bollag (7, 19), *Onyx Pharmaceuticals, Richmond, California 94806*

Michelle A. Booden (4), *Lineberger Comprehensive Cancer Center, CB-7295, University of North Carolina, Chapel Hill, North Carolina 27599*

Janice E. Buss (4), *Department of Biochemistry, Biophysics, and Molecular Biology, Iowa State University, Ames, Iowa 50011*

Andrew D. Catling (28), *Department of Microbiology and Cancer Center, University of Virginia Health Sciences Center, Charlottesville, Virginia 22908-0734*

Meena A. Chellaiah (2), *Renal Division, Barnes-Jewish Hospital, Washington University School of Medicine, St. Louis, Missouri 63110*

Jonathan Chernoff (22), *Division of Basic Science, Fox Chase Cancer Center, Philadelphia, Pennsylvania 19111*

Yong-Jig Cho (18), *Vanderbilt-Ingram Cancer Center, Nashville, Tennessee 37232-6838*

Yun-Jung Choi (7, 19), *Onyx Pharmaceuticals, Richmond, California 94806*

Edwin Choy (3), *Departments of Medicine and Cell Biology, New York University School of Medicine, New York, New York 10016*

Melanie H. Cobb (29), *Department of Pharmacology, University of Texas Southwestern Medical Center, Dallas, Texas 75235-9041*

John Colicelli (10), *Department of Biological Chemistry and Molecular Biology Institute, UCLA School of Medicine, Los Angeles, California 90095*

Adrienne D. Cox (1, 23), *Department of Radiation Oncology and Pharmacology, University of North Carolina School of Medicine, Chapel Hill, North Carolina 27599*

Roger J. Davis (24), *Howard Hughes Medical Institute, Department of Biochemistry and Molecular Biology, University of Massachusetts Medical School, Program in Molecular Medicine, Worcester, Massachusetts 01605*

Channing J. Der (1, 17), *Lineberger Comprehensive Cancer Center, Department of Pharmacology, University of North Carolina, Chapel Hill, North Carolina 27599*

Steven F. Dowdy (2), *Departments of Pathology and Medicine, Howard Hughes Medical Institute, Washington University School of Medicine, St. Louis, Missouri 63110*

DEREK EBERWEIN (27), *Bayer Corporation, West Haven, Connecticut 06516-4175*

SCOTT T. EBLEN (28), *Department of Microbiology and Cancer Center, University of Virginia Health Sciences Center, Charlottesville, Virginia 22908-0734*

JAMES J. FIORDALISI (1), *Departments of Radiation, Oncology, and Pharmacology, University of North Carolina, Chapel Hill, North Carolina 27599*

DANIEL G. GIOELI (26), *Department of Microbiology and Cancer Center, University of Virginia Health Sciences Center, Charlottesville, Virginia 22908*

ERICA A. GOLEMIS (5, 22), *Division of Basic Science, Fox Chase Cancer Center, Philadelphia, Pennsylvania 19111*

SAID A. GOUELI (25), *Signal Transduction Group, Research and Development Department, Promega Corporation, Madison, Wisconsin 53711, and Department of Pathology and Laboratory Medicine, University of Wisconsin School of Medicine, Madison, Wisconsin 53711*

GASTON G. HABETS (19), *Onyx Pharmaceuticals, Richmond, California 94806*

CHRISTIAN HERRMANN (11), *Max Planck Institute for Molecular Physiology, 44227 Dortmund, Germany*

BARBARA HIBNER (27), *Bayer Corporation, West Haven, Connecticut 06516-4175*

KEITH A. HRUSKA (2), *Renal Division, Barnes-Jewish Hospital, Washington University School of Medicine, St. Louis, Missouri 63110*

BRUCE W. JARVIS (25), *Signal Transduction Group, Research and Development Department, Promega Corporation, Madison, Wisconsin 53711*

HAKRYUL JO (18), *Vanderbilt-Ingram Cancer Center, Nashville, Tennessee 37232-6838*

RONALD L. JOHNSON II (1), *Departments of Radiation, Oncology, and Pharmacology, University of North Carolina at Chapel Hill, Chapel Hill, North Carolina 27599*

KIRAN J. KAUR (21), *Department of Cell Biology, University of Texas Southwestern Medical Center, Dallas, Texas 75390*

BRIAN K. KAY (6), *Department of Pharmacology, University of Wisconsin, Madison, Wisconsin 53706-1532*

AKIRA KIKUCHI (9), *Department of Biochemistry, Hiroshima University School of Medicine, Hiroshima 734-8551, Japan*

PAUL T. KIRSCHMEIER (8), *Department of Tumor Biology, Schering Plough Research Institute, Kenilworth, New Jersey 07033*

MARC KNEPPER (19), *Advanced Medicine, Inc., San Francisco, California 94080*

SHINYA KOYAMA (9), *Department of Biochemistry, Hiroshima University School of Medicine, Hiroshima 734-8551, Japan*

PENG LIANG (18), *Vanderbilt-Ingram Cancer Center, Nashville, Tennessee 37232-6838*

DAN LIU (13), *Verna and Marrs McLean Department of Biochemistry and Molecular Biology, Baylor College of Medicine, Houston, Texas 77030*

MARK LYNCH (7), *Bayer Research Center, West Haven, Connecticut 06516*

JOHN F. LYONS (27), *Onyx Pharmaceuticals, Richmond, California 94806*

GWENDOLYN M. MAHON (16), *Department of Microbiology and Molecular Genetics, UMDNJ–New Jersey Medical School, Newark, New Jersey 07103-2714*

MARTIN MCMAHON (30), *Cancer Research Institute and Department of Cellular and Molecular Pharmacology, University of California San Francisco/Mt. Zion Comprehensive Cancer Center, San Francisco, California 94115*

OLGA V. MITINA (22), *Department of Molecular Biology and Medical Biotechnology, Russian State Medical University, Moscow, Russia*

BRION W. MURRAY (32), *Agouron Pharmaceuticals, San Diego, California 92121-1408*

THERESA STINES NAHREINI (31), *Department of Chemistry and Biochemistry, Howard Hughes Medical Institute, University of Colorado, Boulder, Colorado 80309*

MICHAEL NIEDBALA (7), *Bayer Research Center, West Haven, Connecticut 06516*

ANNE K. NORTH (7), *Onyx Pharmaceuticals, Richmond, California 94806*

JIN-KEON PAI (8), *Department of Tumor Biology, Schering Plough Research Institute, Kenilworth, New Jersey 07033*

MARK PHILIPS (3), *Departments of Medicine and Cell Biology, New York University School of Medicine, New York, New York 10016*

SCOTT POWERS (17), *Tularik Genomics, Greenlawn, New York 11740*

KATHERYN A. RESING (31), *Department of Chemistry and Biochemistry, University of Colorado, Boulder, Colorado 80309*

DENNIS T. SASAKI (32), *Signal Pharmaceuticals, Inc., San Diego, California 92121*

TAKEHIKO SASAZUKI (19), *Medical Institute of Bioregulation, Kyushu University, Fukuoka 812, Japan*

HANS J. SCHAEFFER (28), *MDC, Gruppe W. Birchmeier, 13125 Berlin, Germany*

ILYA G. SEREBRIISKII (22), *Division of Basic Science, Fox Chase Cancer Center, Philadelphia, Pennsylvania 19111*

JANIEL M. SHIELDS (17), *Department of Pharmacology, Lineberger Comprehensive Cancer Center, University of North Carolina, Chapel Hill, North Carolina 27599-7295*

SENJI SHIRASAWA (19), *Medical Institute of Bioregulation, Kyushu University, Fukuoka 812, Japan*

ZHOU SONGYANG (12, 13), *Verna and Marrs McLean Department of Biochemistry and Molecular Biology, Baylor College of Medicine, Houston, Texas 77030*

JOHN T. STICKNEY (4), *Department of Cell Biology, Neurobiology, and Anatomy, University of Cincinnati Medical Center, Cincinnati, Ohio 45267-0521*

JAINA SUMORTIN (19), *Onyx Pharmaceuticals, Richmond, California 94806*

MARC SYMONS (7), *The Picower Institute for Medical Research, Manhasset, New York 11030*

GARABET G. TOBY (5), *Division of Basic Science, Fox Chase Cancer Center, Philadelphia, Pennsylvania 19111, and Cell and Molecular Biology Group, University of Pennsylvania School of Medicine, Philadelphia, Pennsylvania 19104-6064*

NICHOLAS S. TOLWINSKI (31), *The Graduate College, Princeton University, Princeton, New Jersey 08544*

L. GERARD TOUSSAINT III (23), *Distinguished Medical Scholar Program, University of North Carolina School of Medicine, Chapel Hill, North Carolina 27599*

AYLIN S. ÜLKÜ (1), *Department of Pharmacology, University of North Carolina at Chapel Hill, Chapel Hill, North Carolina 27599*

LINDA VAN AELST (11), *Cold Spring Harbor Laboratory, Cold Spring Harbor, New York 11724*

ADAMINA VOCERO-AKBANI (2), *Departments of Pathology and Medicine, Howard Hughes Medical Institute, Washington University School of Medicine, St. Louis, Missouri 63110*

YING WANG (10), *Department of Biological Chemistry and Molecular Biology Institute, UCLA School of Medicine, Los Angeles, California 90095*

MICHAEL J. WEBER (26, 28), *Department of Microbiology and Cancer Center, University of Virginia Health Sciences Center, Charlottesville, Virginia 22908-0734*

JOHN K. WESTWICK (23), *Celgene Corporation Signal Research Division, Department of Cell Signaling, San Diego, California 92121*

MICHAEL A. WHITE (21), *Department of Cell Biology, University of Texas Southwestern Medical Center, Dallas, Texas 75390*

IAN P. WHITEHEAD (16), *Department of Microbiology and Molecular Genetics, UMDNJ–New Jersey Medical School, Newark, New Jersey 07103-2714*

ALAN J. WHITMARSH (24), *Howard Hughes Medical Institute, Department of Biochemistry and Molecular Biology, University of Massachusetts Medical School, Program in*

Molecular Medicine, Worcester, Massachusetts 01605

DAVID WHYTE (8), *Sugen Inc., South San Francisco, California 94080*

JULIE L. WILSBACHER (29), *Department of Pharmacology, University of Texas Southwestern Medical Center, Dallas, Texas 75235-9041*

OSWALD WILSON (8), *Department of Tumor Biology, Schering Plough Research Institute, Kenilworth, New Jersey 07033*

MONTAROP YAMABHAI (6), *School of Biotechnology, Suranaree University of Technology, Institute of Agricultural Technology, Nakhon Ratchasima 30000, Thailand*

MAJA ZECEVIC (26), *Department of Microbiology and Cancer Center, University of Virginia Health Sciences Center, Charlottesville, Virginia 22908*

HONG ZHANG (18), *Vanderbilt-Ingram Cancer Center, Nashville, Tennessee 37232-6838*

Preface

As with the Rho and Rab branches of the Ras superfamily of small GTPases, research interest in the Ras branch has continued to expand dramatically into new areas and to embrace new themes since the last *Methods in Enzymology* Volume 255 on Ras GTPases was published in 1995. First, the Ras branch has expanded beyond the original Ras, Rap, and Ral members. New members include M-Ras, Rheb, Rin, and Rit. Second, the signaling activities of Ras are much more diverse and complex than appreciated previously. In particular, while the Raf/MEK/ERK kinase cascade remains a key signaling pathway activated by Ras, it is now appreciated that an increasing number of non-Raf effectors also mediate Ras family protein function. Third, it is increasingly clear that the cellular functions regulated by Ras go beyond regulation of cell proliferation, and involve regulation of senescence and cell survival and induction of tumor cell invasion, metastasis, and angiogenesis. Fourth, another theme that has emerged is regulatory cross talk among Ras family proteins, including both GTPase signaling cascades that link signaling from one family member to another, as well as the use of shared regulators and effectors by different family members.

Concurrent with the expanded complexity of Ras family biology, biochemistry, and signaling have been the development and application of a wider array of methodology to study Ras family function. While some are simply improved methods to study old questions, many others involve novel approaches to study aspects of Ras family protein function not studied previously. In particular, the emerging application of techniques to study Ras regulation of gene and protein expression represents an important direction for current and future studies. Consequently, *Methods in Enzymology,* Volumes 332 and 333 cover many of the new techniques that have emerged during the past five years.

We are grateful for the efforts of all our colleagues who contributed to these volumes. We are indebted to them for sharing their expertise and experiences, as well as their time, in compiling this comprehensive series of chapters. In particular, we hope these volumes will provide valuable references and sources of information that will facilitate the efforts of newly incoming researchers to the study of the Ras family of small GTPases.

Channing J. Der
Alan Hall
William E. Balch

METHODS IN ENZYMOLOGY

VOLUME I. Preparation and Assay of Enzymes
Edited by SIDNEY P. COLOWICK AND NATHAN O. KAPLAN

VOLUME II. Preparation and Assay of Enzymes
Edited by SIDNEY P. COLOWICK AND NATHAN O. KAPLAN

VOLUME III. Preparation and Assay of Substrates
Edited by SIDNEY P. COLOWICK AND NATHAN O. KAPLAN

VOLUME IV. Special Techniques for the Enzymologist
Edited by SIDNEY P. COLOWICK AND NATHAN O. KAPLAN

VOLUME V. Preparation and Assay of Enzymes
Edited by SIDNEY P. COLOWICK AND NATHAN O. KAPLAN

VOLUME VI. Preparation and Assay of Enzymes (*Continued*)
Preparation and Assay of Substrates
Special Techniques
Edited by SIDNEY P. COLOWICK AND NATHAN O. KAPLAN

VOLUME VII. Cumulative Subject Index
Edited by SIDNEY P. COLOWICK AND NATHAN O. KAPLAN

VOLUME VIII. Complex Carbohydrates
Edited by ELIZABETH F. NEUFELD AND VICTOR GINSBURG

VOLUME IX. Carbohydrate Metabolism
Edited by WILLIS A. WOOD

VOLUME X. Oxidation and Phosphorylation
Edited by RONALD W. ESTABROOK AND MAYNARD E. PULLMAN

VOLUME XI. Enzyme Structure
Edited by C. H. W. HIRS

VOLUME XII. Nucleic Acids (Parts A and B)
Edited by LAWRENCE GROSSMAN AND KIVIE MOLDAVE

VOLUME XIII. Citric Acid Cycle
Edited by J. M. LOWENSTEIN

VOLUME XIV. Lipids
Edited by J. M. LOWENSTEIN

VOLUME XV. Steroids and Terpenoids
Edited by RAYMOND B. CLAYTON

VOLUME XVI. Fast Reactions
Edited by KENNETH KUSTIN

VOLUME XVII. Metabolism of Amino Acids and Amines (Parts A and B)
Edited by HERBERT TABOR AND CELIA WHITE TABOR

VOLUME XVIII. Vitamins and Coenzymes (Parts A, B, and C)
Edited by DONALD B. MCCORMICK AND LEMUEL D. WRIGHT

VOLUME XIX. Proteolytic Enzymes
Edited by GERTRUDE E. PERLMANN AND LASZLO LORAND

VOLUME XX. Nucleic Acids and Protein Synthesis (Part C)
Edited by KIVIE MOLDAVE AND LAWRENCE GROSSMAN

VOLUME XXI. Nucleic Acids (Part D)
Edited by LAWRENCE GROSSMAN AND KIVIE MOLDAVE

VOLUME XXII. Enzyme Purification and Related Techniques
Edited by WILLIAM B. JAKOBY

VOLUME XXIII. Photosynthesis (Part A)
Edited by ANTHONY SAN PIETRO

VOLUME XXIV. Photosynthesis and Nitrogen Fixation (Part B)
Edited by ANTHONY SAN PIETRO

VOLUME XXV. Enzyme Structure (Part B)
Edited by C. H. W. HIRS AND SERGE N. TIMASHEFF

VOLUME XXVI. Enzyme Structure (Part C)
Edited by C. H. W. HIRS AND SERGE N. TIMASHEFF

VOLUME XXVII. Enzyme Structure (Part D)
Edited by C. H. W. HIRS AND SERGE N. TIMASHEFF

VOLUME XXVIII. Complex Carbohydrates (Part B)
Edited by VICTOR GINSBURG

VOLUME XXIX. Nucleic Acids and Protein Synthesis (Part E)
Edited by LAWRENCE GROSSMAN AND KIVIE MOLDAVE

VOLUME XXX. Nucleic Acids and Protein Synthesis (Part F)
Edited by KIVIE MOLDAVE AND LAWRENCE GROSSMAN

VOLUME XXXI. Biomembranes (Part A)
Edited by SIDNEY FLEISCHER AND LESTER PACKER

VOLUME XXXII. Biomembranes (Part B)
Edited by SIDNEY FLEISCHER AND LESTER PACKER

VOLUME XXXIII. Cumulative Subject Index Volumes I–XXX
Edited by MARTHA G. DENNIS AND EDWARD A. DENNIS

VOLUME XXXIV. Affinity Techniques (Enzyme Purification: Part B)
Edited by WILLIAM B. JAKOBY AND MEIR WILCHEK

VOLUME XXXV. Lipids (Part B)
Edited by JOHN M. LOWENSTEIN

VOLUME XXXVI. Hormone Action (Part A: Steroid Hormones)
Edited by BERT W. O'MALLEY AND JOEL G. HARDMAN

VOLUME XXXVII. Hormone Action (Part B: Peptide Hormones)
Edited by BERT W. O'MALLEY AND JOEL G. HARDMAN

VOLUME XXXVIII. Hormone Action (Part C: Cyclic Nucleotides)
Edited by JOEL G. HARDMAN AND BERT W. O'MALLEY

VOLUME XXXIX. Hormone Action (Part D: Isolated Cells, Tissues, and Organ Systems)
Edited by JOEL G. HARDMAN AND BERT W. O'MALLEY

VOLUME XL. Hormone Action (Part E: Nuclear Structure and Function)
Edited by BERT W. O'MALLEY AND JOEL G. HARDMAN

VOLUME XLI. Carbohydrate Metabolism (Part B)
Edited by W. A. WOOD

VOLUME XLII. Carbohydrate Metabolism (Part C)
Edited by W. A. WOOD

VOLUME XLIII. Antibiotics
Edited by JOHN H. HASH

VOLUME XLIV. Immobilized Enzymes
Edited by KLAUS MOSBACH

VOLUME XLV. Proteolytic Enzymes (Part B)
Edited by LASZLO LORAND

VOLUME XLVI. Affinity Labeling
Edited by WILLIAM B. JAKOBY AND MEIR WILCHEK

VOLUME XLVII. Enzyme Structure (Part E)
Edited by C. H. W. HIRS AND SERGE N. TIMASHEFF

VOLUME XLVIII. Enzyme Structure (Part F)
Edited by C. H. W. HIRS AND SERGE N. TIMASHEFF

VOLUME XLIX. Enzyme Structure (Part G)
Edited by C. H. W. HIRS AND SERGE N. TIMASHEFF

VOLUME L. Complex Carbohydrates (Part C)
Edited by VICTOR GINSBURG

VOLUME LI. Purine and Pyrimidine Nucleotide Metabolism
Edited by PATRICIA A. HOFFEE AND MARY ELLEN JONES

VOLUME LII. Biomembranes (Part C: Biological Oxidations)
Edited by SIDNEY FLEISCHER AND LESTER PACKER

VOLUME LIII. Biomembranes (Part D: Biological Oxidations)
Edited by SIDNEY FLEISCHER AND LESTER PACKER

VOLUME LIV. Biomembranes (Part E: Biological Oxidations)
Edited by SIDNEY FLEISCHER AND LESTER PACKER

VOLUME LV. Biomembranes (Part F: Bioenergetics)
Edited by SIDNEY FLEISCHER AND LESTER PACKER

VOLUME LVI. Biomembranes (Part G: Bioenergetics)
Edited by SIDNEY FLEISCHER AND LESTER PACKER

VOLUME LVII. Bioluminescence and Chemiluminescence
Edited by MARLENE A. DELUCA

VOLUME LVIII. Cell Culture
Edited by WILLIAM B. JAKOBY AND IRA PASTAN

VOLUME LIX. Nucleic Acids and Protein Synthesis (Part G)
Edited by KIVIE MOLDAVE AND LAWRENCE GROSSMAN

VOLUME LX. Nucleic Acids and Protein Synthesis (Part H)
Edited by KIVIE MOLDAVE AND LAWRENCE GROSSMAN

VOLUME 61. Enzyme Structure (Part H)
Edited by C. H. W. HIRS AND SERGE N. TIMASHEFF

VOLUME 62. Vitamins and Coenzymes (Part D)
Edited by DONALD B. MCCORMICK AND LEMUEL D. WRIGHT

VOLUME 63. Enzyme Kinetics and Mechanism (Part A: Initial Rate and Inhibitor Methods)
Edited by DANIEL L. PURICH

VOLUME 64. Enzyme Kinetics and Mechanism (Part B: Isotopic Probes and Complex Enzyme Systems)
Edited by DANIEL L. PURICH

VOLUME 65. Nucleic Acids (Part I)
Edited by LAWRENCE GROSSMAN AND KIVIE MOLDAVE

VOLUME 66. Vitamins and Coenzymes (Part E)
Edited by DONALD B. MCCORMICK AND LEMUEL D. WRIGHT

VOLUME 67. Vitamins and Coenzymes (Part F)
Edited by DONALD B. MCCORMICK AND LEMUEL D. WRIGHT

VOLUME 68. Recombinant DNA
Edited by RAY WU

VOLUME 69. Photosynthesis and Nitrogen Fixation (Part C)
Edited by ANTHONY SAN PIETRO

VOLUME 70. Immunochemical Techniques (Part A)
Edited by HELEN VAN VUNAKIS AND JOHN J. LANGONE

VOLUME 71. Lipids (Part C)
Edited by JOHN M. LOWENSTEIN

VOLUME 72. Lipids (Part D)
Edited by JOHN M. LOWENSTEIN

VOLUME 73. Immunochemical Techniques (Part B)
Edited by JOHN J. LANGONE AND HELEN VAN VUNAKIS

VOLUME 74. Immunochemical Techniques (Part C)
Edited by JOHN J. LANGONE AND HELEN VAN VUNAKIS

VOLUME 75. Cumulative Subject Index Volumes XXXI, XXXII, XXXIV–LX
Edited by EDWARD A. DENNIS AND MARTHA G. DENNIS

VOLUME 76. Hemoglobins
Edited by ERALDO ANTONINI, LUIGI ROSSI-BERNARDI, AND EMILIA CHIANCONE

VOLUME 77. Detoxication and Drug Metabolism
Edited by WILLIAM B. JAKOBY

VOLUME 78. Interferons (Part A)
Edited by SIDNEY PESTKA

VOLUME 79. Interferons (Part B)
Edited by SIDNEY PESTKA

VOLUME 80. Proteolytic Enzymes (Part C)
Edited by LASZLO LORAND

VOLUME 81. Biomembranes (Part H: Visual Pigments and Purple Membranes, I)
Edited by LESTER PACKER

VOLUME 82. Structural and Contractile Proteins (Part A: Extracellular Matrix)
Edited by LEON W. CUNNINGHAM AND DIXIE W. FREDERIKSEN

VOLUME 83. Complex Carbohydrates (Part D)
Edited by VICTOR GINSBURG

VOLUME 84. Immunochemical Techniques (Part D: Selected Immunoassays)
Edited by JOHN J. LANGONE AND HELEN VAN VUNAKIS

VOLUME 85. Structural and Contractile Proteins (Part B: The Contractile Apparatus and the Cytoskeleton)
Edited by DIXIE W. FREDERIKSEN AND LEON W. CUNNINGHAM

VOLUME 86. Prostaglandins and Arachidonate Metabolites
Edited by WILLIAM E. M. LANDS AND WILLIAM L. SMITH

VOLUME 87. Enzyme Kinetics and Mechanism (Part C: Intermediates, Stereochemistry, and Rate Studies)
Edited by DANIEL L. PURICH

VOLUME 88. Biomembranes (Part I: Visual Pigments and Purple Membranes, II)
Edited by LESTER PACKER

VOLUME 89. Carbohydrate Metabolism (Part D)
Edited by WILLIS A. WOOD

VOLUME 90. Carbohydrate Metabolism (Part E)
Edited by WILLIS A. WOOD

VOLUME 91. Enzyme Structure (Part I)
Edited by C. H. W. HIRS AND SERGE N. TIMASHEFF

VOLUME 92. Immunochemical Techniques (Part E: Monoclonal Antibodies and General Immunoassay Methods)
Edited by JOHN J. LANGONE AND HELEN VAN VUNAKIS

VOLUME 93. Immunochemical Techniques (Part F: Conventional Antibodies, Fc Receptors, and Cytotoxicity)
Edited by JOHN J. LANGONE AND HELEN VAN VUNAKIS

VOLUME 94. Polyamines
Edited by HERBERT TABOR AND CELIA WHITE TABOR

VOLUME 95. Cumulative Subject Index Volumes 61–74, 76–80
Edited by EDWARD A. DENNIS AND MARTHA G. DENNIS

VOLUME 96. Biomembranes [Part J: Membrane Biogenesis: Assembly and Targeting (General Methods; Eukaryotes)]
Edited by SIDNEY FLEISCHER AND BECCA FLEISCHER

VOLUME 97. Biomembranes [Part K: Membrane Biogenesis: Assembly and Targeting (Prokaryotes, Mitochondria, and Chloroplasts)]
Edited by SIDNEY FLEISCHER AND BECCA FLEISCHER

VOLUME 98. Biomembranes (Part L: Membrane Biogenesis: Processing and Recycling)
Edited by SIDNEY FLEISCHER AND BECCA FLEISCHER

VOLUME 99. Hormone Action (Part F: Protein Kinases)
Edited by JACKIE D. CORBIN AND JOEL G. HARDMAN

VOLUME 100. Recombinant DNA (Part B)
Edited by RAY WU, LAWRENCE GROSSMAN, AND KIVIE MOLDAVE

VOLUME 101. Recombinant DNA (Part C)
Edited by RAY WU, LAWRENCE GROSSMAN, AND KIVIE MOLDAVE

VOLUME 102. Hormone Action (Part G: Calmodulin and Calcium-Binding Proteins)
Edited by ANTHONY R. MEANS AND BERT W. O'MALLEY

VOLUME 103. Hormone Action (Part H: Neuroendocrine Peptides)
Edited by P. MICHAEL CONN

VOLUME 104. Enzyme Purification and Related Techniques (Part C)
Edited by WILLIAM B. JAKOBY

VOLUME 105. Oxygen Radicals in Biological Systems
Edited by LESTER PACKER

VOLUME 106. Posttranslational Modifications (Part A)
Edited by FINN WOLD AND KIVIE MOLDAVE

VOLUME 107. Posttranslational Modifications (Part B)
Edited by FINN WOLD AND KIVIE MOLDAVE

VOLUME 108. Immunochemical Techniques (Part G: Separation and Characterization of Lymphoid Cells)
Edited by GIOVANNI DI SABATO, JOHN J. LANGONE, AND HELEN VAN VUNAKIS

VOLUME 109. Hormone Action (Part I: Peptide Hormones)
Edited by LUTZ BIRNBAUMER AND BERT W. O'MALLEY

VOLUME 110. Steroids and Isoprenoids (Part A)
Edited by JOHN H. LAW AND HANS C. RILLING

VOLUME 111. Steroids and Isoprenoids (Part B)
Edited by JOHN H. LAW AND HANS C. RILLING

VOLUME 112. Drug and Enzyme Targeting (Part A)
Edited by KENNETH J. WIDDER AND RALPH GREEN

VOLUME 113. Glutamate, Glutamine, Glutathione, and Related Compounds
Edited by ALTON MEISTER

VOLUME 114. Diffraction Methods for Biological Macromolecules (Part A)
Edited by HAROLD W. WYCKOFF, C. H. W. HIRS, AND SERGE N. TIMASHEFF

VOLUME 115. Diffraction Methods for Biological Macromolecules (Part B)
Edited by HAROLD W. WYCKOFF, C. H. W. HIRS, AND SERGE N. TIMASHEFF

VOLUME 116. Immunochemical Techniques (Part H: Effectors and Mediators of Lymphoid Cell Functions)
Edited by GIOVANNI DI SABATO, JOHN J. LANGONE, AND HELEN VAN VUNAKIS

VOLUME 117. Enzyme Structure (Part J)
Edited by C. H. W. HIRS AND SERGE N. TIMASHEFF

VOLUME 118. Plant Molecular Biology
Edited by ARTHUR WEISSBACH AND HERBERT WEISSBACH

VOLUME 119. Interferons (Part C)
Edited by SIDNEY PESTKA

VOLUME 120. Cumulative Subject Index Volumes 81–94, 96–101

VOLUME 121. Immunochemical Techniques (Part I: Hybridoma Technology and Monoclonal Antibodies)
Edited by JOHN J. LANGONE AND HELEN VAN VUNAKIS

VOLUME 122. Vitamins and Coenzymes (Part G)
Edited by FRANK CHYTIL AND DONALD B. MCCORMICK

VOLUME 123. Vitamins and Coenzymes (Part H)
Edited by FRANK CHYTIL AND DONALD B. MCCORMICK

VOLUME 124. Hormone Action (Part J: Neuroendocrine Peptides)
Edited by P. MICHAEL CONN

VOLUME 125. Biomembranes (Part M: Transport in Bacteria, Mitochondria, and Chloroplasts: General Approaches and Transport Systems)
Edited by SIDNEY FLEISCHER AND BECCA FLEISCHER

VOLUME 126. Biomembranes (Part N: Transport in Bacteria, Mitochondria, and Chloroplasts: Protonmotive Force)
Edited by SIDNEY FLEISCHER AND BECCA FLEISCHER

VOLUME 127. Biomembranes (Part O: Protons and Water: Structure and Translocation)
Edited by LESTER PACKER

VOLUME 128. Plasma Lipoproteins (Part A: Preparation, Structure, and Molecular Biology)
Edited by JERE P. SEGREST AND JOHN J. ALBERS

VOLUME 129. Plasma Lipoproteins (Part B: Characterization, Cell Biology, and Metabolism)
Edited by JOHN J. ALBERS AND JERE P. SEGREST

VOLUME 130. Enzyme Structure (Part K)
Edited by C. H. W. HIRS AND SERGE N. TIMASHEFF

VOLUME 131. Enzyme Structure (Part L)
Edited by C. H. W. HIRS AND SERGE N. TIMASHEFF

VOLUME 132. Immunochemical Techniques (Part J: Phagocytosis and Cell-Mediated Cytotoxicity)
Edited by GIOVANNI DI SABATO AND JOHANNES EVERSE

VOLUME 133. Bioluminescence and Chemiluminescence (Part B)
Edited by MARLENE DELUCA AND WILLIAM D. MCELROY

VOLUME 134. Structural and Contractile Proteins (Part C: The Contractile Apparatus and the Cytoskeleton)
Edited by RICHARD B. VALLEE

VOLUME 135. Immobilized Enzymes and Cells (Part B)
Edited by KLAUS MOSBACH

VOLUME 136. Immobilized Enzymes and Cells (Part C)
Edited by KLAUS MOSBACH

VOLUME 137. Immobilized Enzymes and Cells (Part D)
Edited by KLAUS MOSBACH

VOLUME 138. Complex Carbohydrates (Part E)
Edited by VICTOR GINSBURG

VOLUME 139. Cellular Regulators (Part A: Calcium- and Calmodulin-Binding Proteins)
Edited by ANTHONY R. MEANS AND P. MICHAEL CONN

VOLUME 140. Cumulative Subject Index Volumes 102–119, 121–134

VOLUME 141. Cellular Regulators (Part B: Calcium and Lipids)
Edited by P. MICHAEL CONN AND ANTHONY R. MEANS

VOLUME 142. Metabolism of Aromatic Amino Acids and Amines
Edited by SEYMOUR KAUFMAN

VOLUME 143. Sulfur and Sulfur Amino Acids
Edited by WILLIAM B. JAKOBY AND OWEN GRIFFITH

VOLUME 144. Structural and Contractile Proteins (Part D: Extracellular Matrix)
Edited by LEON W. CUNNINGHAM

VOLUME 145. Structural and Contractile Proteins (Part E: Extracellular Matrix)
Edited by LEON W. CUNNINGHAM

VOLUME 146. Peptide Growth Factors (Part A)
Edited by DAVID BARNES AND DAVID A. SIRBASKU

VOLUME 147. Peptide Growth Factors (Part B)
Edited by DAVID BARNES AND DAVID A. SIRBASKU

VOLUME 148. Plant Cell Membranes
Edited by LESTER PACKER AND ROLAND DOUCE

VOLUME 149. Drug and Enzyme Targeting (Part B)
Edited by RALPH GREEN AND KENNETH J. WIDDER

VOLUME 150. Immunochemical Techniques (Part K: *In Vitro* Models of B and T Cell Functions and Lymphoid Cell Receptors)
Edited by GIOVANNI DI SABATO

VOLUME 151. Molecular Genetics of Mammalian Cells
Edited by MICHAEL M. GOTTESMAN

VOLUME 152. Guide to Molecular Cloning Techniques
Edited by SHELBY L. BERGER AND ALAN R. KIMMEL

VOLUME 153. Recombinant DNA (Part D)
Edited by RAY WU AND LAWRENCE GROSSMAN

VOLUME 154. Recombinant DNA (Part E)
Edited by RAY WU AND LAWRENCE GROSSMAN

VOLUME 155. Recombinant DNA (Part F)
Edited by RAY WU

VOLUME 156. Biomembranes (Part P: ATP-Driven Pumps and Related Transport: The Na,K-Pump)
Edited by SIDNEY FLEISCHER AND BECCA FLEISCHER

VOLUME 157. Biomembranes (Part Q: ATP-Driven Pumps and Related Transport: Calcium, Proton, and Potassium Pumps)
Edited by SIDNEY FLEISCHER AND BECCA FLEISCHER

VOLUME 158. Metalloproteins (Part A)
Edited by JAMES F. RIORDAN AND BERT L. VALLEE

VOLUME 159. Initiation and Termination of Cyclic Nucleotide Action
Edited by JACKIE D. CORBIN AND ROGER A. JOHNSON

VOLUME 160. Biomass (Part A: Cellulose and Hemicellulose)
Edited by WILLIS A. WOOD AND SCOTT T. KELLOGG

VOLUME 161. Biomass (Part B: Lignin, Pectin, and Chitin)
Edited by WILLIS A. WOOD AND SCOTT T. KELLOGG

VOLUME 162. Immunochemical Techniques (Part L: Chemotaxis and Inflammation)
Edited by GIOVANNI DI SABATO

VOLUME 163. Immunochemical Techniques (Part M: Chemotaxis and Inflammation)
Edited by GIOVANNI DI SABATO

VOLUME 164. Ribosomes
Edited by HARRY F. NOLLER, JR., AND KIVIE MOLDAVE

VOLUME 165. Microbial Toxins: Tools for Enzymology
Edited by SIDNEY HARSHMAN

VOLUME 166. Branched-Chain Amino Acids
Edited by ROBERT HARRIS AND JOHN R. SOKATCH

VOLUME 167. Cyanobacteria
Edited by LESTER PACKER AND ALEXANDER N. GLAZER

VOLUME 168. Hormone Action (Part K: Neuroendocrine Peptides)
Edited by P. MICHAEL CONN

VOLUME 169. Platelets: Receptors, Adhesion, Secretion (Part A)
Edited by JACEK HAWIGER

VOLUME 170. Nucleosomes
Edited by PAUL M. WASSARMAN AND ROGER D. KORNBERG

VOLUME 171. Biomembranes (Part R: Transport Theory: Cells and Model Membranes)
Edited by SIDNEY FLEISCHER AND BECCA FLEISCHER

VOLUME 172. Biomembranes (Part S: Transport: Membrane Isolation and Characterization)
Edited by SIDNEY FLEISCHER AND BECCA FLEISCHER

VOLUME 173. Biomembranes [Part T: Cellular and Subcellular Transport: Eukaryotic (Nonepithelial) Cells]
Edited by SIDNEY FLEISCHER AND BECCA FLEISCHER

VOLUME 174. Biomembranes [Part U: Cellular and Subcellular Transport: Eukaryotic (Nonepithelial) Cells]
Edited by SIDNEY FLEISCHER AND BECCA FLEISCHER

VOLUME 175. Cumulative Subject Index Volumes 135–139, 141–167

VOLUME 176. Nuclear Magnetic Resonance (Part A: Spectral Techniques and Dynamics)
Edited by NORMAN J. OPPENHEIMER AND THOMAS L. JAMES

VOLUME 177. Nuclear Magnetic Resonance (Part B: Structure and Mechanism)
Edited by NORMAN J. OPPENHEIMER AND THOMAS L. JAMES

VOLUME 178. Antibodies, Antigens, and Molecular Mimicry
Edited by JOHN J. LANGONE

VOLUME 179. Complex Carbohydrates (Part F)
Edited by VICTOR GINSBURG

VOLUME 180. RNA Processing (Part A: General Methods)
Edited by JAMES E. DAHLBERG AND JOHN N. ABELSON

VOLUME 181. RNA Processing (Part B: Specific Methods)
Edited by JAMES E. DAHLBERG AND JOHN N. ABELSON

VOLUME 182. Guide to Protein Purification
Edited by MURRAY P. DEUTSCHER

VOLUME 183. Molecular Evolution: Computer Analysis of Protein and Nucleic Acid Sequences
Edited by RUSSELL F. DOOLITTLE

VOLUME 184. Avidin–Biotin Technology
Edited by MEIR WILCHEK AND EDWARD A. BAYER

VOLUME 185. Gene Expression Technology
Edited by DAVID V. GOEDDEL

VOLUME 186. Oxygen Radicals in Biological Systems (Part B: Oxygen Radicals and Antioxidants)
Edited by LESTER PACKER AND ALEXANDER N. GLAZER

VOLUME 187. Arachidonate Related Lipid Mediators
Edited by ROBERT C. MURPHY AND FRANK A. FITZPATRICK

VOLUME 188. Hydrocarbons and Methylotrophy
Edited by MARY E. LIDSTROM

VOLUME 189. Retinoids (Part A: Molecular and Metabolic Aspects)
Edited by LESTER PACKER

VOLUME 190. Retinoids (Part B: Cell Differentiation and Clinical Applications)
Edited by LESTER PACKER

VOLUME 191. Biomembranes (Part V: Cellular and Subcellular Transport: Epithelial Cells)
Edited by SIDNEY FLEISCHER AND BECCA FLEISCHER

VOLUME 192. Biomembranes (Part W: Cellular and Subcellular Transport: Epithelial Cells)
Edited by SIDNEY FLEISCHER AND BECCA FLEISCHER

VOLUME 193. Mass Spectrometry
Edited by JAMES A. MCCLOSKEY

VOLUME 194. Guide to Yeast Genetics and Molecular Biology
Edited by CHRISTINE GUTHRIE AND GERALD R. FINK

VOLUME 195. Adenylyl Cyclase, G Proteins, and Guanylyl Cyclase
Edited by ROGER A. JOHNSON AND JACKIE D. CORBIN

VOLUME 196. Molecular Motors and the Cytoskeleton
Edited by RICHARD B. VALLEE

VOLUME 197. Phospholipases
Edited by EDWARD A. DENNIS

VOLUME 198. Peptide Growth Factors (Part C)
Edited by DAVID BARNES, J. P. MATHER, AND GORDON H. SATO

VOLUME 199. Cumulative Subject Index Volumes 168–174, 176–194

VOLUME 200. Protein Phosphorylation (Part A: Protein Kinases: Assays, Purification, Antibodies, Functional Analysis, Cloning, and Expression)
Edited by TONY HUNTER AND BARTHOLOMEW M. SEFTON

VOLUME 201. Protein Phosphorylation (Part B: Analysis of Protein Phosphorylation, Protein Kinase Inhibitors, and Protein Phosphatases)
Edited by TONY HUNTER AND BARTHOLOMEW M. SEFTON

VOLUME 202. Molecular Design and Modeling: Concepts and Applications (Part A: Proteins, Peptides, and Enzymes)
Edited by JOHN J. LANGONE

VOLUME 203. Molecular Design and Modeling: Concepts and Applications (Part B: Antibodies and Antigens, Nucleic Acids, Polysaccharides, and Drugs)
Edited by JOHN J. LANGONE

VOLUME 204. Bacterial Genetic Systems
Edited by JEFFREY H. MILLER

VOLUME 205. Metallobiochemistry (Part B: Metallothionein and Related Molecules)
Edited by JAMES F. RIORDAN AND BERT L. VALLEE

VOLUME 206. Cytochrome P450
Edited by MICHAEL R. WATERMAN AND ERIC F. JOHNSON

VOLUME 207. Ion Channels
Edited by BERNARDO RUDY AND LINDA E. IVERSON

VOLUME 208. Protein–DNA Interactions
Edited by ROBERT T. SAUER

VOLUME 209. Phospholipid Biosynthesis
Edited by EDWARD A. DENNIS AND DENNIS E. VANCE

VOLUME 210. Numerical Computer Methods
Edited by LUDWIG BRAND AND MICHAEL L. JOHNSON

VOLUME 211. DNA Structures (Part A: Synthesis and Physical Analysis of DNA)
Edited by DAVID M. J. LILLEY AND JAMES E. DAHLBERG

VOLUME 212. DNA Structures (Part B: Chemical and Electrophoretic Analysis of DNA)
Edited by DAVID M. J. LILLEY AND JAMES E. DAHLBERG

VOLUME 213. Carotenoids (Part A: Chemistry, Separation, Quantitation, and Antioxidation)
Edited by LESTER PACKER

VOLUME 214. Carotenoids (Part B: Metabolism, Genetics, and Biosynthesis)
Edited by LESTER PACKER

VOLUME 215. Platelets: Receptors, Adhesion, Secretion (Part B)
Edited by JACEK J. HAWIGER

VOLUME 216. Recombinant DNA (Part G)
Edited by RAY WU

VOLUME 217. Recombinant DNA (Part H)
Edited by RAY WU

VOLUME 218. Recombinant DNA (Part I)
Edited by RAY WU

VOLUME 219. Reconstitution of Intracellular Transport
Edited by JAMES E. ROTHMAN

VOLUME 220. Membrane Fusion Techniques (Part A)
Edited by NEJAT DÜZGÜNEŞ

VOLUME 221. Membrane Fusion Techniques (Part B)
Edited by NEJAT DÜZGÜNEŞ

VOLUME 222. Proteolytic Enzymes in Coagulation, Fibrinolysis, and Complement Activation (Part A: Mammalian Blood Coagulation Factors and Inhibitors)
Edited by LASZLO LORAND AND KENNETH G. MANN

VOLUME 223. Proteolytic Enzymes in Coagulation, Fibrinolysis, and Complement Activation (Part B: Complement Activation, Fibrinolysis, and Nonmammalian Blood Coagulation Factors)
Edited by LASZLO LORAND AND KENNETH G. MANN

VOLUME 224. Molecular Evolution: Producing the Biochemical Data
Edited by ELIZABETH ANNE ZIMMER, THOMAS J. WHITE, REBECCA L. CANN, AND ALLAN C. WILSON

VOLUME 225. Guide to Techniques in Mouse Development
Edited by PAUL M. WASSARMAN AND MELVIN L. DEPAMPHILIS

VOLUME 226. Metallobiochemistry (Part C: Spectroscopic and Physical Methods for Probing Metal Ion Environments in Metalloenzymes and Metalloproteins)
Edited by JAMES F. RIORDAN AND BERT L. VALLEE

VOLUME 227. Metallobiochemistry (Part D: Physical and Spectroscopic Methods for Probing Metal Ion Environments in Metalloproteins)
Edited by JAMES F. RIORDAN AND BERT L. VALLEE

VOLUME 228. Aqueous Two-Phase Systems
Edited by HARRY WALTER AND GÖTE JOHANSSON

VOLUME 229. Cumulative Subject Index Volumes 195–198, 200–227

VOLUME 230. Guide to Techniques in Glycobiology
Edited by WILLIAM J. LENNARZ AND GERALD W. HART

VOLUME 231. Hemoglobins (Part B: Biochemical and Analytical Methods)
Edited by JOHANNES EVERSE, KIM D. VANDEGRIFF, AND ROBERT M. WINSLOW

VOLUME 232. Hemoglobins (Part C: Biophysical Methods)
Edited by JOHANNES EVERSE, KIM D. VANDEGRIFF, AND ROBERT M. WINSLOW

VOLUME 233. Oxygen Radicals in Biological Systems (Part C)
Edited by LESTER PACKER

VOLUME 234. Oxygen Radicals in Biological Systems (Part D)
Edited by LESTER PACKER

VOLUME 235. Bacterial Pathogenesis (Part A: Identification and Regulation of Virulence Factors)
Edited by VIRGINIA L. CLARK AND PATRIK M. BAVOIL

VOLUME 236. Bacterial Pathogenesis (Part B: Integration of Pathogenic Bacteria with Host Cells)
Edited by VIRGINIA L. CLARK AND PATRIK M. BAVOIL

VOLUME 237. Heterotrimeric G Proteins
Edited by RAVI IYENGAR

VOLUME 238. Heterotrimeric G-Protein Effectors
Edited by RAVI IYENGAR

VOLUME 239. Nuclear Magnetic Resonance (Part C)
Edited by THOMAS L. JAMES AND NORMAN J. OPPENHEIMER

VOLUME 240. Numerical Computer Methods (Part B)
Edited by MICHAEL L. JOHNSON AND LUDWIG BRAND

VOLUME 241. Retroviral Proteases
Edited by LAWRENCE C. KUO AND JULES A. SHAFER

VOLUME 242. Neoglycoconjugates (Part A)
Edited by Y. C. LEE AND REIKO T. LEE

VOLUME 243. Inorganic Microbial Sulfur Metabolism
Edited by HARRY D. PECK, JR., AND JEAN LEGALL

VOLUME 244. Proteolytic Enzymes: Serine and Cysteine Peptidases
Edited by ALAN J. BARRETT

VOLUME 245. Extracellular Matrix Components
Edited by E. RUOSLAHTI AND E. ENGVALL

VOLUME 246. Biochemical Spectroscopy
Edited by KENNETH SAUER

VOLUME 247. Neoglycoconjugates (Part B: Biomedical Applications)
Edited by Y. C. LEE AND REIKO T. LEE

VOLUME 248. Proteolytic Enzymes: Aspartic and Metallo Peptidases
Edited by ALAN J. BARRETT

VOLUME 249. Enzyme Kinetics and Mechanism (Part D: Developments in Enzyme Dynamics)
Edited by DANIEL L. PURICH

VOLUME 250. Lipid Modifications of Proteins
Edited by PATRICK J. CASEY AND JANICE E. BUSS

VOLUME 251. Biothiols (Part A: Monothiols and Dithiols, Protein Thiols, and Thiyl Radicals)
Edited by LESTER PACKER

VOLUME 252. Biothiols (Part B: Glutathione and Thioredoxin; Thiols in Signal Transduction and Gene Regulation)
Edited by LESTER PACKER

VOLUME 253. Adhesion of Microbial Pathogens
Edited by RON J. DOYLE AND ITZHAK OFEK

VOLUME 254. Oncogene Techniques
Edited by PETER K. VOGT AND INDER M. VERMA

VOLUME 255. Small GTPases and Their Regulators (Part A: Ras Family)
Edited by W. E. BALCH, CHANNING J. DER, AND ALAN HALL

VOLUME 256. Small GTPases and Their Regulators (Part B: Rho Family)
Edited by W. E. BALCH, CHANNING J. DER, AND ALAN HALL

VOLUME 257. Small GTPases and Their Regulators (Part C: Proteins Involved in Transport)
Edited by W. E. BALCH, CHANNING J. DER, AND ALAN HALL

VOLUME 258. Redox-Active Amino Acids in Biology
Edited by JUDITH P. KLINMAN

VOLUME 259. Energetics of Biological Macromolecules
Edited by MICHAEL L. JOHNSON AND GARY K. ACKERS

VOLUME 260. Mitochondrial Biogenesis and Genetics (Part A)
Edited by GIUSEPPE M. ATTARDI AND ANNE CHOMYN

VOLUME 261. Nuclear Magnetic Resonance and Nucleic Acids
Edited by THOMAS L. JAMES

VOLUME 262. DNA Replication
Edited by JUDITH L. CAMPBELL

VOLUME 263. Plasma Lipoproteins (Part C: Quantitation)
Edited by WILLIAM A. BRADLEY, SANDRA H. GIANTURCO, AND JERE P. SEGREST

VOLUME 264. Mitochondrial Biogenesis and Genetics (Part B)
Edited by GIUSEPPE M. ATTARDI AND ANNE CHOMYN

VOLUME 265. Cumulative Subject Index Volumes 228, 230–262

VOLUME 266. Computer Methods for Macromolecular Sequence Analysis
Edited by RUSSELL F. DOOLITTLE

VOLUME 267. Combinatorial Chemistry
Edited by JOHN N. ABELSON

VOLUME 268. Nitric Oxide (Part A: Sources and Detection of NO; NO Synthase)
Edited by LESTER PACKER

VOLUME 269. Nitric Oxide (Part B: Physiological and Pathological Processes)
Edited by LESTER PACKER

VOLUME 270. High Resolution Separation and Analysis of Biological Macromolecules (Part A: Fundamentals)
Edited by BARRY L. KARGER AND WILLIAM S. HANCOCK

VOLUME 271. High Resolution Separation and Analysis of Biological Macromolecules (Part B: Applications)
Edited by BARRY L. KARGER AND WILLIAM S. HANCOCK

VOLUME 272. Cytochrome P450 (Part B)
Edited by ERIC F. JOHNSON AND MICHAEL R. WATERMAN

VOLUME 273. RNA Polymerase and Associated Factors (Part A)
Edited by SANKAR ADHYA

VOLUME 274. RNA Polymerase and Associated Factors (Part B)
Edited by SANKAR ADHYA

VOLUME 275. Viral Polymerases and Related Proteins
Edited by LAWRENCE C. KUO, DAVID B. OLSEN, AND STEVEN S. CARROLL

VOLUME 276. Macromolecular Crystallography (Part A)
Edited by CHARLES W. CARTER, JR., AND ROBERT M. SWEET

VOLUME 277. Macromolecular Crystallography (Part B)
Edited by CHARLES W. CARTER, JR., AND ROBERT M. SWEET

VOLUME 278. Fluorescence Spectroscopy
Edited by LUDWIG BRAND AND MICHAEL L. JOHNSON

VOLUME 279. Vitamins and Coenzymes (Part I)
Edited by DONALD B. MCCORMICK, JOHN W. SUTTIE, AND CONRAD WAGNER

VOLUME 280. Vitamins and Coenzymes (Part J)
Edited by DONALD B. MCCORMICK, JOHN W. SUTTIE, AND CONRAD WAGNER

VOLUME 281. Vitamins and Coenzymes (Part K)
Edited by DONALD B. MCCORMICK, JOHN W. SUTTIE, AND CONRAD WAGNER

VOLUME 282. Vitamins and Coenzymes (Part L)
Edited by DONALD B. MCCORMICK, JOHN W. SUTTIE, AND CONRAD WAGNER

VOLUME 283. Cell Cycle Control
Edited by WILLIAM G. DUNPHY

VOLUME 284. Lipases (Part A: Biotechnology)
Edited by BYRON RUBIN AND EDWARD A. DENNIS

VOLUME 285. Cumulative Subject Index Volumes 263, 264, 266–284, 286–289

VOLUME 286. Lipases (Part B: Enzyme Characterization and Utilization)
Edited by BYRON RUBIN AND EDWARD A. DENNIS

VOLUME 287. Chemokines
Edited by RICHARD HORUK

VOLUME 288. Chemokine Receptors
Edited by RICHARD HORUK

VOLUME 289. Solid Phase Peptide Synthesis
Edited by GREGG B. FIELDS

VOLUME 290. Molecular Chaperones
Edited by GEORGE H. LORIMER AND THOMAS BALDWIN

VOLUME 291. Caged Compounds
Edited by GERARD MARRIOTT

VOLUME 292. ABC Transporters: Biochemical, Cellular, and Molecular Aspects
Edited by SURESH V. AMBUDKAR AND MICHAEL M. GOTTESMAN

VOLUME 293. Ion Channels (Part B)
Edited by P. MICHAEL CONN

VOLUME 294. Ion Channels (Part C)
Edited by P. MICHAEL CONN

VOLUME 295. Energetics of Biological Macromolecules (Part B)
Edited by GARY K. ACKERS AND MICHAEL L. JOHNSON

VOLUME 296. Neurotransmitter Transporters
Edited by SUSAN G. AMARA

VOLUME 297. Photosynthesis: Molecular Biology of Energy Capture
Edited by LEE MCINTOSH

VOLUME 298. Molecular Motors and the Cytoskeleton (Part B)
Edited by RICHARD B. VALLEE

VOLUME 299. Oxidants and Antioxidants (Part A)
Edited by LESTER PACKER

VOLUME 300. Oxidants and Antioxidants (Part B)
Edited by LESTER PACKER

VOLUME 301. Nitric Oxide: Biological and Antioxidant Activities (Part C)
Edited by LESTER PACKER

Volume 302. Green Fluorescent Protein
Edited by P. Michael Conn

Volume 303. cDNA Preparation and Display
Edited by Sherman M. Weissman

Volume 304. Chromatin
Edited by Paul M. Wassarman and Alan P. Wolffe

Volume 305. Bioluminescence and Chemiluminescence (Part C)
Edited by Thomas O. Baldwin and Miriam M. Ziegler

Volume 306. Expression of Recombinant Genes in Eukaryotic Systems
Edited by Joseph C. Glorioso and Martin C. Schmidt

Volume 307. Confocal Microscopy
Edited by P. Michael Conn

Volume 308. Enzyme Kinetics and Mechanism (Part E: Energetics of Enzyme Catalysis)
Edited by Daniel L. Purich and Vern L. Schramm

Volume 309. Amyloid, Prions, and Other Protein Aggregates
Edited by Ronald Wetzel

Volume 310. Biofilms
Edited by Ron J. Doyle

Volume 311. Sphingolipid Metabolism and Cell Signaling (Part A)
Edited by Alfred H. Merrill, Jr., and Yusuf A. Hannun

Volume 312. Sphingolipid Metabolism and Cell Signaling (Part B)
Edited by Alfred H. Merrill, Jr., and Yusuf A. Hannun

Volume 313. Antisense Technology (Part A: General Methods, Methods of Delivery, and RNA Studies)
Edited by M. Ian Phillips

Volume 314. Antisense Technology (Part B: Applications)
Edited by M. Ian Phillips

Volume 315. Vertebrate Phototransduction and the Visual Cycle (Part A)
Edited by Krzysztof Palczewski

Volume 316. Vertebrate Phototransduction and the Visual Cycle (Part B)
Edited by Krzysztof Palczewski

Volume 317. RNA–Ligand Interactions (Part A: Structural Biology Methods)
Edited by Daniel W. Celander and John N. Abelson

Volume 318. RNA–Ligand Interactions (Part B: Molecular Biology Methods)
Edited by Daniel W. Celander and John N. Abelson

Volume 319. Singlet Oxygen, UV-A, and Ozone
Edited by Lester Packer and Helmut Sies

Volume 320. Cumulative Subject Index Volumes 290–319

VOLUME 321. Numerical Computer Methods (Part C)
Edited by MICHAEL L. JOHNSON AND LUDWIG BRAND

VOLUME 322. Apoptosis
Edited by JOHN C. REED

VOLUME 323. Energetics of Biological Macromolecules (Part C)
Edited by MICHAEL L. JOHNSON AND GARY K. ACKERS

VOLUME 324. Branched-Chain Amino Acids (Part B)
Edited by ROBERT A. HARRIS AND JOHN R. SOKATCH

VOLUME 325. Regulators and Effectors of Small GTPases (Part D: Rho Family)
Edited by W. E. BALCH, CHANNING J. DER, AND ALAN HALL

VOLUME 326. Applications of Chimeric Genes and Hybrid Proteins (Part A: Gene Expression and Protein Purification)
Edited by JEREMY THORNER, SCOTT D. EMR, AND JOHN N. ABELSON

VOLUME 327. Applications of Chimeric Genes and Hybrid Proteins (Part B: Cell Biology and Physiology)
Edited by JEREMY THORNER, SCOTT D. EMR, AND JOHN N. ABELSON

VOLUME 328. Applications of Chimeric Genes and Hybrid Proteins (Part C: Protein–Protein Interactions and Genomics)
Edited by JEREMY THORNER, SCOTT D. EMR, AND JOHN N. ABELSON

VOLUME 329. Regulators and Effectors of Small GTPases (Part E: GTPases Involved in Vesicular Traffic)
Edited by W. E. BALCH, CHANNING J. DER, AND ALAN HALL

VOLUME 330. Hyperthermophilic Enzymes (Part A)
Edited by MICHAEL W. W. ADAMS AND ROBERT M. KELLY

VOLUME 331. Hyperthermophilic Enzymes (Part B)
Edited by MICHAEL W. W. ADAMS AND ROBERT M. KELLY

VOLUME 332. Regulators and Effectors of Small GTPases (Part F: Ras Family I)
Edited by W. E. BALCH, CHANNING J. DER, AND ALAN HALL

VOLUME 333. Regulators and Effectors of Small GTPases (Part G: Ras Family II) (in preparation)
Edited by W. E. BALCH, CHANNING J. DER, AND ALAN HALL

VOLUME 334. Hyperthermophilic Enzymes (Part C) (in preparation)
Edited by MICHAEL W. W. ADAMS AND ROBERT M. KELLY

VOLUME 335. Flavonoids and Other Polyphenols (in preparation)
Edited by LESTER PACKER

VOLUME 336. Microbial Growth in Biofilms (Part A: Developmental and Molecular Biological Aspects) (in preparation)
Edited by RON J. DOYLE

VOLUME 337. Microbial Growth in Biofilms (Part B: Special Environments and Physicochemical Aspects) (in preparation)
Edited by RON J. DOYLE

VOLUME 338. Nuclear Magnetic Resonance of Biological Macromolecules (Part A) (in preparation)
Edited by THOMAS L. JAMES, VOLKER DÖTSCH, AND ULI SCHMITZ

VOLUME 339. Nuclear Magnetic Resonance of Biological Macromolecules (Part B) (in preparation)
Edited by THOMAS L. JAMES, VOLKER DÖTSCH, AND ULI SCHMITZ

VOLUME 340. Drug-Nucleic Acid Interactions (in preparation)
Edited by JONATHAN B. CHAIRES AND MICHAEL J. WARING

VOLUME 341. Ribonucleases, Part A (in preparation)
Edited by ALLEN W. NICHOLSON

VOLUME 342. Ribonucleases, Part B (in preparation)
Edited by ALLEN W. NICHOLSON

VOLUME 343. G Protein Pathways (in preparation)
Edited by RAVI SYENGAR AND JOHN D. HILDEBRANDT

Section I

Protein Expression and Protein–Protein Interactions

[1] Mammalian Expression Vectors for Ras Family Proteins: Generation and Use of Expression Constructs to Analyze Ras Family Function

By JAMES J. FIORDALISI, RONALD L. JOHNSON II, AYLIN S. ÜLKÜ, CHANNING J. DER, and ADRIENNE D. COX

Introduction

Cell-based assays are useful for the characterization of Ras family structure–function relationships, identification of upstream regulators and downstream effectors, characterization of signaling inputs and outputs, analysis of the role of Ras family proteins in normal and aberrant cellular metabolism, and evaluation of potential anticancer agents.

Common to all such studies is the need to express the protein(s) of interest within a cell. This is accomplished through the use of plasmid vectors into which are placed the coding sequences of the proteins to be studied, and which can then be introduced into cells by a variety of methods. Protein expression plasmid vectors contain signal sequences required for transcription and translation of the target protein (i.e., promoter elements, polyadenylation sites, etc.) as well as origins of replication for maintenance of the plasmid. Expression vectors have been developed with a variety of features, including selectable markers and sequences encoding epitope tags that are recognized by specific antibodies, which facilitate the subsequent analysis of protein expression and function.

Not all vectors function equally well in different assay systems, even if the sequences being expressed are identical. Similarly, not all proteins are expressed equally well in the same vector. Moreover, the reasons for these differences are not well understood and can be determined only by trial and error. Therefore, choosing the optimum vector for a given protein and assay system can be an empirical and time-consuming endeavor. Undoubtedly, such factors as the identity of the cell line, the gene of interest, the biological readout, as well as others all contribute to variability in the usefulness of the vector.

In this chapter, we attempt to provide readers with a starting point from which to choose the most appropriate vector for their particular proteins of interest and intended uses. We present some observations concerning the strengths and weaknesses of several mammalian protein expression vectors, both commercially available and "homemade." Because there are many vectors currently in use, as well as new vectors and assay systems

0076-6879/00 $35.00

continually being developed, it is not possible to present a comprehensive physical or functional evaluation of all vectors under all circumstances. In this work we identify and discuss most of the major factors that should be considered. In addition to discussing the advantages and disadvantages of particular features of mammalian protein expression vectors, we also compare and contrast them functionally with respect to biological readouts commonly used in the study of Ras protein function, including protein expression, signaling activity in enzyme-linked transcriptional *trans*-activation reporter assays, and transforming ability in fibroblast focus-forming assays. In all cases we use activated, oncogenic Ras proteins as the model system. Because the choice of vector will be influenced by, among other things, the ease with which protein-coding sequences can be introduced into them, we also discuss several techniques for generating and manipulating protein expression constructs. Finally, we discuss several methods for introducing plasmid DNA into mammalian cells, including transfection with a variety of reagents and infection using retroviral packaging vectors.

Properties to Consider in Choosing a Vector

Promoter

In choosing a mammalian protein expression vector (Table I[1–8]), the most important factor to consider is whether the plasmid will express the protein of interest to the desired level in the cell type to be used. Sometimes the highest possible protein expression levels are desired, usually in order to maximize the biological effect being studied. In other cases, lower levels are desired, usually either to achieve more physiologically relevant levels or to minimize toxicity. Protein expression is controlled primarily by the transcriptional promoter region of the vector, which contains elements necessary for transcription (such as binding sites for transcription factors that recruit RNA polymerase) and translation (especially the Kozak se-

[1] M. A. White, C. Nicolette, A. Minden, A. Polverino, L. Van Aelst, M. Karin, and M. H. Wigler, *Cell* **80,** 533 (1995).

[2] R. R. Mattingly, A. Sorisky, M. R. Brann, and I. G. Macara, *Mol. Cell. Biol.* **14,** 7943 (1994).

[3] J. P. Morgenstern and H. Land, *Nucleic Acids Res.* **18,** 1068 (1990).

[4] W. S. Pear, G. P. Nolan, M. L. Scott, and D. Baltimore, *Proc. Natl. Acad. Sci. U.S.A.* **90,** 8392 (1993).

[5] I. Whitehead, H. Kirk, C. Tognon, G. Trigo-Gonzalez, and R. Kay, *J. Biol. Chem.* **270,** 18388 (1995).

[6] C. L. Cepko, B. Roberts, and R. C. Mulligan, *Cell* **37,** 1053 (1984).

[7] J. A. Southern, D. F. Young, F. Heaney, W. K. Baumgartner, and R. E. Randall, *J. Gen. Virol.* **72,** 1551 (1991).

[8] A. Yen, M. Williams, J. D. Platko, C. Der, and M. Hisaka, *Eur. J. Cell Biol.* **65,** 103 (1994).

quence[9]) of the coding sequence. Most promoters found in expression vectors are derived from viral promoters that induce the high rates of protein expression necessary for viral replication. The cytomegalovirus (CMV) promoter, the mouse mammary tumor virus long terminal repeat promoter (MMTV LTR), and the Moloney murine leukemia virus promoter LTR (Mo-MuLV LTR) are commonly used viral promoters.

The CMV promoter generally works well in cell lines derived from primate tissues such as human embryonic kidney cells (HEK293), human breast epithelial cells (T-47D, MCF-7, and MCF-10A), and monkey kidney cells (COS-7), but works less well in cells of rodent origin, such as mouse fibroblasts (NIH 3T3, Rat1, and Rat2) and rat intestinal epithelial cells (RIE-1). The reverse is true of the MMTV LTR and the Mo-MuLV LTR promoters. Naturally, there are always exceptions to such a rule; for example, we have found that pZIP-NeoSV(X)1-based constructs work well in T-47D cells but not in 293 or COS cells. Protein expression levels should always be confirmed directly for each expression construct in the cells of interest, using Western blot analysis or a similar method.

Constitutive versus Inducible Protein Expression

Although most vectors express proteins in a constitutive fashion, protein expression in some vectors is controlled by promoters that contain inducible elements that bind either repressor proteins or inducers that can be inactivated or induced, respectively, by exposure to exogenously added inducing agents. Until then, protein expression does not occur. We have more experience with dexamethasone-inducible vectors[3] (Table I); other common inducible elements are responsive to tetracycline,[10,11] isopropyl-β-D-thiogalactopyranoside (IPTG),[12] and ecdysone (see Ref. 13 and [19] in this volume[14]). Inducible protein expression is desirable if the protein of interest is toxic or otherwise growth inhibitory to the cell, in which case, stable transfection of cells with a vector expressing this protein constitutively would be impossible. Moreover, any transient or temporally distinct cellular phenotype caused by the expression of the protein can be evaluated better if protein expression can be turned on and off relatively rapidly.

[9] M. Kozak, *Nucleic Acids Res.* **9,** 5233 (1981).

[10] L. Chin, A. Tam, J. Pomerantz, M. Wong, J. Holash, N. Bardeesy, Q. Shen, R. O'Hagan, J. Pantginis, H. Zhou, J. W. Horner II, C. Cordon-Cardo, G. D. Yancopoulos, and R. A. DePinho, *Nature (London)* **400,** 468 (1999).

[11] H. S. Liu, C. H. Lee, C. F. Lee, I. J. Su, and T. Y. Chang, *BioTechniques* **24,** 624 (1998).

[12] M. A. Wani, X. Xu, and P. J. Stambrook, *Cancer Res.* **54,** 2504 (1994).

[13] M. J. Calonge and J. Massague, *J. Biol. Chem.* **274,** 33637 (1999).

[14] G. G. Habets, M. Knepper, J. Sumortin, Y.-J. Choi, T. Sasazuki, S. Shirasawa, and G. Bollag, *Methods Enzymol.* **332** [19] 2001 (this volume).

TABLE I
PROPERTIES OF SELECTED MAMMALIAN EXPRESSION VECTORS[a]

Vector	Promoter	Bacterial/ mammalian selection	Epitope tag	Source/Ref.
Constitutive Vectors				
pcDNA3.1	CMV	Amp/Neo, Zeo, or Hyg	Xpress/His_6, Myc/His_6, V5/His_6, or none	InVitrogen/—
pCGN-hygro	CMV	Amp/Hyg	HA	M. Ostrowski[b]/—
pCMV	CMV	Amp/Neo	Myc[c]	InVitrogen/—
pDCR[d]	CMV/Mo-MuLV 5′LTR	Amp/Neo	HA	—/1
pK7-GFP (from pRK7)	CMV	Amp/—	GFP	—/2
pKH3 (from pRK7)	CMV	Amp/—	3 × HA	—/2
pRC/CMV	CMV	Amp/Neo	None	InVitrogen/—
pEGFP	CMV	Kan/Neo	GFP	Clontech/—
pRC/RSV	RSV	Amp/Neo	None	InVitrogen/—
pREP[e]	RSV	Amp/Neo, Hyg, or His_sD	None	InVitrogen/—
pCEP[e]	CMV	Amp/Hyg	None	InVitrogen/—
pRK7	CMV	Amp/—	None	—/2

Vector	Promoter	Bacterial/ mammalian selection	Epitope tag	Inducer	Source/Ref.
Inducible Vectors					
pJ5Ω	MMTV 5′LTR	Amp/—	No	0.1 μM Dex	—/3
pMSG	MMTV 5′LTR	Amp/mycophenolic acid	No	0.1 μM Dex	Pharmacia/—

Vector	Promoter	Bacterial/mammalian selection	Epitope tag	Packaging cell line/Ref.	Source/Ref.
Retroviral Vectors					
pBABE (pBABE-HA)[f]	5′LTR	Amp/Puro	No (HA)	Bosc23/4	—/3
pCTV3H	5′LTR	Tet[g]/Hyg	No	Bosc23/—	—/5
pZIP-NeoSV(X)1 (pZIP B/E[h]; pZBRII[i])[j]	Mo-MuLV 5′LTR	Amp/Neo	No	Bosc23/—	—/6

[a] CMV, Cytomegalovirus; RSV, Rous sarcoma virus; Mo-MuLV LTR, Moloney murine leukemia virus long terminal repeat; Amp, ampicillin; Kan, kanamycin; Neo, neomycin (Geneticin, G418); Zeo, Zeocin (InVitrogen); Hyg, hygromycin; Puro, puromycin; His_6, six contiguous histidines; V5, 14-amino acid sequence (GKPIPNPLLGLDST) derived from simian virus 5 V protein[7]; Xpress, 7-amino acid peptide (DLYDDDDK) containing the 5-amino acid enterokinase cleavage site (DDDDK); Myc, 10-amino acid epitope tag (EQKLISEEDL) derived from the *myc* oncogene protein product; HA, 19-amino acid hemagglutinin epitope tag (MASSYPYDVPDYASLGGPS); 3 × HA, three contiguous HA tags; GFP, 250-amino acid green fluorescent protein; Dex, dexamethasone. All vectors that are Neo^R are also Kan^R.

[b] Generated by M. Ostrowski (Ohio State University, Columbus, OH), personal communication, 1993.

[c] While all three versions of this vector contain an amino-terminal Myc epitope tag, one also contains a targeting sequence for nuclear localization, the second contains a targeting sequence for mitochondrial localization, and the third contains no targeting sequence.

[d] pDCR does not contain a psi (Ψ) sequence and cannot be used for retroviral packaging even though it has both 5′ and 3′ LTR sequences.

[e] pCEP and pREP contain Epstein–Barr virus (EBV) ori (oriP) and nuclear antigen (EBNA-1), allowing them to replicate episomally in primate as well as canine cells [such as Madin–Darby canine kidney (MDCK) epithelial cells], resulting in numerous plasmids per cell. Thus high levels of protein expression may be the result of high per-cell plasmid copy number rather than the strength of the promoter elements.

[f] Generated by J. M. Shields, University of North Carolina, Chapel Hill, NC, unpublished, 1999.

[g] pCTV3 contains the *supF* gene, which overcomes amber mutations in the ampicillin and tetracycline resistance genes of the P3 episomal plasmid. Therefore, *E. coli* strains containing the P3 plasmid (e.g., Top10/P3); InVitrogen) must be used to select pCTV3 on Amp or Tet. The P3 episome is itself Kan resistant.

[h] Generated by G. J. Clark (NIH); unpublished, 1997.

[i] Generated by S. M. Graham.[8]

[j] See Fig. 1.

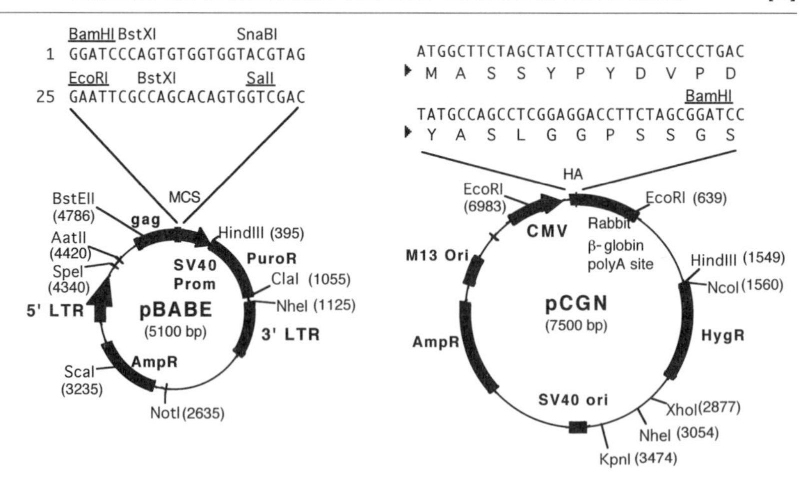

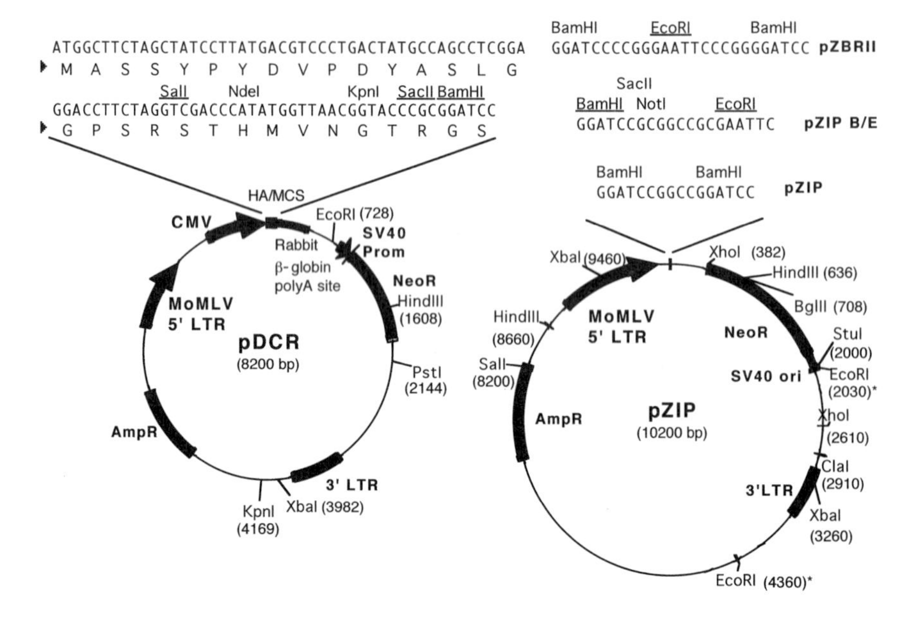

FIG. 1. Restriction maps of noncommercially available mammalian protein expression vectors used routinely in our laboratories. For each plasmid we have identified, when available, the promoter, mammalian origins of replication, bacterial and mammalian selectable markers, retroviral packaging sequences, cloning site sequences, epitope tag coding sequences, and sites for several commonly used restriction enzymes. However, because none of these plasmids has been fully sequenced, to our knowledge, there may be other instances of the restriction

Retroviral Vectors

Retroviral vectors such as pBABE, pCTV3, and pZIP-NeoSV(X)1 offer flexibility in that they can be used either to generate virus for infection of cells (discussed in more detail in Retroviral Vectors for Infection of Mammalian Cells, below) or to transfect directly into cells. Although pDCR contains both a 5′ and 3′ LTR (Fig. 1), it cannot be used as a viral packaging vector because it lacks a psi (Ψ) packaging sequence.

Vectors Containing Epitope Tags

Several vectors (e.g., pcDNA3, pCGN, pDCR, pKH3; see Table I) contain coding sequences for protein motifs that can act as epitope tags for any protein placed into the vector, and that are recognized by commercially available antibodies. Thus, epitope-tagged proteins can be detected by Western blot analysis even if specific antibodies for a novel protein are not available. Also, the expression levels of different proteins containing the same tag can be directly compared without having to determine the relative sensitivities of two different protein-specific antibodies. Antibodies to such tags can also be used to immunoprecipitate proteins and their associated complexes, or to affinity purify proteins for other uses. The hemagglutinin (HA) epitope tag (MASSYPYDVPDYASLGGPS) and the Myc epitope tag (EQKLISEEDL; also sometimes referred to as "9E10," the nomenclature for the monoclonal antibody most commonly used for its detection) are probably the most widely used. Anti-HA and anti-Myc antibodies are available from InVitrogen (Carlsbad, CA), Boehringer-Mannheim/Roche (Indianapolis, IN), Berkeley Antibody Company (BAbCo, Richmond, CA), Affinity BioReagents (Golden, CO), as well as other suppliers. Other common epitope tags for which commercial antibodies are all available (from BAbCo) are those known as His_6 (hexahistidine sequence), FLAG (influenza hemagglutinin, DYKDDDDK), and glu-glu or EE from polyomavirus

---

sites shown. Base pairs in pBABE-Puro and pZIP-NeoSV(X) 1 have been renumbered from Refs. 3 and 6, respectively, to begin at the *Bam*HI site, while base pairs in pDCR are numbered beginning at the *Sal*I site. We have sequenced pCGN-hygro from bp −54 to 3461 and pDCR from bp −60 to 4491, and the sites of several common restriction enzymes in these regions are included. Site locations in pZIP after *Bgl*II (bp 708) are approximate. Restriction sites known to be unique within each plasmid are underlined. In pDCR, the *Nde*I and *Kpn*I sites in the MCS are not unique. Information regarding the construction of these vectors can be found in the indicated references. As shown, variations of pZIP (pZIP B/E and pZBRII) have cloning sites in addition to the single *Bam*HI site of the original pZIP-Neo (see Table I). MCS, Multiple cloning site; gag, Gag viral protein; all others as in Table I. [Created using Gene Construction Kit II (Textco, West Lebanon, NH).]

(EEEEYMPME). The poly(His) epitope tag is also widely used as a tag for affinity purification using solid-phase nickel reagents.

Use of expression vectors containing the coding sequence for the green fluorescent protein (GFP), such as the commercially available pEGFP series (Clontech, Palo Alto, CA), is becoming more common. Although the GFP moiety, like HA and Myc, is detectable with commercially available antibodies and can act as a standard epitope tag, it also permits the direct visualization of the GFP-tagged protein by fluorescence microscopy, making it possible to study the subcellular localization of GFP-tagged proteins in either fixed or live cultured cells.[15–17] Live cell analysis overcomes artifacts introduced by fixation and allows temporal analyses of protein trafficking. Two potential concerns with such a large tag (250 amino acids) are that it may reduce the expression of the tagged protein, or that the tag may affect the biological integrity of the tagged protein. However, when GFP-tagged and endogenous Ras proteins have been directly compared, no differences in posttranslational processing and subcellular localization were noted.[16]

Epitope tags can be located at either the carboxy or amino terminus of a protein; which site is preferred depends on the effect (if any) the tag will have on the function of the protein. For example, most Ras family proteins, such as those of the Ras, Rap, Ral, R-Ras, and Rheb families, undergo extensive posttranslational modifications at the carboxy terminus.[18] These modifications are carried out by enzymes that require the four carboxy-terminal amino acids (CAAX motif) to be exposed. A carboxy-terminal epitope tag would prevent these functionally necessary modifications; thus, only amino-terminal epitope tags should be used with Ras proteins. Although some Ras family proteins, such as Rit and Rin, have no known carboxy-terminal modifications,[19] amino-terminal tagging seems the safer bet here as well because altering the carboxy-terminal sequences alters subcellular localization.[20]

Vector-Specific Considerations

Although the criteria described above are straightforward, there is evidence to suggest that the nature of the vector has other unexpected and as yet unexplained effects on Ras functional assays, including signaling

[15] H. Niv, O. Gutman, Y. I. Henis, and Y. Kloog, *J. Biol. Chem.* **274,** 1606 (1999).

[16] E. Choy, V. K. Chiu, J. Silletti, M. Feoktistov, T. Morimoto, D. Michaelson, I. E. Ivanov, and M. R. Philips, *Cell* **98,** 69 (1999).

[17] H. Yokoe and T. Meyer, *Nat. Biotechnol.* **14,** 1252 (1996).

[18] A. D. Cox and C. J. Der, *Crit. Rev. Oncog.* **3,** 365 (1992).

[19] H. Shao, K. Kadono-Okuda, B. S. Finlin, and D. A. Andres, *Arch. Biochem. Biophys.* **371,** 207 (1999).

[20] C. H. J. Lee, N. G. Della, C. E. Chew, and D. J. Zack, *J. Neurosci.* **16,** 6784 (1996).

TABLE II
FUNCTIONAL ACTIVITY OF Ras IN DIFFERENT VECTORS[a]

Vector	Expression level[b]		Elk-1 activation:	
	Transient	Stable	luciferase activity[c]	Focus formation[d]
pcDNA3.1	+++	+++	++++	0 to +++
pCGN-hygro	+++	+++	+++	+++
pDCR	++	+	++	++
pBABE-puro	+++	+++	++++	+ to +++
pZIP-NeoSV(X)1	+	+++	+	++++

[a] Relative protein expression, transcriptional *trans*-activation, and focus-forming activity of different Ras constructs.

[b] Transient transfection gives more variable results than expression in stably selected cell lines (see text).

[c] Except in pZIP, where all Ras isoforms give similar results, H-Ras generally activates Elk-1 more robustly than N- or K-Ras expressed in the same vector (see Fig. 2).

[d] Note that K-Ras activity is inconsistent in different assays when expressed from pBABE and pcDNA3 but not the other vectors shown (see Figs. 2 and 3).

assays such as enzyme-linked reporter assays and transformation assays such as focus formation. This may explain, in part, some apparent discrepancies in observations seen with the same proteins by different laboratories.

Protein Expression Levels. We have successfully used a variety of vectors (pBABE Puro, pcDNA3, pCGN hygro, pDCR, and pZIP Neo) to generate fibroblast and epithelial cell lines stably expressing many different Ras family proteins (Table II). A representative selection of one such panel of Ras family constructs, made in pZIP-NeoSV(X)1 from H-*ras* mutants with different functional characteristics, is illustrated in Table III.[21–26] To detect Ras proteins, we use anti-pan-Ras antibodies such as OP-40 (pan-Ras Ab-3; Calbiochem, San Diego, CA), isoform-specific antibodies such as the anti-H-Ras antibody 146 (LA069; Quality Biotech, Camden, NJ), or epitope-specific antibodies such as anti-HA (MMS101R; BAbCo) (for details, see Ref. 27). Unlike stable expression, transient transfection into NIH

[21] C. J. Der, B. E. Weissman, and M. J. MacDonald, *Oncogene* **3,** 105 (1988).

[22] C. J. Der, B. T. Pan, and G. M. Cooper, *Mol. Cell. Biol.* **6,** 3291 (1986).

[23] S. Y. Chen, S. Y. Huff, C. C. Lai, C. J. Der, and S. Powers, *Oncogene* **9,** 2691 (1994).

[24] L. A. Quilliam, K. Kato, K. M. Rabun, M. M. Hisaka, S. Y. Huff, S. Campbell-Burk, and C. J. Der, *Mol. Cell. Biol.* **14,** 1113 (1994).

[25] J. E. Buss, P. A. Solski, J. P. Schaeffer, M. J. MacDonald, and C. J. Der, *Science* **243,** 1600 (1989).

[26] A. D. Cox, M. M. Hisaka, J. E. Buss, and C. J. Der, *Mol. Cell. Biol.* **12,** 2606 (1992).

[27] A. D. Cox, P. A. Solski, J. D. Jordan, and C. J. Der, *Methods Enzymol.* **255,** 195 (1995).

TABLE III
STRUCTURAL AND FUNCTIONAL MUTANTS OF H-Ras EXPRESSED IN pZIP-NeoSV(X)1[a]

Name	Mutation	Description	Ref.
H-Ras	None	Wild-type human H-Ras; can cause transformation by overexpression in some cell lines (e.g., NIH 3T3 but not Rat1)	21
H-Ras(12V)	G12V	GTPase-deficient (dominant active) mutant: GAP insensitive, and chronically GTP bound; transforming	A. S. Ülkü, unpublished, 1999
H-Ras(61L)	Q61L	GTPase-deficient (dominant active) mutant: GAP insensitive, increased GDP/GTP exchange, and chronically GTP bound; transforming	22
H-Ras(15A)	G15A	Dominant-negative mutant: Reduced GDP and GTP binding affinity, nucleotide free; forms nonproductive complexes with Ras GEFs and blocks GEF activation of Ras	23
H-Ras(17N)	S17N	Dominant-negative mutant: Reduced GDP binding affinity; forms nonproductive complexes with Ras GEFs and blocks GEF activation of Ras	24
H-Ras(12V, 35S)	G12V, T35S	Effector domain mutant: retains ability to activate Raf, but impaired in PI3-kinase and RalGDS effector pathway activation; reduced transforming activity	A. S. Ülkü, unpublished, 1999
H-Ras(12V, 37G)	G12V, E37G	Effector domain mutant: retains ability to activate RalGDS and related pathways; impaired in Raf and PI3-kinase effector pathway activation; very reduced transforming activity	A. S. Ülkü, unpublished, 1999
H-Ras(12V, 40C)	G12V, Y40C	Effector domain mutant: retains ability to activate PI3-kinase; impaired in Raf and RalGDS effector pathway activation; very reduced transforming activity	A. S. Ülkü, unpublished, 1999
H-Ras(116I)	N116I	Exchange mutant: decreased GDP and GTP binding affinity, and increased GDP/GTP exchange; transforming	21
H-Ras(119N)	D119N	Exchange mutant: decreased GDP and GTP binding affinity, and increased GDP/GTP exchange; transforming	21
H-Ras(61L, 186S)	Q61L, C186S	Prenylation mutant: does not undergo farnesylation or any other CAAX-signaled posttranslational modification, including palmitoylation; cytosolic, nontransforming; dominant inhibitory for activated Ras, by competition for effectors	25
H-Ras(61L, 189L)	Q61L, S189L	Prenylation mutant: modified by geranylgeranylation, still undergoes remaining CAAX-signaled modifications and palmitoylation; retains transforming activity	26

[a] Selected examples of an extensive panel of H-Ras mammalian expression constructs from different functional categories that have been generated by our laboratories. Expression of all proteins in the same background minimizes considerations of vectorology and allows cross-comparison of the functional consequences of different mutations. GAP, Guanosine triphosphatase-activating protein; GEF, guanine exchange factor; PI3-kinase, phosphatidylinositol 3-kinase.

3T3 fibroblasts of a panel of cognate Ras constructs in different vectors shows that expression levels are variable, and depend on both the vector and the insert (Table II). For example, on a transient basis pCGN gives consistently higher Ras protein expression levels than does pZIP in these rodent cells, although the opposite would be expected given the promoter driving each vector. (In a stable population of antibiotic-selected cells, however, pZIP constructs consistently result in high levels, suggesting that transfection efficiency is also important.) Also, although pDCR contains both CMV and LTR promoters, theoretically making this vector ideal for high-level expression in both rodent and primate cells, stable expression of H-Ras(61L), N-Ras(12D), and K-Ras(12V) in NIH 3T3 fibroblasts was comparable to that of the analogous endogenous Ras isoform, and the phenotype characteristic of Ras-induced transformation was not as pronounced as that seen with expression of Ras variants in pZIP (data not shown). This could be considered either a disadvantage or an advantage, depending on the expression level desired.

Finally, K-Ras4B can be expressed stably from pCGN, pDCR, and pZIP at levels similar to those of H-Ras or N-Ras, but is expressed only weakly when the coding sequence is inserted into pBABE and pcCDNA3, in which H-Ras and N-Ras coding sequences are expressed robustly. Transiently, H-Ras is expressed better than N- or K-Ras in the same vector (Table II). The reasons for these differences are not clear, but may have to do with secondary structure considerations in vectors with differing polyadenylation signals. In another example of differing protein expression levels from the same vector, we have found that it is not possible to express the Ras-related proteins Rit or Rin, either stably or transiently, at levels as high as those of Ras (as measured by immunoblotting for the common HA epitope tag) even when the coding sequences are inserted into the same vector, such as pKH3 or pCGN.

Transient Expression Signaling Assays. In enzyme-linked transcriptional *trans*-activation reporter assays, transient transfection of NIH 3T3 fibroblasts with 100 ng of plasmid (per 35-mm dish; see Ref. 28) encoding activated H- and N-Ras produced 10- to 90-fold activation of Elk-1 over empty vector controls in all vectors tested (Fig. 2),[29] with pBABE, pcDNA3, and pCGN producing the highest overall levels and pZIP producing the lowest. Although the 5′ LTR of pBABE gave good activation as expected, it was not expected that the Mo-MuLV LTR promoter of pZIP would have given less activation than the CMV promoter of pCGN in rodent cells. An

[28] C. A. Hauser, J. K. Westwick, and L. A. Quilliam, *Methods Enzymol.* **255,** 412 (1995).
[29] P. J. Casey, P. A. Solski, C. J. Der, and J. E. Buss, *Proc. Natl. Acad. Sci. U.S.A.* **86,** 8323 (1989).

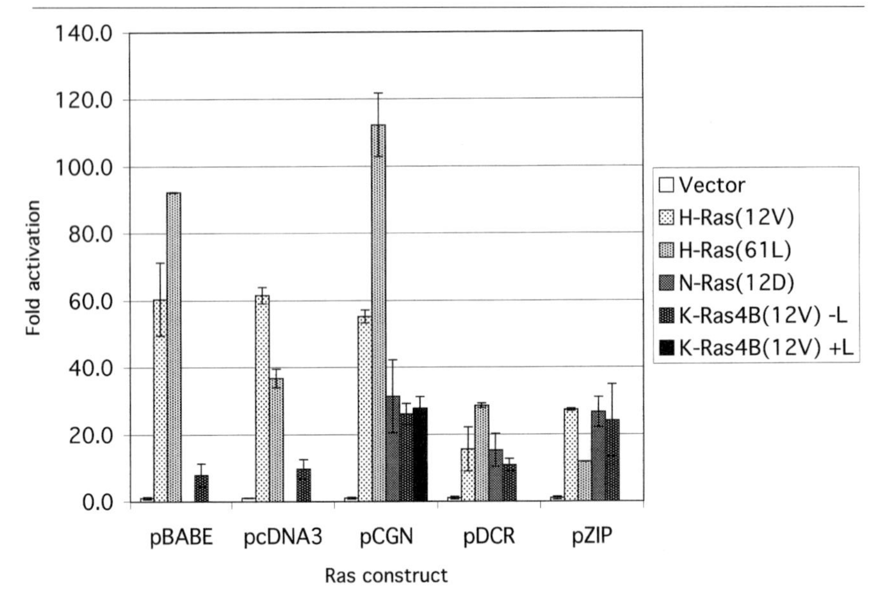

FIG. 2. Activation of Elk-1 by oncogenic Ras variants in pBABE-Puro, pcDNA3, pCGN-hygro, pDCR, and pZIP-NeoSV(X)1. [N-Ras(12D) constructs in pBABE and pcDNA3 were not available for evaluation.] Elk-1 is a substrate of the ERK mitogen-activated protein kinases, and its activation provides an indication of Ras activation of the Raf/ERK pathway. NIH 3T3 fibroblasts were transiently transfected with the indicated plasmid (100 ng/35-mm dish) by calcium phosphate precipitation as described in Transfection of Mammalian Cells. "K-Ras4B(12V) +L" is the version of K-Ras4B that contains a 10-amino acid vector-derived leader sequence,[29] whereas "K-Ras4B(12V) –L" does not contain this leader (pCGN constructs by G. M. Mahon). All dishes were also cotransfected with Gal-Elk-1 plasmid (250 ng/35-mm dish) and Gal-luciferase reporter plasmid (2.5 μg/35-mm dish), which together link Ras activity to expression of luciferase. Three days after transfection, cells were analyzed for luciferase activity (which directly reflects Elk-1 activation by Ras) according to the protocol provided with the enhanced luciferase assay kit (BD-PharMingen, San Diego, CA). Each well was washed with PBS, pH 7.2, and lysed in 150 μl of lysis buffer. Thirty microliters of each cleared lysate was assayed by luminometer. Data are shown as fold activation over empty-vector controls, ±SD for duplicate samples. Data are representative of at least four experiments.

overall pattern of activation similar to that shown in Fig. 2 was also seen with an NF-κB reporter, although the total levels of signal were lower with this reporter construct (3- to 50-fold activation; data not shown).

In general, K-Ras(12V) constructs in all vectors stimulated less Elk-1 activation (only 5- to 25-fold activation) than comparable H- and N-Ras constructs in the same vectors, which correlates with the generally low level of transiently expressed K-Ras that we observed. When expressed from the pCGN vector, the presence or absence in K-Ras4B of a 10-amino acid leader sequence[29] had no effect on its ability to activate Elk-1 transcriptional *trans*-activation. Moreover, levels of Elk-1 activation were greatly reduced for all Ras isoforms when expressed from pEGFP (Clontech; see Table I). Thus, pEGFP containing GFP-tagged H-Ras(61L), N-Ras(12D), and K-Ras(12V) showed only 5-fold activation of Elk-1 even though 20-fold more DNA was transfected (M. Philips, New York University, personal communication, 1999). It is possible that, because of the large size of the GFP moiety (250 residues) compared with Ras (188/189 residues), these proteins were not expressed at levels comparable to those produced by other vectors (we did not assess protein expression levels with pEGFP constructs). However, in other vectors such as pDCR, we have found that fairly low levels of Ras protein expression can support quite robust signaling activity, so there may well be other unknown reasons for the lower activity of the GFP constructs.

Transformation Assays. Perhaps predictably, given that they are the outcome of a complex combination of signaling activities, results in focus-forming assays cannot always be predicted by either protein expression levels or activity of a given construct in specific signaling assays. In general, pZIP constructs give the highest and most consistent activity in all our standard transformation assays.[30,31] All vectors tested with activated Ras (50 ng of plasmid per 60-mm dish) were able to produce transformed foci in NIH 3T3 fibroblasts (Fig. 3 and data not shown), but not at comparable levels and not with every Ras variant. All Ras variants in pZIP produced many large foci, even though overall protein expression levels using this vector for transient transfections were low. It is possible that transfection efficiency with pZIP is lower than with other vectors, resulting in lower overall detectable protein expression, which may mask high expression levels in individual, focus-producing cells. However, we did not observe lower numbers of foci produced by pZIP, suggesting the transfection efficiency is similar with all vectors tested. In any case, the ability of Ras variants to generate foci in pZIP is inconsistent with the relatively low levels of Elk-1 (Fig. 2) and NF-κB activation compared with the same Ras-coding sequences in other vectors. This is likely to be due to transformed phenotype requirements for additional signaling pathways besides those terminating in Elk-1 *trans*-activation. As in pZIP, both H-Ras and K-Ras variants in pBABE also produced many large foci, even though K-Ras4B(12V) activated Elk-1 at only 10% the levels of H-Ras(61L) or H-Ras(12V).

In contrast, striking differences among Ras isoforms were observed with the pcDNA3 vector, in which although H-Ras(12V) and H-Ras(61L) were

[30] A. D. Cox and C. J. Der, *Methods Enzymol.* **238,** 277 (1994).

[31] G. J. Clark, A. D. Cox, S. M. Graham, and C. J. Der, *Methods Enzymol.* **255,** 395 (1995).

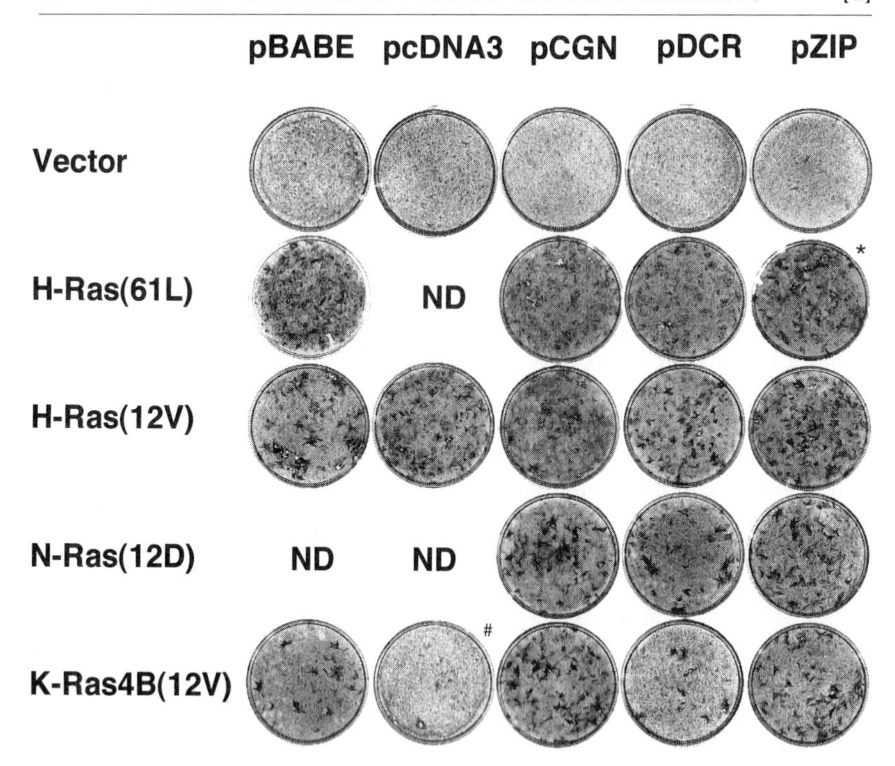

FIG. 3. Focus formation in NIH 3T3 fibroblasts induced by oncogenic Ras variants in different vectors. [N-Ras(12D) constructs in pBABE and pcDNA3 were not available for evaluation.] Except where indicated (*), NIH 3T3 fibroblasts were transfected with 50 ng of the indicated plasmid, by calcium phosphate precipitation (see Transfection of Mammalian Cells). After transfection cells were maintained for 2 weeks under standard growth conditions and fed every 4–5 days until transformed foci appeared (detailed in Ref. 31). Cells were washed once with phosphate-buffered saline, pH 7.2, fixed for 10 min in 75% acetic acid–25% methanol, stained for 1 min with 1% crystal violet in 75% acetic acid–25% methanol, and washed extensively with water. *Only 20 ng of pZIP-H-Ras(61L) was transfected. #Although the correct protein is expressed from pcDNA3-K-Ras4B(12V) and stably expressing transformed cell lines have been generated, this construct is nearly inactive for focus formation, even when transfected at up to 250 ng of DNA per dish (data not shown).

both able to produce foci and induce high levels of Elk-1 activation, K-Ras(12V) activated Elk-1 only 10-fold and was unable to produce foci even at 250 ng per 60-mm dish. Ras in pDCR produced the fewest foci overall, which is consistent with our observation that pDCR produces relatively low levels of protein expression in stable cell lines, resulting in a mildly transformed phenotype in NIH 3T3 cells (data not shown), and with the relatively reduced ability of Ras in pDCR to signal through Elk-1. Similarly, all pEGFP-Ras constructs failed to produce foci at 50 and 200 ng per 60-mm dish, and produced few even at 2–5 μg per 60-mm dish (data not shown). All Ras variants in pCGN produced moderate levels of foci, which is consistent with both the high levels of protein expression and the generally good levels of Elk-1 activation produced by these constructs.

Choosing a Vector

It is clear from these data that, of the vectors tested, pCGN is the preferred vector for reporter assays, while pZIP and pEGFP give much lower activity (in NIH 3T3, HEK293, and COS cells; data not shown). Because they give inconsistent results with different Ras proteins, pBABE and pcDNA3 would not be the first vector of choice for cross-protein comparisons, despite their ability to promote strong activity from H- and N-Ras proteins. For focus-forming assays, pZIP is clearly the preferred vector, although pCGN is also effective at generating foci. pDCR is intermediate in transformation assays, while pEFGP reduces the transforming ability of Ras variants, especially K-Ras, in this assay. Both pBABE and pcDNA3 can generate foci with H- and N-Ras constructs, but give inconsistent results with K-Ras. It is not clear why this is the case, because pBABE and pcDNA3 have been used successfully in our laboratories and others to generate highly transformed NIH 3T3 cell lines stably overexpressing K-Ras(12V), and because K-Ras(12V) expression levels from both stable and transient transfections are comparable to the levels seen with other Ras variants in these vectors (Table II). Overall, if a panel of constructs is to be made in only one vector and will be used for several different readouts, pCGN is the most consistent vector for the assays discussed here. Certainly many other vectors are also available, and widely used, although not discussed here because of our lack of directly comparable experience with them.

Other considerations are also important in vector choice, such as antibiotic resistance for the isolation of stably transfected cells by drug selection. For example, pCGN is hygromycin B resistant; because most of our other vectors are neomycin resistant, we find this convenient for stable selection of multiple constructs in the same cell line. Choosing the best vector for a given study requires matching cell type and promoter, taking into account the levels of expression desired and the biological readout to be used. Unfortunately, all these considerations of "vectorology" means that the ultimate choice of vector remains somewhat empirical. Perhaps the most important point is to realize that each vector–insert combination is potentially different and that, where feasible, results should be confirmed with different vectors.

Generation of Expression Construct

Once an expression vector has been chosen, the gene of interest must be removed from the original vector and placed into the desired vector. The ease with which this can be accomplished depends primarily on whether compatible restriction sites are available both within the vector and flanking the gene of interest. Regardless of which method is used, if an expression vector containing an epitope tag is used (e.g., pCGN or pDCR), it is vital that the coding sequence of the inserted gene be in frame with the coding sequence of the tag, which is usually located just upstream of the cloning site, although some tags are found downstream of the cloning site. The fact that the cloning site may include the same restriction site(s) as that required by the insert does not by itself guarantee that the two coding sequences will be in frame after ligation. If simple subcloning places the two coding sequences out of frame, polymerase chain reaction (PCR) generation of new restriction sites or modification of the multiple cloning site by cassette mutagenesis will likely be necessary, although use of a shuttle vector may be sufficient. These considerations are discussed in detail below.

Simple Subcloning

Preparation of Vector and Insert. Ideally, it will be possible to remove the gene of interest from its current vector with the same restriction enzyme(s) that will be used to insert it into the final vector. To prepare enough insert and final vector for several ligations, we digest 10–20 μg of each purified plasmid with 20–40 units (usually 2–4 μl) of the appropriate restriction enzyme(s) (GIBCO/BRL, Gaithersburg, MD; New England BioLabs, Beverly, MA; Boehringer-Mannheim/Roche; or Promega, Madison, WI) for 1 hr at 37°C in a total volume of 30–50 μl, using the 10× digestion buffer supplied by the manufacturer. This constitutes a 2-fold excess of enzyme, for which 1 unit of activity is defined as the amount of enzyme required to digest 1 μg of DNA at 37°C in 1 hr. Simultaneous digestion with two different enzymes can be done if the digestion buffers required for each are compatible according to the manufacturer information. If two incompatible enzymes are necessary, digestion with one is followed by DNA purification with spin columns [as in the PCR Purification Kit (Qiagen, Valencia, CA) or similar product] and subsequent digestion with the second enzyme. Alternatively, DNA can be purified after the first digestion by phenol–chloroform extraction and ethanol precipitation as follows. The digestion reaction is brought to a total volume of 200 μl with distilled water to ensure that there is enough volume to work with easily. One volume (200 μl) of a mixture of 50% Tris-saturated phenol–48% chloroform–2%

isoamyl alcohol[32] is added to the diluted digestion and vortexed vigorously for 1 min. The sample is microcentrifuged at 16,000*g* for 1 min at room temperature to separate the layers. The top aqueous layer, which contains the DNA, is carefully removed to a clean tube. Usually, one extraction is sufficient. However, if a white precipitate is visible between the aqueous and phenol–chloroform–isoamyl alcohol layers after centrifugation, the extraction should be repeated as many times as necessary to remove it completely, in order to assure a high-quality DNA preparation. The last phenol–chloroform extraction is followed by a single extraction with one volume of 100% chloroform to remove residual phenol that can interfere with subsequent enzyme reactions. To precipitate the DNA, a 1/10 volume (20 μl) of 3 *M* sodium acetate, pH 5.2, and 3–5 volumes (600–1000 μl) of 100% ethanol are mixed with the aqueous layer from the chloroform extraction and kept at −80° for at least 1 hr. DNA is pelleted by centrifugation at 16,000*g* for 15 min at 4°. It is possible that a visible pellet will not be apparent at this stage. After carefully removing and discarding the supernatant, 500 μl of 70% (v/v) ethanol is gently added to the pellet and allowed to stand at room temperature for 5 min. The DNA is then repelleted by centrifugation at 16,000*g* for 5 min at 4°, the supernatant is removed, and the DNA is dried under vacuum. For long-term storage DNA can be resuspended in TE [10 m*M* Tris (pH 7.4), 1 m*M* EDTA, pH 8], which helps prevent DNA degradation by Mg^{2+}-dependent nucleases. However, if DNA is to be used immediately, resuspension in distilled water is recommended. Although phenol–chloroform extraction is a stringent method to ensure the removal of enzymes and other proteins from DNA preparations, we have found generally that DNA purification kits are sufficient for our applications.

Purification of Digested DNA. Because digestion of the insert-containing plasmid DNA will result in two fragments (the insert and the rest of the plasmid), these must be separated from each other before the insert fragment is purified and further manipulated. Likewise, digestion of the final vector may also result in two DNA fragments unless a single enzyme cutting at a single site is used. If digestion of the vector results in two DNA fragments (either with a single enzyme cutting at two sites or with two separate enzymes each cutting at a single site), these must also be separated before the vector fragment is purified. This is accomplished by agarose gel electrophoresis and gel purification, using the Gel Extraction Kit (Qiagen). We run 1% (w/v) agarose gels containing 1× Tris–acetate/EDTA (TAE: 40 m*M* Tris–base, 40 m*M* acetic acid, 1m*M* EDTA), which seems to permit

[32] J. Sambrook, E. F. Fritsch, and T. Maniatis, "Molecular Cloning: A Laboratory Manual." Cold Spring Harbor Press, Cold Spring Harbor, New York, 1989.

more efficient extraction of DNA from the gel than does Tris–borate/EDTA (TBE). After purification from the gel slice, 5% of each purified fragment is run on another agarose gel to confirm purity and estimate concentration.

Dephosphorylation of Digested Vector. If two restriction enzymes are used to create different ends for directional cloning of the insert (e.g., *Sal*I and *Bam*HI for pDCR; see Fig. 1), then the vector and insert preparations are ready for use after agarose gel purification of the digested DNA. However, if only one enzyme is to be used (e.g., *Bam*HI for pCGN and pZIP), it is also necessary to dephosphorylate the digested and compatible ends of the vector to prevent religation without insert. We do this by adding 15 units of calf intestinal alkaline phosphatase (CIAP; GIBCO, Boehringer-Mannheim/Roche, Promega, or New England BioLabs) directly to the restriction digest mixture afte digestion is complete. The vectors we use most often (pZIP and pCGN) do not contain multiple cloning sites; rather, each contains only a single *Bam*HI cloning site, and the digestion buffer for *Bam*HI is compatible with CIAP activity. However, if the digestion buffer used is incompatible with CIAP activity (such as *Kpn*I, *Sac*I, or *Xma*I), the digested vector must first be purified as described above either by the PCR Purification Kit (Qiagen) or by phenol–chloroform extraction and ethanol precipitation, and then treated with CIAP using the buffer supplied with the enzyme. However, CIAP activity is compatible with most commonly used restriction enzymes. After treatment with CIAP, it is once again necessary to purify the DNA by spin column or by phenol–chloroform extraction to ensure that no CIAP carries over to the ligation reaction (see below), where it will dephosphorylate the insert and prevent ligation of the two fragments.

Ligation of Digested Vector and Insert. Two hundred to 500 ng of vector is ligated to a 10-fold molar excess of insert with 5 units of T4 DNA ligase for 1 hr at room temperature in a total volume of 30–50 μl, using the reaction buffer supplied with the enzyme. An identical ligation with vector alone (without insert) is also done as a negative control and to estimate the probability that colonies on the vector plus insert plates are actually likely to contain insert. (If there are five times the number of colonies on the vector plus insert plate as on the vector-only plate, then four of five colonies are likely to contain insert. However, if there are similar numbers on each plate, then few if any of the colonies on the vector plus insert plate are actually likely to contain insert, and new vector and/or insert preparations should be made.) Half of each ligation reaction is transformed into *Escherichia coli* strain DH5α or similar strain and plated onto the appropriate antibiotic selection. The remaining ligation is allowed to continue overnight before transformation and plating in case it is necessary to screen additional colonies because of insufficient yield from the first transformation. Plasmids isolated from several individual bacterial colonies are analyzed for insert by digesting with the same restriction enzyme(s) used to prepare the vector and insert, which should result in the "dropping out" of the insert and its appearance on an agarose gel at the expected molecular weight. Determination of orientation of the insert is discussed below.

Polymerase Chain Reaction Generation of New Restriction Sites

Primer Design. Because of the variety of expression vectors, it is likely that the restriction sites available in the desired vector and insert combinations will not always be compatible for simple subcloning procedures. New sites may be introduced conveniently by amplifying the sequence of interest by PCR, using primers designed to include the appropriate new restriction sequences. The primer encoding the amino terminus of the protein (the 5′ primer) and the primer encoding the carboxyl terminus of the protein (the 3′ primer) should overlap the desired sequence by 18–24 bases to ensure specific priming. The 3′ primer should also include a translation "stop" codon at the end of the desired coding sequence and before the restriction site (unless a vector containing a carboxy-terminal epitope tag is being used, in which case a stop codon would stop translation before the tag is added). Each primer should also include any mutations to be introduced and, of course, the desired restriction sequence(s), followed by an additional 3–5 base pairs (bp) of any sequence. These extra bases will be incorporated into the PCR product, ensuring that the restriction sites will be far enough from the ends of the DNA to be digested efficiently in the next step. Naturally, when adding new restriction sites to an insert, the sites chosen must not be present within the insert itself or else subsequent digestion of the insert for ligation into vector (see below) will destroy the insert. This condition will also limit the choice of vector.

Assuring that Coding Sequences Are in Frame with Epitope Tags. As mentioned above, either the 5′ or the 3′ primer must be designed to place the coding sequence of interest in frame with the coding sequence of any epitope tag that may be present in the vector. To do this, the exact relationship between the restriction site and the coding sequence of the tag within the vector must be known. For example, if *Bam*HI is the restriction site (5′-GGATCC-3′), it must determined whether the coding sequence of the tag utilizes the GGA, the GAT, or the ATC triplet within the *Bam*HI site as a codon. The primer must then be designed to maintain the relationship between the *Bam*HI site triplet/codons and the coding sequence of interest after the fragments are ligated. Failure to keep the coding sequence of the tag and the coding sequence of interest in frame with each other will result in the expression of untagged proteins and possibly low overall levels of expression.

Polymerase Chain Reaction Conditions. Our standard PCR is performed under the following conditions (unless otherwise stated, all reagents are from GIBCO-BRL, Gaithersburg, MD).

- Reaction components (50-μl total volume)
  - *Taq* PCR buffer, 1× (minus Mg^{2+}; supplied by the manufacturer)
  - dNTPs, 200 μM each
  - $MgCl_2$, 1 mM
  - Primer 1, 2 μM
  - Primer 2, 2 μM
  - Template DNA, 0.5–1 μg
  - *Taq* polymerase, 5 units
- Thermocycle programming
  - Program 1 (1 cycle): 95°, 10 min
  - Program 2 (30 cycles)
    - Segment 1: 95°, 1 min
    - Segment 2: 35°, 1.5 min
    - Segment 3: 72°, 2 min
  - Program 3 (1 cycle): 72°, 10 min

Parameters such as [Mg^{2+}], segment length, cycle number, and annealing temperature (segment 2) can be varied to optimize for each amplification,[33] but these conditions have proved reliable in our hands. Although *Taq* is the enzyme of choice for most PCRs, because it is both relatively inexpensive and easy to optimize, other thermostable polymerases such as *Pfu* [which we typically obtain from Stratagene (La Jolla, CA), although there are several other suppliers] are desirable for certain applications. For example, because *Pfu* has higher fidelity than *Taq*, it may be used to amplify larger target sequences (i.e., <1000 bp). Also, unlike *Taq*, which can leave single-base, 3′-adenosine overhangs on each DNA strand, *Pfu* leaves blunt ends, which may be useful for certain subcloning protocols. Finally, we have observed that some Ras family-coding sequences (such as H-*ras*) seldom amplify with any errors, and are therefore quite suitable for *Taq* amplification, whereas others (such as R-*ras*) are more error prone, and best done with *Pfu*.

Cloning Polymerase Chain Reaction Products into Vector. After amplification, the PCR product is separated from template DNA on a TAE–1% (w/v) agarose gel, purified with the Gel Extraction Kit (Qiagen) or similar product, digested, and ligated into digested vector as described in Simple

[33] B. A. White (ed.), "PCR Cloning Protocols: From Molecular Cloning to Genetic Engineering." Humana Press, Totowa, New Jersey, 1997.

Subcloning (above). However, it is sometimes difficult to digest PCR products efficiently, because of the proximity of the restriction sites to the end of the linear DNA strand. Moreover, the efficiency of digestion of PCR products cannot be evaluated with a simple agarose gel, because the product before and after digestion has essentially the same molecular weight. Thus, any problems at this step will not be detected until bacterial colonies have been selected. If digestion of the PCR product does turn out to be inefficient, rather than directly digesting and then ligating the PCR product into the final vector, it is possible to place the gel-purified insert into an intermediate vector without prior digestion. *Taq*-amplified PCR products, which contain 3′-adenosine overhangs on each DNA strand, can be ligated into the TA cloning vector, the TOPO TA cloning vector, or the pBAD TOPO TA cloning vector (InVitrogen), while *Pfu*-amplified products can be ligated into the pCR-Script or pCR-Blunt vectors (Stratagene). (If amplification with a different thermostable polymerase is desired for any of the reasons described above, performing a final single round of PCR with the alternate enzyme will generally add sufficient appropriate ends to permit ligation into the intermediate vector of choice.) Each of these vectors is easily screened for the presence of insert by blue/white color selection. Once the PCR product has been introduced into the intermediate vector, digestion of the insert using the restriction enzymes introduced by PCR occurs efficiently, and allows for easy gel purification and ligation of the insert into the desired final vector as described above (Simple Subcloning).

In addition to introducing new restriction sites for subcloning, this PCR technique has been used extensively by us to generate mutations in the Ras sequence near the carboxy terminus. The maximum length of synthetic oligonucleotide primers ($\approx$100 bases) limits the introduction of mutations to within approximately 25–30 amino acids of the end of the protein with this technique.

Confirmation of Insert Sequence. Although the sequence of all constructs generated should be confirmed by automated sequencing prior to use regardless of which construction method is used, this is especially true of PCR-generated constructs because of the potential introduction of mutations by the polymerase [even by high-fidelity enzymes such as *Pfu* and *Vent* (New England BioLabs)]. Final inserts should be sequenced from both ends, using primers that are located entirely within the vector sequence. Primers within the vector should be designed to assure that the sequence obtained will cross the junction between vector and insert, thus confirming both orientation and frame. The importance of confirming both of these parameters cannot be overstated. Sequencing from both ends has two advantages. First, one sequence can be used to confirm the other. Second, because the quality of sequence data degrades as it proceeds away from the primer (good sequence can usually be obtained up to 600–700 bp from a primer), it guarantees that high-quality data will be available for most or all of the insert. For sequences larger than ≈1200 bp, it may be necessary to use a primer internal to the insert in addition to primers flanking the sequence.

Shuttle Vectors

Because most vectors contain multiple cloning sites (MCS) with several restriction sites available to receive insert, it may be possible to obtain new restriction sites while avoiding PCR, by simple subcloning of an insert into an intermediate shuttle vector. A vector is suitable for use as a shuttle vector if it contains the desired sites flanking the currently available site(s). For example, if the current insert is an *Eco*RI fragment that is to be put into pZIP, which contains only a *Bam*HI site, then the insert can be removed from its current vector by digestion with *Eco*RI as described above and ligated into the MCS of a shuttle vector such as pDUB that has been similarly digested with *Eco*RI. The insert can then be removed from pDUB by digestion with *Bam*HI, sites for which flank the single *Eco*RI site used for cloning into pDUB. The insert now has *Bam*HI ends (with internal *Eco*RI sites) and is ready to be ligated into pZIP. This technique can be used with any enzyme(s) for which there are sites flanking the insert in the shuttle vector (but not in the insert). It is also a useful way of picking up different ends for directional cloning.

The need to keep the coding sequence of interest in frame with an epitope tag in the final vector, the need to discriminate between ligations in either orientation (if a single restriction enzyme is used and a directional result is desired), and the necessity to find a compatible shuttle vector with sites in the right sequence and orientation for the final vector mean that this approach is not always suitable. (For example, pDUB could not be used as a shuttle vector to make a *Bam*HI insert for ligation into pCGN because the resulting insert would be out of frame with the HA tag of the pCGN vector.) Nevertheless, the time and effort saved when such an option is available makes it worthwhile to consider whether a compatible shuttle vector exists for a given insert/final vector combination. If subcloning is done frequently and with a limited number of inserts and vectors, it may even be worthwhile to generate by cassette mutagenesis a multipurpose shuttle vector (see the next section) containing restriction sites that would allow easy shuttling of all inserts among all vectors.

Modification of Vector Multiple Cloning Sites by Cassette Mutagenesis

As described above, it is possible through several techniques to modify the insert to include restriction sites necessary for insertion into the final vector. Alternatively, it is possible to modify the vector to contain the sites required by the unmodified insert. This is accomplished by introducing a DNA "cassette" into the final vector prior to ligation of the insert.

Primer Design. Two synthetic oligonucleotides are designed to anneal with each other to produce a double-stranded cassette containing any sequence desired. In addition to the restriction sites required by the final insert, the oligonucleotides should be designed with 3′ and/or 5′ overhangs that will permit the cassette to be ligated directly into the vector at whatever sites are available without prior digestion. For example, if a vector contains *Bam*HI and *Hin*dIII sites while the insert contains *Eco*RI and *Sal*I sites, a cassette spanning the *Bam*HI–*Hin*dIII region can be designed to include *Eco*RI and *Sal*I sites. It is important to remember that the cassette must not introduce sites that already exist within the vector, either within the MCS or elsewhere in the plasmid. In this example the following oligonucleotides would be synthesized:

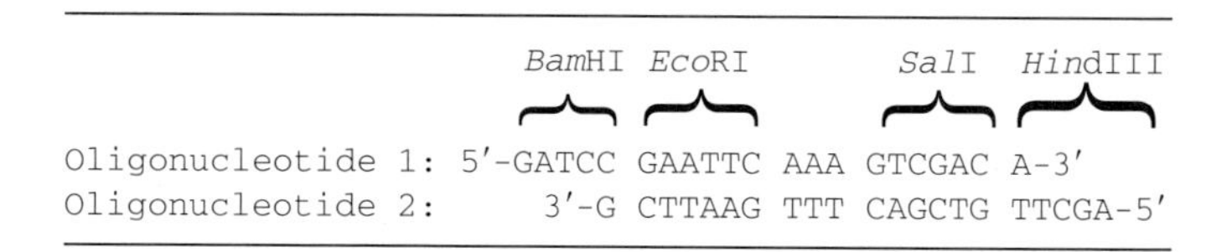

The three A/T pairs in the center of the cassette are necessary to separate the *Eco*RI and *Sal*I sites so that each enzyme will efficiently digest the vector for the final insert.

Generation and Ligation of Cassette. Each oligonucleotide (typically 40 nmol) is resuspended in 200 μl of TE [10 m*M* Tris (pH 7.4), 1 m*M* EDTA, pH 8]. Five microliters of each oligonucleotide, 5 μl of 10× phosphate-buffered saline (PBS: 15 m*M* monobasic potassium phosphate, 80 m*M* dibasic sodium phosphate, 27 m*M* potassium chloride, 1.4 *M* sodium chloride, pH 7.4), and 35 μl of distilled water are mixed and placed in a 500-ml beaker of water at 90°. The annealing reaction is permitted to cool to room temperature and then placed briefly on ice. This cassette can now be treated identically to any other digested DNA insert. A 10-fold molar excess of cassette is ligated into vector that has been appropriately digested and gel purified as described above. After transformation into DH5α (or other bacterial strain), plasmids can be easily screened for insertion of the cassette by digestion with each of the enzymes for which new sites have been introduced (*Eco*RI and *Sal*I in the example above). Plasmid clones that become linearized when so digested (as visualized by agarose gel) are then digested with both enzymes, to receive similarly treated insert as described above.

Other Types of Cassettes. As described in other chapters of this series, such cassettes can also be designed to introduce peptide sequences, such as amino-terminal myristoylation[34] or carboxy-terminal prenylation,[35] into any coding sequence introduced into the vector. The sequences that can be inserted into vectors by this method are limited only by the length of synthetic oligonucleotides that can be obtained. However, even this limitation can be overcome by the sequential insertion of two or more cassettes forming a single continuous sequence.[36]

Determination of Insert Orientation

Some of the vectors discussed above [the original pZIP-NeoSV(X)1 and pCGN] have only a single restriction site (*Bam*HI) for the insertion of the protein-coding sequence. Inserts placed into these vectors can be in two orientations, only one of which will permit protein expression. Two methods can be used to confirm the correct orientation. First, a diagnostic restriction digest can be performed if an appropriate restriction site(s) can be found. Ideally, a single enzyme can be found for which one site exists within the vector and a second site exists within the insert. When possible, inclusion of an appropriate restriction site in the 3′ primer for insert PCR can facilitate this process. For example, pZIP contains a *Sal*I site 2.0 kb upstream of the *Bam*HI cloning site. A *Sal*I site incorporated into the 3′ primer (outside the termination codon but internal to the *Bam*HI site to be used for cloning into the vector) would permit determination of the presence and orientation of insert with a single *Sal*I digest (assuming that *Sal*I is not present in the insert coding sequence). It is imperative that the site within the insert not be located near the center of the insert sequence, but nearer to one end or the other. Digestion with this enzyme will generate fragments of different sizes for the two possible insert orientations, allowing clones with the correct orientation to be identified. Also, the two sites should be appropriately spaced such that the difference in size between the two possible fragments can easily be distinguished by separation on an agarose gel.

Alternatively, a diagnostic PCR can be performed (see above) in which one primer is designed to correspond to a sequence within the correctly oriented insert while the other primer corresponds to a sequence within

[34] P. A. Solski, L. A. Quilliam, S. G. Coats, C. J. Der, and J. E. Buss, *Methods Enzymol.* **250,** 435 (1995).

[35] G. W. Reuther, J. E. Buss, L. A. Quilliam, G. J. Clark, and C. J. Der, *Methods Enzymol.* **327,** 331 (2000).

[36] J. J. Fiordalisi, C. H. Fetter, A. TenHarmsel, R. Gigowski, V. A. Chiappinelli, and G. A. Grant, *Biochemistry* **30,** 10337 (1991).

the vector. With these primers, a product of the appropriate size will be generated only when the insert is in the correct orientation, whereas no product will be generated when it is in backward.

A simple alternative to the orientation problems would be to use vectors capable of accepting directionally cloned inserts. To this end, variations on pZIP have now been generated such that inserts can be cloned in as *Bam*HI–*Eco*RI or as *Eco*RI–*Bam*HI fragments (see Fig. 1).

Importance of Keeping Good Records and Sequence Maps

The preceding discussion demonstrates clearly the potential complexity of DNA manipulation. Prior to designing a subcloning project, we strongly advise accumulating all available information concerning the sequences and restriction sites of the vectors and inserts to be used. Of course, many vectors have not been completely sequenced, and indeed only rough restriction maps may be available for "homemade" vectors.

A computer-based database of vector, insert, and construct sequences that includes complete restriction maps (when available) in a format that can be easily searched is of tremendous benefit in planning subsequent manipulations and for keeping long-term records. Moreover, this information can be provided to other researchers whenever vectors or constructs change hands, ensuring that this information is not lost over time. In our laboratories, we use a program called Gene Construction Kit II (GCK II; Textco, West Lebanon, NH) that is extremely helpful in documenting the generation of new constructs both graphically and as specific, searchable DNA sequences. Figure 1 was generated using GCK II.

Transfection of Mammalian Cells

Several techniques can be used to introduce plasmid DNA into mammalian cells. These include (1) transfection by calcium phosphate precipitation of DNA, (2) transfection by lipid–DNA complexing, (3) infection by retrovirus, and (4) electroporation. Of these, we have found the first three to be satisfactory for all our applications. (Retroviral infection of mammalian cells is discussed separately below.) To maximize the desired biological readout and reduce the need for large amounts of DNA, the efficiency of the transfection method chosen is of prime importance. The simplicity of the method and the cost of reagents are also factors in the selection of a procedure. Several variables, especially the cell type to be transfected, can affect the efficiency of each of the methods discussed below and, if necessary, each technique should be optimized for a given application. Each technique can be used either to generate cell lines stably expressing the protein of interest or to produce transient protein expression for assays of limited duration. First we discuss each transfection method and then we discuss how each can be used for transient or stable transfection.

Transfection by Calcium Phosphate Precipitation of DNA

NIH 3T3 cells are particularly efficient at the uptake of calcium phosphate–DNA precipitates.[30,31] For the generation of stable cell lines, using NIH 3T3 cells and a plasmid with a selectable marker, we typically transfect 50–100 ng of DNA onto a 60-mm dish.[30,31] If no selectable marker is available, we cotransfect with a 40- to 100-fold molar excess of a second, selectable plasmid containing an antibiotic resistance gene but no expressed insert. After selection, several thousand colonies per microgram of DNA are typically obtained. Transient transfections are done with 10 ng to 2 μg of plasmid DNA per dish (with or without selectable marker), depending on the sensitivity of the biological readout and the expression levels achieved by the plasmid. Focus-forming assays using activated Ras proteins require 10–100 ng of DNA per dish, whereas enzyme-linked transcription factor reporter assays require 50–200 ng of plasmid per dish.

For focus assays and for generation of stable cell lines, NIH 3T3 cells are plated the day before transfection at 2.5×10^5 cells per 60-mm dish, or are plated 2 days before transfection at 1.25×10^4 cells per dish. For enzyme-linked transcriptional activation reporter assays, NIH 3T3 cells are plated at 1×10^5 per 35-mm dish the day before transfection, or at 5×10^4 cells 2 days before transfection. Just prior to transfection, high molecular weight carrier DNA (calf thymus DNA; Boehringer-Mannheim/Roche) is made up to 40 μg/ml in *N*-2-hydroxyethylpiperazine-*N'*-2-ethanesulfonic acid (HEPES)-buffered saline (HBS), *exactly* pH 7.05 (20 m*M* HEPES, 140 m*M* sodium chloride, 5 m*M* potassium chloride, 1.3 m*M* dibasic sodium phosphate, 5.5 m*M* glucose). Make enough of this mixture for 500 μl per 35- or 60-mm dish. Aliquot this mixture into separate polystyrene tubes (polypropylene will bind the DNA and reduce transfection efficiency) to receive each plasmid to be transfected. To each tube, add the appropriate amount of plasmid DNA followed by a 1/10 volume of 1.25 *M* calcium chloride. Vortex briefly and let stand for 15 min at room temperature. Precipitated DNA should appear as a fine white powder. Add 500 μl per 35- or 60-mm dish of the DNA dropwise to each dish so that it is evenly spread over the plate. Cells should be kept in complete (i.e., serum-containing) medium during the procedure. Return the cells to standard incubation conditions for 3 to 5 hr. To ensure efficient DNA uptake, cells should be glycerol shocked after incubation as follows. Aspirate DNA-containing medium from the cells and wash once with fresh medium. Add 1–2 ml of 15% (v/v) glycerol in HBS to each plate and rock to cover the cells with an even layer. Immediately aspirate the glycerol. Cells must be exposed to glycerol for a total of no more than 3 min from the initial addition, as glycerol is toxic to cells. After 3 min, add 2–3 ml of fresh medium to each dish to stop glycerol shock and aspirate. Replace with complete growth medium (with serum, without selection antibiotic) and return to normal growth conditions. Subsequent treatment of cells will depend on the assay being performed.

Transfection by Lipid–DNA Complexing

Unlike NIH 3T3 cells, most cell lines do not take up calcium phosphate–DNA complexes efficiently, and are better transfected by liposome-mediated transfer. Several lipid-based transfection reagents are commercially available. We routinely use LipofectAMINE (GIBCO-BRL/Life Technologies, Gaithersburg, MD), SuperFect (Qiagen), or the expensive but efficient reagent FuGENE (Boehringer-Mannheim/Roche), which form liposome–DNA complexes in which the DNA is contained within a lipophilic vesicle that can fuse with and penetrate the cell membrane. We have also successfully used Effectene (Qiagen), which the manufacturer describes as a "non-liposomal lipid."

With all reagents, high transfection efficiency depends on achieving optimum ratios of DNA to reagent, reagent to cell number, and DNA to cell number and will depend on the cell line as well as several other factors. In our hands all have performed adequately for generation of stable cells lines using such diverse cell types as human embryonic kidney cells (HEK293), human breast epithelial cells (T-47D), and rat intestinal epithelial cells (RIE-1), which are not as easily transfected by the much less expensive calcium chloride methods. We have also observed greater activity in enzyme-linked reporter assays when these reagents are used, compared with calcium chloride transfection, suggesting that transient transfection assays can also benefit from their use. In particular, consistent, high transfection efficiencies are required for reporter assays, and in cells such as RIE-1 this is not achievable in our hands without use of Superfect or FuGENE. In contrast, NIH 3T3 fibroblasts transfect well with calcium and we do not use lipid reagents with these cells. For experiments requiring the highest possible transfection efficiency and consistency, we suggest that a side-by-side comparison be done before choosing one reagent over another. We have found a surprising degree of variation, such that even different individuals within the same laboratory, after using the same reagents to transfect the same expression constructs into the same cells, swear by different optimal liposomal reagents!

Transfection with each reagent is performed essentially as described in the protocols provided by the manufacturers. In all cases 1–2 μg of DNA is mixed with the lipid reagent and allowed to complex. Adherent, cultured cells are then exposed to the lipid–DNA complexes for 2–24 hr, during which time they take up the DNA. Cells are then returned to normal growth conditions until needed for the assay being performed.

Effect of Transfection Procedure on Biological Activity. We have observed in certain instances that the transfection procedure used can influence the outcome of an assay. For example, transient transfection of NIH 3T3 fibroblasts with activated R-Ras(38V) or R-Ras (87L), using FuGENE, resulted in suppression of NF-κB transcriptional *trans*-activation activity in a reporter assay, whereas an identical, simultaneous transfection using calcium chloride precipitation resulted in a reproducible 2.5-fold activation of NF-κB activity, even when assays were normalized to β-galactosidase (β-Gal) to account for transfection efficiency. Although we have not explored these unusual and unexpected observations in depth, they clearly suggest that care must be taken in choosing a transfection method and that results should be confirmed by a different method.

Stable versus Transient Transfection

Any of the described methods can be used for stable or transient transfection of cells. If cells stably expressing the protein of interest are not needed, then after transfection the cells are simply subjected to whatever treatment is appropriate for the assay being performed (i.e., cells may be lysed, fixed, and stained, etc.). A detailed discussion of various techniques for analyzing Ras proteins and their functions can be found in Refs. 27 and 31. However, many assays for Ras function, such as for anchorage-independent growth or migration, require a population of cells that are all expressing the protein of interest, so stable expression in selected cells is required.

Stable Transfection

Forty-eight hours after transfection, medium is aspirated from the transfected cells. Cells are washed once with HBS and exposed to 1 ml of trypsin–ethylenediaminetetraacetic acid (EDTA) (75 units of trypsin/ml, 1.5 m*M* EDTA, pH 8.0) for 5 min at 37° to remove the cells from the dish. The nonadherent cells are triturated briefly with a pipet to ensure a single-cell suspension, which is then split to two (or more) 100-mm plates containing complete growth medium with the appropriate selective antibiotic [i.e., G418/geneticin (400 μg/ml) or hygromycin B (200 μg/ml)]. The amount of the trypsinized cell suspension passaged into selection will depend on the efficiency of transfection, the density of the transfected culture, and the number of stably expressing clones desired. Normally we split the cell suspension from the original 60-mm dish such that one-third and two-thirds of the total volume are each plated onto a 100-mm dish, resulting in 1 : 10 and 1 : 5 splits, respectively, which yields at least 1 plate with 50–70 colonies. If a polyclonal population of cells is desired to avoid the problem of clonal variation, these colonies can be pooled. However, if individual clones are desired, cells should be passaged into selection at a lower density to ensure that individual colonies can be isolated (see below). Also, if transfection efficiency is high, then we typically will passage only one-sixth to one-tenth of the original culture. The selective antibiotics will kill nonresistant cells only if the cells are actively dividing. Thus it is important to passage cells into selection at a density that will remain subconfluent for 3–5 days, at which point cell death should be well underway.

After passage into selective medium, cells are returned to standard growth conditions and fed with fresh selective medium every 3–4 days. After 2–3 days in hygromycin B, the vast majority of cells will begin to die, leaving the plate apparently almost completely empty. Selection in G418 takes significantly longer, from 5 to 10 days. After 10–30 days (depending on the cell line and transfection efficiency), growth will be sufficient that colonies will be clearly visible even without microscopic examination. Selection is complete when there is no longer evidence of cell death and well-separated colonies are apparent. Selected cells may be removed from selective medium and returned to normal growth medium if desired. We generally find it unnecessary to maintain selective pressure on Ras-expressing cells. An exception is RIE-1 cells, in which Ras expression is often much less well tolerated than that of other oncogenes.

Retroviral Vectors for Infection of Mammalian Cells

As an alternative method to transfection, retroviral infection of mammalian cells offers a number of advantages. First, infection is generally much more efficient in delivering DNA to cells. This results in a higher percentage of cells expressing the desired protein, commonly 70–90% of a population when viral particle number is not the limiting factor in infection. For transient assays, achieving a higher percentage of cells expressing a construct can increase the intensity of a desired signal and provide a more accurate representation of the behavior of a population of cells in response to the exogenous protein. For stable expression, infection delivers the DNA construct to a majority of exposed cells, making drug selection to establish a population much faster. Second, infection can be used for certain cell types that are difficult to transfect, such as rat intestinal epithelial cells (RIE-1) and human epithelial cell types including DLD-1 and HCT116 colon cancer cells and MCF-10A breast cells.

Retroviral Vectors and Viral Packaging

To infect cells, cDNA sequences must first be shuttled into retroviral vectors. As described above, these vectors have 5′ and 3′ LTRs flanking a mammalian selection marker and an MCS into which the coding sequence of interest can be cloned. pBABE-Puro, pCTV3H, and pZIP-Neo are common retroviral vectors that provide puromycin, hygromycin, and neomycin resistance, respectively (Table I). One advantage to these three vectors is that using virus produced with different vectors, cells can be simultaneously infected with two constructs that provide different resistance during double drug selection. Also, pBABE and pCTV3H are high-yield vectors and easy to purify in large quantities when propagated in bacteria. Moreover, they each provide several restriction sites within the MCS, facilitating directional insertion of coding sequences. In contrast, pZIP-Neo is a low-copy vector and its yield in bacteria should be enhanced by the use of chloramphenicol in the growth medium to enhance plasmid DNA production.[6]

DNA cloned into retroviral vectors must be packaged into virus by first transfecting packaging cell lines with the desired construct. Packaging lines are human or murine cell lines that stably express the Gag, Pol, and Env viral proteins necessary for packaging viral particles to form infectious virus. Depending on the identity of these proteins expressed in the packaging lines, the viral particles produced may infect murine or human cells. Ecotropic viruses, regardless of whether they are produced in murine- or human-derived packaging lines, infect rodent cells. Amphotropic viruses can infect human cells in addition to rodent cells. Examples of packaging lines are Bosc23 cells,[4] which produce ecotropic virus, and Phoenix[37] or Bing[4] cells, which produce amphotropic virus. Each of these cell lines is derived from 293T human embryonic kidney epithelial cells.

We routinely infect NIH 3T3 and rat intestinal epithelial (RIE-1) cells with activated Ras family members [H-Ras(61L) and H-Ras (12V), K-Ras(12V), and N-Ras(12D) and N-Ras (61K)] (Table III) to establish stable cell lines. Similar methods using amphotropic virus can be applied to infect human cells; however, extreme caution must be taken in making and handling amphotropic virus encoding dominant positive oncogenes. These viruses have only 1- to 4-hr half-lives at 37°, but the potential for transmission from aerosol or liquid contamination necessitates first determining the requirements of each institution for working with viruses. Although eco-

[37] T. M. Kinsella and G. P. Nolan, *Hum. Gene Ther.* **7,** 1405 (1996).

tropic virus has not been shown to infect human cells, precautions must be taken to limit exposure to retrovirus by deactivating virus in solutions or on plastic surfaces before removal from the hood by using 10% (v/v) bleach. Solutions treated with 10% (v/v) bleach can be safely poured down the drain after 5 min.

An alternative to using amphotropic virus with activated oncoproteins is to establish human cells that express the ecotropic virus receptor and use ecotropic virus for infection. Regardless of the virus type being used, all cell lines to be infected should be tested for the presence of helper virus that would allow infected cells to produce virus themselves. Helper virus can be detected by assaying the target cell line supernatant for reverse transcriptase (RT) activity, using either a chemiluminescent or colorimetric RT activity kit (available from Boehringer-Mannheim/Roche).

Production of Virus

To package ecotropic and amphotropic virus, respectively, we use Bosc23 and Phoenix cells maintained in Dulbecco's modified Eagle medium (DMEM) supplemented with 10% fetal calf serum (FCS). Cells are plated at 2×10^6 per 60-mm dish and allowed to grow to confluence overnight. The following day (day 1) the medium is changed and the DNA to be transfected is prepared. Five micrograms of DNA is diluted in 0.9 ml of HBS with 100 μl of 1.25 M $CaCl_2$. The mixture is vortexed, incubated at room temperature for 1–2 min, and added to the cells. The medium is changed on day 2 and virus is collected on day 3. Harvested virus is filtered through a 0.45-μm syringe filter and stored in 0.5- to 1-ml aliquots at $-80°$ for up to 6 months. Growth medium can also be placed on cells on day 3 for viral collection on day 4; however, titers for the second collection are much lower. In addition, virus infectivity drops with each successive freeze–thaw of stored aliquots, so storage of smaller aliquots is advisable. All solutions and plastic material in contact with virus must be bleached for at least 5 min before removal from the tissue culture biohazard hood. The hood surfaces should also be wiped down with 10% (v/v) bleach.

Infection of Cells

For efficient infection, cells should be plated at 10–20% confluency in 60-mm dishes. On day 1, fresh or stored virus is diluted in an equal or excess volume of medium to a total volume of 1 ml. Polybrene (hexadimethrine bromide; Sigma, St. Louis, MO) is added to a final concentration of 8 μg/ml, and the viral solution is mixed gently. The medium is aspirated, and the cells are incubated for 3–4 hr with the viral mixture at 37°. After incubation, virus is aspirated and fresh medium added. Cells can be split on day 2 or day 3 depending on growth rate, and should be maintained at or below 70% confluence. On day 3, drug is added to the growth medium to begin selection of infected cells for stable protein expression. For transient assays, cells can be collected when protein expression is believed to peak, at ~48 hr. As described above, selection may take 4–10 days or longer, depending on the drug used and the cell type, but cell death should be apparent after 2–5 days. Selection is more rapid when cells have been infected than when they have been transfected. It has been our experience that splitting cells into drug on day 3 increases initial cell death. In addition, when cell death seems to have stopped, splitting cells to maintain subconfluent cultures can also lead to more death.

Titering of Virus

When using virus to establish cell lines stably expressing a desired protein, viral titering is not necessary because the process of drug selection eliminates uninfected cells, making the relative number of infected cells unimportant. However, when using virus to transiently express protein in cells, it is important to determine the percentage of cells that become infected or else overall protein expression levels in an infected cell population will not necessarily reflect equivalent expression levels on a per-cell basis. For example, if the overall protein expression levels in two separate infections are equal, this may reflect either an equal number of cells each expressing equivalent levels of protein or it may reflect fewer cells in one infection each expressing higher levels of protein than cells in the other infection. To achieve relatively equal protein expression in cells infected with different viruses, the titer of each virus should be determined in order to estimate what percentage of cells are infected. In this way overall protein expression levels can be normalized to the number of infected cells. Each new preparation of virus needs to be titered separately, because variations in packaging cells and transfected DNA can alter viral production.

Titering is achieved by infecting cells with known volumes of virus and proceeding with drug selection as described above. On day 3, cells are split at different ratios (e.g., 1:5, 1:10, 1:20). Within 10–12 days, when drug selection is complete, colonies are counted before they spread. The number of colonies is divided by the fraction of cells plated, and also divided by the volume (ml) of virus used to infect, giving the titer in colony-forming units (CFU) per volume of virus [e.g., 20 colonies in 1:5 dilution from 100 μl virus = $20/(0.2 \times 0.1)$ = 1000 CFU/ml]. If infected cells have vastly different replication rates, a replication factor should be added to the titering equation. However, replication rates over the 2 days prior to splitting seldom differ significantly and are often ignored. Titers that vary by a factor of 3 or less are considered to be equal; titers varying >3-fold are significantly different. It is also important to be aware that different cells take up viruses to different extents, so infecting normal versus Ras-transformed cells with the same viral titer can give different levels of infection and subsequent protein expression.

In our experience, titers can be expected to range from 10^4 to 10^5 CFU/ml using Bosc23 cells, although titers as high as 10^6 CFU/ml have been achieved. Viral titer can vary on the basis of the packaging system used and the vector being packaged. For example, larger vectors routinely give lower titers. Moreover, the quality of the packaging cell line also influences viral titer, and we routinely replenish packaging cell lines with low-passage cell stocks if titers fall below 10^3 CFU/ml.

Conclusion

The use of mammalian expression vectors to express dominant activated or dominant negative mutants of Ras and Ras superfamily proteins has provided powerful and versatile approaches to assess their signaling and biological properties. The ectopic overexpression of Ras GTPases in a range of cell-based assays has yielded much of the information that provides the foundation for our current understanding of the role of Ras proteins in regulating signaling as well as normal and malignant cell physiology. In this chapter, we have summarized experiences with the application of a variety of mammalian expression vectors for transient and stable expression studies in a range of rodent and primate cell lines. We have provided some general guidelines for the choice, construction, and application of expression vectors. A major conclusion from these analyses is that no one vector is ideally suited for all applications or cell lines. Those useful for stable expression may be inadequate for transient expression. Species and cell type differences can greatly influence the usefulness of a particular vector. Finally, the vector-specific nature of some assays also prompts a caution regarding observations made from exogenous overexpression of Ras proteins from heterologous promoters. The general tendency is to utilize a particular expression vector because it yields a positive response for a specific assay. If vast overexpression is required to achieve a response, consideration should be given to whether such a response represents a physiologically relevant function of Ras. Nevertheless, despite the obvious concerns associated with ectopic protein expression studies, the importance of this approach to the study of Ras family function in mammalian cells, where genetic manipulation options are limited, will continue to be considerable.

Acknowledgments

We thank all members of the Cox and Der laboratories for sharing their experiences with different combinations of vector, insert, and readouts. We also thank Gwen Mahon, Todd Palmby, and Ben Rushton for generation of expression constructs. Our research is supported by NIH grants to C.J.D. (CA42978, CA59577, CA63071), to ADC (CA76092) and to both C.J.D. and A.D.C. (CA67771).

[2] Protein Transduction: Delivery of Tat–GTPase Fusion Proteins into Mammalian Cells

By ADAMINA VOCERO-AKBANI, MEENA A. CHELLAIAH, KEITH A. HRUSKA, and STEVEN F. DOWDY

Introduction

Small GTPases, such as CDC42, Rac, and Rho, regulate the cytoskeletal architecture of the cell depending on the type of extracellular signals received.[1–3] For instance, Rho plays an important role in the formation of actin stress fibers and focal adhesions. In addition, Rho has been implicated in the formation of podosomes in osteoclasts required for bone adsorption.[4–6] Because osteoclasts are essentially resistant to the introduction of expression constructs by transfection or retroviral delivery, it is difficult to address specific questions involving Rho and/or other small GTPases. Although Rho proteins can be microinjected into osteoclasts, the number of injected cells is extremely limiting and excludes most biochemical analyzes. Therefore, to dissect the requirement for Rho function in podosome development, we applied the method of protein transduction to deliver constitutively active and dominant-negative forms of Rho to ~100% of osteoclasts.

The proof of concept for the transduction of proteins into cells was first described in 1988 independently by Green and Loewenstein[7] and Frankel and Pabo[8] with the discovery that human immunodeficiency virus (HIV)

[1] A. J. Ridley and A. Hall, *Cell* **70,** 389 (1992).
[2] C. D. Nobes and A. Hall, *Cell* **81,** 53 (1995).
[3] A. J. Ridley, *Nat. Cell Biol.* **1,** E64 (1999).
[4] P. C. Marchisio, D. Cirillo, L. Naldini, M. V. Primavera, A. Teti, and A. Zambonin-Zallone, *J. Cell Biol.* **99,** 1696 (1984).
[5] D. Zhang, N. Udagawa, I. Nakamura, H. Murakami, S. Saito, K. Yamasaki, Y. Shibasaki, N. Morii, S. Narumiya, N. Takahashi, *et al. J. Cell Sci.* **108,** 2285 (1995).
[6] M. Chellaiah, C. Fitzgerald, U. Alvarez, and K. Hruska, *J. Biol. Chem.* **273,** 11908 (1998).
[7] M. Green and P. M. Loewenstein, *Cell* **55,** 1179 (1988).
[8] A. D. Frankel and C. O. Pabo, *Cell* **55,** 1189 (1988).

0076-6879/00 $35.00

Tat protein can cross cell membranes. In 1994, Fawell *et al.*[9] expanded on this observation by demonstrating that heterologous proteins chemically cross-linked to a 36-amino acid domain of Tat were able to transduce into cells. However, these reports were not followed by a method to generate and efficiently transduce proteins of interest into cells. Subsequent to the Tat discovery, other transduction domains have been identified that reside in the Antennapedia (Antp) protein from *Drosophila*[10] and in herpes simplex virus (HSV) viral protein 22 (VP22).[11] The exact mechanism of transduction across cellular membranes remains unclear; however, Tat peptides and fusion proteins have been shown to transduce into cells at 4° in a receptor less fashion.[12] Thus, in principle and practice, all cell types are potentially targetable by protein transduction.

The exact mechanism of transduction across lipid bilayers is currently unknown; however, we hypothesized that, because of reduced structural constraints, higher energy (ΔG), denatured proteins may transduce more efficiently into cells than lower energy, correctly folded proteins. Once inside the cell, transduced denatured proteins would be correctly refolded by chaperones[13] such as heat shock protein 90 (HSP90).[14] Indeed, an analysis of Tat–p27^{Kip1} protein revealed that urea-denatured proteins elicit biological phenotypes more efficiently than soluble, correctly folded protein.[15] However, correctly folded proteins are capable of transducing into cells and therefore, denaturation, while apparently advantageous in most situations and results in dramatic yield increases, is not an absolute obligatory step.

We describe here the methodology to generate and transduce full-length Tat fusion proteins that can be applied to a broad spectrum of proteins independent of size or function. Briefly, bacterially expressed N′-terminal in-frame Tat fusion proteins are isolated from bacteria by sonication in 8 *M* urea. The use of 8 *M* urea achieves two goals. First, the majority of recombinant proteins in bacteria are present in inclusion bodies as denatured insoluble proteins, especially full-length proteins. Sonication in urea solubilizes this material, thus allowing for its isolation. Second, denatured

[9] S. Fawell, J. Seery, Y. Daikh, C. Moore, L. L. Chen, B. Pepinsky, and J. Barsoum, *Proc. Natl. Acad. Sci U.S.A.* **91,** 664 (1994).

[10] D. Derossi, A. H. Joliot, G. Chassaings, and A. Prochiantz, *J. Biol. Chem.* **269,** 10444 (1994).

[11] G. Elloit and P. O'Hare, *Cell* **88,** 223 (1997).

[12] E. Vives, P. Brodin, and B. Leblus, *J. Biol. Chem.* **272,** 16010 (1997).

[13] S. Gottesman, S. Wickmer, and M. R. Maurizi, *Genes Dev.* **11,** 815 (1997).

[14] C. Schneider, L. Sepp-Lorenzino, E. Nimmesgern, O. Ouerfelli, S. Danishefsky, N. Rosen, and F. U. Hartl, *Proc. Natl. Acad. Sci. U.S.A.* **93,** 14536 (1996).

[15] H. Nagahara, A. Vocero-Akbani, E. L. Snyder, A. Ho, D. G. Latham, N. A. Lissy, M. Becker-Hapak, S. A. Ezhevsky, and S. F. Dowdy, submitted (1998).

Tat fusion proteins have an enhanced potential to elicit biological responses.[15] The denatured Tat fusion proteins are made soluble in an aqueous buffer and added directly to the medium of cells in tissue culture. We have transduced Tat fusion proteins into a variety of primary and transformed cell types, including peripheral blood lymphocytes (PBLs), diploid human fibroblasts, keratinocytes, bone marrow stem cells, osteoclasts, fibrosarcoma cells, leukemic T cells, osteosarcoma, glioma, hepatocellular carcinoma, renal carcinoma, NIH 3T3 cells, and all cells present in whole blood, including both nucleated and enucleated cells.[15–20] We have transduced proteins into all cells and tissues present in mice, including across the blood–brain barrier.[21,22] To date, we used this strategy to generate and transduce more than 60 full-length proteins and domains from 15 to 120 kDa from a variety of classes, suggesting that many, if not most, proteins may be transduced into cells.[15–22]

Materials and Reagents

Buffers

Buffer Z: 8 *M* urea–100 m*M* NaCl–20 m*M* HEPES (pH 8.0)
Buffer A: 50 m*M* NaCl–20 m*M* HEPES (pH 8.0)
Buffer B: 1–2 *M* NaCl–20 m*M* HEPES (pH 8.0)
Phosphate-buffered saline (PBS)
Paraformaldehyde, 4% (w/v)
Imidazole, 5 *M*
Guanidine hydrochloride, 6 *M*
Urea, 8 *M*
Coomassie Brillant Blue

Reagents

pTAT-HA plasmid
BL21(DE3)LysS bacteria (Novagen, Madison, WI)

[16] S. A. Ezhevsky, H. Nagahara, A. Vocero-Akbani, D. R. Gius, M. Wei, and S. F. Dowdy, *Proc. Natl. Acad. Sci. U.S.A.* **94,** 10699 (1997).
[17] N. A. Lissy, L. Van Dyk, M. Becker-Hapak, J. H. Mendler, A. Vocero-Akbani, and S. F. Dowdy, *Immunity* **8,** 57 (1998).
[18] A. Vocero-Akbani, N. Vander Heyden, N. A. Lissy, L. Ratner, and S. F. Dowdy, submitted (1998).
[19] A. Ho, S. Schwarze, G. Waksman, S. J. Mermelstein, and S. F. Dowdy, submitted (2000).
[20] M. A. Chellaiah, N. Soga, S. Swanson, S. McAllister, U. Alvarez, D. Wang, S. F. Dowdy, and K. A. Hruska, *J. Biol. Chem.* **275,** 11993 (2000).
[21] S. Schwarze, A. Ho, A. Vocero-Akbani, and S. F. Dowdy, *Science* **285,** 1569 (1999).
[22] S. Schwarze and S. F. Dowdy, *Trends Pharmacol.* in press (2000).

12CA5 Anti-hemagglutinin (HA) antibodies (BAbCo, Berkeley, CA)
Protein fluorescein isothiocyanate (FITC) labeling kit (Pierce, Rockford, IL)
Ni-NTA resin (Qiagen, Valencia, CA)
Resource Q and S resin (optional; Pharmacia, Piscataway, NJ)
LB medium (Sigma, St. Louis, MO)

Equipment

Gravity columns (Bio-Rad, Hercules, CA)
Mono Q and S columns (5/5 or 10/10; Pharmacia)
PD10 (Sephadex G-25) columns (Pharmacia)
Sonicator
Fast protein liquid chromatograph (FPLC, optional; Pharmacia)
Sodium dodecyl sulfate–polyacrylamide gel electrophoresis (SDS–PAGE) and protein transfer units
Fluorescence-activated cell sorter (FACS)
Fluorescence microscope (preferably confocal)

Generation of Transducible Proteins: In-Frame Tat fusion Proteins

Transducible proteins are generated by cloning the cDNA of interest into the pTAT-HA bacterial expression vector (Fig. 1A). The pTAT-HA vector contains an N′-terminal 6-histidine leader followed by the 11-amino acid Tat protein transduction domain[15] flanked by glycine residues for free bond rotation, a hemagglutinin (HA) tag, and a polylinker (Fig. 1A). To obtain a genetic N′-terminal in-frame fusion with the Tat leader, the 5′ untranslated regions (UTRs) of the cDNA must be deleted. In general, we clone into the *Nco*I restriction site, with the ATG being in-frame with the upstream leader. The pTAT-HA-cDNA plasmid is then transformed into a bacterial strain that yields a high copy plasmid number, such as DH5α. Individual clones are isolated and analyzed for the correct insert by standard molecular biology techniques. The pTAT-HA-cDNA plasmid is then transformed into the high-expressing BL21(DE3)LysS bacterial strain.

Six to 12 independent isolates are grown as 1-ml overnight cultures in the presence of 100 μM isopropyl-β-D-thiogalactopyranoside (IPTG). Pellet and then boil the bacteria in 2× SDS gel loading buffer and analyze by SDS–PAGE. One gel is stained with Coomassie blue and expression of the recombinant protein is compared with that of a nonexpressing BL21(DE3)LysS control lysate. Please note that not all Tat fusion proteins are expressed at a substantial enough level to be viewed in this manner. Therefore, the second gel, if required, is transferred to a filter and probed with anti-HA antibodies or antibodies directed against the cDNA-encoded

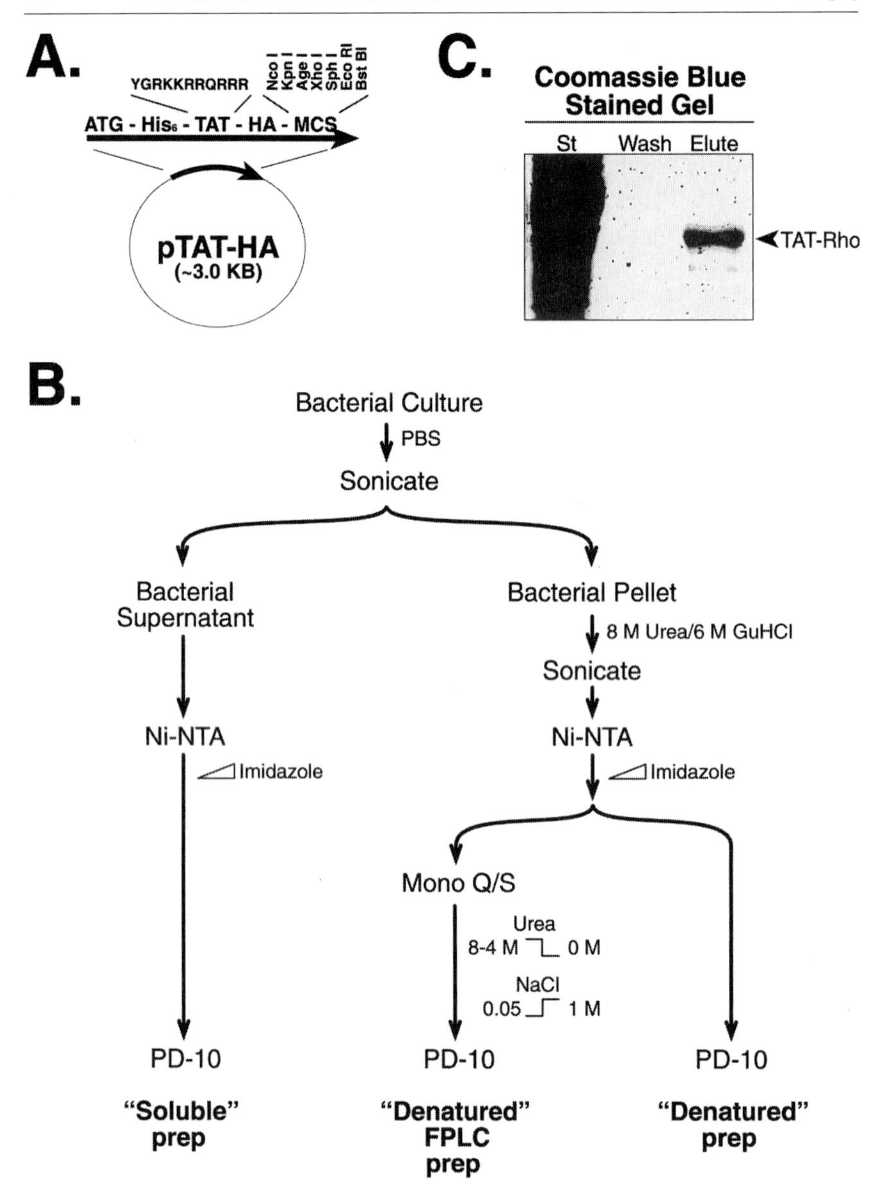

FIG. 1. Characterization of pTAT vector and purification protocol. (A) pTAT expression vector. (B) Purification protocols. (C) Tat–Rho fusion protein is purified over Ni-NTA resin, resolved by SDS–PAGE, and stained with Coomassie blue. St, Start.

product by immunoblot analyses. Although the Tat leader confers a ~3.5-kDa increase in molecular mass, we consistently detect a 5- to 10-kDa increase by denaturing SDS–PAGE. Make glycerol stocks of three to five of the highest expressing BL21(DE3)LysS isolates and choose a single isolate to focus on.

Large-Scale Purification of Tat Fusion Proteins

We have devised a denaturing protocol, using either 8 *M* urea or 6 *M* guanidine hydrochloride, to purify Tat fusion proteins to increase both the transduction potential[15] and the yield by recovering all the recombinant protein present in bacterial inclusion bodies (Fig. 1B). However, we have observed that several proteins produced superior results when purified under native conditions. In addition, although the protocol was originally designed for use with an FPLC protein purification system, because of the use of single steps to perform buffer changes, commonly available resins combined with gravity flow columns can be substituted for FPLC columns. Morever, the purification can be performed in a batch method.

Start a 100- to 200-ml LB medium overnight culture of high-expressing pTAT-HA-cDNA BL21(DE3)LysS isolate. The following morning, inoculate the entire volume into 1 liter of LB plus 100–500 μM IPTG and shake the mixture for 5 (or up to 10) hr at 37°. Isolate and wash the cell pellet in PBS, and then resuspend it in 10–20 ml of buffer Z and sonicate on ice (three 15-sec pulses) or until turbid. Clarify by centrifugation (5000 rpm at 4° for 10 min) and save the supernatant. Note that some Tat fusion proteins, including Tat–caspase-3, Tat–pRB, and Tat–Bid, are toxic to log-phase bacterial cells. Therefore, we inoculate a 1-liter overnight culture and isolate the cell pellet immediately in the morning. The theory is that stationary-phase cells (which would occur sometime in the night) are either more resistant to toxic proteins or more capable of transporting them into inclusion bodies than log-phase cells.

The sonicate is equilibrated in 10–20 m*M* imidazole, then applied at 4° or room temperature to a preequilibrated 3- to 10-ml Ni-NTA column in buffer Z plus 10–20 m*M* imidazole. Allow the column to proceed by gravity or apply slight air pressure via syringe as required. Save some of the start (St) and flowthrough for SDS–PAGE analysis. Wash the column in ~50 ml of buffer Z plus 10–20 m*M* imidazole. Elute the Tat fusion protein by stepwise addition of 5–10 ml of buffer Z containing 100 m*M*, 250 m*M*, 500 m*M*, and 1 *M* imidazole. Analyze the start, flowthrough, and each column fraction by Coomassie blue (Fig. 1C) and immunoblot analysis as outlined above. Pool appropriate fractions together.

For Tat fusion proteins purified under native conditions, sonicate in

PBS, apply to Ni-NTA as outlined above, elute with increasing imidazole, then desalt via PD10 column into the medium of choice.

Troubleshooting

If the Tat protein is detected primarily in the flowthrough, decrease the imidazole concentration in loading buffer Z and if a high background of contaminating bacterial proteins is detected, increase the imidazole concentration of the loading buffer. In addition, because of the "head" binding nature of the Ni-NTA column, removing the urea (see below) on this column results in aggregation and precipitation of the protein on the column. If binding to Ni-NTA remains problematic, change buffer Z from 8 *M* urea to 6 *M* guanidine hydrochloride with 10–20 m*M* imidazole. Because 6 *M* guanidine hydrochloride is a more stringent denaturant than 8 *N* urea, the protein becomes further unfolded and exposes the hexahistidine (His_6) purification leader. However, removal of guanidine hydrochloride requires additional steps than 8 *M* urea; therefore, we routinely use 8 *M* urea unless otherwise required.

Solubilization of Purified Tat Fusion Proteins into Aqueous Buffer

To treat tissue culture cells and animal models with Tat fusion proteins, the denaturant, 8 *M* urea or 6 *M* guanidine hydrochloride, must be rapidly removed. This also achieves the goal of obtaining high energetic (ΔG) denatured Tat fusion proteins. Several choices for this procedure are described below. In our experience, use of Mono Q or S ion-exchange chromatography yields superior results compared with removal by desalting columns. However, we have generated transducible Tat fusion proteins by both of these approaches (Fig. 1B).

Ion-Exchange Chromatography: Fast Protein Liquid Chromatography

Although the purification of most proteins by ion-exchange chromatography (Q and S resins) is dependent in large measure on the isoelectric point (p*I*) of the fusion protein, it has been our experience that in 4–8 *M* urea Tat fusion proteins will bind to Mono Q or S resin regardless of p*I* predictions.

If FPLC is at 4°, to avoid crystallization of 8 *M* urea, the pooled Ni-NTA fractions from above are diluted 1:1 with 20 m*M* HEPES, pH 8.0 for Mono Q and pH 6.5 for Mono S. This will result in 50 m*M* NaCl and 4 *M* urea final concentrations. If FPLC is at room temperature, dilute in buffer Z with no NaCl to obtain a final 50 m*M* NaCl. Inject the sample into a 5/5 or 10/10 (preferred) Mono Q/S column attached to an FPLC equilibrated in buffer A plus urea (4 or 8 *M* depending on temperature). Wash with ~20–50 ml of buffer A (no urea) and elute with a single 1–2 *M* NaCl step in buffer A. Analyze start, flowthrough, and eluate fractions by SDS–PAGE as outlined above and pool appropriate fractions. The sample is then desalted on a PD10 (Sephadex G-25) desalting column and analyzed by SDS–PAGE. We routinely equilibrate the PD10 in sterile PBS or tissue culture medium minus the serum; however, some full-length proteins and domains are more stable to precipitation when desalted in 10 m*M* Tris–HCl–0.1 m*M* EDTA. Check start, flowthrough, and column fractions as outlined above by SDS–PAGE.

By reducing the urea from 4 to 0 *M* in a single step, the denatured proteins are forced to become soluble in an aqueous environment. Because of a mixed population of protein configurations and biophysical properties, use of a single 1–2 *M* NaCl step is preferred to a linear NaCl gradient and will result in a sharper protein peak resolution and an increased concentration.

Troubleshooting. If failure to bind or weak binding of protein is observed, try the other column type regardless of predicted p*I*. If strong binding is observed (i.e., no protein present in flowthrough), but no or poor yield in the eluate, several modifications to reduce the effective avidity of the protein to the resin can be applied (Fig. 2). As an example, if protein X binds the Q resin at 50 m*M* NaCl, but fails to elute, clean the column with HCl/NaOH (see manufacturer directions). Equilibrate the sample and column with 100 m*M* NaCl, inject the sample, and elute as described above (see Fig. 2). Check start, flowthrough, and column fractions as outlined above by SDS–PAGE. Continue to increase the NaCl concentration by 50 m*M* steps until the recovery yield is 70–90% of input or protein is observed in the flowthrough. Alternatively, decrease (Q resin) or increase (S resin) the pH of buffer A by steps of 0.5 pH units until a small amount of protein is detected in the flowthrough fraction. The goal is to decrease the avidity of the protein for the column in urea to the point of obtaining a reversal binding. We have used both of these "blasphemous" strategies to isolate several proteins, including injecting samples at 200 m*M* NaCl/pH 7.0 into a Mono Q column.

Because of the ionic nature of guanidine hydrochloride, if 6 *M* guanidine hydrochloride is used to denature and purify the Tat fusion protein on Ni-NTA it must be desalted into 8 *M* urea on a PD10 column prior to addition to the Q or S resin.

Gravity Columns and Batch Preparations

Because of the design protocol of using single elution steps, FPLC or high-performance liquid chromatography (HPLC) is not required. Set up

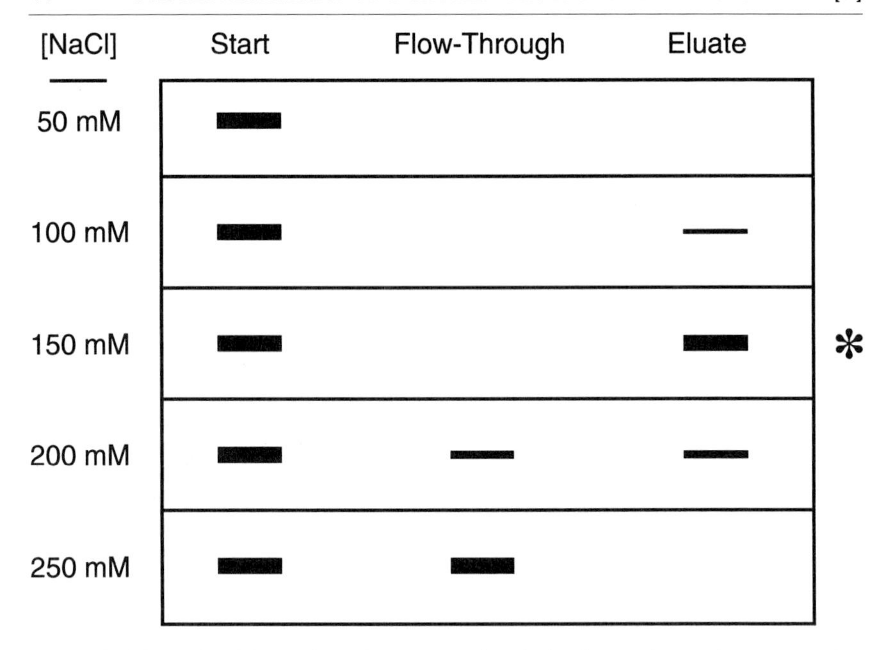

FIG. 2. Schematic of increasing salt concentration during ion-exchange chromatography. Tat fusion protein present at 8–4 *M* urea and 50 m*M* NaCl binds Mono Q or S resin, and is absent from the flow-through; however, it fails to elute with 1–2 *M* NaCl. After cleaning the column with HCl-NaOH, equilibrate the sample and column with 100 m*M* NaCl, inject, and elute. Repeat as required by increasing the NaCl concentration as indicated. As indicated by the asterisk, at some point during the increase in NaCl concentration the protein will have high enough avidity to bind the resin, thus failing to appear in the flowthrough, but low enough avidity so that it will be competed off with 1–2 *M* NaCl.

a 1- to 5-ml ion-exchange column with 30-μm Resource Q or S resin (Pharmacia). The sample is diluted as described above and, because of a low back pressure, is injected into the preequilibrated column via syringe. Wash the column with ~50 ml of buffer A plus imidazole and elute Tat fusion protein with a single step of 1–2 *M* NaCl. The sample is then desalted on a PD10 desalting column, collected, and analyzed by SDS–PAGE. Troubleshoot as described above.

For batch preparations, add 1–5 ml of Resource Q/S resin directly to the 10–20 ml of sonicate and place in a wheel at room temperature or 4° for 15–60 min. Pellet the beads at 2000–5000 rpm, aspirate the supernatant, and wash three times in 10–20 ml of buffer A. Elute by addition of 1–10 ml of buffer B and analyze as described above. This method has the added benefit that by mixing equal amounts of Q and S resin, predetermination of resin binding by the protein is not required.

Desalting Column

The theory behind rapid desalting is that passage of a denatured protein from 8 *M* urea or 6 *M* guanidine hydrochloride through the interface into PBS on the other side forces the protein to rapidly hide its hydrophobic residues and become aqueously soluble (Fig. 1B). PD10 desalting columns have a 1:1.4 dilution factor. Therefore, denatured proteins are separated from each other, helping to avoid aggregation of the proteins and subsequent precipitation on the column. Because of column volume size and consequential restraints, do not add more than 1.0 ml of the Tat fusion protein in 8 *M* urea to a disposable PD10 desalting column. Equilibrate the column in PBS or tissue culture medium minus serum as described above. One-milliliter, column fractions are isolated and analyzed by SDS–PAGE as described above. Tat fusion proteins usually elute in fractions 6 and 7. Although we have used this strategy to remove the urea from many Tat fusion proteins, in general, it also produces protein preparations that are more susceptible to precipitation and freeze–thaw problems. However, it is a rapid and inexpensive procedure.

Troubleshooting. The biggest issue with desalting directly from 8 *M* urea or 6 *M* guanidine hydrochloride is precipitation on the PD10 column. This can be alleviated by dilution of the protein sample or by use of equilibrated columns with 10–20% (v/v) glycerol and/or 1% (w/v) bovine serum albumin (BSA) to help stabilize the protein. In addition, some proteins are more stable when desalted into 10 m*M* Tris–0.1 m*M* EDTA.

After solubilization in an aqueous buffer or tissue culture medium, determine the concentration by comparison with a standard, such as BSA, in an SDS–polyacrylamide gel, and/or by Bio-Rad protein concentration analysis. The purified Tat fusion protein is then flash frozen in 100- to 200-μl volumes in 10% (v/v) glycerol and store at −80°. After freezing, test a vial for loss of protein due to freeze–thaw-induced precipitation by thawing a vial, spinning it at 12,000–100,000 rpm at 4° for 10 min, and analyzing it by SDS–PAGE. In addition, we never refreeze thawed protein preparations. Frozen Tat fusion proteins, such as Tat–E1A, are capable of retaining transduction potential on thawing when stored at −80° for 2 years.

Labeling Tat Fusion Proteins with Fluorescein Isothiocyanate

Because of the ability of Tat fusion proteins to enter cells in a receptor- and transporter-independent fashion, in theory and in practice, all cell types are susceptible to protein transduction. Indeed, we have previously generated more than 60 Tat fusion proteins and demonstrated their ability to transduce into a wide variety of primary and transformed mammalian cells. Moreover, we have shown the ability of several proteins to transduce into all cells present in mouse models.[21]

To test the ability of Tat fusion proteins to transduce into cells of interest, label the protein with fluorescein isothiocyanate (FITC), which will conjugate to lysine residues. Twenty to 50 μg of Tat fusion protein is placed in a 300-μl reaction mix as per the manufacturer instructions for 2 hr at room temperature. The nonconjugated FITC is then removed from the FITC-labeled Tat protein either by gel-filtration chromatography or by use of a PD10 desalting column. We routinely use a PD10 column equilibrated in PBS. Isolate 1-ml column fractions and check each fraction by Bio-Rad protein concentration analysis and/or SDS–PAGE. If available, check the efficiency of FITC cross-linking by use of a fluorometer. Poorly labeled Tat fusion proteins will result in a low signal-to-noise ratio, thereby requiring alteration of the labeling conditions by increasing/decreasing the amount of protein to FITC.

To analyze for transduction of FITC-labeled Tat fusion protein into cells, add 100–400 μl of purified FITC-labeled Tat fusion protein to 10^5 to 10^6 cells in 0.5–1 ml of medium. Analyze for transduction by flow cytometry (FACS) analysis at 10, 20, and 30 min postaddition of the FITC-labeled Tat fusion protein. In our hands, FITC-labeled Tat fusion proteins rapidly transduced into ~100% of cells, achieving maximum intracellular concentration in less than 5 min. In addition, we note a narrow intracellular concentration range of the transduced protein within the population as supported by the narrow FACS peak width between control and transduced cells.

To analyze for transduction of FITC-labeled TAt fusion proteins by fluorescence confocal microscopy, treat cells on a glass coverslip with the FITC-labeled Tat fusion protein for 15–60 min, then rapidly wash individual coverslips two or three times in PBS buffer and fix in 4% (w/v) paraformaldehyde. Mount the coverslips and view them under a fluorescent microscope. At early time points (less than 2–4 hr), while the FITC-labeled Tat fusion protein is in excess, fluorescence is seen in both the nuclear and cytoplasmic compartments, without the merely perimembrane staining that is observed with free FITC negative control (Fig. 3).

Troubleshooting. Remember that protein transduction is concentration dependent. Therefore, if the extracellular concentration is changed, such as by slowly washing the cells, the intracellular FITC-labeled Tat fusion protein will attempt to reach chemical equilibrium and transduce back out of the cells. For FACS this is generally not an issue; however, for fluorescence confocal microscopy, this can make a significant difference in the ability to detect intracellular FITC-labeled Tat fusion proteins.

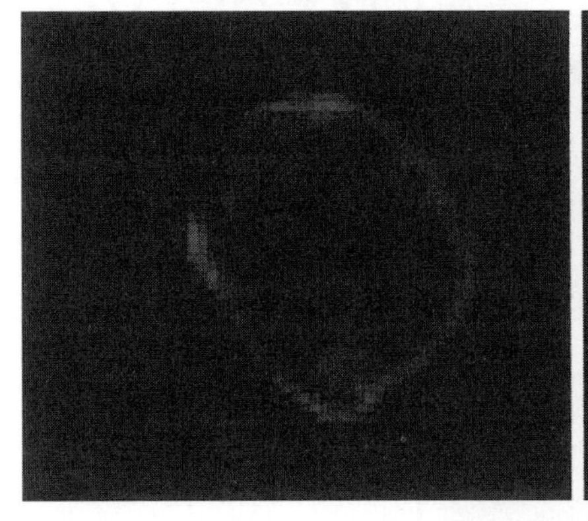

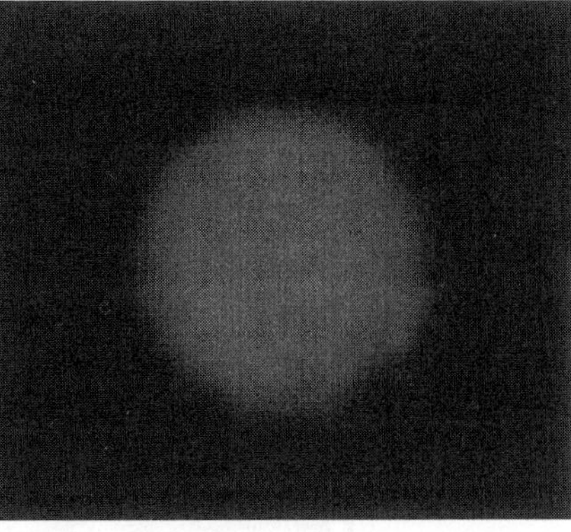

FIG. 3. Analysis of FITC-labeled Tat fusion proteins added to cells. Jurkat T cells were treated with free FITC control (*left*) or FITC-labeled Tat–β-Gal (*right*), rapidly washed, fixed in 4% (w/v) paraformaldehyde, and analyzed by fluorescence confocal microscopy. Note the perimembrane-only staining of control free FITC and the appearance of Tat–β-Gal–FITC in all regions of the T cell.

Induction of Podosome Formation in Osteoclasts by Tat–Rho

Rho has previously been implicated in the formation of stress fibers and the disassembly of podosomes in osteoclasts.[1–3] However, osteoclasts are resistant to genetic manipulations.[4–6] Therefore, to test directly for the involvement of Rho in regulating the actin architecture, Tat–Rho V14 (constitutively active form) and Tat–Rho N19 (dominant-negative form) fusion proteins are generated and purified as outlined above.[20] Human osteoclasts are treated with 100 n*M* Tat–Rho V14 or Tat–Rho N19 proteins or control Tat leader and Tat–HSV-TK proteins (NAGAHARA) for 15 and 30 min and fixed in 4% (w/v) paraformaldehyde. The cellular architecture is then assayed by staining the fixed osteoclasts with rhodamine-conjugated phalloidin (Molecular Probes, Eugene, OR), an actin-binding protein (Fig. 4). Fluorescence confocal microscopy shows the maintenance of peripheral, small ring like actin structures of podosomes in control Tat leader- and Tat–HSV-TK protein-treated osteoclasts (Fig. 4A and B). However, treatment with Tat–Rho V14 shows dissassembly of podosomes as early

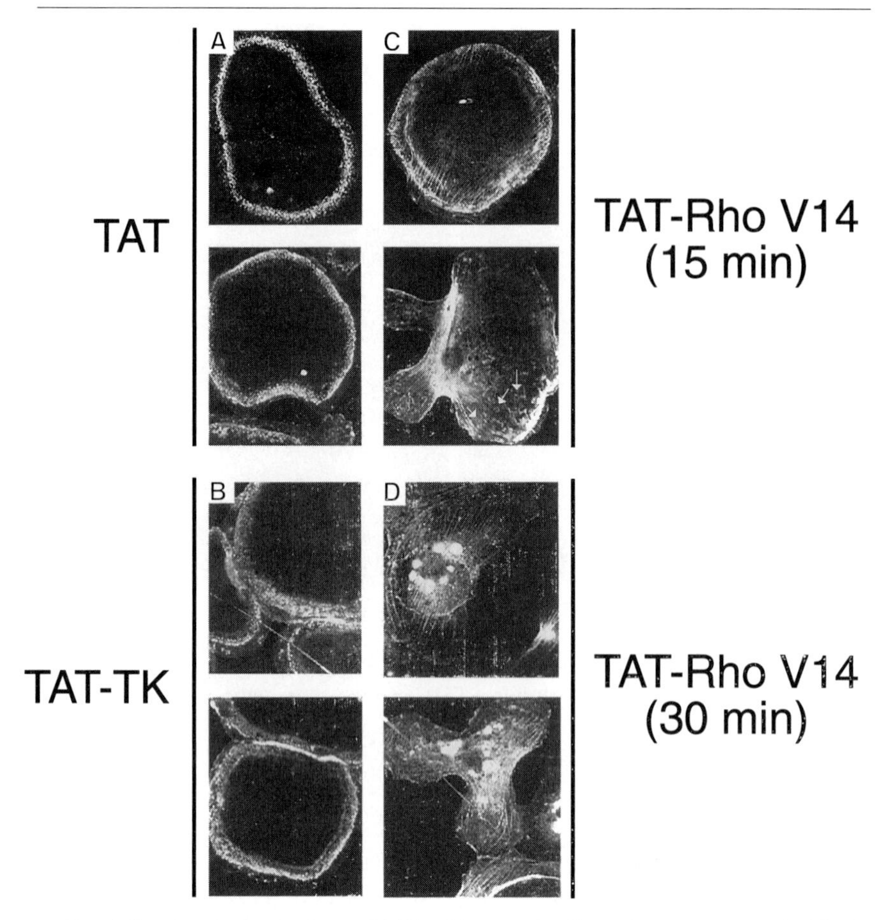

FIG. 4. Treatment of osteoclasts with Tat–Rho protein. Osteoclasts were treated with the Tat leader control protein (A), Tat–HSV-TK control protein (B), or Tat–Rho V14 protein for 15 min (C) or 30 min (D). Treated cells were fixed, stained with rhodamine-conjugated phalloidin, and analyzed by fluorescence confocal microscopy. Oseoclasts treated with Tat-Rho V14 proteins showed rapid disassembly of podosome actin structures and the appearance of stress fibers, whereas the control-treated osteoclasts maintained their podosomes.

as 15 min postaddition (Fig. 4C) and the appearance of actin stress fibers. Stress fibers are maintained at 30 min postaddition (Fig. 4D) and for as long as 3 hr (data not shown). Thus, transduced, constitutively active Rho not only reaches intracellular equilibrium, but also refolds and induces the rapid disassembly of podosomes and the appearance of stress fibers in less than 15 min in ~100% of osteoclasts.

Moreover, treatment of cells with Tat–CDC42 V12 protein results in the formation of filopodia in <15 min (data not shown).

These observations demonstrate the utility of transducing proteins, such as Rho, into ~100% of primary cells that are highly resistant to other forms of manipulation, to dissect both biological and *in vivo* biochemical processes.

Discussion

We have successfully utilized the procedure described here to generate and transduce more than 60 Tat fusion proteins from 15 to 120 kDa into ~100% of all cells assayed thus far, including primary cells, transformed cells, and all cells/tissues present in mice.[15–21] Although we have not gone beyond 120 kDa, at this point in our limited understanding of the mechanism of proteins transduction,[22] there appears to be no size constraint.

The use of a genetic in-frame fusion combined with denaturation of the proteins achieves several goals, including isolation of the bulk of recombinant proteins that are usually present in inclusion bodies, increased efficiency of biological response, and ease of use. Once inside the cell, transduced denatured proteins appear to be rapidly refolded by chaperones[13] such as HSP90,[14] and are capable of binding their cognate intracellular targets and performing biochemical functions, such as cell cycle arrest, cell migration, protection from and induction of apoptosis, transcriptional inhibition and activation, and various enzymatic activities.[15–21] Indeed, every Tat wild-type fusion protein we have generated thus far has given us a known or new biological or intracellular biochemical phenotype. Furthermore, because of the rapid rate of transduction, real-time kinetic experiments can be performed in primary cells, an undertaking not possible by any other means.

We conclude that transduction of full-length Tat fusion proteins directly into ~100% of primary or transformed cells has broad implications for manipulating intracellular processes in both experimental *in vitro* tissue culture systems and animal models.

[3] Green Fluorescent Protein-Tagged Ras Proteins for Intracellular Localization

By EDWIN CHOY and MARK PHILIPS

Introduction

On the basis of immunofluorescence and immunoelectron microscopy, Ras protein was reported to be localized at the cytosolic leaflet of the plasma membrane.[1] The basis for its affinity for membranes was elucidated when it was recognized that all Ras proteins are modified by a farnesyl lipid and that some are additionally modified by palmitic acid.[2] Ras proteins proved to be members of a large family of proteins that terminate with a CAAX motif that directs sequential posttranslational modifications by three enzymes: a prenyltransferase (e.g., farnesyltransferase), a prenylprotein AAX endopeptidase, and a prenylcysteine-directed carboxyl methyltransferase.[3] Palmitoylation is dependent on CAAX processing, occurs at cysteine residues in close proximity to the CAAX motif, and represents one of two types of secondary plasma membrane-targeting signals, the alternative being a polybasic domain proximal to the CAAX motif.[4]

Since its introduction into cell biology, green fluorescent protein (GFP) has revolutionized our ability to localize proteins in tissues, cells, and organelles.[5] The basis of this revolution is the fact that the cDNA for GFP, incorporated into an appropriate expression plasmid and transfected, is sufficient for fluorescence in virtually any cell without additional cofactors. Moreover, GFP expression is well tolerated by cells and other proteins may be fused to either end of GFP without affecting its fluorescence, allowing for simple construction of chimeric cDNAs that direct the expression of GFP fusion proteins that report the localization of the fused protein. Most important, because GFP forms its fluorophore autocatalytically *in vivo,* GFP-tagged proteins can be observed in living cells and organisms.

Analysis of Ras localization by GFP technology has forced a reevalua-

[1] M. C. Willingham, I. Pastan, T. Y. Shih, and E. M. Scolnick, *Cell* **19,** 1005 (1980).

[2] J. F. Hancock, A. I. Magee, J. E. Childs, and C. J. Marshall, *Cell* **57,** 1167 (1989).

[3] S. Clarke, *Annu. Rev. Biochem.* **61,** 335 (1992).

[4] J. F. Hancock, H. Paterson, and C. J. Marshall, *Cell* **63,** 133 (1990).

[5] K. F. Sullivan and S. A. Kay, "Methods in Cell Biology," Vol. 58: "Green Fluorescent Proteins" Academic Press, San Diego, California, 1999.

0076-6879/00 $35.00

tion of long-held assumptions about Ras trafficking.[6] Whereas it was previously assumed that Ras translocated directly from the cytosol, where it is synthesized, to the plasma membrane by virtue of CAAX processing, analysis of GFP–Ras fusion proteins revealed endomembrane trafficking en route to the plasma membrane. Rather than promoting nonspecific membrane association, prenylation targets CAAX proteins specifically to the endomembrane, where they are further processed (proteolyzed and methylated) and then sent to the plasma membrane only if they possess a secondary targeting sequence (palmitoylation sites or a polybasic domain). Moreover, the pathway to the plasma membrane utilized by N- and H-Ras (palmitoylated) appears to differ from that used by K-Ras (polybasic). This difference can be observed with GFP-tagged Ras protein in that, whereas GFP–N-Ras and GFP–H-Ras are visualized on rapidly motile peri-Golgi vesicles, GFP–K-Ras is not. This chapter details the methods used for tagging Ras proteins with GFP and imaging the fusion protein in live cells to observe subcellular localization and membrane trafficking.

Expression Constructs

The original clone of the green fluorescent protein (GFP) of the jellyfish *Aequorea victoria* isolated in 1992[7] has been widely disseminated and extensively mutated. GFP variants are commercially available from seveal vendors. The most extensive selection of GFP expression vectors is available from Clontech (Palo Alto, CA) as their Living Colors line of products and from Invitrogen (San Diego, CA). In addition to GFP, Clontech offers both red- and blue-shifted mutants of GFP that allow for simultaneous tracking of two molecules in either live or fixed cells, although the filter sets or confocal laser lines required for some of these variants are not standard. The discovery of a red fluorescent protein from the reef coral *Anemonia majano* and its commercialization by Clontech will greatly facilitate two-color protein tagging because conventional imaging configurations will suffice.

Although GFP has been utilized successfully to tag proteins at the N terminus, C terminus, or internally, N-terminal tagging of Ras and other CAAX proteins is required to leave the C-terminal membrane-targeting domain intact. We have utilized pEGFP (Clontech) for our Ras constructs. This plasmid directs the expression of enhanced green fluorescent protein

[6] E. Choy, V. K. Chiu, J. Silletti, M. Feoktistov, T. Morimoto, D. Michaelson, I. E. Ivanov, and M. R. Philips, *Cell* **98,** 69 (1999).

[7] D. C. Prasher, V. K. Eckenrode, W. W. Ward, F. G. Prendergast and J. J. Cormier, *Gene* **111,** 229 (1992).

TABLE I
PCR PRIMERS FOR AMPLIFICATION FROM Ras cDNAs OF INSERTS FOR pEGFP-C3 TO PRODUCE EGFP-TAGGED Ras CONSTRUCTS

Primer	Sequence
Forward N-Ras	5′-CTT CGA ATT CTG ACT GAG TAC AAA CTG GTG-3′
Reverse N-Ras	5′-TTA GGG CCC TTA CAT CAC CAC ACA TGG CAA-3′
Reverse N-Ras C186S	5′-TTA GGG CCC TTA CAT CAC CAC AGA TGG CAA-3′
Reverse N-Ras C181S	5′-GGA GGG CCC TTA CAT CAC CAC ACA TGG CAA TCC CAT TGA ACC-3′
Forward K-Ras	5′-CTT CGA ATT CTG ACT GAA TAT AAA CTT GTG-3′
Reverse K-Ras	5′-GCG GGG CCC TTA CAT AAT TAC ACA CTT TGT-3′
Forward H-Ras	5′-CTT CGA ATT CTG ACC GAA TAC AAG CTT GTT-3′
Reverse H-Ras	5′-GCA GGG CCC TCA GGA GAG CAC ACA CTT GCA-3′

(EGFP) under the control of a cytomegalovirus (CMV) immediate-early promoter and enhancer. The construct is "enhanced" in two ways. First, two amino acid substitutions (F64L and S65T) result in a red-shifted excitation peak (488 nm) and a fluorophore that is 35 times brighter and more photostable than the wild type. Second, extensive silent mutations introduced to optimize codon usage for mammalian cells leads to higher protein expression. pEGFP variants are available (Clontech) with a multiple cloning site (MCS) 5′ or 3′ of the EGFP sequence, each in three reading frames. To tag Ras proteins or fragments thereof we have utilized almost exclusively pEGFP-C3, which allows for insertion of the Ras coding sequence 3′ of the EGFP sequence. Our inserts are derived from polymerase chain reaction (PCR) amplification, allowing for adjustment of the reading frame to the C3 variant by primer design. Similarly, we take advantage of the PCR primers to introduce distinct 5′ and 3′ restriction sites compatible with the pEGFP MCS (usually 5′ *Eco*RI and 3′ *Apa*I), assuring unidirectional insertion into the double-digested vector. Although not essential, it is good practice to eliminate the Ras initiation methionine or avoid a Kozsak sequence by appropriate PCR primer design, to reduce the chance of directing the expression of untagged protein. A stop codon must be included in the 3′ primer.

Using the cDNA of choice (e.g., wild-type or mutant N-, H-, or K-Ras, Ras-related GTPases, other CAAX proteins) the coding sequence or a 3′ fragment thereof can be PCR amplified with appropriate restriction sites designed into the primers (Table I) using a polymerase of moderate fidelity (e.g., *Taq*). Mutations of the hypervariable membrane-targeting domain of Ras proteins (e.g., C → S) can be introduced by appropriate modification of the 3′ primer (Table I). The PCR product is then double digested and inserted into the MCS of pEGFP.

Primer Design

In designing the 5′ primer, the following should be included: (1) a 5′ terminal extension of 2–4 nucleotides to facilitate restriction enzyme activity, (2) a restriction site compatible with the MCS, (3) 10 to 15 nucleotides of the coding sequence of the Ras insert in frame with the EGFP coding sequence, and (4) if the initiation codon of the Ras insert is retained, exclusion of a Kozsak sequence that may facilitate initiation.

In designing the 3′ primer, the following should be included: (1) a 5′ terminal extension of three nucleotides to facilitate restriction enzyme activity, (2) a restriction site found in the MCS 3′ of that utilized for the 5′ primer, (3) a stop anticodon in frame with the coding sequence of Ras, and (4) 10 to 15 nucleotides of the anticoding sequence of Ras.

Cloning Methods

We have utilized the following PCR conditions to amplify inserts for EGFP–Ras constructs, using a Perkin-Elmer (Norwalk, CT) thermal cycler.

Reactants and Buffer

5′ Primer, 20 μM	5μl
3′ Primer, 20 μM	5μl
Ras cDNA plasmid, 1 μg/μl	1μl
dNTPs, 10mM	2.5 μl
Taq polymerase	0.5 μl
Polymerase buffer, 10×	5 μl
Doubly distilled H_2O	31 μl

Cycles

Cycle number	Melting	Annealing	Extension
1	5 min at 98°	1 min at 55°	1 min at 72°
2–30	1 min at 98°	1 min at 55°	1 min at 72°
31			10 min at 72°
32: Hold at 4°			

Amplification products should be stored at 4° and then purified by 1.2% (w/v) agarose gel electrophoresis. PCR-amplified Ras should migrate at ~0.6 kb (visualized by ethidium bromide). The electrophoresed amplification product can be excised from the gel with a sterile razor and the DNA extracted by kit (Qiagen, Valencia, CA) according to the manufacturer instructions and brought up at 1 μg/μl in 10 mM Tris–HCl, pH 7.4.

Restriction Digest

(Restriction digest must be applied to both PCR product and GFP expression vector):

PCR product or pEGFP DNA, 1 mg/ml	1.0 μl
Restriction enzyme A (to cut 5′ end)	0.5 μl
Restriction enzyme B (to cut 3′ end)	0.5 μl
Restriction buffer, 10× (as per manufacturer suggestion)	2.0 μl
Doubly distilled H_2O	16 μl

Incubate the restriction digest at 37° for 2 hr (or as per manufacturer suggestion). Purify the cut DNA by kit (Qiagen) as per manufacturer instructions. The PCR product should be brought up to 50 ng/μl and the vector to 1 μg/μl, each in 10 mM Tris–HCl, pH 7.4.

Ligation Reactions

Set up six ligation conditions in 1.5-ml Eppendorf tubes as specified in Table II. Incubate the ligation reactions at 14° for >16 hr.

Bacterial Transformation

Thaw competent bacterial cells (e.g., DH5α treated with $MgCl_2$) and heat shock at 65° for 2 min followed by incubation over ice for 3 min. Add 100 μl of the competent bacteria to the products of each of the ligation reactions described above. Incubate over ice for 30 min. Add 1 ml of prewarmed (37°) LB broth and incubate the tubes, with end-over-end rotation, at 37° for 1 hr. Centrifuge the tubes at 10,000 rpm for 10 sec at room temperature and remove 0.9 ml of supernatant. Resuspend the cells in the remaining 0.1 ml of LB broth and spread on LB plates containing selection antibiotics (e.g., kanamycin at 50 μg/ml for pEGFP). Incubate at 37° for 16 hr. Condition 1 (Table II) should give >100 colonies per plate if the bacteria are competent. Condition 2 should give 50–200 colonies if the ligase is active. Condition 3 should give <5 colonies per plate if the vector is completely digested by both restriction enzymes. Conditions 4–6 should give at least 2-fold the number of colonies observed from condition 3 if insertion of the PCR product is successful. Choose from among conditions 4–6 the one that gives the most colonies per plate and pick 5–20 colonies. Prepare small-scale plasmid DNA by kit (Qiagen) from each of these

TABLE II
LIGATION CONDITIONS FOR CLONING GFP–Ras CONSTRUCTS

Condition	Volume required (μl)							
	Uncut vector (1 mg/ml)	Single cut vector (1 mg/ml)	Double cut vector (1 mg/ml)	Double cut PCR product (50 ng/ml)	T4 ligase[a]	10× buffer with ATP[a]	Doubly distilled H_2O	Purpose
1	1	0	0	0	0	2	17	Bacterial competence
2	0	1	0	0	1	2	16	Ligase activity
3	0	0	1	0	1	2	16	Vector digestion
4	0	0	1	1	1	2	15	Cloning
5	0	0	1	2	1	2	14	Cloning
6	0	0	1	4	1	2	12	Cloning

[a] Life Technologies (Rockville, MD).

colonies. Double-digest 1 μg of the plasmid DNA and analyze the insert size by 1.2% (w/v) agarose gel electrophoresis. Check the insert sequence of plasmids with inserts of the expected size. Prepare large-scale plasmid DNA by kit (Qiagen) from one or more plasmids that contain the correct insert.

C-Terminal Constructs

We have found that the hypervariable C-terminal membrane-targeting domains of Ras proteins, when used alone to extend GFP, target the fluorescent protein in a manner identical to that of the full-length Ras protein. Furthermore, we have found that mutated versions of the hypervariable domain and the CAAX motif alone can be useful in analyzing membrane targeting.[6] To prepare these short constructs we have utilized synthetic complementary oligonucleotides in place of the PCR products discussed above. DNA inserts of 45 bp or less that are used to produce C-terminal constructs are generated by synthesizing complementary positive and negative oligonucleotides (with appropriate restriction site overhangs) that have a 5′ *Bgl*II site, a 3′ stop codon, and a 3′ *Apa*I site in their sequence (Table III). These single-stranded oligonucleotides are purified by agarose gel electrophoresis and resuspended at 10 m*M* in 10 m*M* Tris–HCl, pH 7.4. Twenty nanomoles (2 μl) of each complementary strand is mixed in 1.5-ml Eppendorf tubes with 2 μl of 10× ligation buffer (Life Technologies, Rockville, MD) and 14 μl of doubly distilled H_2O and heated to 100° for 5 min. The tubes are placed on ice and 1 U of T4 ligase (Life Technologies) and 1 μg of pEGFP-C3 vector linearized by double digestion with *Bgl*II and *Apa*I are added to each tube and incubated at 14° for 12 hr. The complementary strands anneal to form linear double-stranded DNA (dsDNA) and are subsequently ligated into the *Bgl*II and *Apa*I sites of

TABLE III
COMPLEMENTARY OLIGONUCLEOTIDES FOR CONSTRUCTION OF SHORT DOUBLE-STRANDED DNA INSERTS[a]

Primer	Sequence
Forward CVVM	5′-GAT CTC TGT GTG GTG ATG TAA GGG CC-3′
Reverse CVVM	5′-C TTA CAT CAC CAC ACA GA-3′
Forward CVLS	5′-GAT CTC TGC GTT CTG TCT TAA GGG CC-3′
Reverse CVLS	5′-C TTA ACA CAG AAC GCA GA-3′
Forward CVIM	5′-GAT CTC TGT GTA ATT ATG TAA GGG CC-3′
Reverse CVIM	5′-C TTA CAT AAT TAC ACA GA-3′

[a] pEGFP-C3 5′ *Bgl*II and 3′ *Apa*I sites.

pEGFP-C3. Transformation of competent bacteria with these constructs is accomplished as described above.

Transfection

Although GFP-tagged Ras constructs expressed by transfection can be analyzed in fixed cells, the ability to observe their subcellular localization in living cells offers tremendous advantages in studying membrane targeting. These include increased brightness, lower background, higher resolution, avoidance of fixation artifacts, and the ability to dynamically observe organelles. Currently available high-efficiency transfection methods (e.g., LipofectAMINE and SuperFect) permit a wide variety of cell lines to be utilized. We have used successfully COS-1, CHO, HEK293, MDCK, ECV309, NIH 3T3, PAE, and HeLa cells for studying the subcellular localization of CAAX proteins.

Choice of Cell Type

Each cell line has unique properties that can facilitate specific studies. For example, because COS-1 cells are efficiently transfected and express large T antigen, pEGFP vectors that contain a simian virus 40 (SV40) origin of replication induce high levels of expression of GFP or a fusion protein. Because COS-1 cells are relatively large and spread symmetrically they afford excellent views of GFP–Ras in peripheral lamellipodia (Fig. 1A).

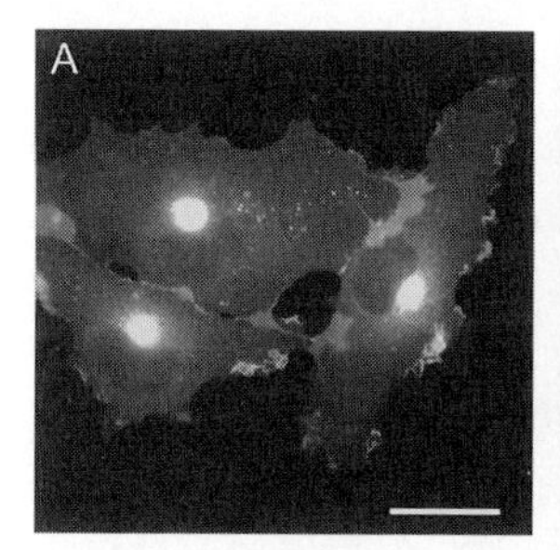

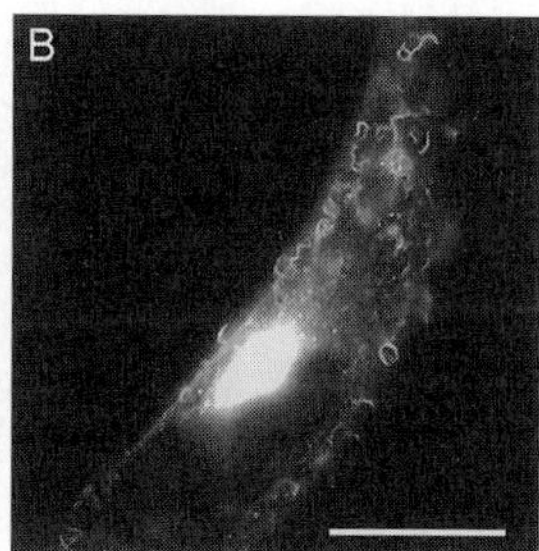

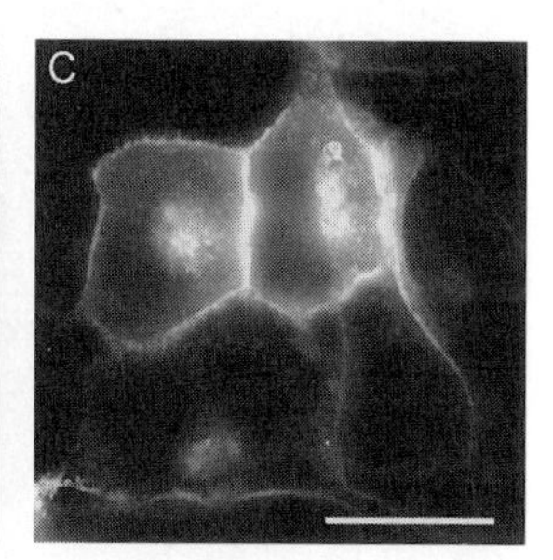

FIG. 1. Steady state localization of GFP–N-Ras in various transiently transfected cells. pEGFP-N-ras was transfected with SuperFect into COS-1 (A), CHO (B), or MDCK (C) cells. The cells were imaged alive with a Zeiss Axiovert 100S microscope 24 hr after transfection, and images were captured digitally with a Princeton Instruments cooled CCD camera and MetaMorph imaging software. Note that in COS-1 and CHO cells the brightest structure, the Golgi, was intentionally overexposed to reveal expression in peripheral lamellipodia in COS-1 cells and dorsal lamellipodia in CHO cells. In the columnar confluent MDCK cells the expanse of plasma membrane in the *z* axis results in a structure that is as bright as the Golgi. Bars: 10 μm.

In addition, their morphologically discrete Golgi (Fig. 2A) and extended endoplasmic reticulum (ER) (Fig. 2B) make them an excellent choice for visualizing CAAX constructs in the endomembrane. However, COS-1 cells are relatively pleiomorphic, they form foci spontaneously, and can have bizarre multilobar nuclei confounding some observations. MDCK cells, especially those allowed to grow to confluence, are much more homogeneous than COS-1 cells. Because they become cuboidal or columnar on

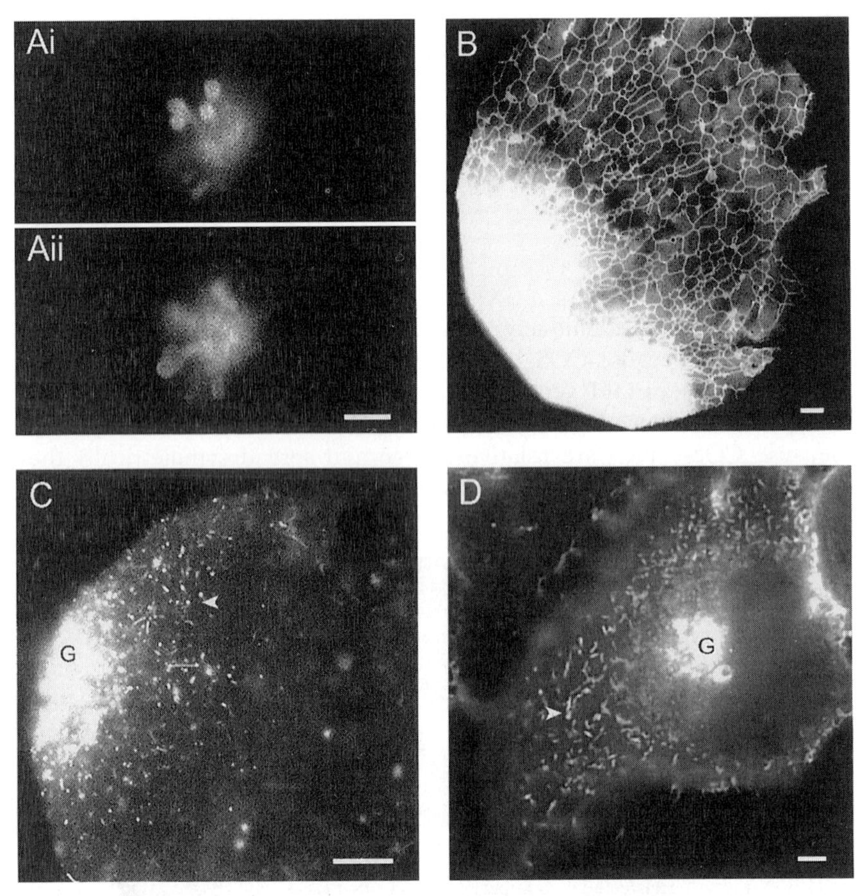

FIG. 2. Localization of GFP–N-Ras in organelles. (A) Views of two focal planes of the same Golgi in a COS-1 cell expressing GFP–N-Ras. (B) Endoplasmic reticulum localization of palmitoylation-deficient GFP–N-RasC181S (the grossly overexposed Golgi is partially obscured by the mercury lamp diaphragm). (C) Rapidly motile peri-Golgi vesicles (arrowhead) in a COS-1 cell transfected with GFP–N-Ras. (D) GFP–N-Ras expressed in the apical microvilli (arrowhead) of an MDCK cell (although the bright Golgi is deeper in the cell it can still be seen in this nonconfocal image). G, Golgi. Bars: 1 μm.

confluent growth, the dorsolateral plasma membrane that often forms a polygonal shape in the *xy* dimension is extended in the *z* axis and offers an excellent structure with which to assess plasma membrane targeting (Fig. 1C). In addition to the continuous basolateral membrane, GFP–Ras can been seen in discrete microvilli on the apical surface (Fig. 2D). Like COS-1 cells, CHO cells are easy to transfect but in contrast to COS-1 cells grow with a characteristic spindle shape in which GFP–N-Ras and GFP–H-Ras can be observed to localize in an elongated Golgi nestled at the hilum of the kidney-shaped nucleus (Fig. 1B). Whereas plasma membrane-associated GFP–Ras is seen in peripheral lamellipodia in COS-1 cells, dorsal lamellipodia are better seen in CHO cells (Fig. 1B). CHO cells, especially the occasional giant cell, also reveal extensive ER networks when transfected with CAAX constructs that lack a second membrane-targeting signal. Endothelial-like cell lines such as ECV309 and PAE cells are extremely sessile and reveal distinctive endomembrane but limited plasma membrane fluorescence. We have reported that whereas the subcellular localizations of GFP–N-Ras and GFP–H-Ras are similar in that they are highly expressed at steady state in the Golgi, GFP–K-Ras is expressed at a lower level in the Golgi area.[6] This differential expression pattern is cell-type dependent in that whereas GFP–K-Ras is observed not at all in the Golgi of confluent MDCK cells, in COS-1 cells a perinuclear fluorescence is revealed that coincides with the Golgi but is less intense and more diffuse than the Golgi illuminated by GFP–N-Ras and GFP–H-Ras (Fig. 3).

Culture and Transfection Techniques

COS-1, CHO, MDCK, HEK293, NIH 3T3, and HeLa cells can be maintained in 5% CO_2 at 37° in Dulbecco's modified Eagle's medium

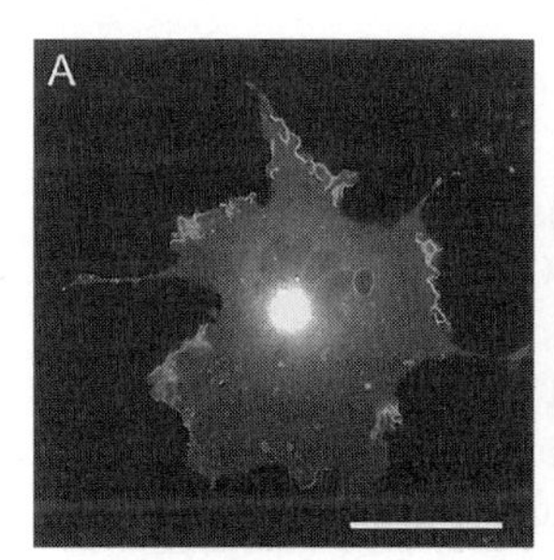

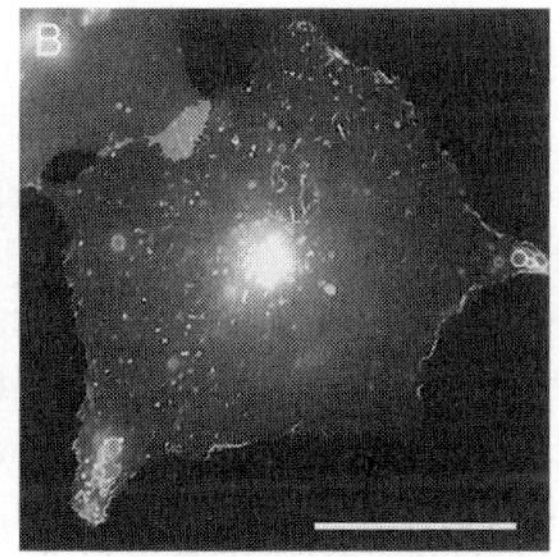

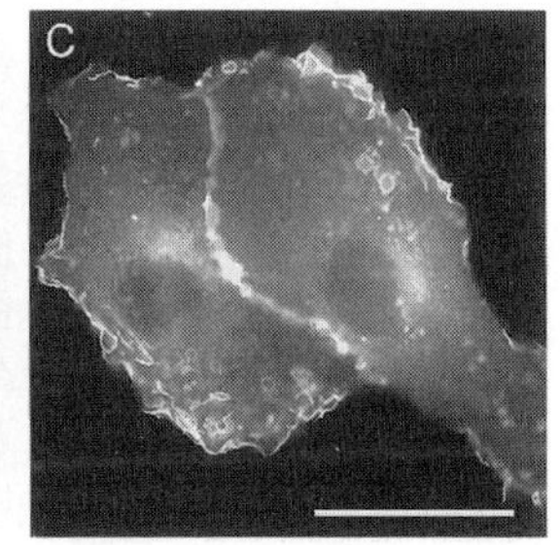

FIG. 3. Differential steady state localization of GFP-tagged N-, H-, and K-Ras in COS-1 cells. COS-1 cells were transiently transfected with GFP-tagged N-Ras (A), H-Ras (B), or K-Ras (C) and imaged alive after 24 hr. Note that whereas GFP–N-Ras and GFP–H-Ras illuminate a tight perinuclear structure identifiable as Golgi, GFP–K-Ras gives a dimmer and more diffuse perinuclear fluorescence. Bars: 10 μM

(DMEM) containing 10% (v/v) fetal bovine serum (FBS). ECV cells can be maintained in Media 199 (Life Technologies) containing 10% FBS. If cells are to be transfected and then fixed prior to imaging (e.g., for colocalization of GFP–Ras with an immunofluorescently stained molecule) they can be grown on alcohol-washed coverslips (No. 1.5 thickness) placed in 24-well plates or commercially available chambered coverslips (Nunc, Roskilde, Denmark). Whereas fixation, solubilization, and washing can be accomplished in the same 24-well plate, to conserve reagents, immunostaining is best performed by inverting the coverslips onto 50- to 100-μl droplets beaded on Parafilm. Fixed cells on coverslips can be mounted inverted on glass microscope slides, using a mounting agent that retards photobleaching (e.g., Mowiol; Calbiochem, La Jolla, CA).

The fluorescence of tissue culture plastic precludes its use for high-quality imaging of live cells unless there is access to water immersion objectives. Cells grown on glass coverslips, particularly 22-mm circles, can be used for visualizing GFP-tagged molecules in live cells, using a custom chamber [e.g., Harvard Apparatus, *www.harvardapparatus.com* (South Natick, MA)] or by inverting and mounting with silicon grease on a silicon rubber gasket applied to a microscope slide such that medium can be placed in the well formed by a central cutout. Cells grown in chambered microscope slides (e.g., Lab-Tek from Nunc) can be used with an inverted microscope; however, the thickness of the glass slide prohibits the use of high-powered objectives with short working distances. We have found that the most practical systems, provided there is access to an inverted microscope, utilizes commercially available 35-mm culture dishes that have a 14-mm cutout at the bottom that forms a miniculture well sealed by a No. 1.5 glass coverslip (MaTek, Ashland, MA).

Transfection Method

1. One day prior to transfection, seed 1.5×10^5 cells into 35-mm MaTek dishes.

2. Twenty-four hours after seeding, cells should be approximately 30% confluent and ready to be transfected with a pEGFP-ras fusion plasmid, using one of several transfection techniques. Although the traditional calcium phosphate or DEAE-dextran methods will suffice, we have found either LipofectAMINE (Life Technologies) or SuperFect (Qiagen) to be efficient and reliable. Both reagents are best used according to the manufacturer instructions. Control cells should be transfected with pEGFP that directs the expression of unfused GFP that illuminates the cytosol and, in some cell types (e.g., COS-1), accumulates in the nucleus.

3. Fluorescence appears over several hours. In addition to requiring the biosynthesis of a threshold amount of fluorescent protein, the cyclic

tripeptide that forms the fluorophore of GFP must be created, a process that proceeds autocatalytically with a time constant of up to 4 hr. Fluorescence can be observed as early as 5 hr posttransfection although full expression revealing steady state localization of GFP–Ras is best observed after 16–48 hr.

Stable Transformants

Many of the commercially available GFP vectors, including pEGFP, incorporate selectable markers and are therefore useful in establishing cell lines that stably express GFP-tagged proteins. We have found that it is relatively easy to establish cell lines that stably express GFP-tagged Ras proteins using pEGFP and combining aminoglycoside (G418) selection and fluorescence-activated cell sorting (FACS). For this purpose we have used CHO, MDCK, ECV, and NIH 3T3 cells. A G418 killing curve (0.1 to 3 mg/ml) must be performed for each cell type prior to selection to establish the appropriate dose. In addition, a nontransfected control should be treated in parallel with transfected cells to verify selection of resistant clones. We have used the following method.

1. At 5% confluence, cells can be transfected by any method (e.g., calcium phosphate, DEAE-dextran, LipofectAMINE, SuperFect).
2. Both control and transfected plates of cells are grown in appropriate medium containing the concentration of G418 established by the killing curve.
3. After 3–9 days, whereas all cells in the nontransfected control plate will have died (there should be <1 viable cell per low-power field) the plates with transfected cells should have numerous colonies of fluorescent cells.
4. The selected cells are harvested by addition of trypsin–EDTA and reseeded onto 10-cm plates at serial dilutions of 10^{-4}, 10^{-5}, 10^{-6}, 10^{-7}, and 10^{-8} of the initial plate. This wide range of dilution is useful to account for the variabilities in expression and transfection efficiencies between different samples.
5. These cultures are allowed to grow in G418 for 1–3 weeks until individual colonies (clones) of fluorescent cells can be identified.
6. Lubricant-sealed sterile steel cylinders can be placed around individual clones to isolate them for trypsin–EDTA harvesting. Ten to 20 clones should be selected on the basis of growth and fluorescence.
7. These clones are then grown in six-well plates in medium without antibiotic selection.
8. After 5–10 days these six-well plates should achieve 50% confluence of fluorescent cells.

9. Despite clonal growth, cell lines often manifest great heterogeneity in GFP fusion protein expression. Relatively high expressors can be selected and the resulting cell line made more homogeneous in expression by a single round of FACS.

In addition to cell lines that constitutively express GFP–Ras, we have succeeded in producing cell lines that inducibly express GFP–N-Ras. Although there is no reason to believe that one of the commercially available inducible expression systems (e.g., Tet-On from Clontech or GeneSwitch from InVitrogen) would not be well suited for this task, we chose to engineer our own isopropyl-β-D-thiogalactopyranoside (IPTG)-inducible expression system. To accomplish this we subcloned EGFP–N-Ras into pCMV3R, placing the fusion protein under the control of a cytomegalovirus (CMV) promoter that has been modified to incorporate three binding sites for the *lac* repressor. This plasmid also incorporates a neomycin resistance gene driven by a constitutive promoter. This pCMV3R-EGFP–N-ras plasmid was then transfected into a CHO cell line (obtained from M. Roth, University of Texas Southwestern Medical Center, Dallas, TX) stably transfected (using hygromycin selection) with a plasmid that constitutively drives the expression of the *lac* repressor. After transfection, G418 selection (as described above) allowed for the isolation of cell lines that stably incorporate pCMV3R-EGFP–N-ras. This system proved to be both tight and efficient in that, whereas in the absence of IPTG no cellular fluorescence was observed, high-level expression of GFP–N-Ras was seen after induction with IPTG. This system allowed for synchronous initiation of GFP fusion protein expression that facilitated trafficking studies.[6]

Imaging

The requirements for imaging GFP-tagged Ras proteins in fixed cell preparations are identical to those for conventional immunofluorescence studies. EGFP-tagged proteins can be imaged with conventional fluorescein isothiocyanate (FITC) filter sets, although custom sets optimized for GFP are also available [Chroma Technology (Brattleboro, VT), *www.chroma.com*]. Although cultured coverslips can be mounted inverted on a glass slide over medium-containing wells created by silicon rubber gaskets such that an upright microscope can be used, for imaging GFP–Ras in live cells an inverted microscope has many advantages. Whether conventional or confocal, combining an inverted microscope with the MaTek culture dishes described above allows for repeated imaging of the same culture without contamination and, when combined with a heated stage or a microincubator (Harvard Apparatus), kinetic analysis and directly observed pharmacologic manipulation.

We have found that a laser-scanning confocal microscope, although indispensable for certain applications, is not required to obtain membrane-targeting information from GFP–Ras expressed in live cultured cells. Although epifluorescence microscopes equipped with film cameras will suffice, digital image acquisition and manipulation offers tremendous advantages. These include high light sensitivity [with cooled charge coupled device (CCD) cameras], instantaneous acquisition setting adjustment for optimized exposure, time-lapse capability, digital image enhancement and analysis, and digital storage and transfer. Such systems are available from a variety of vendors [e.g., Leica (Bensheim, Germany), Nikon (Garden City, NY), and Zeiss (Thornwood, NY)] and can be purchased as complete systems or assembled with a microscope, digital camera, and imaging software from individual vendors. We have used a Zeiss Axiovert 100 microscope equipped with a Princeton Instruments (Trenton, NJ; *www.prinst.com*) cooled CCD camera optimized for GFP (model RTE/CCD-1300Y/HS). The halogen and mercury lamps, filter wheels, shutters, and camera are all controlled by MetaMorph software [Universal Imaging (West Chester, PA), *www.image1.com*] that also offers state-of-the-art image enhancement and analysis. This system is also ideal for capturing time-lapse image sequences that can be converted to QuickTime or AVI movies. We have used this technology to capture the movement of peri-Golgi vesicles illuminated by GFP–N-Ras and GFP–H-Ras.

GFP–Ras proteins expressed with pEGFP as described above and visualized in living cells tend to be so bright that attenuation of the mercury lamp is often required (this will also cut down on bleaching). After transient transfection one usually observes a range of expression levels. Often the brightest cells are rounded, phase-dense, and presumably apoptotic and should be excluded from analysis. Cells with low to midrange expression usually manifest normal morphology. Objectives with magnifications of ×10–×40 are best suited for imaging clusters of cells. Oil immersion objectives with magnifications of ×63 or ×100 and relatively high numerical apertures (1.2–1.4) are required for analyzing subcellular localization at the organelle level. Although cooled CCD cameras have an impressive dynamic range, they do not compare with the human eye. Thus, by directly observing a cell transfected with GFP–N-Ras or GFP–H-Ras it is often possible to simultaneously appreciate localization in both peripheral plasma membrane and central Golgi. However, because the Golgi tends to be so much brighter than the peripheral membrane, for digital imaging it is necessary to choose an exposure that is appropriate for one structure or the other. We often purposefully overexpose the Golgi in order to record plasma membrane localization. This is especially true for capturing ER localization of GFP-tagged Ras variants that lack secondary targeting sequences (i.e., have CAAX motifs alone). To avoid overwhelming glare from the markedly overexposed Golgi we close the diaphragm of the mercury lamp, leaving only the central field illuminated, and position the cell of interest such that the Golgi is obscured by the diaphragm, thus allowing long exposures (Fig. 2B and C).

Confocal microscopy, particularly when an inverted confocal microscope is available, offers a few advantages over digital imaging with a cooled CCD camera in analyzing the membrane localization of GFP–Ras in live cells. These include the ability to scan in the *z* axis, the capacity for three-dimensional reconstructions and rotations, and more precise colocalization of GFP–Ras with other molecules. In addition, high-end laser-scanning confocal microscopes such as the Zeiss 510 allow for sophisticated fluorescence recovery after photobleaching (FRAP) and fluorescence loss in photobleaching (FLIP),[8] technologies that hold great promise in analyzing membrane trafficking of GFP–Ras.

Acknowledgment

This work was supported by National Institutes of Health Research Grants GM55279 and AI36224.

[8] J. Lippincott-Schwartz, J. F. Presley, K. J. M. Zaal, K. Hirschberg, C. D. Miller and J. Ellenberg, *Methods Cell Biol* **58,** 261 (1999).

[4] Targeting Proteins to Membranes, Using Signal Sequences for Lipid Modifications

By JOHN T. STICKNEY, MICHELLE A. BOODEN, and JANICE E. BUSS

Introduction

Attachment of acylation signals to heterologous proteins has been widely used either to substitute for the normal membrane-binding mechanisms of a protein or to relocate a cytosolic protein to membranes. New designs for attachment of sequences that produce palmitoylation without myristate or isoprenoid attachment, and the use of the increasingly popular C*aa*X plus polybasic sequence from the K-Ras4B protein, are presented. Criteria for selecting a signal sequence and additional signals such as dual acylation sites and polybasic domains that can be used to improve targeting of acylated and prenylated proteins to membranes are included.

0076-6879/00 $35.00

Considerations for Using Lipidation Signals

Numerous examples of proteins with artificial membrane-targeting sequences that acquire constitutive biological function or become independent of normally required protein interactions have been produced. H-Ras proteins have been targeted to membranes by means of myristoylation signal sequences derived from the p60*src* protein kinase,[1,2] retroviral Gag proteins,[3] and even a transmembrane domain.[4] These unnaturally targeted forms of H-Ras have remarkably good biological activity in transforming NIH 3T3 cells. Ras proteins appear to depend heavily on their lipid's contribution for membrane localization, a situation in which a surrogate lipid can substitute quite well. The protein kinase Raf-1 normally utilizes Ras as a way to gain access to its (as yet ill-defined) activators in membranes, but can become Ras-independent if targeted to membranes by a C-terminal C*aa*X motif.[5,6] Membrane targeting can even be used to inhibit activity, by preventing the membrane-tethered chimeric protein from moving to its proper location. Prenylated forms of the mitogen-activated protein (MAP) kinases, extracellular signal-regulated kinase 1 (ERK1) and ERK2 fail to enter the nucleus and can interfere with transcriptional activation of c-*fos*.[7] This ability to change and study how membrane binding is coupled to biological activity is the most notable strength of this adoptive membrane-binding approach.

However, attachment of a lipidation sequence does not always activate function. Some proteins [e.g., some of the ADP-ribosylation factor (ARF) proteins involved in vesicular transport] may use the lipid for a more specific role, such as for participating at an interface with another protein.[8] In these cases, alternative lipids, especially at distant sites, will not suffice. Importantly, the technique of swapping lipids can help distinguish these two roles.

A central property of myristate and farnesyl lipids is that although they become attached permanently to a protein, neither is particularly good at

[1] P. M. Lacal, C. Y. Pennington, and J. C. Lacal, *Oncogene* **2,** 533 (1988).

[2] B. M. Willumsen, A. D. Cox, P. A. Solski, C. J. Der, and J. E. Buss, *Oncogene* **13,** 1901 (1996).

[3] J. E. Buss, P. A. Solski, J. P. Schaeffer, M. A. MacDonald, and C. J. Der, *Science* **243,** 1600 (1989).

[4] K. C. Hart and D. J. Donoghue, *Oncogene* **14,** 945 (1997).

[5] D. Stokoe, S. G. MacDonald, K. Cadwallader, M. Symons, and J. F. Hancock, *Science* **264,** 1463 (1994).

[6] S. J. Leevers, H. F. Paterson, and C. J. Marshall, *Nature* (*London*) **369,** 411 (1994).

[7] F. Hochholdinger, G. Baier, A. Nogalo, B. Bauer, H. H. Grunicke, and F. Uberall, *Mol. Cell. Biol.* **19,** 8052 (1999).

[8] C. D'Souza-Schorey and P. D. Stahl, *Exp. Cell Res.* **221,** 153 (1995).

sustaining membrane binding of that protein on its own.[9] Basic residues near the site of lipid attachment are apparently utilized to increase the interaction of a myristoylated or farnesylated protein with membranes.[10,11] For the experimenter, if high-efficiency, stable membrane binding is desired, a permanent (myristoyl or prenyl) lipid and multiple basic residues will likely be a good choice.

In contrast, palmitates are quite good at binding to lipid bilayers, yet are ephemeral, with the lipid being added and removed multiple times during the life span of a protein.[12,13] This gives palmitate a kinetic opportunity to influence and change protein hydrophobicity and behavior. There are now several examples of situations in which the rate of palmitate attachment to and hydrolysis from a signaling protein can vary.[14–20] As more experience is gained with use of palmitoylation of heterologous proteins we will learn whether palmitates may be useful for actively regulating interfaces between the acyl–protein and membranes or partner proteins.

Selecting Lipidation Signals

The only characterized signal sequences for lipid modification are located at the ends of the native protein. The investigator must thus decide if the chosen acceptor protein can best tolerate addition of a new domain at the N or C terminus without disruption of an essential region or activity of the target protein. To minimize disruption of acceptor protein structure the signal sequences described here introduce the smallest number of amino acids necessary for effective lipid modification and membrane binding.

Signals for N-Myristoylation

N-Terminal sequences for myristoylation are notably brief: a glycine to which the myristoyl group is attached, followed by three residues with relatively few limitations on character, and then a fifth residue that is

[9] S. Shahinian and J. R. Silvius, *Biochemistry* **34,** 3813 (1995).
[10] W. van't Hof and M. D. Resh, *J. Cell Biol.* **136,** 1023 (1997).
[11] J. F. Hancock, H. Paterson, and C. J. Marshall, *Cell* **63,** 133 (1990).
[12] A. I. Magee, L. Gutierrez, I. A. McKay, C. J. Marshall, and A. Hall, *EMBO J.* **6,** 3353 (1987).
[13] S. M. Mumby, C. Kleuss, and A. G. Gilman, *Proc. Natl. Acad. Sci. U.S.A.* **91,** 2800 (1994).
[14] J. T. Dunphy and M. E. Linder, *Biochim. Biophys. Acta* **1436,** 245 (1998).
[15] S. M. Mumby, *Curr. Opin. Cell Biol.* **9,** 148 (1997).
[16] P. B. Wedegaertner and H. R. Bourne, *Cell* **77,** 1063 (1994).
[17] M. Y. Degetyarev, A. M. Spiegel, and T. L. Z. Jones, *J. Biol. Chem.* **268,** 23769 (1993).
[18] L. Adams, M. Bouvier, and T. L. Z. Jones, *J. Biol. Chem.* **274,** 26337 (1999).
[19] L. J. Robinson, L. Busconi, and T. Michel, *J. Biol. Chem.* **270,** 995 (1995).
[20] S. Bhamre, H. Y. Wang, and E. Friedman, *J. Pharmacol. Exp. Ther.* **286,** 1482 (1999).

TABLE I
SEQUENCES FOR ATTACHING LIPIDS TO PROTEINS[a]

Lipid and amino acid sequences producing lipid modifications	Source protein
Myristate only	
[1]Met-**Gly(M)**-Ser-Ser-Lys-Ser-Lys-Pro-Lys-Asp-Pro-Ser-Gln-Arg-Arg[15]-	Src tyrosine kinase
[1]Met-**Gly(M)**-Gln-Ser-Leu-Thr-Thr-Pro-Leu-Ser-Leu[11]-	Rasheed SV Gag
Dual N-terminal acyl groups	
[1]Met-**Gly(M)**-**Cys(P)**-Thr-Val-Ser-Ala-Glu-Asp-Lys-Ala-Ala[12]-	G_i α subunit
[1]Met-**Gly(M)**-**Cys(P)**-Val-**Cys(P)**-Ser-Ser-Asn-Pro-Glu[10]-	Lck tyrosine kinase
Palmitate (N terminal)	
[1]Met-Val-**Cys(P)**-**Cys(P)**-Met-Arg-Arg-Thr-Lys-Gln-Val[11]-	GAP43
Palmitate plus polybasic (C terminal)	
-[181]**Cys(P)**-Met-Ser-**Cys(P)**-Lys-Cys*-Val-Leu-Lys-Lys-Lys-Lys-Lys-Lys (stop)	ExtRas mutant
Palmitate plus farnesyl (C terminal)	
-[181]**Cys(P)**-Met-Ser-**Cys(P)**-Lys-**Cys(F)**-Val-Leu-Ser (stop)	H-Ras
Farnesyl plus polybasic (C terminal)	
-[171]Ser-Lys-Asp-Gly-Lys-Lys-Lys-Lys-Lys-Lys-Ser-Lys-Thr-Lys-**Cys(F)**-Val-Ile-Met (stop)	K-Ras4B

[a] Lipidated residues are set in boldface type. Attached lipids are designated (M), myristate; (P), palmitate; or (F) farnesyl. Palmitoylation status of Cys* is not known.

most favorable if serine or threonine.[21] The signal sequences for myristate addition are exceptionally well conserved from plants to fungi to animals, allowing successful acylation of proteins bearing signals derived from mammalian proteins in a wide selection of expression systems. When N-terminal residues of a myristoylated protein are appended to a heterologous protein, a leader of 12–16 amino acids is often used, and may also contain additional features such as basic residues or cysteines that can be palmitoylated (Table I).

Signals for Isoprenoid Attachment

The canonical sequence for isoprenoid addition is a *CaaX* motif, in which the C is the cysteine to which the isoprenoid will be attached, followed by two generally (but not always) aliphatic residues and the final residue of the protein, X.[22] Two lengths of isoprenoid can be attached to proteins, the 15-carbon farnesyl or the 20-carbon geranylgeranyl group. The X residue is one of the major features that determine which size isoprenoid will

[21] L. Knoll, D. Johnson, M. Bryant, and J. Gordon, *Methods Enzymol.* **250,** 405 (1995).
[22] F. L. Zhang and P. J. Casey, *Annu. Rev. Biochem.* **65,** 241 (1996).

be attached. The geranylgeranyltransferase (GGTase I) prefers X to be leucine or phenylalanine, while the farnesyltransferase (FTase) accepts a broader range of residues, including methionine, serine, glutamine, and alanine. It is possible for the prenyltransferases to modify a four-residue C*aa*X peptide, but membrane targeting of a normally nonprenylated protein has usually been performed with a longer sequence, encompassing either a polybasic domain or additional cysteine residues conferring palmitoylation on the chimeric product.

Lack of Signal Sequences for S-Palmitoylation

In frustrating contrast to the well-defined signals for myristate and isoprenoid attachment, palmitoylation signal sequences have proved enigmatic.[2] This may be because palmitoyltransferases(s) appear to require little more than proximity of a cysteine to a lipid bilayer, and largely disregard the primary character of amino acids in its vicinity as long as these residues are compatible with membrane interactions. This ambiguity of context occurs in palmitoylated cysteines at either the N or C terminus.[14,15] Although prenylation occurs prior to palmitate attachment, the presence of isoprenoid does not appear to be a necessary part of a recognition motif for a Ras palmitoyltransferase, as substitution of a polybasic domain for the C*aa*X motif permits palmitoylation in the absence of prenylation in both yeast Ras2[23] and mammalian H-Ras.[24] Conversely, palmitoylation appears to require that the protein possess a method for transient membrane interaction, such as the presence of basic residues to support ionic interactions with acidic membrane phospholipids.[23]

Attachment of C-Terminal Signals for Isoprenoid Modification

Strategy

Selection of C-Terminal CaaX-Containing Sequence. The most common motifs used for prenylation-dependent plasma membrane targeting of proteins are derived from the C termini of the N-, H-, and K-Ras GTPases. Because inclusion of a second signal in addition to prenylation is usually necessary for efficient membrane association, amino acids N terminal to the C*aa*X sequence are often added to provide a polybasic region (as is the case for K-Ras4B) or signal for palmitoylation (for N- and H-Ras) that

[23] D. A. Mitchell, L. Farh, T. K. Marshall, and R. J. Deschenes, *J. Biol. Chem.* **269,** 21540 (1994).
[24] M. A. Booden, T. L. Baker, P. A. Solski, C. J. Der, S. G. Punke, and J. E. Buss, *J. Biol. Chem.* **274,** 1423 (1999).

allows for enhanced membrane interaction. However, data suggest that these regions of the various Ras isoforms may also play a role in signaling events[24] and/or intracellular trafficking[25] in addition to their role in plasma membrane interaction. In addition, work in our laboratory suggests that the mere presence of a *CaaX* motif does not assure that a protein will in fact be prenylated to a significant extent.[25a] Thus, once a C-terminal motif has been chosen, the prenylation status and subcellular location of the chimeric protein must be examined.

Choosing a Method. A previous *Methods in Enzymology* chapter[26] described the use of two sequential polymerase chain reaction (PCR) reactions to create a chimeric mutant protein possessing a C-terminal prenylation/membrane association signal. However, this method requires the possession of both target and signal sequence genes. The following methods describe how to create a chimeric protein possessing a C-terminal K-Ras4B-derived signal, using a one-step PCR, and to test whether it is isoprenoid-modified and membrane-associated.

Methods

One-Step PCR. The example shown here is from our studies to identify why the murine guanylate binding protein 1 (mGBP1), which possesses a usable *CaaX* motif, incorporates little [^{3}H]mevalonate. Figure 1A shows the nucleotide sequence of the mutagenic primer B used to create this mGBP1 : K-Ras4B chimera. Both 5′ (primer A, not shown) and 3′ (primer B) oligonucleotides possess a unique restriction site to allow directional cloning of the PCR product and four flanking nucleotides that serve as a molecular "handle" to facilitate restriction enzyme digestion of the PCR product. Primer A includes only 18 nucleotides for the 5′ end of mGBP1 because no mutations are introduced in this region. Primer B has 24 nucleotides used to anneal to the mGBP1 template, the 18 codons (54 nucleotides) encoding the K-Ras4B polybasic domain and *CaaX* motif plus a stop codon and restriction enzyme site. During the design of these primers, the program Oligo 4.0 (National Biosciences, Plymouth, MN) was used to determine the potential of and propose corrections for primer self-association and homo- and heterodimerization of primers.

Samples for PCR contain 100–500 ng of template DNA, 200 μM

[25] E. Choy, V. K. Chiu, J. Silletti, M. Feoktistov, T. Morimoto, D. Michaelson, I. E. Ivanov, and M. R. Philips, *Cell* **98,** 69 (1999).

[25a] J. T. Stickney and J. E. Buss, *Mol. Biol. Cell* **11,** 2191 (2000).

[26] P. A. Solski, L. A. Quilliam, S. G. Coats, C. J. Der, and J. E. Buss, *Methods Enzymol.* **250,** 435 (1995).

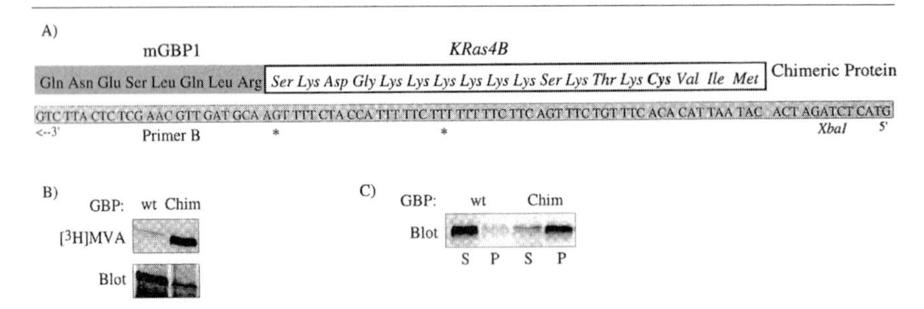

FIG. 1. Attachment of prenylation signal from K-Ras4B to mGBP1. (A) Nucleotide sequence of the 3′ primer used to create the prenylation signal for the mGBP1 : K-Ras4B chimera. Asterisks indicate where alternative codons were used to decrease undesirable primer interactions. (B) [^{3}H]MVA incorporation into wild-type (wt) and chimeric GBP : K4B (Chim). Proteins were expressed in COS-1 cells, labeled overnight with [^{3}H]mevalonate, and then isolated with a polyclonal anti-GBP serum (D. Paulnock, Madison, WI). Immunoprecipitates were resolved by SDS–PAGE, transferred to PVDF, and exposed to film for 2 weeks, and then the blot was probed with the anti-GBP serum. (C) Distribution of the chimeric GBP : K4B protein between cytoplasmic and membrane-containing fractions. Subcellular fractions were prepared by centrifugation at 100,000*g* and resolved by SDS–PAGE, and the chimera was detected by immunoblot detection with anti-GBP serum. The isoprenoid dependence of this association can be tested by repeating this experiment, using compactin to inhibit prenylation of mGBP1Chim and observing a decrease in the amount of chimera in the P100 fraction (not shown).

of each of the four deoxynucleotide triphosphates (dNTPs), 400 n*M* of each primer, proofreading DNA polymerase (1 unit of Deep Vent; New England BioLabs, Beverly, MA), and vendor-supplied DNA polymerase buffer in a 50-μl reaction volume. A "hot start" of 95° is used before addition of DNA polymerase to reduce the amount of mispriming that can occur during the early cycles of PCR. After 5 min, the polymerase is added and 28 cycles of 95° for 30 sec, 54° for 1 min, and 72° for 4 min are used to amplify and incorporate the mutated sequences into the template gene. These times and temperatures have been used successfully for amplification of mGBP1 and Ras proteins (below).

Ligation of Polymerase Chain Reaction Products and Verification of Sequence. After the PCR is completed, 5 μl is separated on an agarose gel. If a single easily visible band of the correct size is detected, the remaining PCR product is purified. We find the QIAquick PCR purification kit (Qiagen, Valencia, CA) to be one of several successful methods. For ligation, 10–25 μl of the clean PCR product and 1–1.5 μg of pcDNA3 are individually digested with the appropriate restriction enzymes and digestion products are purified on a 0.8% (w/v) agarose gel. The amplified insert and cut vector are excised from the gel, and the DNA is extracted by placing the gel pieces in a filtered P200 pipette tip cut to fit inside a 1.5-ml microcentrifuge tube and spinning them for 10 min at 4000 rpm at 4°.[27] Recovery of DNA is generally ~50% by this method. The volume of eluate varies on the basis of the size of the gel slice, but usually ranges from 20 to 40 μl. The actual amount should be measured as this will influence the total volume of the subsequent ligation reaction.

If the PCR gives an easily detectable amount of product, one-half of the eluted insert DNA is mixed with 3 μl of cut vector (usually corresponding to 50–120 ng of plasmid) for ligation in a minimum volume, with T4 DNA ligase (1–3 U/μl; Promega, Madison, WI) and 10× ligase buffer [Promega; 30 m*M* Tris-HCl (pH 7.8), 10 m*M* $MgCl_2$, 10 m*M* dithiothreitol (DTT), 1 m*M* ATP]. Because glycerol inhibits the ligase enzyme, the amount of T4 DNA ligase [in 50% (v/v) glycerol] used must be less than 10% of the total volume (typically 3–4 μl) to keep the final concentration $<5\%$. After addition of water and ligation buffer, we also add additional ATP to a final concentration of 1 m*M*. This mixture is then heated to 65° for 5 min to allow for dissociation of any remaining DNA that may be base paired to the restriction sites of the insert. The tube is cooled on ice, ligase is added, and the reaction is incubated overnight at 14°.

After the overnight ligation another 1 μl of ligase is added at room temperature for 1 hr to "boost" any remaining ligation events that may have stalled because of loss of enzyme activity. The entire mixture is used to transform *Escherichia coli* and DNA from individual bacterial colonies is isolated and screened for the presence of the insert by digesting the DNA with the appropriate enzymes. Because all PCR-generated sequences are subject to random mutations, DNAs meeting the size criteria are further purified and the presence of the desired mutation, and the absence of unintended mutations, in the DNA sequence are confirmed by sequencing from both ends. If pcDNA3 is used, then T7-1 and SP6 primers can be used to sequence the 5′ and 3′ ends of the insert, respectively.

Testing Chimeric Protein for Prenylation. To verify that the resultant chimeric protein is, as planned, actually isoprenoid modified, the most common method used is metabolic labeling of transfected cells with [^{3}H]mevalonate (usually 100–200 μCi/60-mm dish) overnight (16–24 hr) in the presence of 10–50 μM compactin or lovastatin.[28] These compounds inhibit cells from synthesizing mevalonate and force them to use the [^{3}H]mevalonate supplied exogenously. Cell lysates or immunoprecipitates

[27] A. D. Dean and J. E. Greenwald, *BioTechniques* **18,** 980 (1995).

[28] S. M. Mumby and J. E. Buss, *Methods Companion Methods Enzymol.* **1,** 216 (1990).

are prepared and proteins are separated by sodium dodecyl sulfate–polyacrylamide gel electrophoresis (SDS–PAGE) and subsequently transferred to polyvinylidene difluoride (PVDF) membrane. The proteins bearing radiolabel are visualized by exposure to film after the PVDF membrane has been treated with a fluorographic enhancer (Fig. 1B). We prefer to use aerosol enhancers such as En^{3}Hance (New England Nuclear Life Science Products, Boston, MA) rather than liquid enhancing solutions. The PVDF membrane is dried, sprayed lightly with En^{3}Hance, dried, sprayed again in an orientation 90° to the first, and dried again. Typical fluorographic exposure times range from 10 days to 3 weeks. After film development, enhancing reagent is removed from the PVDF membrane to allow immunoblot analysis of the same membrane. The membrane is rewet with methanol, which also extracts the organic enhancer, and then rinsed with a Tween 20/Tris-buffered saline (TTBS) solution. Alternating rinses of methanol and TTBS are performed until no cloudy solution is formed when the methanol rinse contacts an aqueous solution. The blot is then transferred to a blocking solution and immunoblot detection can be performed as usual.

Biochemical Verification of Membrane Association, Using Subcellular Fractionation. Because lipid signals are chosen as a means to redirect a protein to membranes, it is important to determine just how efficiently the lipid and adjoining sequences confer this association to the chimeric protein. Biochemical separations of P100 "membranes" are a common means of detecting membrane binding, but must be interpreted with caution, as proteins that are either insoluble or legitimately bound to nuclei, cytoskeletal elements, or internal membranes can also be present, in addition to proteins associated with plasma membranes.

Cells expressing the chimeric protein are rinsed with phosphate-buffered saline (PBS) and allowed to swell for 5–15 min in a hypotonic buffer [1 m*M* Tris, (pH 7.4), 1 m*M* $MgCl_2$, 1 m*M* Pefabloc, 1 μM leupeptin, 2 μM pepstatin, and 0.1% (w/v) aprotinin] on ice. The swollen cells are broken in a Dounce homogenizer, NaCl is added to adjust the ionic strength to 0.15 *M*, and the membranes are pelleted by centrifugation at 100,000*g* for 30 min at 4°. The supernatant (S100) is removed and the pellet (P100) is resuspended in a volume of hypotonic buffer equal to that of the S100 fraction. Proteins in both fractions are then precipitated by the addition of 4 volumes of cold acetone on ice for 30 min and collected by centrifugation at 3000 rpm at 4° for 30 min. Acetone is removed and the pellets are air dried before addition of electrophoresis sample buffer and SDS–PAGE. The distribution of the chimeric protein in the cytoplasmic and membrane fractions is then determined by immunoblotting (Fig. 1C).

Methods to Produce Palmitoylated Protein

Strategy

The most common examples of palmitoylation occur in concert with either myristate or farnesyl attachment, and require prior attachment of the stable acyl or prenyl group. Signals for myristoylation or dual myristoyl/palmitoyl lipid groups can also be attached by the techniques described here, and additional details have been described in previous volumes of *Methods in Enzymology.*[26,29] Several proteins that have N-terminal palmitates but lack myristate have been identified, including GAP43/neuromodulin and the α subunits of some heterotrimeric GTP-binding proteins such as G_q and G_{11}.[15] Importantly, the palmitates of the GAP43 N-terminal leader show the same active palmitate cycling that occurs on the intact protein.[30] Thus, these palmitoylation signals, although having imperfectly described motifs, can be used to produce dynamic acylation of a target protein.

The examples described here can be used to produce H-Ras proteins that have either (1) an N-terminal palmitoylation signal, mimicking the first 11 amino acids of the GAP43 protein[30] or (2) a unique C-terminal palmitoylation signal, derived from H-Ras, that can be modified in the absence of isoprenoid.[24]

N-Terminal Palmitoylation. Two strategies can be used to attach an amino-terminal palmitoylation signal to H-Ras. The first utilizes a three-piece ligation method in which an oligonucleotide encoding the first 11 amino acids of the GAP43 protein is used to link the acceptor protein and vector sequences.[26] The major limitation to this method is that the acceptor protein must have a unique restriction site near the N terminus of the protein that can be used as a splice junction, or an artificial site must be engineered in the acceptor, using site-directed mutagenesis or PCR. This approach may be necessary for larger proteins. However, for proteins up to (and increasingly beyond) ~60 kDa (equivalent to ~2 kb of cDNA), attachment of a palmitoylation sequence can be accomplished directly by PCR-directed mutagenesis.

This second strategy utilizes a 5′ oligonucleotide primer encoding the first 11 amino acids of the GAP43 protein and approximately 18 nucleotides that match the target H-Ras sequence (Fig. 2A). It is important for the target sequence in the primer to begin with the second codon of the protein,

[29] G. Reuter, J. E. Buss, L. A. Quilliam, G. J Clark, and C. J. Der, *Methods in Enzymol.* **327,** 331 (2000).

[30] S. G. Coats, M. A. Booden, and J. E. Buss, *Biochemistry* **38,** 12926 (1999).

A) N-terminal leader

5'-Primer for attachment of GAP-43 residues 1 - 11 to HRas

5'-GGATCCACC*ATGGTGTGCTGTATGAGAAGAACCAAACAGGTT***ACAGAATACA**-3'

BamH1

3'-Primer for mutation of HRas Cys186 to Ser

5'-CTCGAGTCA**GGATAACACA***CT***CTTGCAGCTCATGCA**-3'

Xho1

B) C-terminal extension

5'-Primer for HRas N-terminus

5'-GGGGGGATCCACC**ATGACAGAATACAAGCTT**-3'

BamH1

3'-Primer for C-terminal extension on HRas

5'-GGGGGGATCCTCA*CTTCTTCTTCTTCTTCTT***GAGCACACACTTGCAGCT**-3'

BamH1

C)

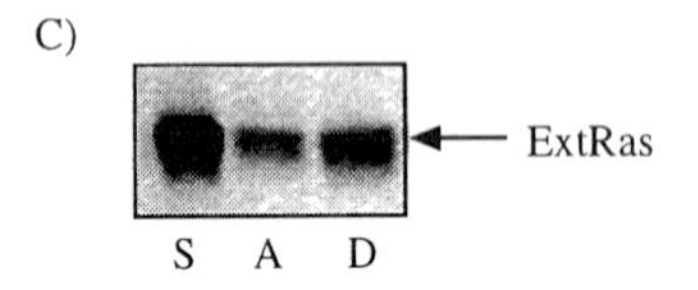

FIG. 2. Attachment of N-terminal or C-terminal sequences for palmitoylation. (A) Mutagenic oligonucleotide for attaching N-terminal GAP43 residues to H-Ras. The 5′ primer contains a *Bam*HI site, the first 11 codons of GAP43 (italics), and codons 2–5 of H-Ras (boldface). The 3′ primer introduces a C186S substitution (italic) to prevent prenylation, has 27 nucleotides complementary to 3′ sequences of H-Ras (bold-face), a termination codon, and a *Xho*I site. (B) Design of C-terminal polybasic extension to induce palmitoylation of the C-terminal cysteines of H-Ras. The 5′ primer is fully complementary to the 5′ sequences of human H-Ras (boldface). The 3′ mutagenic oligonucleotide contains nucleotides complementary to the 3′ sequences of H-Ras (boldface), a Ser189Lys substitution followed by five additional lysine residues (italics), a termination codon, and a *Bam*HI site. (C) Distribution of ExtRas between soluble and membrane fractions. After separation of S100 and P100 fractions, the P100 fraction was further subjected to Triton X-114 partitioning. All fractions were precipitated in cold acetone and resuspended in 100 μl of electrophoresis sample buffer, and equal portions of cytosolic (S), aqueous (A), or detergent (D) phase were resolved by SDS–PAGE. H-Ras proteins were detected by immunoblotting with the anti-H-Ras monoclonal antibody 3E4-146 (Quality Biotech, Camden, NJ).

because retention of the initiating methionine of the target allows its occasional, unintended use during protein synthesis. This can result in production of a mixed population of chimeric and nonchimeric protein that begins at the original Met-1 and lacks the desired signal sequence.[30] The 3′ oligonucleotide primer should contain 12 bases of sequence complementary to the target coding sequence, followed by a termination (TGA) codon, and a restriction site convenient to perform directional cloning.

N-Terminal Dual Acylation. Importantly, investigators designing custom versions of an N-terminal leader should remember that if a codon for glycine is generated at the second position, the protein may also become myristoylated, especially if the amino acid at position 6 is a serine or threonine. The sequences for just such a leader, deliberately producing a dually modified myristoyl–palmitoyl chimeric target protein, can be modeled after the Lck or G_i α sequences shown in Table I. On the basis of reports using green fluorescent protein (GFP) bearing these types of signals and on the basis of analyses of native proteins with dual acylation motifs, these signals may do more than confer general plasma membrane targeting; they may confer an increased presence of the chimeric protein in lipid "rafts" at the plasma membrane.[31,32]

C-Terminal Palmitoylation. The remodeling of the C terminus of H-Ras, to permit palmitoylation of Cys-181, -184, and -186, in the absence of an isoprenoid modification, can also be accomplished by PCR. The 5′ oligonucleotide primer (Fig. 2B) should contain a convenient restriction site for cloning, the initiating ATG codon of the target gene, followed by ~18 bases of the 5′ sequence of the target gene (in this case, H-Ras). The 3′ oligonucleotide primer should contain ~18 bases of matched sequence, the nucleotides encoding H-Ras amino acids 181–188 with the sites for palmitate attachment, bases altering the final codon from a TCC (serine) to CTT (lysine), and codons adding 5 additional lysines, followed by a termination codon, and a restriction site convenient to perform directional cloning (total length, ~70 nucleotides). The lysine in the "X" position of the C*aa*X motif is important, as it prevents farnesylation of the protein even if the other five lysines should be removed by proteolysis. The positively charged lysines are envisioned to substitute for the hydrophobic farnesyl and provide an ionic platform for interaction of the C terminus with acidic phospholipids of the membrane and enable the cysteines to become palmitoylated. The yeast Ras2 protein has been successfully modified and palmitoylated using a variation on this basic theme (-Lys-Leu-Ile-Lys-Arg-Lys).

[31] P. Zlatkine, B. Mehul, and A. Magee, *J. Cell Sci.* **110,** 673 (1997).

[32] M. D. Resh, *Biochim. Biophys. Acta* **1451,** 1 (1999).

Methods

Construction of Chimeric DNA by Polymerase Chain Reaction. The PCR and ligation conditions described in the previous section have been used successfully for construction of the GAP43 : H-Ras or ExtRas[24] chimeras; however, it was necessary to remove the template from the vector by digestion with the appropriate enzymes and to gel purify the 0.6-kb Ras DNA in order to reduce mispriming of the oligonucleotides on vector sequences.

Checking Chimeric Protein for Palmitoylation. To examine the palmitoylation status of the expressed chimeric protein, cells expressing the chimera are labeled for 4 hr with [^{3}H]palmitate (1 mCi/ml; New England Nuclear Life Science Products) in medium containing cycloheximide (25 μg/ml).[28] The chimeric protein is immunoprecipitated, resolved by SDS–PAGE, and detected by fluorography as described above. Electrophoresis sample buffer for protein bearing [^{3}H]palmitate must contain a minimal amount of reducing agent (no mercaptoethanol, only 50 m*M* dithiothreitol) to prevent hydrolysis of the palmitoylcysteine thioester bond. The presence of cycloheximide in the labeling medium is also needed to ensure that radiolabel is incorporated only into those proteins that have been synthesized previously and is not derived from [^{3}H]palmitate that has been metabolized into ^{3}H-labeled amino acids or to [^{3}H]myristate, and incorporated into myristoylation sites. Fluorographic exposure times of 1 week to 1 month are not uncommon, unless the protein has been highly overexpressed.

Combination Subcellular Fractionation and Hydrophobic Partitioning. The level of membrane association supported by palmitoylation alone is significantly less than the highly efficient combinations of myristate plus palmitate, or isoprenoid plus basic residues. Because ~60% of the ExtRas protein is cytosolic, the following variation on Triton X-114 phase separation was developed to separate this large pool of soluble protein and quickly learn if the membrane-bound protein displays any hydrophobic character suggestive of the presence of the desired lipid modification. Cytosolic (S100) and membrane-containing (P100) fractions are prepared by the technique described in the previous section, soluble proteins are removed, and the crude P100 fraction is resuspended in 0.5 ml of ice-cold buffer containing 12.5% (v/v) Triton X-114 in Tris-buffered saline (additional details in Ref. 11) and thoroughly mixed. The sample is then warmed to 37° for 2 min to trigger phase separation and subjected to a brief centrifugation in a room-temperature microcentrifuge. The upper (aqueous) phase, containing proteins with predominantly hydrophilic character, is removed to a fresh tube. The remaining detergent-enriched fraction contains proteins with exposed hydrophobic features (e.g., lipids). The proteins in the cytosolic S100 fraction can be similarly partitioned if it is suspected that the signal for lipid modification confers only inefficient or weak membrane binding. The proteins in each separated fraction are precipitated with acetone and then resuspended in electrophoresis sample buffer, displayed by SDS–PAGE, and detected by immunoblotting (Fig. 2C).

Summary

Changing an existing lipid or appending a lipid to a cytosolic protein has emerged as an important technique for targeting proteins to membranes and for constitutively activating the membrane-bound protein. The potential for more precise or regulated interactions of lipidated proteins in membrane subdomains suggests that this method for membrane targeting will be of increasing usefulness.

Acknowledgments

This work was supported by the Roy J. Carver Charitable Trust, the Elsa Pardee Foundation, and an award from the NSF POWRE program.

[5] Targeting Proteins to Specific Cellular Compartments to Optimize Physiological Activity

By GARABET G. TOBY and ERICA A. GOLEMIS

Introduction

For most proteins, localization to specific cellular compartments is generally a prerequisite for appropriate physiological function. Different cellular compartments provide different environmental conditions that affect the physiological and enzymatic activities of proteins, and, in addition, house interaction partners such as other proteins or nucleic acids essential for biological activity. Targeting of proteins to specific intracellular locations is highly regulated, and in many cases has been shown to rely on short peptide motifs. For instance, all newly synthesized proteins destined to be secreted or membrane bound initially enter the endoplasmic reticulum (ER). In the ER, nascent peptides in an unfolded state undergo posttranslational modifications such as glycosylation, disulfide bond formation, or assembly into oligomers. Further carbohydrate modifications occur in the various cisternae of the Golgi complex. Proteins that are to become residents of the ER are retained in the lumenal compartment by way of a recognition-retention carboxy-terminal tetrapeptide signal (KDEL).[1] Membrane-bound proteins contain the information necessary to direct them to the cell surface, mainly through posttranslational fatty acid modifications. Nuclear proteins utilize discrete nuclear localization sequences (NLS), often rich in lysine residues, to target them to the cell nucleus. The short peptide sequences that perform such targeting are in some cases recognized by chaperones (e.g., importin a and b for nuclear transport, reviewed in Ref. 2) that direct the protein to its final destination. In other cases, the posttranslational modifications (e.g., fatty acylation) render the protein more hydrophobic, enhancing the tendency to anchor in the phospholipid membranes.[3] Alternatively, the targeting sequence is recognized by specific receptors, leading to the docking of the protein and its subsequent translocation into its specific cellular compartment [e.g., peroxisome-targeting sequence I (PTS I) and PTS II receptors, peroxins Pex5p and Pex7p, reviewed in Ref. 4].

The issue of appropriate cellular compartmentalization is of particular relevance to small GTPases of the Ras superfamily.[5] At present there are more than 700 defined members of the superfamily across species, including 29 members in the completely sequenced organism *Saccharomyces cerevisiae,* and at least 66 in humans.[6] Members of the superfamily fall into four or five groups, represented by the Ras, Rho, ADP-ribosylation factor (ARF), Ypt/Sec/Rab, and Ran families. Members of these groups are specialized for distinct functions, executed in distinct cellular locations. For instance, the Rho proteins (Cdc42, Rac, Rho, and others) control actin cytoskeleton and cell movement, and are intermediates in cell signaling related to oncogenic transformation and apoptosis,[7,8] while Rab proteins control vesicular transport, and are colocalized with components of the Golgi and endosomal compartments.[9] Studies of the interaction of Ras family molecules have indicated that many of these proteins are involved in extremely complex webs of association with overlapping sets of effector

[1] H. R. Pelham, *Cell Struct. Funct.* **21,** 413 (1996).
[2] D. Gorlich, *EMBO J.* **17,** 2721 (1998).
[3] M. D. Resh, *Biochim. Biophys. Acta* **1451,** 1 (1999).
[4] E. H. Hettema, B. Distel, and H. F. Tabak, *Biochim. Biophys. Acta* **1451,** 17 (1999).
[5] H. R. Bourne, D. A. Sanders, and F. McCormick, *Nature (London)* **348,** 125 (1991).
[6] J. A. Garcia-Ranea and A. Valencia, *FEBS Lett.* **434,** 219 (1998).
[7] I. M. Zohn, S. L. Campbell, R. Khosravi-Far, K. L. Kossman, and C. J. Der, *Oncogene* **17,** 1415 (1998).
[8] L. Van Aelst and C. D'Souza-Schorey, *Genes Dev.* **11,** 2295 (1997).
[9] P. Novick and P. Brennwald, *Cell* **75,** 597 (1993).

molecules,[10] implying that studies intended to analyze functional properties of superfamily proteins should take particular heed to minimize factors that might contribute to promiscuous protein interactions.

Structure–function analysis usually entails making an extensive series of deletions or mutations in the amino acid sequence of a protein to identify residues critical for protein activity. If such sequence modification results in disruption of a protein motif necessary for targeting a protein to an appropriate cellular compartment, this will result in an apparent loss of function, even if a protein remains capable of normal activities (interaction with substrates, catalysis, adaptor function, etc.), complicating interpretation of findings. Alternatively, the construction of epitope tag fusions of the protein [e.g., influenza hemagglutinin (FLAG), hemagglutinin (HA), Myc] can in some cases lead to a change in the conformation of a protein that may prevent it from localizing to the location that matches the endogenous protein localization. For these reasons, and others noted below (see Applications), it is desirable to be able to control protein localization by use of an invariant standard targeting motif that can be routinely added to a series of proteins under evaluation, to eliminate or at least to reduce questions of localization as a variable in analysis.

Expression of Proteins as Fusions to Localization Signals

Cellular Compartmental Localization Signals

To overcome localization problems, one strategy is to develop a vector series allowing the convenient expression of a fusion of a cDNA that encodes the protein of interest and a localization signal of choice for specific subcellular compartments. The following section provides a brief review of a number of common signals that are known to direct proteins to specific compartments within the cell and can be utilized to develop targeting vectors. Targeting motifs discussed here are summarized in Table I.

Endoplasmic Reticulum/Golgi Complex. ER protein residents encompass a specific sequence of amino acids (KDEL) at their carboxy terminus that makes their sorting and retention in the ER possible. Proteins that contain such a signal are ones that help in the correct folding and modifications of nascent peptides. Such proteins include BiP [binding protein, a homolog of heat shock protein 70 (hsp-70)], protein disulfide isomerase (PDI), and calreticulin. This retention is mediated by a process known as retrograde transport (reviewed in Ref. 1), in which proteins containing the

[10] S. L. Campbell, R. Khosravi-Far, K. L. Rossman, G. J. Clark, and C. J. Der, *Oncogene* **17,** 1395 (1998).

TABLE I
TARGETING SIGNALS AND DOMAINS FOR DISCRETE CELLULAR COMPARTMENTS

Cellular compartment	Consensus signal	Fuse to protein terminal	Origin	Ref.
Er/Golgi	KDEL[a]	C	BiP	1
	K(X)KXX[a]	C	Emp47	12
	FF	C	Emp24	13
Peroxisomal matrix	PTS I (SKL)[a]	C	Catalase A	4
	PTS II [(R/K)(L/V/I)X_5 (H/Q)(L/A)]	N		4
Membranes	MGSSKSK[a]	N	Src family	16
	*Caa*X[a]	C	Ras family	15
	Ph domain[a]	N	Akt	19
	PDZ domain		MAGUK	21
Nuclei	PKKKRKV[a]		SV40	22
Mitochondria	MSVLTPLLLRGLTGSA-RRLPVPRAKISL[a]	N	COX VIII	23
Cytoskeleton				
Focal adhesions	FAT (1–159)[a]	C	FAK C terminus	24
	TBS (167–208)[a]	N	Vinculin N terminus	27
F-actin	C1 domain (223–228)		εPKC	29
Chloroplast	(S/T)RRXFLK	N		32

Abbreviations: MAGUK, membrane-associated guanylate kinase; FAK, focal adhesion kinase.
[a] Signal has been shown to be sufficient to direct fused reporters or cDNAs to corresponding cellular compartment.

KDEL sequence advance through the ER to the cisternae of the Golgi complex, and then are recognized and sent back to the ER compartments.

In addition to the KDEL retention signal, a dilysine signal [K(X)KXX] at the C terminus of a number of proteins also plays part in their retention in the ER lumen. Whereas KDEL results in a specific enrichment of proteins in the ER, K(X)KXX-fused proteins tend to be concentrated in the intermediate compartment and *cis*-Golgi complex.[11,12]

Finally, a double phenylalanine (FF) motif present in the cytoplasmic domain of major constituents of the *cis*-Golgi network (p24 family) has been shown to be involved in Golgi retention of some proteins, based on interaction with coat protein II (COP II.)[13]

[11] L. V. Lotti, G. Mottola, M. R. Torrisi, and S. Bonatti, *J. Biol. Chem.* **274,** 10413 (1999).
[12] S. Schroder, F. Schimmoller, B. Singer-Kruger, and H. Riezman, *J. Cell Biol.* **131,** 895 (1995).
[13] M. Dominguez, K. Dejgaard, J. Fullekrug, S. Dahan, A. Fazel, J. P. Paccaud, D. Y. Thomas, J. J. Bergeron, and T. Nilsson, *J. Cell Biol.* **140,** 751 (1998).

Peroxisomes. Proteins are targeted to peroxisomes because of the presence of a peroxisome-targeting sequence (PTS) at either their C terminus (PTS I) or N terminus (PTS II).[4] Specific targeting of most proteins into the peroxisome matrix is associated with the presence of an evolutionarily conserved C-terminus tripeptide SKL (e.g., acyl-CoA oxidase) or a conserved derivative (S/C/A, K/R/H, L) designated PTS I. PTS I has been shown to be necessary and sufficient to direct reporter proteins to peroxisomes.[14] Although PTS II is found in fewer peroxisomal matrix proteins, in some cases it is the only detectable localization motif. The defined PTS II is (R/K)(L/V/I)X_5(H/Q)(L/A).

Membranes. Fatty acylation of proteins is a conserved mechanism of targeting proteins to the cytoplasmic surface of cellular membranes. These modifications of proteins for targeting include the attachment of myristate, palmitate, and isoprenoid moieties. The consensus sequences for most of these fatty acyl modifications have been characterized (see Ref. 15 and references therein).

Src family members translocate to the membranes by encoding an N-terminal myristate attachment sequence (MAS) that enables *N*-myristoyltransferase (NMT) to catalyze myristate attachment.[16] The common sequence for NMT substrate is MGXXX(S/T). The requirement for G is absolute; in general S/T is preferred at position 6, although not absolute. Although myristoylation is necessary for targeting, an additional polybasic cluster of amino acids (R or K at positions 7 and 8) adjacent to the fatty acyl modification enhances the attachment of proteins to the membrane by establishing electrostatic bonds between the positively charged amino acids and the negatively charged phospholipids.[3] Alternatively, a second fatty acyl modification can attribute to better insertion of proteins in the phospholipid membrane (e.g., the dual N-terminal myristate and palmitate attachment of some Src family members).[3,17]

In addition, C-terminal lipid modifications are common among a large number of proteins. One of the most well-characterized modifications includes the C-terminal prenylation of Ras. Prenylation requires the presence of a C*aa*x motif at the C terminus of proteins (C is a cysteine, *a* is an aliphatic amino acid, and X is any amino acid). The cysteine is necessary for the modification because it is the site of attachment of the lipid moiety. Attachment of either the C_{15}-farnesyl group or the C_{20}-geranylgeranyl group

[14] S. J. Gould, G. A. Keller, and S. Subramani, *J. Cell Biol.* **107,** 897 (1988).

[15] P. A. Solski, L. A. Quilliam, S. G. Coats, C. J. Der, and J. E. Buss, *Methods Enzymol.* **250,** 435 (1995).

[16] M. D. Resh, *Cell* **76,** 411 (1994).

[17] W. van't Hof and M. D. Resh, *J. Cell Biol.* **136,** 1023 (1997).

is highly influenced by the nature of the aliphatic amino acid.[15] Like N-terminal myristoylation, additional polybasic motifs or a second lipid modification (e.g., dual palmitoylation and isoprenylation of Ras) enhances the attachment of proteins to the intracellular surface of cytoplasmic membranes.[18]

Direct fatty acylation is not the only way by which proteins are directed and anchored to the cytoplasmic membrane. Pleckstrin homology (PH) domains have the ability to bind to phospholipids and hence proteins that contain PH domains within their amino acid sequences have the ability to be membrane associated. For example, the ability of the protooncogene product Akt to translocate and attach to the plasma membrane depends on its PH domain in addition to its phosphorylation on receiving growth cues.[19] Although different PH domains from different proteins have variable affinities for lipid binding, the PH domain of phospholipase (PLC)-δ1 fused to green fluorescent protein (GFP) predominantly localizes at the plasma membrane,[20] demonstrating sufficiency of the motif for localization. In addition, PSD-95, Dlg, and ZO-1 (PDZ) domains, which occur in diverse molecules including protein tyrosine phosphatases and serine/threonine kinases, enable these molecules to bind to the carboxy terminus of a subset of receptors and with Shaker-type K^+ channels.[21] One potential advantage in using these domains as a fusion module might be the targeting of proteins to specific clusters of membrane-associated signaling proteins, in proximity to other proteins of interest.

Nuclei. The first well-characterized amino acid nuclear targeting sequence (NLS) was identified in nucleoplasmin and the simian virus 40 (SV40) large T antigen.[22] This sequence, PKKKRKV, is sufficient to direct fusion proteins to the nucleus. While this sequence is regarded as a model, a bipartite NLS is also found in many nuclear proteins,[22] again enriched in basic amino acids.

Mitochondria. Directing proteins to the mitochondria can be achieved by fusing a cDNA of interest in frame with a mitochondrial targeting element derived from subunit VIII of human cytochrome *c* oxidase. Chimeric proteins have been shown to localize efficiently to the mitochondria when a 29-amino acid presequence (see Table I) is fused at the N terminus of the cDNA of interest.[23]

[18] S. G. Coats, M. A. Booden, and J. E. Buss, *Biochemistry* **38,** 12926 (1999).

[19] B. A. Hemmings, *Science* **275,** 628 (1997).

[20] M. Fujii, M. Ohtsubo, T. Ogawa, H. Kamata, H. Hirata, and H. Yagisawa, *Biochem. Biophys. Res. Commun.* **254,** 284 (1999).

[21] C. P. Ponting, C. Phillips, K. E. Davies, and D. J. Blake, *BioEssays* **19,** 469 (1997).

[22] C. Dingwall and R. A. Laskey, *Trends Biochem. Sci.* **16,** 478 (1991).

[23] R. Rizzuto, A. W. Simpson, M. Brini, and T. Pozzan, *Nature* (*London*) **358,** 325 (1992).

Cytoskeleton. Structure–function analysis of the domain structure of the focal adhesion kinase revealed a stretch of 159 amino acids at the carboxy terminus of the protein[24] that is essential for its targeting to the focal adhesion sites, points of cellular attachment to the extracellular matrix through its integrins.[25] This stretch of amino acids is termed a focal adhesion targeting (FAT) domain. Alternatively, the talin-binding sequences from vinculin have also been utilized to target proteins to focal adhesions.[26] Fusion of the N-terminal 45 kDa of vinculin to v-Src targets the latter to focal adhesions. Within this region, the talin-binding sequence (TBS) has been identified as amino acids 167–208 (GMTKMAKMIDERQQELTH-QEHRVMLVNSMNTVKELLPVLIS).[27]

Finally, other domains that enable association of proteins to other cytoskeletal components have been identified. As one example, the C1 domain of protein kinase C isoform ε (εPKC) encompasses an isozyme-unique sequence that mediates the binding of this PKC family member to F-actin (see Ref. 28; discussed with other examples in Ref. 29).

Chloroplasts. The plant chloroplast targeting signal is a bipartite domain. An import motif (the transit peptide) occurs at the N terminus of proteins destined to reach the chloroplasts. This signal is cleaved during transport, exposing a second N-terminal signal (the export signal) that directs proteins across the thylakoid membrane. The export signal encompasses three distinct domains: a positive domain (N domain), a hydrophobic domain (H domain), and a more polar domain (C domain). The N domain encompasses a double arginine (RR) motif. Mutating these residues to KK abolishes targeting.[30] A consensus of (S/T)RRXFLK, in which the twin arginines are invariable, is shared among a set of membrane proteins that bind to complex redox factors.[31,32]

Experimental Design

Most of the above-described targeting motifs can readily be incorporated into a compatible expression vector in such a way as to facilitate

[24] J. D. Hildebrand, M. D. Schaller, and J. T. Parsons, *J. Cell Biol.* **123,** 993 (1993).
[25] K. Burridge and M. Chrzanowska-Wodnicka, *Annu. Rev. Cell. Dev. Biol.* **12,** 463 (1996).
[26] E. C. Liebl and G. S. Martin, *Oncogene* **7,** 2417 (1992).
[27] P. Jones, P. Jackson, G. J. Price, B. Patel, V. Ohanion, A. L. Lear, and D. R. Critchley, *J. Cell Biol.* **109,** 2917 (1989).
[28] R. Prekeris, M. W. Mayhew, J. B. Cooper, and D. M. Terrian, *J. Cell Biol.* **132,** 77 (1996).
[29] D. Mochly-Rosen and A. S. Gordon, *FASEB J.* **12,** 35 (1998).
[30] A. M. Chaddock, A. Mant, I. Karnauchov, S. Brink, R. G. Herrmann, R. B. Klosgen, and C. Robinson, *EMBO J.* **14,** 2715 (1995).
[31] B. C. Berks, *Mol. Microbiol.* **22,** 393 (1996).
[32] A. M. Settles and R. Martienssen, *Trends Cell Biol.* **8,** 494 (1998).

subsequent in-frame fusion with cDNAs of interest. In constructing such vectors, several points should be considered. A first issue is that of the strength of the promoter utilized to express targeted chimeric proteins. A strong promoter will generally ensure that substantial levels of the protein of interest are produced, and in conjunction with a targeting signal, will lead to robust levels of the protein at a desired intracellular location, a favored situation for dominant negative applications (see below). A weak promoter may be preferred when the experimental aim is to closely approximate physiological protein levels. A second issue is the evaluation of whether the targeting motif selected is effective in providing the desired localization to fused proteins. The functionality of novel targeting vectors is most readily assayed with a reporter gene product fused in frame to the targeting signal. Although a wide variety of reporters can be used (e.g., β-galactosidase and luciferase), the use of GFP[33] is advantageous in this case. Localization of the targeting signal–GFP fusion is assayed by transfecting/transforming cells with the targeted reporter and directly observing them under a fluorescence microscope, in conjunction with antibody-based detection of markers resident in specific cellular compartments. Alternatively, cell fractionation can be utilized to establish localization to some cellular compartments. A third point is that in designing a fusion vector, particularly if the targeting motif is small, it is possible to modify the fusion cassette to include additional features of interest. For example, the two-hybrid activation-domain fusion vector pJG4-5[34] includes an NLS and an epitope tag to facilitate antibody detection, in addition to the transcriptional activation moiety required for its function. Finally, it is important to be alert for possible pitfalls arising from the use of targeting sequences, and hence creation of a novel protein. The new chimera may not be appropriately folded, and hence may demonstrate anomalous activity or be subject to enhanced proteolytic degradation. Hypothetically, targeting sequence-fused proteins, if highly expressed, could compete for the normal receptor of the target sequence, and hence block other cellular proteins utilizing that sequence from their normal intracellular locale. Controls should be performed to ensure the fused protein is appropriately expressed and modified, and not inducing nonspecific toxicity.

Available Targeting Vectors

We have used this strategy and modified the widely used mammalian expression vector pcDNA3 (InVitrogen, Carlsbad, CA) to generate three

[33] M. Chalfie, Y. Tu, G. Euskirchen, W. W. Ward, and D. C. Prasher, *Science* **263,** 802 (1994).
[34] J. Gyuris, E. A. Golemis, H. Chertkov, and R. Brent, *Cell* **75,** 791 (1993).

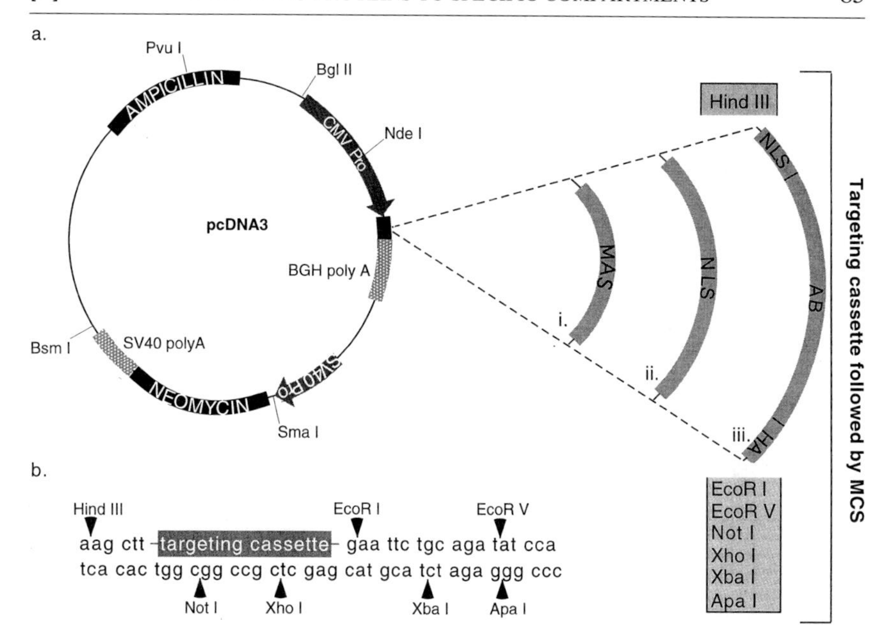

FIG. 1. (a) Vector map for plasmids pGTM, pGTN, and pGTNAT. The plasmid backbone is that of pcDNA3 (InVitrogen). Expression of cDNA inserts is driven by the cytomegalovirus (CMV) promoter, with the bovine growth hormone (bGH) polyadenylation transcription termination signal sequence incorporated 3′ to the cDNA insertion site to enhance high-level mRNA expression. The multiple cloning site (MCS) provides a variety of unique restriction sites for convenient cloning. pcDNA3 also encompasses a neomycin resistance gene that allows the selection of stable mammalian cell lines, while the ampicillin resistance gene is used for selection in bacteria. Targeting cassettes were introduced into the MCS using *Hin*dIII and *Eco*RI restriction sites in order to generate (i) pGTM, (ii) pGTN, and (iii) pGTNAT (see text for further details). The complete nucleotide sequence of pcDNA3 is available from the InVitrogen website (*www.invitrogen.com*). (b) The targeting cassettes of pGTM, pGTN, and pGTNAT allow fusion to cDNAs at the *Eco*RI or downstream sites (indicated), using the "gaa ttc" frame.

different targeting vectors that direct proteins and protein domains to different compartments of the cell.[35] The resulting vectors utilize a cytomegalovirus promoter in conjunction with the bovine growth hormone polyadenylation transcription termination signal sequences to drive expression of inserted genes. In the next section, we describe the following three targeting vectors: go to nucleus (pGTN), go to nucleus activate transcription (pGTNAT), and go to membrane (pGTM). All three vectors retain a wide selection of restriction sites in the polylinker (Fig. 1).

[35] G. Toby, S. F. Law, and E. A. Golemis, *BioTechniques* **24,** 637 (1998).

pGTN expresses a protein as a fusion to an NLS derived from SV40. pGTNAT is a derivative of pGTN that includes, in addition to the NLS, an acid blob (AB) that acts as a transcriptional activation domain and a hemagglutinin epitope tag. This vector is designed for use with mammalian forms of the two-hybrid system,[36,37] but may also be used for deletion studies of transcription factors, to normalize transcriptional activation potential of expressed proteins. Finally, pGTM expresses the inserted protein as a fusion to a myristoylation attachment sequence (MAS) derived from the N terminus of the protooncogene product Src. The function of these targeting vectors was assayed by fusing GFP with the incorporated cassette and studying the GFP distribution in HeLa cells under a fluorescence microscope, compared with GFP expressed from the parental pcDNA3 vector. In cells transfected with GFP/pcDNA3, GFP does not show a discrete localization, and the green staining observed in the fluorescein isothiocyanate (FITC) channel is throughout the cell. In contrast, when cells are transfected with either GFP/pGTN or GFP/pGTNAT, GFP is mainly localized to the nucleus; when transfected with GFP/pGTM, GFP staining is seen mainly at the plasma membrane.

Some vectors are commercially available; for example, InVitrogen has developed "pShooter" vectors that include targeting sequences for the nucleus, mitochondria, and cytoplasm Clontech (Palo Alto, CA) has developed a series of vectors in which targeting sequences are used to direct GFP to different intracellular compartments, to facilitate immunofluorescence analysis; these could be readily modified to express specific proteins of interest.

Applications

Beyond the applications discussed above, there are other situations in which targeting vectors may prove useful. For example, targeting signal fusion proteins can be used to generate constitutively active mutants of proteins. Expression of the oncoprotein Akt as a fusion to a myristate attachment signal renders the protein constitutively active. On receiving growth signals, Akt is phosphorylated at specific residues and translocates to the cell membrane. Akt constructs with an amino-terminal myristoylation signal (Akt–Myr) renders Akt constitutively active even in the absence of growth cues, demonstrated by the ability of Akt–Myr to induce focus formation of transformed cells within 10 days of transfection.[38] Similarly,

[36] Y. Luo, A. Batalao, H. Zhou, and L. Zhu, *BioTechniques* **22,** 350 (1997).

[37] S. Fields and O. Song, *Nature (London)* **340,** 245 (1989).

[38] M. Aoki, O. Batista, A. Bellacosa, P. Tsichlis, and P. K. Vogt, *Proc. Natl. Acad. Sci. U.S.A.* **95,** 14950 (1998).

Ras p21 proteins exert their biological functions when associated with the inner surface of the plasma membrane, dependent on dual palmitate and farnesyl modification at the C terminus.[39,40] One critical biological function of Ras is the activation of the Raf signaling kinase; in studies in which a membrane localization signal from Ras is appended to the carboxy terminus of Raf, the modified Raf protein becomes constitutively active and can be further activated by epidermal growth factor, independently of Ras.[41]

Alternatively, a common approach to probing gene function is expressing defined dominant negative forms of a protein of interest "X," that retain the ability to compete with the wild-type version of the protein for interaction with partner proteins, but lack normal function.[42] In such studies, ability to express the dominant negative forms so that they are concentrated at the site of intended action is helpful in minimizing nonspecific interactions and nonphysiological effects.

Finally, although this review has focused on the use of targeting motifs for intracellular localization, such motifs can also be incorporated in strategies involving traversal of the plasma membrane. One situation in which this may be useful is in the development of refined therapeutic approaches in gene therapy. In one example, addition of an NLS multimer to cationic lipid and a reporter construct resulted in enhanced transport of transfection complex to the nucleus, and better gene expression.[43] In a second example, it may be of interest to use cell lines to express secreted (toxic) proteins containing extracellular targeting motifs, to direct fusion proteins to specific cell populations for removal. Currently, the integrin-binding sequence RGD has been used in this capacity to direct the drug doxirubicin toward breast cancer cells in an animal model system.[44] In summary, the use of specific motifs to circumscribe spheres of protein actions represents an increasingly valuable tool in refining biological expression systems.

Acknowledgments

We are grateful to Margret Einarson for critical comments on the manuscript. E.A.G. and G.G.T. are supported by American Cancer Society Grant RPG-94-025-06-CCG (to E.A.G.), and by core funds CA-06927 (to Fox Chase Cancer Center).

[39] B. M. Willumsen, K. Norris, A. G. Papageorge, N. L. Hubbert, and D. R. Lowy, *EMBO J.* **3,** 2581 (1984).

[40] B. M. Willumsen, A. Christensen, N. L. Hubbert, A. G. Papageorge, and D. R. Lowy, *Nature* (*London*) **310,** 583 (1984).

[41] S. J. Leevers, H. F. Paterson, and C. J. Marshall, *Nature* (*London*) **369,** 411 (1994).

[42] I. Herskowitz, *Nature* (*London*) **329,** 219 (1987).

[43] A. I. Aronsohn and J. A. Hughes, *J. Drug Target.* **5,** 163 (1998).

[44] R. Pasqualini, E. Koivunen, and E. Ruoslahti, *Nat. Biotechnol.* **15,** 542 (1997).

[6] Mapping Protein–Protein Interactions with Alkaline Phosphatase Fusion Proteins

By MONTAROP YAMABHAI and BRIAN K. KAY

Introduction

Mediation of protein–protein interaction through modular binding domains, such as SH2, SH3, PTB, PDZ, WW, or EH domains, is an important mechanism for intracellular signaling.[1] The physiological importance of one such module, the SH3 domain, is evidenced by its presence in both the regulators and effectors of the Ras-mediated mitogen-activated protein (MAP) kinase signaling pathway.[2,3] SH3 domain-containing proteins have also been shown to function in a variety of diverse biological events including cytoskeletal organization,[4] subcellular localization of proteins,[5] and endocytosis.[6]

SH3 domains are known to bind to short proline-rich peptide sequences within interacting proteins. Typically, the ligands have the consensus sequence, PxxP, and adopt polyproline type II (PPII) helical structures that fit into shallow hydrophobic pockets, formed by aromatic residues on the surface of SH3 domains.[7] Furthermore, SH3 domains bind the peptide ligands in one of two opposite orientations, N to C terminus or C to N terminus.[8,9] The orientation and specificity of peptide ligands for SH3 domains is heavily influenced by interaction between the nonproline residues in the PPII helix and the residues flanking the main hydrophobic binding surface of the SH3 domain.[10,11]

In this chapter, we describe a convenient and versatile method for determining the molecular recognition properties of SH3 domains, using bacterial alkaline phosphatase as a fusion protein. A variety of proteins or

[1] T. Pawson and J. D. Scott, *Science* **278,** 2075 (1997).
[2] G. B. Cohen, R. Ren, and D. Baltimore, *Cell* **80,** 237 (1995).
[3] T. Pawson, *Nature (London)* **373,** 573 (1995).
[4] K. L. Carraway and C. A. Carraway, *BioEssays* **17,** 171 (1995).
[5] D. Bar-Sagi, D. Rotin, A. Batzer, V. Mandiyan, and J. Schlessinger, *Cell* **74,** 83 (1993).
[6] P. S. McPherson, *Cell Signal.* **11,** 229 (1999).
[7] B. Kay, M. Williamson, and M. Sudol, *FASEB J.* **14,** 231 (2000).
[8] S. Feng, J. Chen, H. Yu, J. Simmon, and S. Schreiber, *Science* **266,** 1241 (1994).
[9] W. A. Lim, F. M. Richards, and R. Fox, *Nature (London)* **372,** 375 (1994).
[10] T. Kapoor, A. Andreotti, and S. Schreiber, *J. Am. Chem. Soc.* **120,** 23 (1998).
[11] J. T. Nguyen, C. W. Turck, F. E. Cohen, R. N. Zuckermann and W. A. Lim. *Science* **282,** 2088 (1998).

0076-6879/00 $35.00

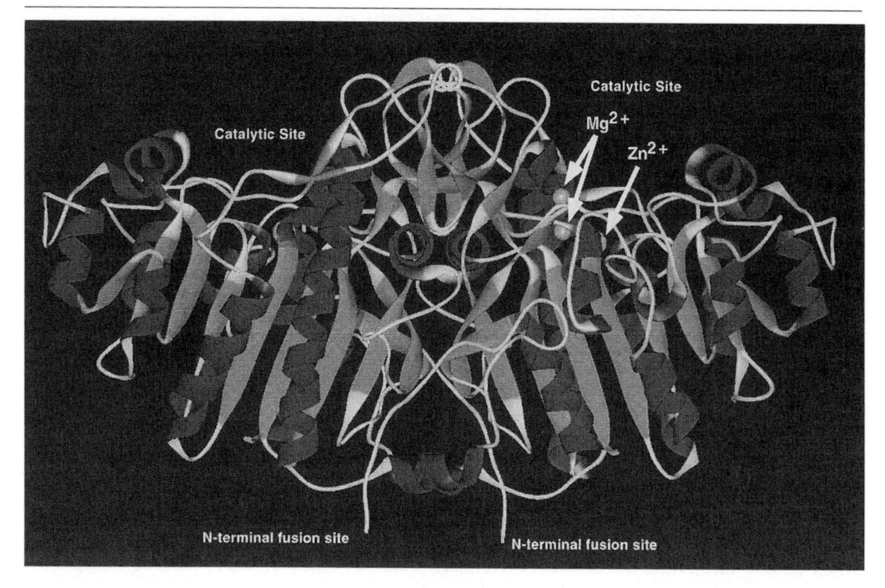

FIG. 1. Structure of bacterial alkaline phosphatase. The three-dimensional structure of a homodimer of *Escherichia coli* alkaline phosphatase is shown.[37] The structure of the protein (PDB database accession number 1ALK) is shown as modeled with the program WebLab ViewerLite 3.2 (*http://www.msi.com*). The catalytic sites with their Zn^{2+} and Mg^{2+} ions and the N-terminal fusion sites are indicated.

peptides have been fused to the N terminus of alkaline phosphatase (AP) by genetic engineering, including antibody chains (i.e., single-chain and Fab antibody segments),[12–15] epitopes,[16,17] and the extracellular domain of the Steel receptor.[18] Figure 1 highlights the three-dimensional structure of a homodimer of *E. coli* AP. Because fusions to AP occur at its N terminus, which protrudes away from the globular body of the protein, many protein segments or peptides can be fused to the AP without interfering with the

[12] W. Wels, I. M. Harwerth, M. Zwickl, N. Hardman, B. Groner, and N. E. Hynes, *Bio/Technology* **10,** 1128 (1992).

[13] E. Weiss and G. Orfanoudakis, *J. Biotechnol.* **33,** 43 (1994).

[14] A. Carrier, F. Ducancel, N. B. Settiawan, L. Cattolico, B. Maillere, M. Leonetti, P. Drevet, A. Menez, and J. C. Boulain, *J. Immunol. Methods* **181,** 177 (1995).

[15] B. H. Muller, D. Chevrier, J. C. Boulain, and J. L. Guesdon, *J. Immunol. Methods* **227,** 177 (1999).

[16] R. Kerschbaumer, S. Hirschl, C. Schwager, M. Ibl, and G. Himmler, *Immunotechnology* **2,** 145 (1996).

[17] Q. Yuan, J. J. Pestka, B. M. Hespenheide, L. A. Kuhn, J. E. Linz, and L. P. Hart, *Appl. Environ. Microbiol.* **65,** 3279 (1999).

[18] J. Flanagan and P. Leder, *Cell* **63,** 185 (1990).

catalytic activity of the enzyme. In our hands, we have used the culture media as one-step detection probes for detecting protein–protein interactions on a variety of surfaces including, nitrocellulose or PVDF membranes, microtiter plate wells, and plastic pins.

Description of Bacterial Alkaline Phosphatase Fusion System

Generation of Fusion Proteins

Alkaline Phosphatase Fusion Vector. The major components of the expression plasmid include (1) a bacterial signal peptide derived from OmpA that directs secretion of the fusion protein into the periplasmic space, where it can leak into the culture medium, (2) the FLAG epitope at the N terminus of the mature AP protein, (3) multiple cloning sites downstream of the epitope for insertion of coding segments, (4) a regulated promoter to permit induction of the fusion protein, and (5) an antibiotic resistance gene to maintain the plasmid by selection in bacterial cells. A map of the AP fusion vector, pMY101, is shown in Fig. 2. We have observed

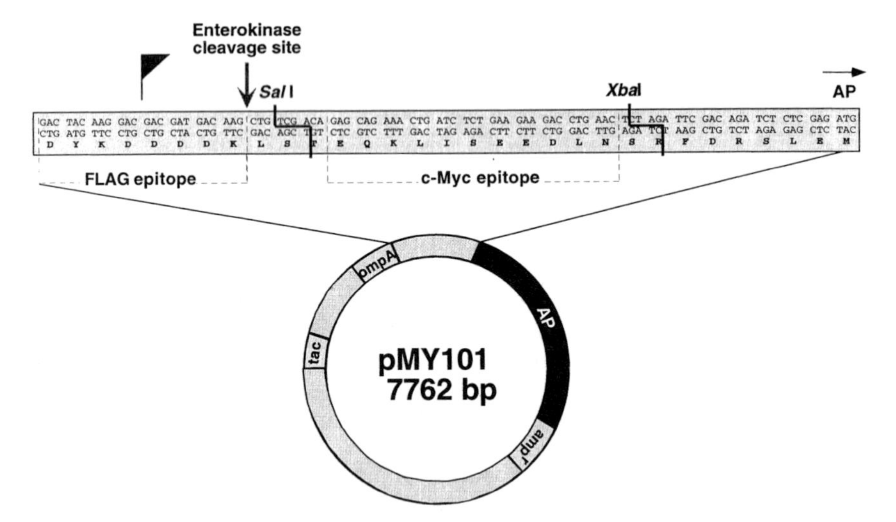

FIG. 2. Map of the bacterial alkaline phosphatase fusion vector, pMY101. The coding region of a segment of the *E. coli* alkaline phosphatase (AP) gene is shown with the FLAG and c-Myc epitope sequences below. The *Sal*I and *Xba*I restriction sites flank the c-Myc epitope in the same reading frame as cloned peptides that were displayed by a bacteriophage M13 combinatorial peptide library. The 7762-base pair (bp) vector was derived from the pFLAG-1-AP vector in which the OmpA signal/leader sequence is upstream of the FLAG and c-Myc epitopes and the mature AP-coding sequence (black). The vector also carries genes for ampicillin resistance (Amp^r) and the *lac* repressor (*lac*I), which regulates the *tac* promoter, upstream of the AP gene.

that polypeptides ranging from 8 to 300 amino acids in length can be fused to the N terminus of AP, without loss of either the phosphatase activity or binding properties of the fused segment.

Preparation of Alkaline Phosphatase Fusion Protein. After the DNA fragment encoding the peptide or protein of interest is cloned in frame into the pMY101 vector, the construct is transformed into *E. coli.* Recombinant plasmids can be identified by polymerase chain reaction (PCR) or restriction digestion analysis, and verified by DNA sequencing. Ampicillin-resistant colonies are used to inoculate 5–15 ml of LB medium (containing ampicillin at 100 μg/ml) and allowed to grow overnight at 37°. Addition of isopropyl-β-D-thiogalactopyranoside (IPTG) to 50–100 μM final concentration in the culture medium during log-phase growth of the bacteria increases the yield of most fusion proteins. The AP fusion protein, which is secreted into the culture media, can be separated from the bacterial cells by centrifugation (500*g* for 5 min at 4°) and stored at 4°.

The AP fusion protein should be kept at 4° for no longer than 2 weeks, as the integrity of the fused protein or the AP activity is often lost with time. Protease inhibitors such as Complete (Boehringer Mannheim, Indianapolis, IN) can be added to increase the shelf life of the fusion protein, although longer storage may require purification. To confirm that the AP fusion protein is active, transfer 50 μl of culture supernatant to an Eppendorf tube containing 50 μl of *p*-nitrophenyl phosphate (pNPP) (Sigma Fast; Sigma, St. Louis, MO); typically, the solution becomes bright yellow within 20 min. As the culture medium does not interfere with the interaction of the AP fusion protein with other proteins, no additional purification is necessary prior to its use.

Troubleshooting

If little or no AP activity is detected in the culture medium, several considerations are suggested for improving the yields of the AP fusion protein of interest.

1. *Lower the temperature of the bacterial culture:* Growing the cells at 30° sometimes leads to improved secretion of the fusion proteins into the culture medium. This is especially important if the N-terminal fusion to AP is larger than 90 amino acids.

2. *Start with a fresh transformant:* In a number of cases, we have observed that bacteria lose their ability to secrete AP fusion protein over time. Presumably this is due to the negative selective pressure on the bacteria caused by long-term overexpression of AP, which may be detrimental to growth. This problem can be solved by using bacteria from a newly transformed colony to inoculate the culture medium.

3. *Break the cell membrane:* On rare occasions, the fusion proteins are retained in the periplasmic space or cytoplasm instead of leaking out into the culture medium. In this situation, the protein can be recovered by harvesting the contents of the periplasmic space with an osmotic shock.

4. *Concentrate or purify the protein:* The fusion protein can be concentrated from the culture medium by centrifugation. Transfer 10–20 ml of culture medium containing the AP fusion protein of interest into a Centriprep YM-30 (Millipore, Bedford, MA) filtration device. Centrifuge at 1500*g* for 30–45 min at 4°, according to the manufacturer protocol. If desired, exchange the buffer of the retained liquid by adding Tris-buffered saline [TBS; 25 m*M* Tris–HCl (pH 7.5), 145 m*M* NaCl, 3 m*M* KCl] containing 0.1% (v/v) Tween 20 to the concentrate and centrifuge a second time. Conversely, the AP fusion proteins can be affinity purified with an anti-FLAG antibody (M1 monoclonal antibody; Sigma) coupled to agarose (Sigma). We have also found that a run of six histidines can be added to the N terminus of AP, thereby permitting the purification of the fusion protein by chromatography over a resin containing nickel-loaded iminodiacetic acid.[19]

Binding Strength and Specificity of Fusion Proteins

To utilize AP fusion proteins to study SH3 domain–ligand interaction, either the peptide ligand or the SH3 domain can be fused to the N terminus of the enzyme. As illustrated in Fig. 3 the AP fusion proteins have been used to study binding properties of the SH3 domains of Src and Abl. While the two SH3 domains are 52% similar in primary structure, their ligand specificities are quite different: the optimal ligand preferences for the Src SH3 domain are R(A/P)LPPLP or PPVPPR, while the Abl SH3 domain binds PPPVPLP.[20] This specificity can be demonstrated with AP fusion proteins by immobilizing either peptide ligands or SH3 domains on the surface of microtiter plate wells.

In Fig. 3A, individual wells of a 96-well high-binding microtiter plate (Costar, Cambridge, MA) are incubated with 1 μg of streptavidin (Sigma) in 100 m*M* $NaHCO_3$ (pH 8.3) for 30 min at room temperature. The wells are then washed three times with TBS containing 0.1% (v/v) Tween 20. For each triplicate set of wells, 0.1 m*M* biotinylated peptide is added and incubated for 30 min. To eliminate nonspecific binding of the wells, 10 μg of bovine serum albumin (BSA) is then added to each well and incubated for an additional 30 min.

[19] A. Kurakin, N. G. Hoffman, and B. K. Kay, *J. Peptide Res.* **52,** 331 (1998).

[20] A. Sparks, J. Rider, N. Hoffman, D. Fowlkes, L. Quilliam, and B. Kay, *Proc. Natl. Acad. Sci. U.S.A.* **93,** 1540 (1996).

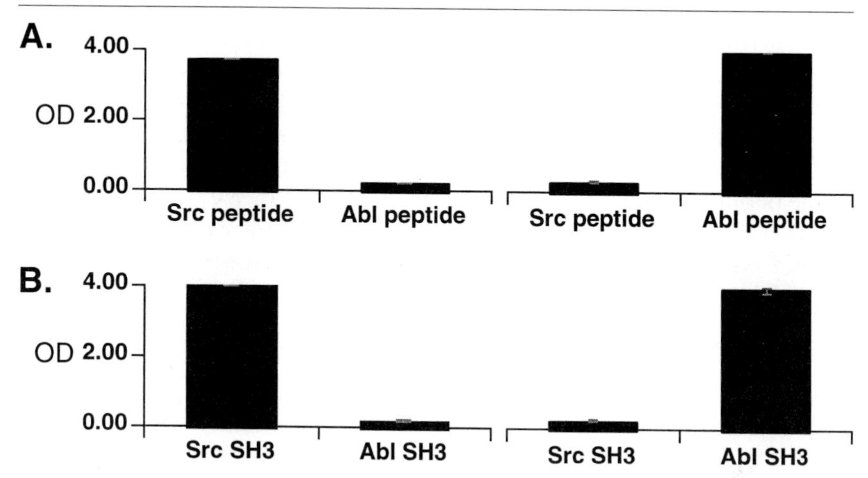

FIG. 3. Binding strength and specificity of SH3 domain–ligand interaction, using the AP fusion system. (A) Biotinylated Src and Abl SH3 domain peptide ligands were added to duplicate wells of a microtiter plate coated with streptavidin, washed, and incubated with AP fusions to either Src (*left*) or Abl (*right*) SH3 domains. (B) Binding properties of Src (*left*) or Abl (*right*) SH3 peptide ligands (fused to AP) to either Src SH3 or Abl SH3 domains (fused to GST) that were immobilized onto triplicate wells of microtiter plate. Bound AP fusion proteins were detected by incubation with pNPP; average optical density values and standard error are shown. The synthetic peptide ligands of the Src and Abl SH3 domains are SGSGVLKRPLPIPPVTR and SGSGSRPPRWSPPPVPLPTSLDSR, respectively. The peptide ligands of the Src and Abl SH3 domains that were fused to the N terminus of AP are ISQRALPPLPLMSDPA and GPRWSPPPVPLPTSLD, respectively. The enzymatic conversion of pNPP was allowed to proceed overnight to demonstrate that this type of assay yields not only a robust signal, but also a low background.

In Fig. 3B, glutathione *S*-transferase (GST)–SH3 fusion proteins are immobilized by adding 1 μg of protein in 100 μl of 0.1 m*M* $NaHCO_3$ (pH 8.3) into microtiter wells, and incubating for 1 hr at room temperature. To eliminate nonspecific binding of the wells, 10 μg of BSA is added to each well and incubated for an additional 30 min. Glutathione *S*-transferase (GST) fusion proteins are prepared according to the manufacturer protocol (Pharmacia, Piscataway, NJ).

After the peptide ligands or GST fusion proteins are immobilized on the microtiter plate wells, 100 μl of culture supernatant containing SH3–AP or peptide–AP fusions is added to the appropriate wells that have been washed and contain 50 μl of TBS. After a 1-hr incubation the wells are washed five times with TBS–0.1% (v/v) Tween 20. The amount of AP fusion protein retained in the microtiter plate wells is estimated by adding 150 μl of pNPP, incubating the solution for 20–120 min at room temperature, and quantitating the absorbance of the wells with a plate spectrophotometer (Molecular Devices, Sunnyvale, CA), at a wavelength of 405 nm. If necessary, the color reactions can be extended overnight at room temperature to maximize the absorbance values.

The results in Fig. 3 demonstrate that the interactions between the Src and Abl SH3 domains and their peptide are specific, whether the domain or the peptide ligand is fused to the N terminus of AP. We typically achieve a higher degree of specificity and lower background with the AP fusion protein system than with soluble synthetic or phage-displayed peptide ligands as probes.[21]

Notes

1. In every incubation step, the wells of the microtiter plates must be sealed with plastic wrap (or tape) to avoid evaporation.

2. The FLAG antibody can serve as a useful positive control to ensure that equal amounts of AP fusion protein have been added per microtiter plate well. Coat separate wells with the antibody and verify that the amounts of AP enzyme activity retained in the different wells are comparable.[21]

Types of Assays

AP fusion proteins can be used to detect SH3 domain and ligand interactions in a number of different experimental formats.

Detection of Protein–Protein Interactions on Membranes

The AP fusion system has been used to detect interacting proteins that have been immobilized on nitrocellulose or PVDF membranes (i.e., dot or Western blots). Nonspecific protein-binding sites on the membranes can be blocked by incubating them in 1.5% (w/v) BSA in TBS–0.1% (v/v) Tween 20, for 1 hr at room temperature, or overnight at 4°. The membranes are then incubated with culture supernatant containing the AP fusion protein in a small plastic bag, and after 1–2 hr of incubation at room temperature, the membranes are washed three times with agitation in TBS–0.1% (v/v) Triton X-100 for 10 min. AP fusion proteins that have been retained on the membranes are finally detected by either colorimetric or chemiluminescent reactions. For colorimetric detection of the AP fusion protein, the membranes are incubated in nitroblue tetrazolium–5-bromo-4-chloro-3-indolylphosphate toluidinium (NBT–BCIP; Sigma Fast) solution for 10 min, washed with deionized water,

[21] M. Yamabhai and B. K. Kay, *Anal. Biochem.* **247,** 143 (1997).

and allowed to dry. For chemiluminescent detection of the fusion protein, the membranes are incubated in TBS for 10 min, drained of excess buffer, and covered with a solution of disodium 3-(4-methoxyspiro{1,2-dioxetane-3,2′-(5′-chloro)tricyclo[3.3.1.1]decan}-4-yl)phenyl phosphate (CSPD; Tropix, Bedford, MA). The chemiluminescent reaction is allowed to proceed for 5 min in the dark before the excess solution is drained away, and the membranes are exposed to X-ray film (Eastman Kodak, Rochester, NY) for 15–20 min.

Figure 4 demonstrates that peptide–AP fusion proteins can be used to probe for particular proteins present on dot or Western blots. The Src SH3 peptide ligand–AP fusion binds only to the Src SH3 domain, whereas the Abl SH3 ligand–AP fusion binds only to the Abl SH3 domain. Thus, AP fusion proteins can be used to detect specific SH3 domain-mediated protein–protein interactions on blots. SH3 domains appear to retain their three-dimensional structure (or renature efficiently) once affixed to a mem-

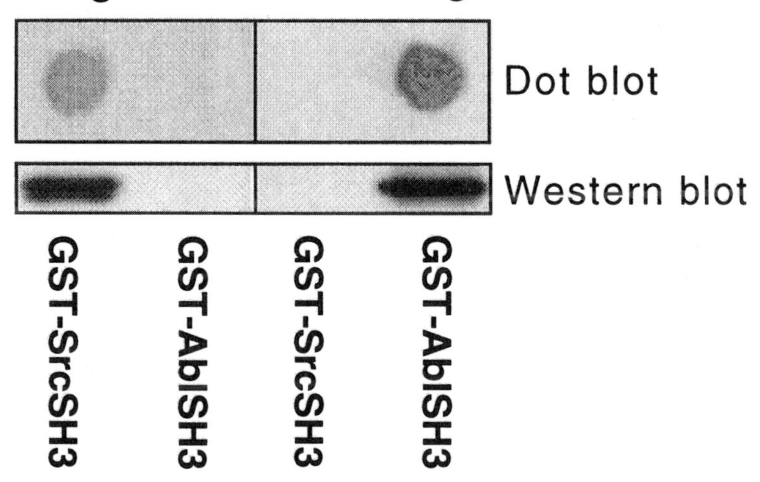

FIG. 4. Dot-blot and Western blot analyses with AP fusion proteins. One microgram each of GST–Src SH3 and GST–Abl SH3 fusion proteins were dot blotted onto nitrocellulose membrane (*top*) or fractionated by SDS–PAGE and transferred to PVDF membrane (*bottom*). The immobilized proteins were incubated either with the Src SH3 ligand–AP fusion protein (*left*) or Abl SH3 ligand–AP fusion protein (*right*). Colorimetric (NBT–BCIP) or chemiluminescent (CSPD) reagents were used for detection of the bound AP fusion protein in the dot (*top*) and Western (*bottom*) blots, respectively. The Src SH3 peptide ligand–AP fusion bound only to the GST–Src SH3 domain, whereas the Abl SH3 ligand–AP fusion bound only to the GST–Abl SH3 domain.

brane, presumably because of their innate thermal stability and lack of disulfide bonds.[22,23]

The AP fusion system can also be used to probe cDNA expression libraries in bacteriophage λ. We have used AP fusions to different protein interaction modules (i.e., SH3 domains of Src and Abl, EH domains of intersectin) to screen a mouse 16-day embryo cDNA expression library (Novagen, Madison, WI). Intersectin is a protein containing two N-terminal EH domains, an α-helical structure, and five C-terminal SH3 domains, and has been shown to function in endocytosis and cellular signaling.[24–26] The EH domain is a protein interaction module that has been shown to bind peptides containing the peptide sequence asparagine-proline-phenylalanine (NPF).[24,27,28]

In the primary screen, the library was plated at a density of 3×10^4 recombinant phage per 90-mm plate. Thirty plates were screened with each of the three AP fusion proteins. After four rounds of screening, individual plaques were isolated and plasmids bearing the cDNA inserts were rescued from the isolated λ phage by Cre-mediated excision, according to the manufacturer protocols (Novagen). Inserts were sequenced by automated fluorescent dideoxynucleotide sequencing and are summarized in Table I. All the isolated proteins contain short peptide sequences that have been previously shown to bind each domain, and in many cases have been demonstrated to be interacting partners in cells. Thus, the AP fusion system is a useful tool in identifying potential interacting proteins from a cDNA expression library.

Detection on Plastic Pins

H. M. Geysen and colleagues have devised a convenient method for synthesizing 96 peptides at a time on plastic pins, using microtiter plate wells as reaction chambers.[29] This technique can be easily duplicated in

[22] W. A. Lim, R. O. Fox, and F. M. Richards, *Protein Sci.* **3,** 1261 (1994).

[23] M. C. Parrini and B. J. Mayer, *Chem. Biol.* **6,** 679 (1999).

[24] M. Yamabhai, N. G. Hoffman, N. L. Hardison, P. S. McPherson, L. Castagnoli, G. Cesareni, and B. K. Kay, *J. Biol. Chem.* **273,** 31401 (1998).

[25] N. K. Hussain, M. Yamabhai, A. R. Ramjaun, A. M. Guy, D. Baranes, J. P. O'Bryan, C. J. Der, B. K. Kay, and P. S. McPherson, *J. Biol. Chem.* **274,** 15671 (1999).

[26] X. K. Tong, N. K. Hussain, E. de Heuvel, A. Kurakin, E. Abi-Jaoude, C. C. Quinn, M. F. Olson, R. Marais, D. Baranes, B. K. Kay, and P. S. McPherson, *EMBO J.* **19,** 1263 (2000).

[27] A. E. Salcini, S. Confalonieri, M. Doria, E. Santolini, E. Tassi, O. Minenkova, G. Cesareni, P. G. Pelicci, and P. P. Di Fiore, *Genes Dev.* **11,** 2239 (1997).

[28] S. Paoluzi, L. Castagnoli, I. Lauro, A. E. Salcini, L. Coda, S. Fre, S. Confalonieri, P. G. Pelicci, P. P. Di Fiore, and G. Cesareni, *EMBO J.* **17,** 6541 (1998).

[29] H. M. Geysen, *Southeast Asian J. Trop. Med. Public Health* **21,** 523 (1990).

TABLE I
IDENTITY OF PROTEINS ISOLATED FROM SCREENS OF cDNA EXPRESSION LIBRARY WITH VARIOUS AP FUSION PROTEINS[a]

AP fusion partner	Amino acid sequence of cDNA clones isolated	Interacting protein	Ref.
Src SH3 domain	QHRPRLPSTESLSRRPLPALPVSEAPAPSPAPSP APGRKGSIQDRPLPPPPPCLPGYGGLKPEGD	Efs	38
	TSEAPPLPPRNAGKGPTGPPSTLPLGTQTSSGSS TLSKKRPPPPPPPGHKRTLSDPPSPLPHGPPN	S19	39, 40
Abl SH3 domain	PPAPPPPPPLPSGPAYASALPPPPGPPPPPPLPS TGPPPPPPPPPPLPNQAPPPPPPPPAPPLP	Mena	41, 42
Intersectin EH domain	NPFLPSGAPPTGPSVTNPFQPAPPATLTLNQLRL SPVPPVGAPPTYISPLGGGPGLPPMMPPGPPAP NTNPFLL	Epsin 1	24, 25, 43–45
	NPFLAPGAAAPAPVNPFQVNQPQPLTLNQLRGSP VLGSSASFGSGPGVETVAPMTSVAPHSSVGASGS SLTPLGPTAMNMVGSVGIPPSAAQSTGTTNPFLL	Epsin 2	24, 25, 43–45

[a] A λ library expressing cDNA segments prepared from a 16-day-old mouse embryo was probed with various domain–AP fusions. Putative peptide ligands of the SH3 and EH domains are underlined in the isolated cDNA segments.[21,24] Additional information regarding either the identified proteins (right-hand column) or evidence of protein–protein interactions can be found in the references noted above.

most laboratories and blocks of plastic pins, software, and reagents can be purchased from Mimotopes (Clayton Victoria, Australia; *http://www.mimotypes.com*). To explore the possible utility of the pin format in dissecting the specificity of SH3 domains, we have synthesized two sets of peptides on pins and examined their binding to SH3 domain fusions to AP. After the synthesis was complete, the nonspecific binding of proteins to the pins was blocked by incubating them in TBS–0.2% (w/v) BSA at room temperature. After a 30-min incubation, the pins were washed three times by dipping them in a tray containing TBS–0.1% (v/v) Tween 20 for 10 min each time. Next, the pins were placed into wells of a 96-well microtiter plate containing 100 μl of the conditioned culture medium (containing the secreted AP fusion protein) and 50 μl of TBS to adjust the pH. After 1 hr of incubation, the pins were washed by dipping them three times (10 min each) in a tray containing TBS–0.1% (v/v) Tween 20. The amount of AP fusion protein retained on the wells was then revealed by incubating the pins in a microtiter plate containing 100 μl of pNPP solution in each well. The enzymatic reaction was allowed to proceed until the optical density (OD) of the wells reached values of 1.0 or greater. To reuse the pins, bound AP fusions can be stripped off by incubating the block of pins at 50° in

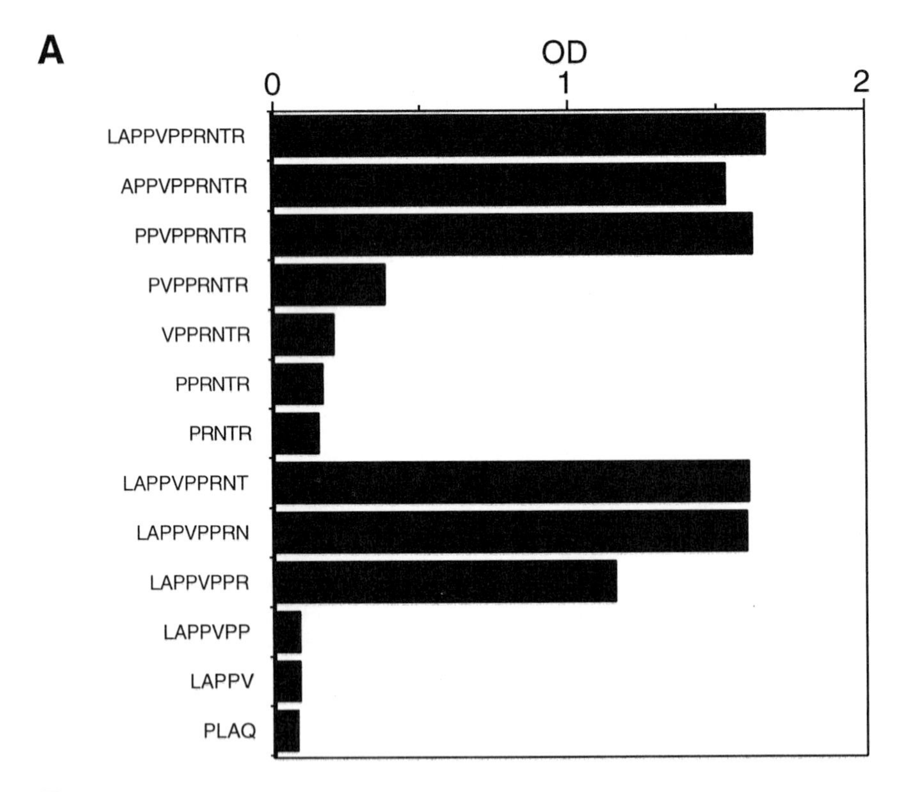
A
OD
0
1
2
LAPPVPPRNTR
APPVPPRNTR
PPVPPRNTR
PVPPRNTR
VPPRNTR
PPRNTR
PRNTR
LAPPVPPRNT
LAPPVPPRN
LAPPVPPR
LAPPVPP
LAPPV
PLAQ

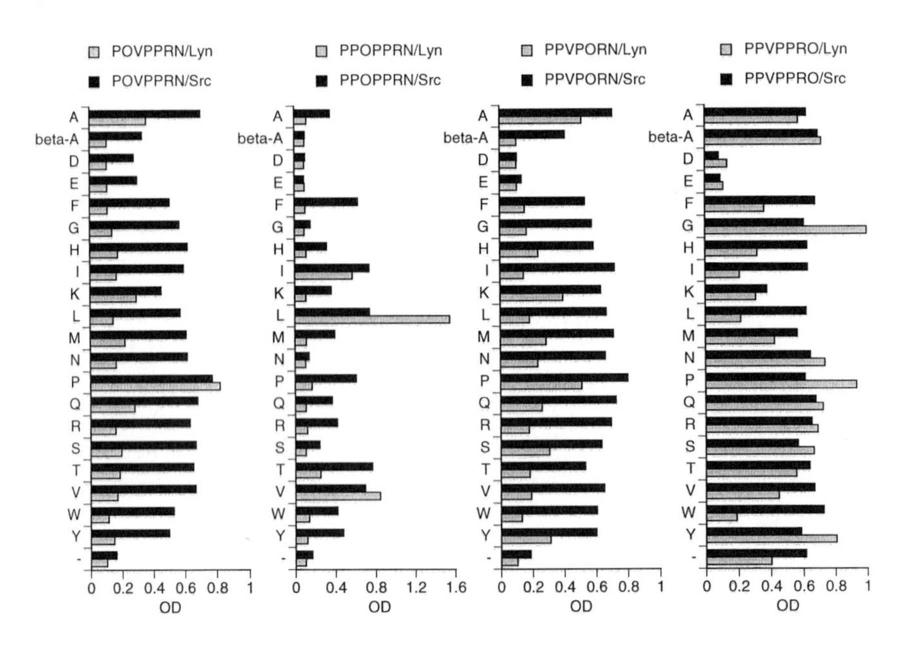
B
POVPPRN/Lyn
POVPPRN/Src
PPOPPRN/Lyn
PPOPPRN/Src
PPVPORN/Lyn
PPVPORN/Src
PPVPPRO/Lyn
PPVPPRO/Src
A
beta-A
D
E
F
G
H
I
K
L
M
N
P
Q
R
S
T
V
W
Y
-
0 0.2 0.4 0.6 0.8 1
0 0.4 0.8 1.2 1.6
OD

a solution of 10% (w/v) sodium dodecyl sulfate (SDS) with 30 min of constant sonication.

A series of peptides, corresponding to truncations from the N and C termini of the parent sequence LAPPVPPRNTR, were synthesized on plastic pins, to define the boundaries of a class II peptide ligand for the Src SH3 domain.[30] The pins were then incubated with the Src SH3–AP fusion and the bound AP fusion was detected as described above. As seen in Fig. 5A, the Src SH3–AP fusion protein can bind to peptides synthesized on plastic and the minimal peptide ligand of the Src SH3 domain is PPVPPR.

A second set of peptides on pins was synthesized and tested for their binding to two different SH3 domains. In this set, each residue within the minimal ligand PPVPPRN was systematically replaced with 20 different amino acids, to test the importance of each amino acid at each position (Fig. 5B). When the pins were probed with the Src SH3–AP fusion, the optimal ligand was deduced to be PxnPx′Rx′ (where x is any amino acid, n represents all but β-A, D, E, G, and N, and x′ means all but acidic amino acids). When the same set of pins was reprobed with an AP fusion to the Lyn SH3 domain, a member of the Src family of tyrosine kinases, the optimal ligand was observed to be PP(L/V/I)PoRo′ (where o represents A, H, K, M, N, P, Q, S, T and Y, and o′ represents all but D, E, I, L, and W). Thus, even though the Src and Lyn SH3 domains are 70% similar in primary structure, their binding preferences are distinct, in agreement with previously published phage display experiments.[31,32]

[30] A. B. Sparks, L. A. Quilliam, J. M. Thorn, C. J. Der, and B. K. Kay, *J. Biol. Chem.* **269,** 23853 (1994).

[31] R. J. Rickles, M. C. Botfield, Z. Weng, J. A. Taylor, O. M. Green, J. S. Brugge, and M. J. Zoller, *EMBO J.* **13,** 5598 (1994).

[32] T. P. Stauffer, C. H. Martenson, J. E. Rider, B. K. Kay, and T. Meyer, *Biochemistry* **36,** 9388 (1997).

FIG. 5. Mapping the ligand preferences of SH3 domains with peptides synthesized on pins. A set of peptides on pins (*y* axis) was designed to define the boundaries of a particular peptide ligand for the Src SH3 domain that was isolated from a phage display combinatorial peptide library (A). The block of pins was incubated with the Src SH3–AP fusion protein and the bound protein was detected by incubation of the pins with pNPP. Optical density at 405 nm is shown. Binding of the two different SH3 domains to a set of pins in which a number of residues in the peptide sequence PPVPPRN were systematically replaced with different amino acids. The binding of Src SH3–AP to the pins was measured first, before sonicating and washing the pins, and testing the binding of Lyn SH3–AP. O represents the residue replaced one at a time with β-A, A, D, E, F, G, H, I, K, L, M, N, P, Q, R, S, T, V, W, Y, or nothing. All optical density signals below 0.2 are considered background.

Determination of Modular Domain–Ligand Binding Properties on Microtiter Plates

The affinity of a particular domain for its ligand is commonly measured by fluorescence polarization, equilibrium dialysis, or surface plasmon resonance. We have found that competition assays with AP fusion proteins are a less time-consuming and expensive method for estimating the binding strength between various domains and their peptide ligands. We have utilized displacement of AP fusion proteins from binding target proteins immobilized onto microtiter plate wells with soluble compounds (i.e., peptides, domains) as a convenient means of estimating their relative binding strengths.

We have utilized this assay format to evaluate the binding strength of peptide ligands for the Src SH3 domain and the EH domains of intersectin. In this type of competition assay, 1 μg of GST–SH3 or GST–EH fusion protein prepared according to standard protocols[33] is added to microtiter plate wells in 100 μl of 0.1 m*M* $NaHCO_3$ (pH 8.3). After the protein has been allowed to bind to the well surface at room temperature for 1 hr, nonspecific protein-binding sites are blocked by adding 100 μl of TBS–0.2% (w/v) BSA to the wells and incubating for 1 hr at room temperature (or overnight at 4°). The wells are then washed three times with TBS–0.1% (v/v) Tween 20 and various amounts of synthetic peptides are added to the wells along with 100 μl of peptide ligand–AP fusion protein. After 2 hr of incubation at room temperature, the wells are then washed five times, and the relative amount of AP fusion protein retained in the wells is detected by reaction with 100 μl of pNPP.

Figure 6 demonstrates a competition assay in which increased amounts of soluble Src-1–AP and EH-1–AP fusion proteins, were added to microtiter plate wells containing immobilized GST fused to EH or SH3 domain. The soluble peptides inhibited only the binding of their respective peptide–AP fusion proteins. The 50% inhibitory concentration (IC_{50}) values of the peptides for both the Src SH3 and intersectin EH domains are ~10 μM and ~3 μM, respectively. Thus, the IC_{50} value obtained from this type of assay can be used to predict the binding strength of different peptide ligands to protein interaction modules.

Discussion and Conclusions

With the ability to display a wide range of polypeptides at the N terminus of bacterial AP, it is possible to identify or map protein–protein interactions

[33] D. B. Smith and K. S. Johnson, *Gene* **67,** 31 (1988).

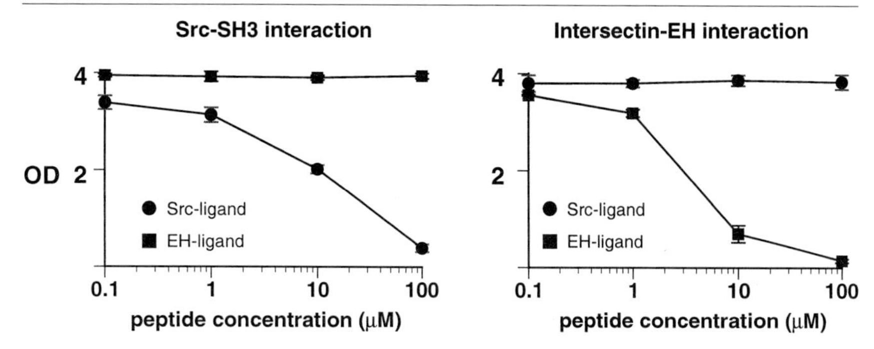

FIG. 6. Estimation of the binding strength of the Src SH3 and intersectin EH domain–ligand interactions. One microgram of GST–Src SH3 or GST–intersectin EH fusion proteins was immobilized in microtiter plate wells. AP fusions to either the Src SH3 ligand (*left*) or the EH ligand (*right*) were incubated in the presence of increased amount of either soluble Src peptide (RPLPIPPVTR) or EH peptide ligand (DCTNPFSCWR, cyclized through the cysteines). Average $OD_{405\ nm}$ values are shown for triplicate wells, along with standard error.

in a number of formats. There are several kinds of alkaline phosphatase substrates available (i.e., soluble or insoluble, chemiluminescent or colorimetric), so a particular substrate can be chosen to suit specific needs. Furthermore, the utility of this system is enhanced by the fact the AP fusion protein is secreted from *E. coli* and the culture medium can be used directly in experiments; once the appropriate fusions have been constructed, the cells are simply grown overnight, and the fusion protein is harvested with the culture supernatant. One limitation of the current vector, however, is that this system is not appropriate for studying protein–protein interactions that require C-terminal sequences, because the fusions to AP occur at its N terminus.

The successful application of the AP fusion system to examine binding properties of the SH3 domain and other protein-binding modules, as described in this chapter, suggests that this system may prove useful for the general study of protein–protein interactions. Protein interaction modules may be well suited to display as AP fusion proteins, because the modules are small, typically fold properly in *E. coli,* lack disulfide bonds, and their interaction surface is opposite the N and C termini of the module. The AP system may be useful in studying the peptide ligand preferences of protein interaction modules, because it permits the detection of modest (i.e., 10 μM) interactions by eliminating additional incubation and washing steps as a one-step detection reagent. Peptides synthesized on pins or spots[34] can

[34] A. Kramer, U. Reineke, L. Dong, B. Hoffmann, U. Hoffmuller, D. Winkler, R. Volkmer-Engert, and J. Schneider-Mergener, *J. Peptide Res.* **54,** 319 (1999).

be quickly surveyed for binding to a particular protein interaction module expressed as a fusion with AP.

One interesting application of the AP fusion system is its potential use in the drug discovery process. Libraries of natural products or small molecules can be screened for inhibitors of protein–protein interactions that either block the binding of an AP fusion protein and an interacting protein in a microtiter plate well,[35] or bind an AP fusion protein directly.[36] The identification of compounds that can specifically antagonize specific protein–protein interactions should prove useful in analyzing the physiological consequences of these molecular interactions, as well as suggest leads for drug development.

Acknowledgments

We appreciate Mario Geysen and Dan Kinder at the GlaxoWellcome Research Institute (Research Triangle Park, NC) for help in synthesizing peptides on plastic pins. We thank Paul Hamilton, Jeremy Kasanov, Stephen Knight, Dennis Prickett, Jeffrey Rubin, and Bernard Weisblum for comments on the manuscript.

[35] B. K. Kay, A. Kurakin, and R. Hyde-DeRuyscher, *Drug Discov. Today* **3,** 370 (1998).

[36] D. Wagner, C. Markworth, C. Wager, F. Schoenen, C. Rewerts, B. Kay, and H. Geysen, *Combinatorial Chem. High Throughput Screening* **1,** 143 (1998).

[37] E. E. Kim and H. W. Wyckoff, *J. Mol. Biol.* **218,** 449 (1991).

[38] M. Ishino, T. Ohba, H. Sasaki, and T. Sasaki, *Oncogene* **11,** 2331 (1995).

[39] M. T. Brown, J. Andrade, H. Radhakrishna, J. G. Donaldson, J. A. Cooper, and P. A. Randazzo, *Mol. Cell. Biol.* **18,** 7038 (1998).

[40] P. Lock, C. L. Abram, T. Gibson, and S. A. Courtneidge, *EMBO J.* **17,** 4346 (1998).

[41] F. B. Gertler, A. R. Comer, J. L. Juang, S. M. Ahern, M. J. Clark, E. C. Liebl, and F. M. Hoffmann, *Genes Dev.* **9,** 521 (1995).

[42] F. B. Gertler, K. Niebuhr, M. Reinhard, J. Wehland, and P. Soriano, *Cell* **87,** 227 (1996).

[43] H. Chen, S. Fre, V. Slepnev, M. Capua, K. Takei, M. Butler, P. Di Fiore, and P. De Camilli, *Nature (London)* **394,** 793 (1998).

[44] H. Chen, V. I. Slepnev, P. P. Di Fiore, and P. De Camilli, *J. Biol. Chem.* **274,** 3257 (1999).

[45] E. Santolini, A. E. Salcini, B. K. Kay, M. Yamabhai, and P. P. Di Fiore, *Exp. Cell Res.* **253,** 186 (1999).

[7] Assays of Human Postprenylation Processing Enzymes

By YUN-JUNG CHOI, MICHAEL NIEDBALA, MARK LYNCH, MARC SYMONS, GIDEON BOLLAG, and ANNE K. NORTH

Introduction

Ras proteins are localized to the cytoplasmic membranes of most eukaryotic organisms, ranging from yeasts to humans.[1] Membrane localization is dependent on a series of posttranslational processing events. The first of these events is the farnesylation of Cys-186 in the C*aa*X (Cys-aliphatic-aliphatic-arbitrary residue) motif. The farnesylated protein is then a substrate for a prenyl-directed endoprotease that liberates the carboxy-terminal *aa*X tripeptide. Proteolytic cleavage results in a free carboxyl group on the terminal farnesylated cysteine, and this carboxylate is subsequently modified by a prenylcysteine-directed carboxylmethyltransferase. Both proteolysis and methylation are important for proper membrane localization.

Although the human farnesyltransferase has been studied for many years,[1] primary structures of the human protease and methyltransferase have only more recently been explored. The yeast gene encoding the prenyl-directed endoprotease (named RCE1, for Ras-converting enzyme) was discovered by elegant genetic techniques.[2] While yeast cells lacking RCE1 are viable, the function of activated Ras proteins is suppressed. On the basis of the yeast RCE1 sequence, the human enzyme (hRCE1) has been cloned, expressed, and characterized.[3] Furthermore, the RCE1 homolog has been deleted from the mouse genome; while the resulting strain of mice invariably dies during development, the observation that these mice appear free of obvious pathologies at birth indicates a nonessential function for RCE1 during many physiological processes.[4] This finding is even more surprising, because it appears that RCE1 is the only postprenylation endoprotease: both geranylgeranylated and farnesylated proteins of many subfamilies (e.g., Ras, Rho, Rab, and lamins) are substrates. Figure 1A displays the homology among phylogenetically diverse endoprotease genes, indicating an important conservation of function.

[1] A. D. Cox and C. J. Der, *Biochim. Biophys. Acta.* **1333,** F51 (1997).
[2] V. L. Boyartchuk, M. N. Ashby, and J. Rine, *Science* **275,** 1796 (1997).
[3] J. C. Otto, E. Kim, S. G. Young, and P. J. Casey, *J. Biol. Chem.* **274,** 8379 (1999).
[4] E. Kim, P. Ambroziak, J. C. Otto, B. Taylor, M. Ashby, K. Shannon, P. J. Casey, and S. G. Young, *J. Biol. Chem.* **274,** 8383 (1999).

0076-6879/00 $35.00

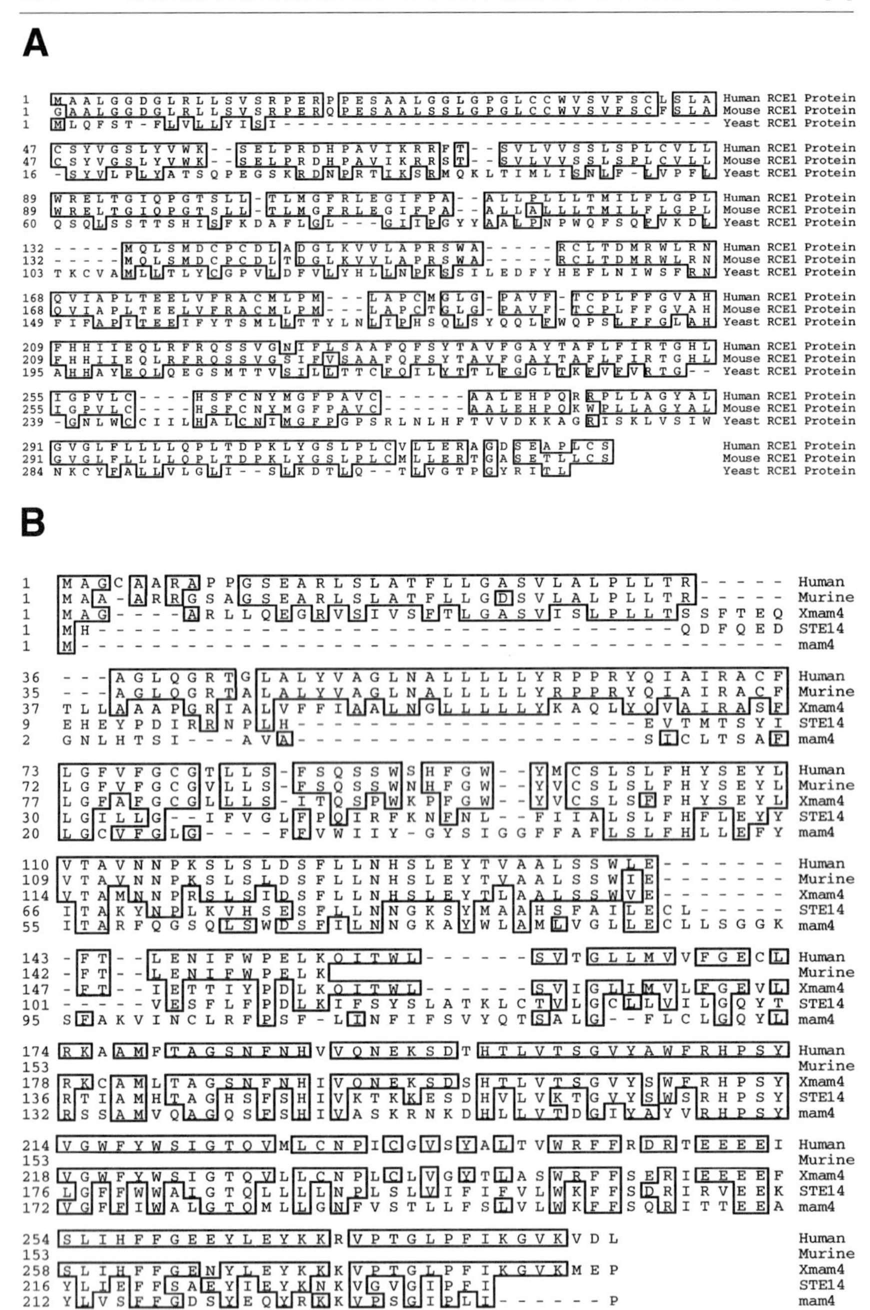

FIG. 1. Protein sequence alignments. (A). Comparison of prenyl-directed endoproteases. (B) Comparison of prenylcysteine methyltransferases.

The role of the prenylcysteine carboxylmethyltransferase appears to mirror that of the prenyl-directed endoprotease. Although disruption of the yeast gene (STE14) results in yeast sterility, the function of the endogenous Ras pathway is not detectably altered.[5] As also observed after disruption of RCE1 in yeast or mouse, Ras proteins are substantially mislocalized in STE14 mutant yeast. On the basis of the yeast and *Xenopus laevis* methyltransferase sequences, the human enzyme has been cloned, expressed, and characterized.[6] Again, homology among methyltransferases of diverse species suggests the importance of postprenylation processing (Fig. 1B). While hRCE1 and hSTE14 share homologies with paralogs in other species, no homology with other proteases and methyltransferases are apparent. Both enzymes are integral membrane proteins, spanning the membrane with five to seven hydrophobic segments.

To explore the roles of the postprenylation processing enzymes, we have cloned hRCE1 and hSTE14, and have found the cDNAs to be identical to the published sequences. The corresponding mRNAs are ubiquitously expressed in human tissues, and both epitope-tagged recombinant proteins localize to the endoplasmic reticulum. Both recombinant enzymes are active when expressed in human or insect cell lines, and substrate analogs inhibit the native and recombinant activities with similar potency and specificity. Overexpression of either protein stimulates the complementary activity, suggesting these enzymes may collaborate in cells. Using a human cell line engineered to allow inducible expression of K-RasD12,[7] we show that the rates of K-RasD12 protein synthesis and degradation correlate with the rates of methylation and demethylation, respectively. Finally, screening of a random small molecule library has revealed a class of specific inhibitors. Using the most potent of these inhibitors, we confirm that Ras is mislocalized. Here, we detail the methods involved.

Materials

Isolation of Human STE14 and RCE1 cDNAs

The expressed sequence tag (EST) cDNA database (NCBI) is screened for sequences related to the *Saccharomyces cerevisiae* STE14 and RCE1 protein, using the BLAST program. The hRCE1 partial cDNA (missing the 5′ end) is assembled from ESTs W96411 and T97242. To obtain full-

[5] C. A. Hrycyna, S. K. Sapperstein, S. Clarke, and S. Michaelis, *EMBO J.* **10,** 1699 (1991).

[6] Q. Dai, E. Choy, V. Chiu, J. Romano, S. R. Slivka, S. A. Steitz, S. Michaelis, and M. R. Philips, *J. Biol. Chem.* **273,** 15030 (1998).

[7] G. G. M. Habets, M. Knepper, J. Sumortin, Y.-J. Choi, T. Sasazuki, S. Shirasawa, and G. Bollag, *Methods Enzymol.* in preparation **332** [19] (2001) (this volume).

length RCE1 a human pancreatic cDNA library (Stratagene, La Jolla, CA) is screened and 5′ rapid amplification of cDNA ends (RACE) is performed. To obtain the hSTE14 cDNA, the human pancreatic cDNA library (Stratagene) is screened with a probe design based on the ESTs AA143108 and T19783. The hSTE14 and hRCE1 genes are cloned into pcDNA3.1-His containing an N-terminal "Xpress" epitope tag (InVitrogen, Carlsbad, CA).

Reagents

Synthetic peptides are obtained from Synpep (Dublin, CA). The specific methyltransferase substrate is biotin-Lys-Lys-Ser-Lys-Thr-Lys-(farnesyl) Cys, and the substrate for the coupled protease/methyltransferase assay is biotin-Lys-Lys-Ser-Lys-Thr-Lys-(farnesyl)Cys-Val-Leu-Ser. These peptides are designed to mimic the C terminus of K-Ras4B. Protease inhibitors (Pefabloc, leupeptin, and pepstatin) are from Boehringer-Roche (Indianapolis, IN). [^{3}H]Methyl-*S*-adenosylmethionine ([^{3}H]SAM), [^{3}H]methylmethionine, and [^{35}S]methionine are from Dupont-NEN (Boston, MA). Streptavidin-coated scintillation proximity assay (SPA) beads are from Amersham-Pharmacia Biotech (Piscataway, NJ).

Methods

Immunofluorescence

Porcine aortic endothelial cells are grown on coverslips and microinjected in the nucleus with Xpress-tagged hRCE1 or hSTE14 plasmids (Ted Pella, Inc., Redding, CA) (25 μg/ml). After 15 hr of incubation, cells are fixed in 4% (w/v) formaldehyde dissolved in phosphate-buffered saline (PBS) and further processed for indirect double immunofluorescence, using a mouse anti-Xpress monoclonal antibody (Invitrogen, Carlsbad, CA) and a rabbit polyclonal anti-calreticulin antibody (ABR). Immunofluorescence procedures and photography are essentially as previously described.[8] For examining Myc epitope-tagged K-RasD12 localization, a human cell line with ecdysone-inducible expression of K-RasD12 is constructed.[7] This cell line grown on coverslips is pretreated with compound for 1 hr, and then induced for K-RasD12 expression for 8 hr using 10 μ*M* ponasterone A in the presence of compound. Cells are then fixed in formaldehyde and visualized with anti-Myc (9E10) antibody and fluorescein isothiocyanate (FITC)-conjugated anti-mouse antibody as previously described.[8]

[8] D. Stokoe, S. G. Macdonald, K. Cadwallader, M. Symons, and J. F. Hancock, *Science* **264,** 1463 (1994).

Membrane Preparation for Enzymatic Assays

Membranes are isolated by a modification of published methods. Nonadherent tissue culture cells are washed twice in PBS containing 20 m*M* EDTA and then subjected to hypotonic lysis on ice in a buffer containing 5 m*M* Tris (pH 8), 5 m*M* EDTA, 1 m*M* Pefabloc, leupeptin (10 μg/ml) and aprotinin (10 μg/ml). The samples are homogenized with a Dounce homogenizer (Wheaton, Millville, NJ) and subjected to a low-speed spin (4000*g* at 4° for 5 min) to remove unbroken cells and insoluble cell debris. Micosomal membranes are then collected by centrifuging the resulting supernatant at 100,000*g* at 4° for 45 min and resuspending the pellet in buffer as described above and containing 0.25 *M* sucrose. Assays of endogenous protease and methyltransferase enzymes from a selection of human tumor cell lines suggest robust activities from each.

Protease and Methyltransferase Assays by Scintillation Proximity Assay

Methyltransferase or coupled protease/methyltransferase activities *in vitro* with peptide substrates are determined by measuring the transfer of ^{3}H from [^{3}H]methyl-*S*-adenosylmethionine ([^{3}H]SAM) to biotinylated peptides. The standard mixture contains 100 m*M* HEPES (pH 7.3), 125 n*M* biotinylated peptide, 0.125 μCi of [^{3}H]SAM, and 1 μg of membranes in a final volume of 100 μl. Assays are conducted at room temperature for 60 min. Enzymatic activity is terminated by the addition of 5 m*M* EDTA and 0.1% (v/v) Tween 20. The streptavidin-coated SPA beads are then allowed to settle overnight at ambient temperature before quantitation by scintillation counting.

Inhibition of Activities

A potent and selective peptidomimetic inhibitor (named RPI) of prenyl-directed endoprotease activity has been described.[9] Prenylcysteine carboxylmethyltransferase activity has been shown to modify small molecule analogs of a single C-terminal prenylcysteine residue such as AFC (*N*-acetyl-*S*-*trans,trans*-farnesyl-L-cysteine) and AGGC (*N*-acetyl-*S*-*trans, trans*-geranylgeranyl-L-cysteine).[10] These compounds also function as selective inhibitors of prenylcysteine carboxylmethyltransferase, whereas a structurally similar compound, AGC (*N*-acetyl-*S*-*trans*-geranylcysteine), in which the isoprenoid is the C_{10}-geranyl group, is not an effective inhibitor of methylation. Membrane fractions (1 μg) are incubated with various

[9] Y. T. Ma, B. A. Gilbert, and R. R. Rando, *Biochemistry* **32,** 2386 (1993).

[10] C. Volker, M. H. Pillinger, M. R. Philips, and J. B. Stock, *Methods Enzymol.* **250,** 216 (1995).

concentrations of inhibitor (diluted into 100 m*M* HEPES) at room temperature for 20 min before addition of substrate and label (125 n*M* biotinylated peptide; 0.125 μCi of [^{3}H]SAM). Reactions are then incubated for a further 60 min at room temperature before stopping. All the inhibitors discussed above behaved as published in this assay format. This assay format is used to screen a random small molecule library.

Labeling of Cells with Radioactive Methionine

The assay for determination of base-labile counts incorporated into K-RasD12 is a modification of previously published protocols.[11] The ecdysone-inducible K-RasD12 cells[7] are grown in medium lacking methionine for 1 hr. Cells are then treated with [^{3}H]Met (50 μCi/ml) and 10 μM ponasterone A for 24 hr. Lysis is achieved by treating the cells with lysis buffer [20 m*M* Tris (pH 8), 150 m*M* NaCl, 1.5 m*M* $MgCl_2$, 10% (v/v) glycerol, 1% (v/v) Triton X-100, 1 m*M* EGTA] on ice for 5 min. The lysates are clarified by centrifugation at 10,000*g* at 4° for 10 min and Y13-259 resin (20 μg/ml) (Oncogene Science, Uniondale, NY) is added. After rotation at 4° for 16 hr, the resin is pelleted by brief centrifugation and washed three times with PBS. The immunopurified Ras proteins are then resolved on 4–20% (w/v) SDS–polyacrylamide gels, and the gels are sliced. Gel slices are then treated with 1 *N* NaOH at 37° for 16 hr and measured for counts that are able to diffuse into the surrounding scintillant (Ras methylation), or the gel slices are neutralized with 1 *N* HCl and dissolved in scintillant and counted (total Ras). For measurement of protein degradation and demethylation kinetics, cells that have been treated with [^{3}H]Met (50 μCi/ml) and 10 μM ponasterone A for 24 hr are washed and incubated with 100 μM unlabeled methionine, before processing for Ras methylation and total Ras as described above.

Membrane Partitioning of K-RasD12

Cells are labeled with either [^{3}H]Met or [^{35}S]Met and membranes are separated from cytosols as described above, except for using buffer containing 5 m*M* Tris (pH 7.5), 1 m*M* EGTA, 25 m*M* NaF, 1 m*M* dithiothreitol (DTT). Membranes and cytosols are then solubilized with lysis buffer and immunopurified with Y13-259 resin as described above. After resolution on 4–20% (w/v) SDS–polyacrylamide gels, the samples are subjected to autoradiography ([^{35}S]Met samples), or Ras methylation ([^{3}H]Met samples) is determined as described above.

[11] C. Volker and J. B. Stock, *Methods Enzymol.* **255,** 65 (1995).

Results

Sequence Comparisons

Figure 1A shows an alignment of hRCE1 with homologs of various species. The cDNA encodes a protein of 329 residues with a predicted molecular mass of 36 kDa, with at least 5 hydrophobic stretches of greater than 20 residues that putatively span the membrane. One clone that was identified from the cDNA library differs from the EST sequences in that there is a deletion of 21 residues from positions 231 to 251. This could represent a splice variant of the RCE1 gene, and preliminary polymerase chain reaction (PCR) analysis of cDNA libraries from various tissues suggests expression of both isoforms. The mouse RCE1 sequence was derived from an assortment of ESTs, with some editing to correct putative sequence mistakes. Figure 1B shows an alignment of hSTE14 with homologs from other species. The open reading frame (ORF) encodes a protein of 284 amino acids of predicted molecular mass 32 kDa, again containing multiple membrane-spanning domains. The partial mouse STE14 sequence was derived from the EST clone AA022288 with a nucleotide arbitrarily inserted to correct a frame shift.

Intracellular Localization of RCE1 and STE14

To study the intracellular localization of the human endoprotease (hRCE1) and the human methyltransferase (hSTE14), we performed microinjection studies in porcine aortic endothelial (PAE) cells (Fig. 2). Indirect immunofluorescence for epitope-tagged hRCE1 and hSTE14 revealed a fine reticular staining throughout the cell, in addition to the nuclear envelope (Fig. 2A, C, and E), suggesting that it localizes to the endoplasmatic reticulum (ER) compartment. Double staining for calreticulin, a calcium-binding protein that resides in the lumen of the ER,[12] shows striking colocalization of hSTE14 and hRCE1 with that of calreticulin (compare Fig. 2C and D with Fig. 2E and F) confirming that these postprenylation processing enzymes localize to the ER compartment. These findings are consistent with previous experiments that localize RCE1[13] and hSTE14[6] to the endoplasmic reticulum.

Kinetics of Postprenylation Processing

The methods described above allow determination of the rates of K-RasD12 synthesis, degradation, methylation, and demethylation. Be-

[12] M. Michalak, R. Milner, K. Burns, and M. Opas, *Biochem. J.* **285,** 681 (1992).

[13] W. K. Schmidt, A. Tam, K. Fujimura-Kamada, and S. Michaelis, *Proc. Natl. Acad. Sci. U.S.A.* **95,** 11175 (1998).

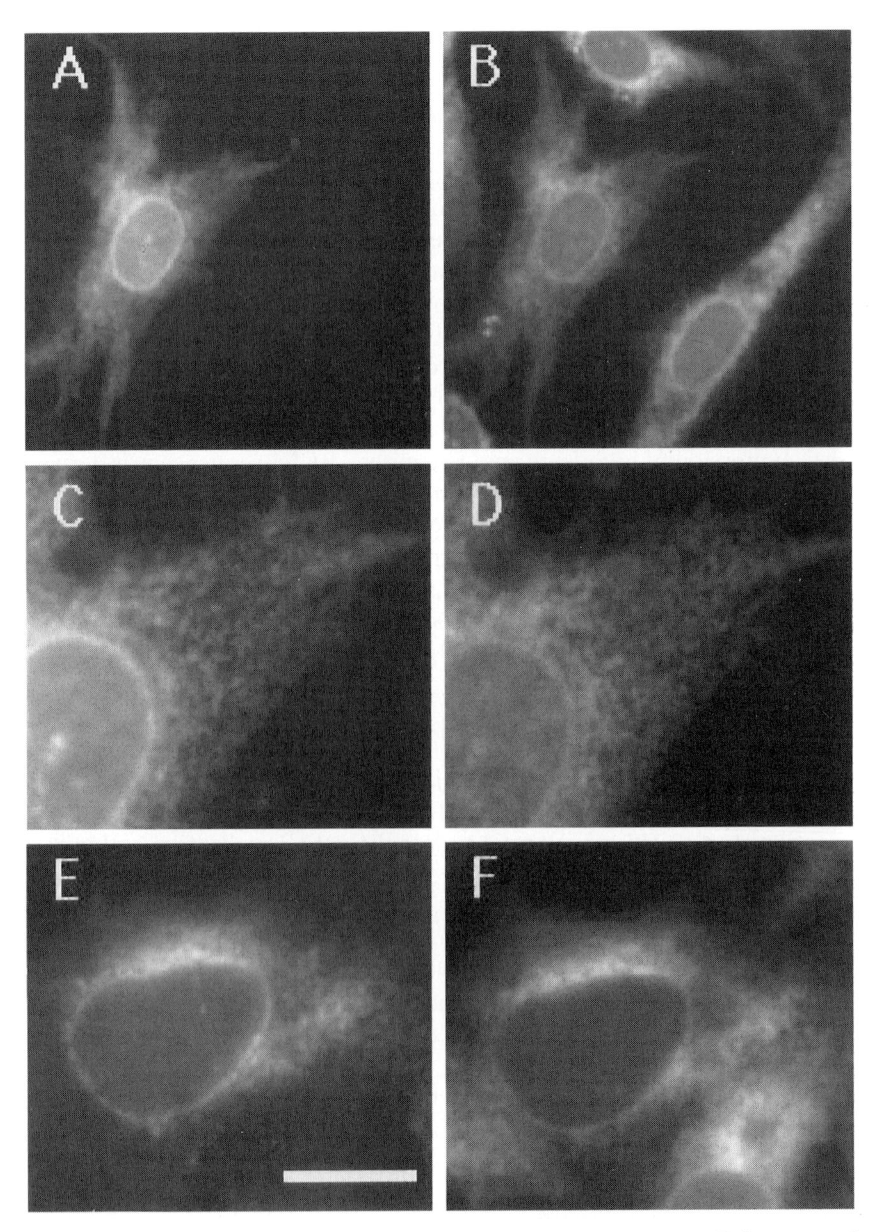

FIG. 2. Subcellular localization of prenyl-directed processing enzymes. Epitope-tagged human cDNAs encoding hSTE14 and hRCE1 were microinjected into PAE cells and visualized by indirect immunofluorescence. Staining of the endoplasmic reticulum marker calreticulin is shown for reference. Both hSTE14 and hRCE1 colocalize with calreticulin to the endoplasmic reticulum. Fluorescence micrographs show double indirect immunofluoresence for hSTE14 (A and C), hRCE1 (E), and endogenous calreticulin (B, D, and F) in porcine aortic endothelial cells. Bar 25 μm in the low-magnification (×40) micrographs (A and B); 10 μm in the high-magnification (×100) micrographs (C–F).

cause methylation requires prior endoproteolysis and because this step is irreversible, the methylation rates allow an appreciation of the overall kinetics of postprenylation processing. An example of the robust assay available for determining the methylation state of K-RasD12 is shown in Fig. 3A. Using this assay, the rates of K-RasD12 protein synthesis and methylation could be compared (Fig. 3B). During this time course, it is clear that no distinction can be made between the rates of protein synthesis and methylation: methylation does not appear to be a rate-limiting step of Ras biosynthesis. This is consistent with the localization studies; postpre-

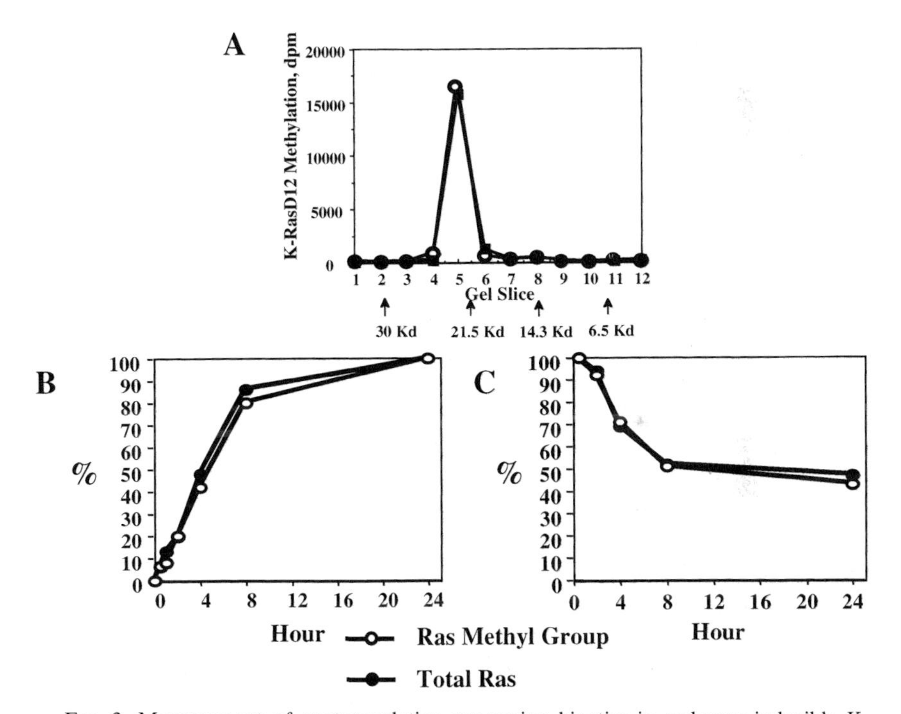

FIG. 3. Measurement of postprenylation processing kinetics in ecdysone-inducible K-RasD12-expressing cells. (A) Induced K-RasD12 from cells labeled with [^{3}H]methionine was immunoprecipitated and run on SDS-polyacrylamide gels, gel slices were treated with 1 *N* NaOH, and base-labile ^{3}H was counted. Results of duplicate measurements are shown. (B) Comparison of K-RasD12 synthesis and methylation. Cells were labeled with [^{3}H]methionine and time courses of K-RasD12 protein synthesis (total counts in gel slice) and methylation (base-labile counts) were compared. (C) Comparison of K-RasD12 degradation and demethylation. Cells were labeled with [^{3}H]Met and time courses of K-RasD12 protein synthesis (total counts in gel slice) and methylation (base-labile counts) were compared after "chase" with a large excess of unlabeled methionine.

nylation processing in the endoplasmic reticulum occurs as part of the pathway leading to Ras membrane localization.[14]

Methylesterification is putatively a reversible process, and the existence of a methylesterase for certain proteins has been suggested.[15] Nonetheless, early experiments demonstrated that carboxylmethylation of low molecular weight proteins is quite stable.[16] Using the methods described, we have examined the stability of K-RasD12 methylesterification. As shown in the pulse–chase experiment of Fig. 3C, K-RasD12 demethylation occurs at essentially the same rate as K-RasD12 protein degradation. These results argue against the presence of a methylesterase. Postprenylation processing appears to be a rapid, irreversible event.

Roles of Postprenylation Processing in Ras Membrane Localization

Biochemical characterization of postprenylation processing has demonstrated that complete processing is essential for proper membrane affinity.[17] In yeast, loss of STE14[5] or RCE1[2] results in improper cytosolic localization of Ras, even though effects on Ras pathway function are subtle. Furthermore, the selective STE14 inhibitors AFC and AGGC cause mislocalization of Ras and inhibition of Ras signaling events.[10,18] Mice lacking RCE1 have again confirmed that incompletely processed Ras no longer localizes to the plasma membrane.[4]

Given the importance of Ras function during tumorigenesis, we have embarked on experiments to determine if inhibition of postprenylation processing can inhibit tumor formation. We have screened a large collection of small molecules for inhibitors of hRCE1 and hSTE14. One screening hit of particular interest was identified as a 300 n*M* hSTE14 inhibitor. This compound appears to be selective, as other methyltransferases such as the *Eco*RI and Dam DNA methyltransferases are not inhibited by 10 μM concentrations of the screening hit compound. This compound is able to inhibit K-RasD12 methylation in intact cells (Fig. 4A) and causes displacement of K-RasD12 from the membrane fraction (Fig. 4B). Indirect immunofluorescence analysis of these cells demonstrates significant loss of K-RasD12 plasma membrane association (Fig. 4C). Preliminary results suggest that the screening hit compound inhibits the proliferation of human tumor

[14] E. Choy, V. K. Chiu, J. Silletti, M. Feoktistov, T. Morimoto, D. Michaelson, I. E. Ivanov, and M. R. Philips, *Cell* **98,** 69 (1999).

[15] D. Pérez-Sala, E. W. Tan, F. J. Canada, and R. R. Rando, *Proc. Natl. Acad. Sci. U.S.A.* **88,** 3043 (1991).

[16] D. Chelsky, B. Ruskin, and D. E. Koshland, Jr., *Biochemistry* **24,** 6651 (1985).

[17] J. F. Hancock, K. Cadwallader, and C. J. Marshall, *EMBO J.* **10,** 641 (1991).

[18] Y. T. Ma, B. A. Gilbert, and R. R. Rando, *Methods Enzymol.* **250,** 226 (1995).

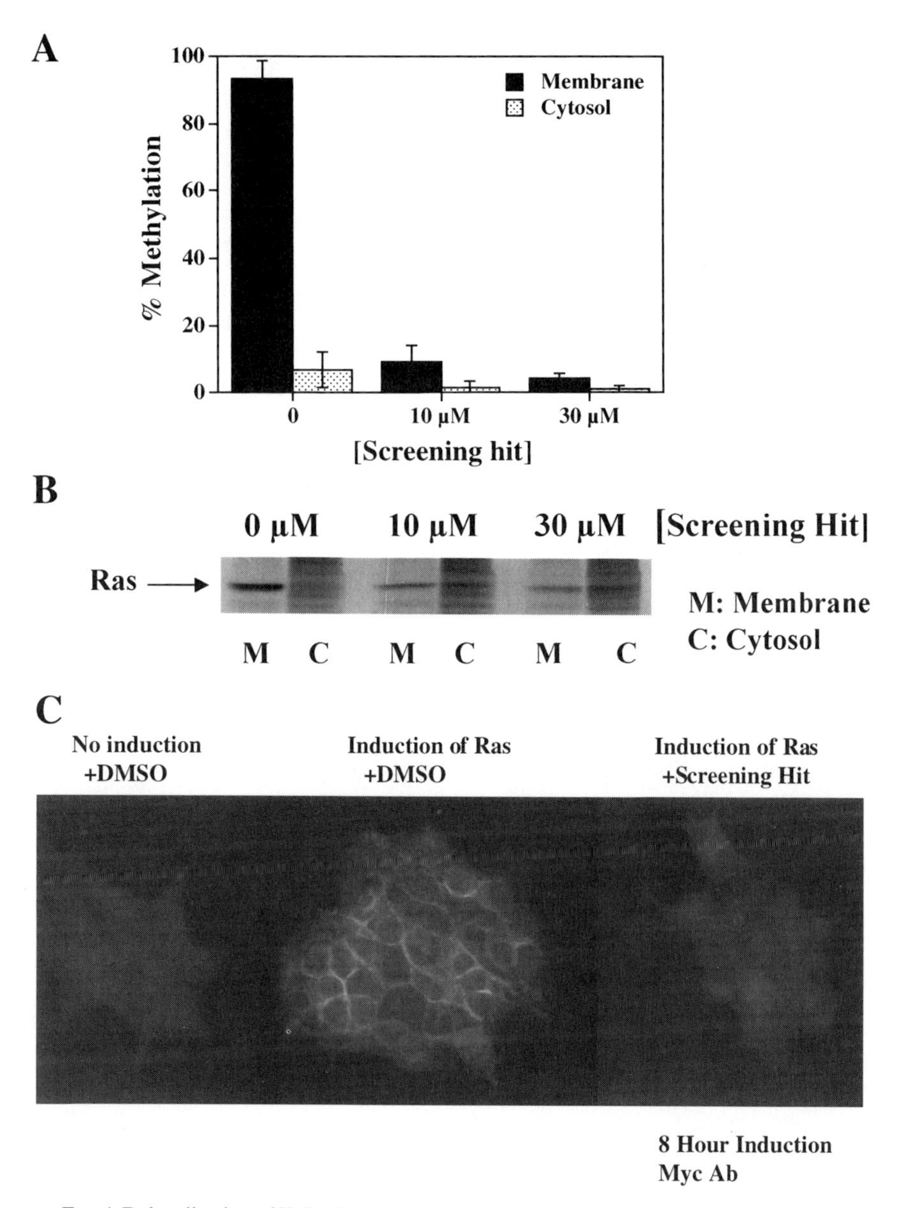

FIG. 4. Relocalization of K-RasD12 from membrane to cytosol on inhibition of methylation. (A) Comparison of K-RasD12 methylation in membrane and cytosolic fractions of cells labeled with [^{3}H]Met and treated with the indicated levels of the methyltransferase inhibitor (screening hit). (B) Comparison of K-RasD12 levels in membrane and cytosolic fractions of cells labeled with [^{35}S]Met and treated with the indicated levels of the methyltransferase inhibitor (screening hit). (C) Indirect immunofluorescence of cells either uninduced (*left*), induced for 8 hr with 10 μM ponasterone A (*middle*), or induced for 8 hr with 10 μM ponasterone A in the presence of 10 μM screening hit (*right*). Visualization is achieved via the Myc epitope on K-RasD12.

cells. Therefore, we are optimistic that targeting postprenylation processing could lead to anticancer therapeutics.

Conclusions

Maturation of many proteins that are targeted for cellular membranes involves prenylation, endoproteolysis and carboxymethylation. Prenylation is achieved primarily by two distinct geranylgeranyltransferases and one farnesyltransferase. In contrast, postprenylation processing appears to require a single endoprotease (hRCE1) and a single methyltransferase (hSTE14). While hRCE1 encodes several isoforms, apparently via alternative splicing, there is currently no evidence for additional genes for either the endoprotease or the methyltransferase. [The cloning of the human ortholog of the yeast mating **a**-factor protease STE24 (AFC1) that can mediate the amino- and carboxyl-terminal cleavages has been reported.[19] While the human STE24 was shown to cleave yeast **a**-factor, no **a**-factor orthologs have been identified in mammals. To date, there has been no role assigned for STE24 in postprenylation Ras processing.]

Kinetic analysis of K-Ras biosynthesis reveals a tight correlation between the rates of protein expression and complete posttranslational processing. Furthermore, demethylation of K-Ras is not detected. These observations suggest that membrane localization of K-Ras (and perhaps many other prenylated proteins) is a biosynthetic, and not a regulatory, event.

While deletion of the RCE1 and STE14 genes in yeast results in viable cells, knockout of the mouse RCE1 gene results in embryonic lethality. This points to a role for proper processing of certain GTPases during mouse development. Nonetheless, these mice are completely devoid of postprenylation Ras processing, yet a few of the mice survive postnatally. This surprising observation predicts a lack of general toxicity caused by inhibitors of postprenylation processing. Availability of the human enzymes and enhanced knowledge of their roles should expedite the development of novel Ras-targeted cancer therapies.

[19] A. Tam, F. J. Nouvet, K. Fujimura-Kamada, H. Slunt, S. S. Sisodia, and S. Michaelis, *J. Cell Biol.* **142,** 635 (1998).

[8] *In Vivo* Prenylation Analysis of Ras and Rho Proteins

By PAUL T. KIRSCHMEIER, DAVID WHYTE, OSWALD WILSON, W. ROBERT BISHOP, and JIN-KEON PAI

Introduction

Farnesyltransferase inhibitors (FTIs) were designed to inhibit oncogenic forms of the small GTPase (smGTPase), Ras.[1] During the development of these compounds experimental evidence showed that NIH 3T3 cells transformed by H-*ras* were more sensitive to FTIs than NIH 3T3 cells transformed with K-*ras*.[2,3] Further reports suggested a mechanism for differential sensitivity to FTIs in cells transformed by the different *ras* isoforms. One report showed that FTIs block mevalonate incorporation into Ras proteins immunoprecipitated from H-*ras*-transformed cells but did not block mevalonate incorporation into Ras proteins immunoprecipitated from K-*ras*-transformed cells.[4] Other reports showed that of the Ras proteins only H-Ras was exclusively a substrate for farnesyl protein transferase (FPT), whereas K-Ras4A/4B and N-Ras were substrates for both FPT and geranylgeranyl protein transferase 1 (GGPT-1).[4,5] These results suggested that H-Ras was exclusively farnesylated; however, the type of modification of K- and N-Ras in the presence of FTIs remained uncharacterized. Methodology was therefore developed to distinguish the prenylation of the individual Ras isoforms and other smGTPases in cells.[6]

Two systems were used to examine the prenylation of Ras proteins and other smGTPases under different growth or treatment conditions. The first involved heterologous expression of individual proteins in a cellular system; labeling the cells overexpressing the protein with [^{3}H]mevalonate and determining the prenyl group associated with the overexpressed protein by chemical analysis. This analysis provided data on an individual Ras isoform; it did not provide direct information about the endogenous protein under

[1] A. D. Cox and C. J. Der, *Biochim. Biophys. Acta* **1333,** F51 (1997).

[2] T. Nagasu, K. Yoshimata, C. Rowell, M. D. Lewis, and A. M. Garcia, *Cancer Res.* **55,** 5310 (1995).

[3] L. Sepp-Lorenzino, Z. Ma, E. Rands, N. E. Kohl, J. B. Gibbs, A. Oliff, and N. Rosen, *Cancer Res.* **55,** 5302 (1995).

[4] G. L. James, M. S. Brown, and J. L. Goldstein, *Proc. Natl. Acad. Sci. U.S.A.* **93,** 4454 (1996).

[5] F. L. Zhang, P. Kirschmeier, D. Carr, L. James, R. W. Bond, L. Wang, R. Patton, W. T. Windsor, R. Syto, R. Zhang, and W. R. Bishop, *J. Biol. Chem.* **272,** 10232 (1997).

[6] D. B. Whyte, P. Kirschmeier, T. N. Hockenberry, I. Nunez-Oliva, L. James, J. J. Catino, W. R. Bishop, and J.-K. Pai, *J. Biol. Chem.* **272,** 14459 (1997).

0076-6879/00 $35.00

normal expression conditions. A second system was designed to examine protein prenylation of endogenously expressed proteins. This system relied on construction of a human cell line transfected with a mevalonate transporter expression vector to enhance the uptake and incorporation of radiolabeled mevalonolactone into cellular proteins.

Prenyl Analysis of Small GTPases in COS Cells

Ras Family Members

The method involves the prenyl analysis of the Ras/Rho family of smGTPases. The approach, however, is extended to study the prenylation of other proteins such as HK33 and HDJ-2. In general it is applicable to the prenyl analysis of all proteins with *Caa*X sequences (where *a* indicates an aliphatic amino acid). COS cells are used for these experiments and are obtained from the American Type Culture Collection (ATCC, Manassas, VA).[7] The cells are maintained in Dulbecco's modified Eagle's medium (DMEM; GIBCO, Grand Island, NY) that is supplemented with glucose (4500 mg/liter) (w/v), glutamine (0.89 mg/liter) (w/v), 0.1 m*M* nonessential amino acids, gentamicin (10 μg/liter), and 10% (v/v) fetal bovine serum (FBS; HyClone, Logan, UT).

The cDNAs encoding the Ras or Rho proteins are generally obtained by polymerase chain reaction (PCR) amplification. Mutant H-*ras* (G12V) is obtained by reverse transcriptase (RT)-PCR amplification of RNA isolated from the human bladder carcinoma cell line T24. Mutant (Q61H) K-*ras*4A is obtained by RT-PCR amplification of RNA isolated from the lung carcinoma cell line NCI-H460. Mutant (G12V) K-*ras*4B is obtained by RT-PCR amplification of RNA isolated from the colon carcinoma cell line SW-620. N-*ras* (G12N) is a gift from A. Wolfman (Cleveland Clinic, OH). Because a monoclonal antibody, Y13-259, is widely used for Ras immunoprecipitation, epitope tags are not added to the Ras proteins during the construction of expression vectors.[8] PCR-amplified cDNAs are cloned into mammalian expression vectors: pSV Sport (Life Technologies, Rockville, MD) for H-*ras* and pCIneo (Promega, Madison, WI) for K- and N-*ras*. In our experience K-*ras* is much more highly expressed with the pCIneo vector than with the pSV Sport vector. The reasons for differences in expression of these two *ras* isoforms are not clear; however, for practical reasons, we have chosen vectors that result in high levels of expression for each Ras isoform.

[7] J. Sambrook, E. M. Fritsch, and T. Maniatis, "Molecular Cloning: A Laboratory Manual." Cold Spring Harbor Laboratory Press, Cold Spring Harbor, New York, 1989.
[8] M. E. Furth, L. J. Davis, B. Fleurdelys, and E. M. Scolnick, *J. Virol.* **43,** 294 (1982).

Rho Family Members

The strategy for constructing expression vectors for members of the Rho family is different from that for the Ras family because specific antibodies for the Rho proteins are not always available. The SwissProt database has been searched with the FINDPATTERNS algorithm (GCG) to identify human proteins with C*aa*X sequences potentially recognized by FPT. The search reveals several smGTPases (RhoD, TC10, RhoE, Rho6, and Rho7) in which the X in the C*aa*X sequence is either T or M. The coding sequences for these proteins have been PCR amplified and cloned into pFLAG-CMV-2 (Sigma, St. Louis, MO) expression vectors. These vectors are convenient and have worked well for expression of these Rho family members. Other methods for epitope tagging proteins would be expected to work as well.[9]

After construction, the coding sequence for each Ras or Rho protein is verified by sequence analysis. The expression plasmids are isolated and purified in large scale by either $CsCl_2$ gradient centrifugation or affinity column (Qiagen, Valencia, CA) methods. Both methods yield plasmids satisfactory for transfection.

Transfection and Labeling of Small GTPases

A cotransfection strategy is used to efficiently label Ras proteins in the COS cell system. Individual Ras or Rho expression vectors are mixed with an equal amount of the mevalonate transporter expression plasmid, pMEV.[10] Inclusion of the mevalonate transporter plasmid results in greater uptake of the exogenously supplied mevalonolactone. It is possible to do the experiment in the absence of the pMEV expression plasmid, but lower levels of incorporation of label should be expected.

Step 1. Reducing Volume of [^{3}H]Mevalonolactone. Preparation for labeling involves reducing the volume of the [^{3}H]mevalonolactone. Typically, this reagent is supplied at 40 Ci/mmol (1 m Ci/ml) as a solution in 100% ethanol (DuPont-NEN, Boston, MA). The volume of the stock solution is slowly reduced under a stream of nitrogen to increase the concentration to 100 m Ci/ml. The [^{3}H]mevalonolactone is not completely dried under these conditions. It is essential to remember that [^{3}H]mevalonolactone is volatile and proper containment practices must be strictly observed. DuPont-NEN sells a charcoal trap cartridge to be used during the concentration step; it significantly aids proper containment of the radioactivity. The concentrated [^{3}H]mevalonolactone is stored at $-20°$ prior to use.

[9] J. W. Jarvik and C. A. Telmer, *Annu. Rev. Genet.* **32,** 601, (1998).
[10] C. M. Kim, J. L. Goldstein, and M. S. Brown, *J. Biol. Chem.* **267,** 23113 (1992).

Step 2. Cell Culture. To obtain larger quantities of radiolabeled protein for chemical analysis two 100-mm dishes of COS cells are used for each condition. Approximately 2×10^6 COS cells are plated per dish and allowed to attach and grow overnight. The following morning the medium is changed with fresh DMEM warmed to 37°.

Step 3. Transfection. After 2 hr of incubation the cells are prepared for transfection with the cationic lipid adjuvant LipofectAMINE (Life Technologies). Briefly, plasmid DNA is prepared by mixing 15 μg of a *ras* expression plasmid with 15 μg of pMEV in a volume of 1600 μl of Opti-MEM-1 reduced serum medium (Life Technologies). LipofectAMINE (150 μl) is also added to 1600 μl of Opti-MEM. The DNA and lipid solutions are gently mixed and allowed to stand for 30 min. At this time 12.8 ml of Opti-MEM is added to the DNA–lipid complexes and mixed thoroughly with a pipette. The solution contains sufficient transfection mixture for two 100-mm dishes of cells. The COS-7 cells are washed twice in Opti-MEM-1 prewarmed to 37°. Half the DNA–lipid complex is added to each of the two 100-mm dishes and the cells are returned to the incubator for 4–5 hr. The DNA–lipid mixture is aspirated off the cells and replaced by fresh DMEM containing 10% (v/v) FBS. FTIs are added to the medium at the indicated concentrations. The cells are placed back in the incubator for about 20–22 hr.

Step 4. Mevalonate Starvation and Compound Addition. Mevalonate starvation medium is prepared prior to removing the cells from the incubator. This medium is DMEM supplemented as described above except that dialyzed FBS (HyClone) is used instead of the standard serum supplement. This substitution enhances the reproducibility and consistency of the labeling procedure. Mevastatin [stock solution of 20 m*M* in phosphate-buffered saline (PBS)] is added to prewarmed DMEM supplemented with dialyzed FBS such that the final concentration is 20 μM. Mevastatin in this incubation blocks endogenous mevalonate synthesis. At the same time FTIs are prepared and added to the culture medium. The tricyclic FTIs, such as SCH44342, SCH56582, or SCH66336, are dissolved in tissue culture-grade dimethyl sulfoxide (DMSO) (Sigma) at 1000× stock concentrations. Compounds are added directly to the tissue culture medium such that the final concentration of DMSO is 0.1% (v/v). The transfected COS cells are removed from the incubator and the medium is replaced with 10 ml per dish of the mevalonate starvation medium plus the indicated concentration of FTI. The cells are returned to the incubator for approximately 2 hr.

Step 5. Mevalonate Labeling. The labeling medium is prepared during the mevalonate starvation process. The concentrated [^{3}H]mevalonolactone prepared during step 1 is removed from the freezer. The label is added to warm mevalonate starvation medium such that the final concentration of

[^{3}H]mevalonolactone is 100 μCi/ml. DMSO and compounds are next added to the labeling medium. The cells are removed from the incubator and the starvation medium is replaced by labeling medium. Cells are incubated for an additional 18 hr to label the cellular proteins.

Step 6. Cell Lysis and Immunoprecipitation. Labeled cells are washed twice with ice-cold PBS and lysed in radioimmunoprecipitation assay (RIPA) buffer [150m*M* NaCl, 1% (v/v) Nonidet P-40 (NP-40), 0.5% (w/v) deoxycholate, 0.1% (w/v) sodium dodecyl sulfate (SDS), 50m*M* Tris (pH 7.5), 50 μM leupeptin, 1 m*M* Pefabloc, SC, aprotinin (2 μg/ml), soybean trypsin inhibitor (2 μg/ml), pepstatin (1 μg/ml), 2 m*M* benzamidine, and 2 m*M* EDTA].[10] Typically, 1 ml of lysis buffer is used per 100-mm dish. The viscosity of the lysate is reduced before immunoprecipitation.[11] The total amount of radioactivity incorporated is estimated by trichloroacetic acid (TCA) precipitation and liquid scintillation counting of 10 μl of lysate. Proteins are immunoprecipitated from approximately 20 million cpm of lysate diluted to 1 ml in RIPA buffer in a 1.5-ml microcentrifuge tube. For the Ras proteins an agarose conjugate of the pan-Ras rat monoclonal antibody, Y13-259 (Santa Cruz Biotechnologies, Santa Cruz, CA), is used. Thirty microliters of a well-suspended slurry of Y13-259 agarose conjugate is added to each milliliter of lysate and rotated for 3–4 hr at 4°. FLAG-tagged Rho proteins are immunoprecipitated with anti-FLAG M2 agarose (Sigma) under the same conditions used for the Ras proteins. After this incubation the tubes are centrifuged at 6000 rpm for 2 min at 4° in a microcentrifuge to pellet the agarose and immunoprecipitate the proteins. Care must be taken not to disturb the pellet while removing the supernatant containing unprecipitated proteins. The agarose pellet is washed four times with ice-cold RIPA buffer. The Ras or Rho proteins are now in a suitable form to initiate analysis of their associated prenyl groups. They can be stored at −80° until use.

Methyl Iodide (Iodomethane) Cleavage

Structural analysis of the prenyl groups attached to the immunoprecipitated proteins is performed by methyl iodide cleavage and high-performance liquid chromatography (HPLC) as initially described by Casey *et al.*[12]

Step 1. Acetone Extraction. Noncovalently bound lipids are first removed from immunoprecipitated proteins. The labeled protein immunoprecipitate is suspended in 0.8 ml of ice-cold acetone in a 1.5-ml microcentrifuge tube

[11] E. Harlow and D. P. Lane, "Antibodies: A Laboratory Manual." Cold Spring Harbor Laboratory Press, Cold Spring Harbor, New York, 1988.

[12] P. J. Casey, P. A. Solski, C. J. Der, and J. E. Buss, *Proc. Natl. Acad. Sci. U.S.A.* **86,** 8323 (1989).

and left at 2° for >4 hr. Carrier protein, 5 μl of bovine serum albumin [BSA, fatty acid free (Sigma), 10 mg/ml in PBS], is added to each tube. The mixture is centrifuged for 15 min at 14,000 rpm at 2° to pellet the protein. The acetone is removed under house vacuum. The acetone wash of the pellet is repeated. After the second wash, the protein pellet is air dried in the fume hood.

Step 2. Trypsin Digestion. The delipidated proteins are digested with trypsin to produce peptide fragments. The dried pellet from the acetone wash is suspended in 190 μl of 100 m*M* Tris–HCl, pH 7.7, containing 5% (v/v) acetonitrile (buffer A). Twenty microliters of trypsin (Calbiochem, La Jolla, CA), 100 μg/100 μl in trypsin buffer [50 m*M* HEPES (pH 8.0), 1 m*M* EDTA, 1 m*M* dithiothreitol (DTT), 10 m*M* $MgCl_2$], is also added and the sample is incubated overnight at 37°. After digestion the amount of radioactivity in 2 μl of the sample is determined. Best results are obtained with counts per minute >500 in this sample.

Step 3. Methyl Iodide Cleavage. The sample is transferred to a 4-ml amber vial containing a small stir bar, and 0.2 ml of 25 m*M* Tris–HCl, pH 7.7, containing 80% (v/v) acetonitrile (buffer B) is added. Finally, 100 μg each (10 μl of a solution consisting of 0.5 mg/0.5 ml of ethanol) of *N*-acetylfarnesylcysteine and *N*-acetylgeranylgeranylcysteine (Biomol, Plymouth Meeting, PA) are added. The isoprenyl alcohols released during the cleavage of the added prenylcysteines act as carrier. Next, the sample is diluted by the addition of 0.8 ml of 3% (w/v) formic acid, and the cleavage reaction is initiated by the addition of 0.1 ml of methyl iodide (iodomethane; Sigma). Low light conditions are maintained because of the light sensitivity of methyl iodide. The vial is sealed with a Teflon-lined cap and stirred in the dark overnight.

Step 4. Prenyl Alcohol Extraction. Unreacted methyl iodide is removed by evaporation under nitrogen for 1 hr over a warm water bath. The sample is neutralized by the addition of ~0.15 ml of 30% (w/v) sodium carbonate. The final pH of 7.0 to 8.5 is confirmed by spotting a small sample onto pH indicator paper. The neutralized solution is incubated with stirring for >6 hr at room temperature. The released prenyl groups are extracted twice with 1.2 ml each of chloroform–methanol (9:1, v/v). The organic (lower) layer is combined, dried under a stream of nitrogen, and dissolved in a small volume (100–150 μl) of 50% (v/v) acetonitrile containing 0.1% (w/v) trifluoroacetic acid (TFA) (solvent A). An aliquot of this final sample is checked by scintillation counting to determine the efficiency of the cleavage and extraction.

High-Performance Liquid Chromatography Analysis

The example used here is the HPLC analysis of the prenyl alcohol released from the Rho family member TC10, and the data are shown in Fig. 1.

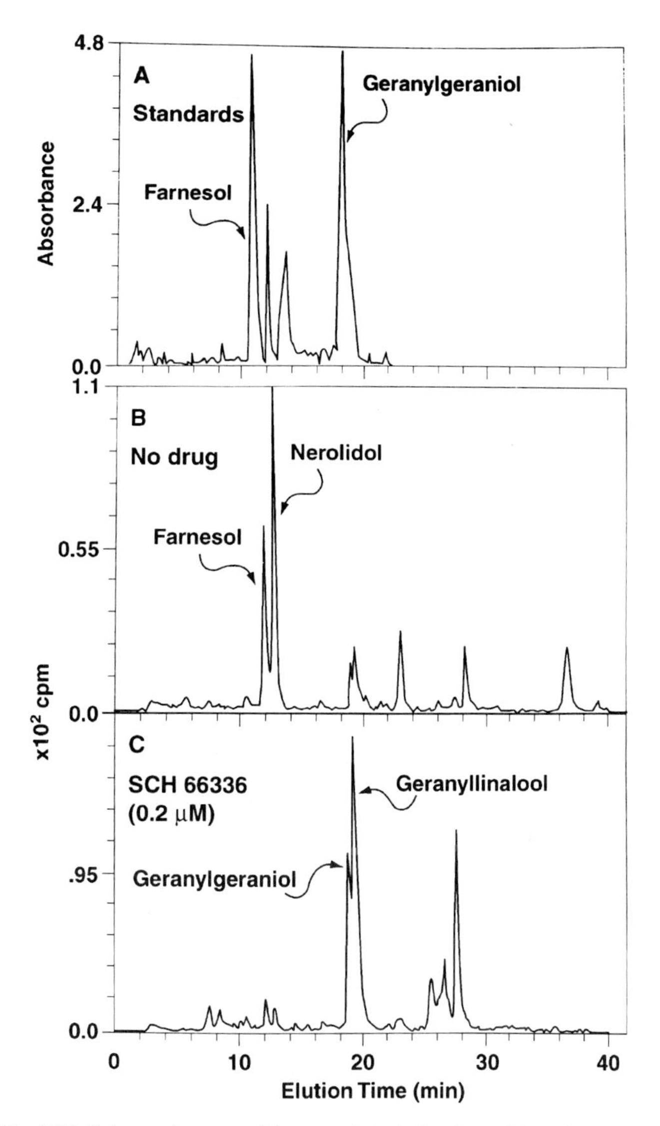

FIG. 1. The HPLC chromatograms of the prenyl alcohols released from immunoprecipitated TC10 protein. Epitope-tagged TC10 was expressed in COS-7 cells and immunoprecipitated with anti-FLAG M2 agarose. The prenyl alcohols were cleaved from the immunoprecipitated protein as described and separated by HPLC. The elution profile of the nonradioactive standards, farnesol and geranylgeraniol, is shown in (A). Under the gradient conditions used farnesol eluted at ~12 min and geranylgeraniol eluted at ~18.5 min. In untreated COS-7 cells the isoprene alcohols cleaved from immunoprecipitated TC10 eluted at ~12 min as shown in (B). However, in COS-7 cells incubated with 0.2 μM SCH66336, the isoprenyl alcohols eluted with a retention time of ~19 min as shown in (C). Significant amounts of radioactivity in (C) elute with a retention time of 25 to 29 min. These products appear to be related to C_{20}-isoprenyl alcohols but rigorous product identification was not done.

Step 1. Standards. Standards are used to identify the retention times of the isoprenyl alcohols on the HPLC system. Nonradioactive *trans,trans*-farnesol (Fluka, Ronkonkoma, NY) and *cis*-nerolidol (Fluka) are used as the standard for the 15-carbon isoprenyl alcohols. Nonradioactive all-*trans*-geranylgeraniol (American Radiolabeled Chemicals, St. Louis, MO) and geranyllinalool (Fluka) are used as standards for the 20-carbon isoprenyl alcohols. For radioactive standards *trans,trans*-[1-^{3}H]farnesol (American Radiolabeled Chemicals) and all-*trans*-[1-^{3}H]geranylgeraniol are used.

Step 2. Reversed-Phase High-Performance Liquid Chromatography. The isoprenoid alcohols are analyzed by reversed-phase HPLC (Waters, Milford, MA) on a C_{18} column (Symmetry C_{18}, 3.9 × 150 mm; Waters). The system is equipped with a flow scintillation analyzer (Radiomatic 525TR; Packard, Downers Grove, IL) to detect radioactive fractions and a UV detection system from HPLC separation. The prenyl alcohols are dissolved previously in a small volume of solvent A [50% (v/v) acetonitrile containing 0.1% (w/v) TFA]. The sample is loaded on the column in solvent A and further developed with an additional 2 ml of solvent A. The prenyl alcohols are eluted with a 15-ml linear gradient of 50% solvent A/50% solvent B [100% (v/v) acetonitrile containing 0.1% (w/v) TFA] and 10% solvent A/90% solvent B at a flow rate of 1 ml/min. After the gradient the column is washed with 20 ml of solvent B at the same flow rate. Under these conditions the farnesol standard elutes with a retention time of ~12 min and the geranylgeraniol standard elutes at ~19 min.

Using these gradient conditions two peaks corresponding to the C_{15} and the C_{20} peak are commonly observed. This is most likely due to a carbocationic rearrangement of the cleaved prenyl alcohols. A clear separation of the peaks is observed. In separate experiments the standards neriolodol and geranyllinallool, which represent the C_{15} and C_{20} rearrangement products, respectively, elute with retention times consistent with this observation. The percentage of radioactivity found in each peak can be used to calculate the amount of protein that is farnesylated and compare that with the amount that is geranylgeryanylated. This approach is particularly useful for doing dose–response curves for prenylation of Ras proteins in the presence of farnesyltransferase inhibitors.[6]

The results from prenyl group analysis of the Rho family member TC10 are shown in Fig. 1. The C*aa*X sequence of TC10 is CLIT, which predicts that the protein should normally be farnesylated. In control cells the radioactivity cleaved from the immunoprecipitated TC10 elutes with the same retention time as the C_{15} standard. This result is consistent with the motif prediction. However, in the presence of FTIs the cleaved radioactivity elutes with the same retention time as the C_{20} standard. Using these gradient conditions two peaks corresponding to the C_{15} and the C_{20} peak are commonly observed. The doublets at C_{15} and C_{20} are most likely due to a carbocationic rearrangement of the cleaved prenyl alcohols. Changes in the gradient used would result in improved resolution of each double peak.

The experiment shows that the farnesylation of TC10 is completely inhibited by the FTI, but rather than remaining unprenylated it is geranylgeranylated.[13] The carboxy terminus of TC10 is basic, which is believed to be one feature contributing to recognition of normally farnesylated substrates by geranylgeranyltransferase 1. Furthermore, these experiments show that the threonine at the carboxy terminus does not prevent alternative prenylation by geranylgeranyltransferase. Similar results have been observed with the K- and N-Ras proteins whereas H-Ras is exclusively farnesylated.[6] The exact structural determinants at the carboxy terminus of a farnesylated protein that permit alternative prenylation still need to be investigated. The use of mevalonate labeling in heterologous expression systems provides a fundamental methodology for addressing this question.

Prenyl Analysis of Endogenous Small GTPases

The second methodology involves determining the prenyl groups attached to endogenous Ras and Rho proteins. Rather than using an overexpression system, a cell line has been developed that enhances incorporation of exogenously added mevalonate. The human colon tumor cell line DLD-1 is used for these experiments. The cell line is mutant for p53 and is heterozygous at the K-*ras* locus, expressing both a mutated activated K-Ras protein and a wild-type protein.[14] N-Ras is also relatively highly expressed in this cell line. While we concentrate on Ras for these experiments, this cell system has been useful for characterizing the prenylation of other endogenous proteins as well. This system is particularly useful if proteins are difficult to express in the heterologous system.

Construction of DLD pMEV-1

Step 1. Transfection of DLD-1. The DLD-1 colon tumor cell line is obtained from the ATCC and grown in DMEM supplemented with glucose (4500 mg/liter), glutamine (0.89 mg/liter), 0.1 *M* nonessential amino acids, gentamicin (10 μg/liter), and 10% (v/v) fetal bovine serum. To construct stable transformants expressing the mevalonate transporter and appropriate control cells the DLD-1 cells are plated at 3.0×10^5 cells in 60-mm culture dishes and incubated overnight. The following day the medium is

[13] O. Wilson, W. H. Jin, S. Black, P. Kirschmeier, H. Ashar, W. R. Bishop, and J.-K. Pai, in preparation.

[14] S. Shirasawa, M. Furuse, N. Yokoyama, and T. Sasazuki, *Science* **260,** 85 (1993).

changed 4 hr prior to transfection and the cells are returned to the incubator. DNA (10 μg of pMEV-1 plasmid DNA plus 1.0 μg of pCIneo plasmid DNA) is diluted into 300 μl of Opti-MEM-1 reduced serum medium (Life Technologies). LipofectAMINE (40 μl) is also added to 300 μl of Opti-MEM. The DNA and lipid solutions are gently mixed and allowed to stand for 30 min. At this time 2.4 ml of Opti-MEM is added to the DNA–lipid complexes and mixed thoroughly with a pipette. A control transfection solution is prepared by omitting the pMEV plasmid DNA. The DLD-1 cells are washed twice in Opti-MEM-1 prewarmed to 37°. The DNA–lipid complex is added to the dishes and the cells are returned to the incubator for 4–5 hr. The DNA–lipid mixture is aspirated off the cells and replaced with fresh DMEM containing 10% (v/v) FBS. After 40 hr of incubation the contents of each 60-mm dish is split into five 100-mm dishes containing DMEM and 10% (v/v) FBS containing G418 (800 μg/ml). The DLD-1 cells require higher levels of G418 than NIH 3T3 cells for effective cell killing. Colonies are allowed to grow for 12 days with several medium changes.

Step 2. Characterization of Stable Transformants. Ten clones from each transfection are isolated. The control and pMEV-transfected clones are initially characterized by their sensitivity to 20 μM mevastatin. The control and pMEV-transfected cells as well as the parental DLD-1 cell line are killed within several days after continuous exposure to medium containing 20 μM mevastatin. The clones are further characterized by growth in medium containing 20 μM mevastatin plus 100 μM mevalonate (Sigma). Under these conditions the parental DLD-1 cell line and the control transfected clones are all killed and are therefore equally sensitive to mevastatin in the presence or absence of exogenously added mevalonate. In contrast, several of the pMEV-transfected clones grow in medium containing 20 μM mevastatin in the presence of 100 μM mevalonate.

These clones are then tested for their ability to incorporate [^{3}H]mevalonate into a TCA-precipitable fraction. Results from one experiment, shown in Fig. 2, demonstrate a large enhancement of incorporation of exogenously supplied mevalonate in this transfected cell line. It is important to note that this level of enhancement is not achieved with all the transfectants tested. Each transfectant needs to be individually tested by mevalonate incorporation.

Labeling and Determination of Prenylation in DLD pMEV-1 Cells

Labeling and determination of the prenyl groups attached to the Ras proteins in DLD pMEV-1 cells are performed under growth and mevalonate starvation conditions similar to those used for COS-7 cell labeling.

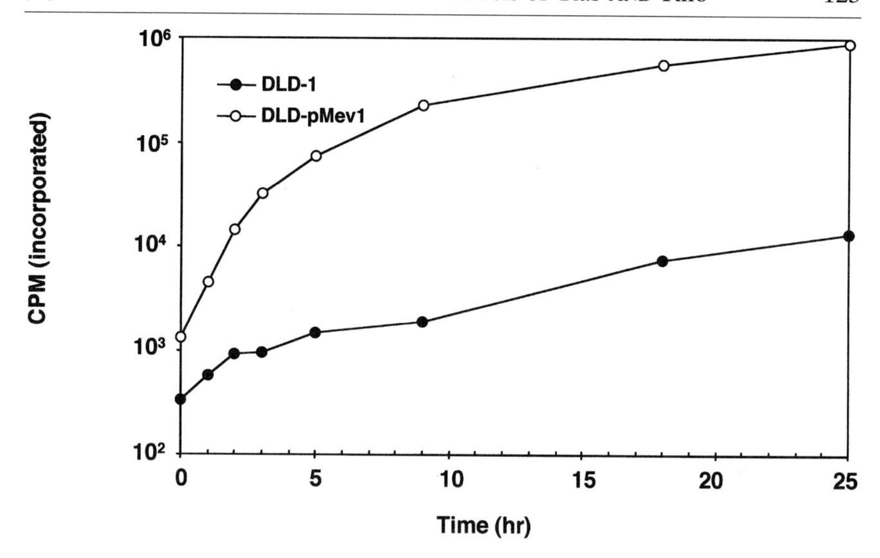

FIG. 2. Mevalonate incorporation into DLD-1 and DLD pMEV-1 cells was compared. Cells (3×10^6) were plated in 60-mm dishes. The following day the cells were starved for mevalonate and then incubated in labeling medium as described in text. Cells were washed in cold PBS and lysed in RIPA buffer at the indicated time points. Ten percent of the lysate was TCA precipitated and counted by liquid scintillation counting.

Step 1. Labeling Conditions. DLD pMEV-1 cells are plated at 2×10^6 cells per 100-mm dish. After overnight incubation the medium is replaced with the same mevalonate starvation medium as described above. At the same time compounds and DMSO are also added to the culture medium. After a 2-hr incubation the starvation medium is replaced with labeling medium prepared as described above. The cells are incubated with the compounds and [^{3}H]mevalonolactone for an additional 18 to 20 hr.

Step 2. Cell Lysis and Immunoprecipitation. The cells are washed twice with ice-cold PBS and lysed in RIPA buffer as described above. For the experiments done with Ras proteins immunoprecipitation is done with Y13-259 agarose. Because DLD-1 expresses both K- and N-Ras both proteins are represented in the immunoprecipitate when using Y13-259. For experiments with other proteins, specific antibodies are required to immunoprecipitate the endogenous protein. The conditions for cell lysis and Ras immunoprecipitation are described above. The immunoprecipitates can be stored at $-80°$ or can be processed to cleave the prenyl groups. The methyl iodide cleavage procedure is done exactly as described above.

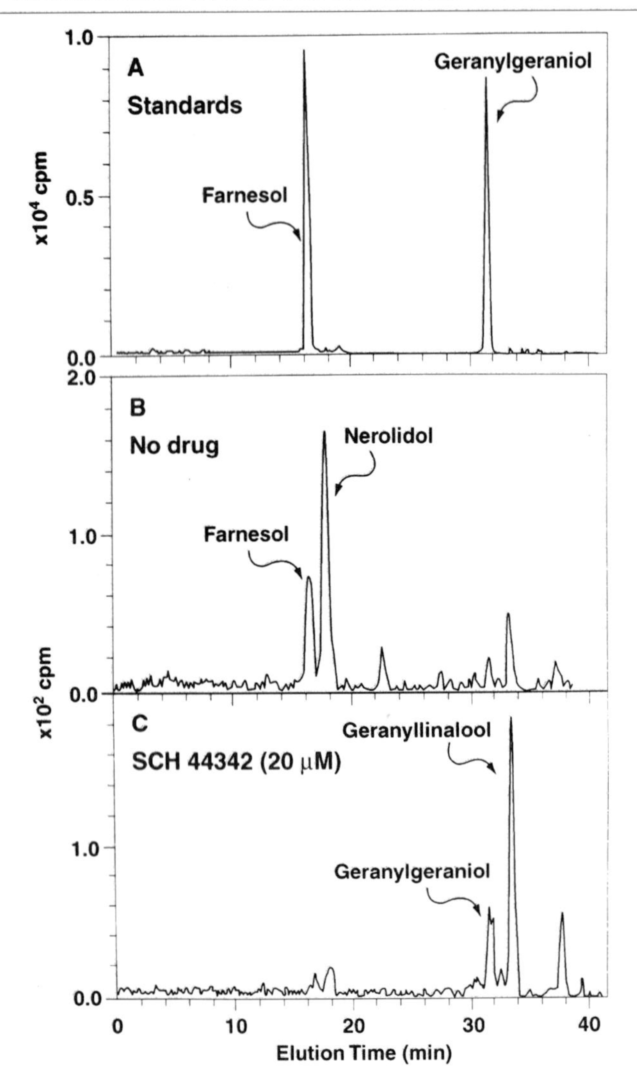

FIG. 3. The HPLC chromatograms of the prenyl alcohols released from immunoprecipitated Ras proteins from DLD-1 pMEV cells are shown along with the radioactive standards. For this experiment a shallower gradient was used than in the experiment shown in Fig. 1. The radioactive farnesol standard elutes with a retention time of ~18 min and the radioactive geranylgeraniol standard elutes at ~31 min as shown in (A). In (B) where no FTI was incubated with the cells, the cleaved isoprenyl alcohols from the immunoprecipitated Ras proteins eluted at ~18 min. When the cells were incubated with 20 μM SCH44342 the cleaved isoprenyl alcohols eluted at ~31–32 min. The gradient conditions used in this experiment resolved the primary isoprenyl alcohol cleavage products and rearranged isoprenyl alcohols produced during the cleavage procedure.

Step 3. High-Performance Liquid Chromatography Analysis. The HPLC conditions used for these experiments are slightly different from the conditions described above and the results are shown in Fig. 3. HPLC is performed on a C_{18} column (Symmetry C_{18}, 3.9 × 150 mm; Waters). The column is developed with 3 ml of solvent A, and then with a 30-ml linear gradient of 90% solvent A/10% solvent B to 100% acetonitrile/0.1% (v/v) TFA (solvent B), followed by 7 ml of solvent B at a flow rate of 1 ml/min. In this experiment radioactive isoprenyl alcohol standards are used to identify the retention times for the farnesol (C_{15}) or geranylgeraniol (C_{20}) cleavage products. The radioactive standards are diluted 1:50 in solvent A before they are applied to the reversed-phase column. The farnesol (C_{15}) standard elutes at about 15 min and the geranylgeraniol (C_{20}) standard elutes at about 29 min under these HPLC conditions. These conditions spread out the peaks on the gradient and clearly allow the separation of farnesol and geranylgeraniol from their rearrangement products nerolidol and geranyllinalool. Figure 3B shows the major isoprenyl group associated with the endogenous Ras proteins eluted with the retention time of the C_{15}-prenyl alcohol. This peak represents products cleaved from both K- and N-Ras proteins because both are expressed in the cell and both are immmunoprecipitated by Y13-259. When the cells are treated with the FTI SCH44342,[15] the cleaved isoprenyl alcohol elutes with the same retention time as the C_{20} standards (Fig. 3C). This result shows that the FTIs inhibit farnesylation of the endogenous K- and N-Ras isoforms but that prenylation of these proteins is not inhibited. In the case of the Ras proteins the results from the DLD pMEV system corroborate the results observed in the overexpression system.[6]

[15] W. R. Bishop, R. Bond, J. Petrin, L. Wang, R. Patton, R. Doll, G. Njoroge, Catino, J. Schwartz, W. Windsor, R. Syto, D. Carr, L. James, and P. Kirschmeier, *J. Biol. Chem.* **270,** 30611 (1995).

[9] Ras Interaction with RalGDS Effector Targets

By SHINYA KOYAMA and AKIRA KIKUCHI

Introduction

RalGDS is a GDP/GTP exchange protein for Ral, a member of the small GTP-binding protein superfamily. RalGDS was originally isolated by polymerase chain reaction (PCR) using regions conserved between *CDC25* and *ste6*, which encode GDP/GTP exchange proteins of Ras in

0076-6879/00 $35.00

Saccharomyces cerevisiae and *Schizosaccharomyces pombe,* respectively.[1] However, RalGDS does not stimulate GDP/GTP exchange of Ras. Among various small G proteins, RalGDS stimulates GDP/GTP exchange of RalA and RalB. Ral was originally isolated by probing with an oligonucleotide corresponding to one of the GTP-binding domains of Ras.[2] Although the function of Ral is not well understood, its effector protein, Ral-binding protein 1 (RalBP1), has been identified.[3] RalBP1 has a Rho GTPase-activating protein (GAP) domain and indeed stimulates the GTPase activity of Rac and CDC42. In *Drosophila,* genetic analyses demonstrate that Ral regulates Jun N-terminal kinase activity, which is activated by Rac and CDC42,[4] but in mammals there is no report to show that Ral regulates the function of Rac and CDC42. Instead, POB1 and Reps have been identified as RalBP1-binding proteins.[5,6] Both proteins share the EH domain, which was originally identified in Eps15, that is important for regulating receptor-mediated endocytosis.[7] The EH domain of POB1 interacts with Eps15 and Epsin, which is also involved in endocytosis. Although the functions of the Ral signaling pathway have long been unknown, Ral, RalBP1, and POB1 are reported to regulate the receptor-mediated endocytosis for epidermal growth factor (EGF) and insulin.[8,9]

There is a RalGDS family consisting of RalGDSa, RalGDSb, RGL, and Rlf (Fig. 1).[1,10,11] Yeast two-hybrid screening has shown a link between Ras and Ral by demonstrating that RalGDS family members bind to Ras. They share a CDC25 homology domain in the central region and a Ras-interacting domain (RID) in the C-terminal region. CDC25 homology domain stimulates the GDP/GTP exchange of Ral. RID interacts with the GTP-bound form of Ras. Therefore, RalGDS is an effector protein of Ras

[1] C. F. Albright, B. W. Giddings, J. Liu, M. Vito, and R. A. Weinberg, *EMBO J.* **12,** 339 (1993).

[2] P. Chardin and A. Tavitian, *EMBO J.* **5,** 2203 (1986).

[3] S. B. Cantor, T. Urano, and L. A. Feig, *Mol. Cell. Biol.* **15,** 4578 (1995).

[4] K. Sawamoto, P. Winge, S. Koyama, Y. Hirota, C. Yamada, S. Miyao, S. Yoshikawa, M. H. Jin, A. Kikuchi, and H. Okano, *J. Cell Biol.* **146,** 361 (1999).

[5] A. Yamaguchi, T. Urano, T. Goi, and L. A. Feig, *J. Biol. Chem.* **272,** 31230 (1997).

[6] M. Ikeda, O. Ishida, T. Hinoi, S. Kishida, and A. Kikuchi, *J. Biol. Chem.* **273,** 814 (1998).

[7] P. P. Di Fiore, P. G. Pelicci, and A. Sorkin, *Trends Biochem. Sci.* **22,** 411 (1997).

[8] S. Nakashima, K. Morinaka, S. Koyama, M. Ikeda, M. Kishida, K. Okawa, A. Iwamatsu, S. Kishida, and A. Kikuchi, *EMBO J.* **18,** 3629 (1999).

[9] K. Morinaka, S. Koyama, S. Nakashima, T. Hinoi, K. Okawa, A. Iwamatsu, and A. Kikuchi, *Oncogene* **18,** 5915 (1999).

[10] A. Kikuchi, S. D. Demo, Z.-H. Ye, Y.-W. Chen, and L. T. Williams, *Mol. Cell. Biol.* **14,** 7483 (1994).

[11] R. M. F. Wolthuis, B. Bauer, L. J. V. Veer, A. M. M. Vries-Smith, R. H. Cool, M. Spaargaren, A. Wittinghofer, B. M. T. Burgering, and J. L. Bos, *Oncogene* **13,** 353 (1996).

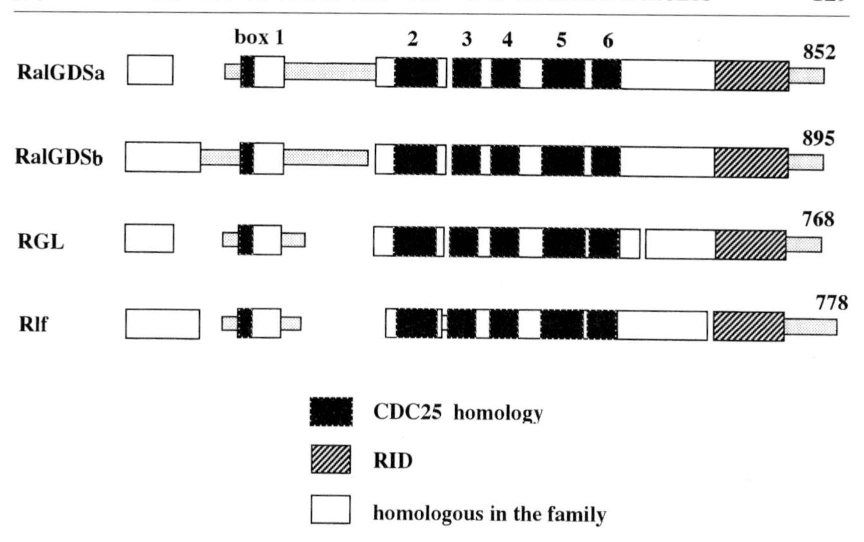

FIG. 1. RalGDS family. Boxes 1–6 represent CDC25 homology domains and RID represents the Ras-binding domain. White boxes are homologous regions in the family. The numbers at the carboxyl terminus (right end) indicate amino acid numbers of the proteins.

and transmits the signal from Ras to Ral. This chapter describes assays for the binding of RalGDS to Ras in intact cells and *in vitro,* and the GDP/GTP exchange of Ral by Ras through RalGDS in intact cells and *in vitro.*

Materials

Guanosine 5′-(3-*O*-thio)triphosphate (GTPγS) is purchased from Roche Diagnostics (Mannheim, Germany). 3-[(3-Cholamidopropyl)dimethylammonio]-1-propanesulfonic acid (CHAPS) and *n*-octyl-β-D-glucopyranoside (*n*-OG) are from Dojindo Laboratories (Kumamoto, Japan). Bovine serum albumin (BSA, fraction V), guanosine 5′-triphosphate (GTP), and guanosine 5′-diphosphate (GDP) are from Sigma (St. Louis, MO). L-α-Dimyristoylphosphatidylcholine (DMPC), phosphatidylcholine (PC), phosphatidylethanolamine (PE), phosphatidylserine (PS), and phosphatidylinositol (PI) are from Wako Pure Chemicals (Osaka, Japan). [^{35}S]GTPγS, [^{3}H]GDP, glutathione–Sepharose 4B (GSH–Sepharose), protein A–Sepharose CL-4B, and Cy2- or Cy3-labeled anti-mouse IgG and anti-rabbit IgG are from Amersham Pharmacia Biotech (Uppsala, Sweden). Alexa488- or Alexa594-labeled anti-mouse IgG and anti-rabbit IgG are from Molecular Probes (Eugene, OR). The anti-Ras monoclonal antibody (F235) is from Santa Cruz Biotechnology (Santa Cruz, CA). Polyethyleneimine (PEI) cellulose thin-layer chromatography (TLC) plates are from Merck (Darmstadt, Germany). All other chemicals are of reagent grade.

Buffers and Reaction Mixtures

Lysis buffer: 20 m*M* Tris–HCl (pH 7.5), 1% (w/v) Nonidet P-40 (NP-40), 137 m*M* NaCl, 10% (w/v) glycerol, and 1 m*M* dithiothreitol (DTT)

Protease inhibitors: 1 m*M* phenylmethylsulfonyl fluoride (PMSF), aprotinin (20 μg/ml), and leupeptin (10 μg/ml)

Buffer A: 20 m*M* Tris–HCl (pH 7.5), 1 m*M* EDTA, 5 m*M* $MgCl_2$, and 1 m*M* DTT

G protein-labeling buffer: 50 m*M* Tris–HCl (pH 7.5), 10 m*M* EDTA, 5 m*M* $MgCl_2$, 1 m*M* DTT, and BSA (1 mg/ml)

Reaction buffer: 50 m*M* Tris–HCl (pH 7.5), 20 m*M* $MgCl_2$, 10 m*M* EDTA, 1 m*M* DTT, and BSA (1 mg/ml)

Western Blotting

Protein samples are boiled in Laemmli sample buffer, resolved by sodium dodecyl sulfate–polyacrylamide gel electrophoresis (SDS–PAGE), and transferred to nitrocellulose filters. The filters are incubated with the anti-Ras antibody (F235), anti-RalGDS antibody, anti-hemagglutinin (HA) antibody 12CA5, or anti-Myc antibody 9E10 as the primary antibody, and an alkaline phosphatase-conjugated anti-mouse or rabbit IgG as the secondary antibody, and developed with nitro blue tetrazolium and 5-bromo-4-chloro-3-indolylphosphate as a substrate. To achieve higher sensitivity, horseradish peroxidase-conjugated anti-mouse or rabbit IgG is used as the secondary antibody and developed with an enhanced chemiluminescence (ECL) system (Amersham Pharmacia Biotech).

Complex Formation of RalGDS and Ras in Intact Cells

Transient Expression of RalGDS and Ras in COS Cells

COS cell are cultured at 37° in Dulbecco's modified Eagle's medium (DMEM) containing 10% (v/v) calf serum, penicillin (50 μg/ml), and streptomycin (50 μg/ml). COS cells (60–70% confluent on 10-cm-diameter dishes) are transfected with various plasmids to express Ras and RalGDS by the DEAE-dextran method.[12] Sixty hours after transfection, the cells

[12] C. Gorman, *in* "DNA Cloning: A Practical Approach" (D. M. Glover, ed.), p. 143. IRL Press, Oxford, 1985.

are lysed for 1 hr at 4° in 0.5 ml of the lysis buffer containing protease inhibitors. Insoluble material is removed by centrifugation for 30 min at 4° at 13,000*g*.

Immunoprecipitation Assay

The lysates (160 μg of protein) expressing HA-RalGDS and Ras^{G12V}, an active form in which Gly-12 is mutated to valine, are prepared and the proteins of the lysates are immunoprecipitated with the anti-Ras antibodies. An anti-Ras monoclonal antibody, Y13-238, precipitates any form of Ras, while Y13-259 reacts with Ras in a region neighboring the effector loop and interfers with the interaction between Ras and its effector proteins. Y13-259 does not precipitate Ras that is complexed with its effector protein Raf, whereas Y13-238 does.[13,14] The anti-Ras antibody and 10 μl of protein A–Sepharose are added to the lysate and rocked at 4° for 1 hr. The immunoprecipitates are washed once with the lysis buffer, twice with 100 m*M* Tris–HCl (pH 7.5)–0.5 *M* LiCl, and once with 10 m*M* Tris–HCl (pH 7.5). The precipitates are probed with the anti-Ras (F235) antibody. Coprecipitated HA–RalGDS is detected by the anti-HA antibody.

HA–RalGDS is precipitated with Y13-238 but not with Y13-259.[10,15] Although Ras^{G12V} forms a complex with RalGDS, c-Ras or Ras^{S17N}, a dominant negative form, does not.[10,15] Furthermore, effector loop mutants of Ras recognize their specific effector proteins. $Ras^{G12V/T35S}$ forms a complex with RalGDS but not with Raf, whereas $Ras^{G12V/E37G}$ exhibits the opposite binding.[16,17]

Extracellular Signal-Dependent Complex Formation of RalGDS and Ras

COS cells expressing c-Ras and HA–RalGDS are incubated in DMEM containing 2% (v/v) dialyzed serum for 24 hr. The cells are then washed with DMEM and stimulated with various concentrations (0–300 ng/ml) of EGF for various periods (0–60 min).[15] After the stimulation, the cells are lysed and the lysates are immunoprecipitated with the anti-Ras antibody, and then the immunoprecipitates are probed with the anti-HA antibody.

[13] P. H. Warne, P. R. Viciana, and J. Downward, *Nature* (*London*) **364,** 352 (1993).

[14] X. F. Zhang, J. Settleman, J. M. Kyriakis, E. Takeuchi-Suzuki, S. J. Elledge, M. S. Marshall, J. T. Bruder, U. R. Rapp, and J. Avruch, *Nature* (*London*) **364,** 308 (1993).

[15] A. Kikuchi and L. T. Williams, *J. Biol. Chem.* **271,** 588 (1996).

[16] M. A. White, C. Nicolette, A. Minden, A. Polverino, L. Van Aelst, M. Karin, and M. H. Wigler, *Cell* **80,** 533 (1995).

[17] M. Okazaki, S. Kishida, T. Hinoi, T. Hasegawa, M. Tamada, T. Kataoka, and A. Kikuchi, *Oncogene* **14,** 515 (1997).

HA–RalGDS is coprecipitated with c-Ras, dose dependent on EGF, with maximum binding observed at 20 min after stimulation.[15]

Regulation of Subcellular Localization of RalGDS by Ras

COS cells expressing HA–RalGDS and Ras^{G12V} or $Ras^{G12V\Delta}$ are washed with ice-cold phosphate-buffered saline (PBS) and suspended in 1 ml of buffer A containing protease inhibitors. This suspension is sonicated and centrifuged at 700*g* for 5 min at 4° to remove unbroken cells and nuclei. The homogenate is centrifuged at 100,000*g* for 30 min at 4°. The supernatant is used as the cytosol fraction. The pellet is resuspended in 1 ml of buffer A containing 1% (w/v) NP-40 and protease inhibitors, rocked for 1 hr, and then centrifuged at 100,000*g* for 30 min at 4°. The supernatant is used as the membrane fraction. Aliquots (20 μg of protein) of the cytosol and membrane fractions are probed with the anti-HA or anti-Ras (F235) antibody. For the immunoprecipitation assay, aliquots (200 μg of protein) of the cytosol and membrane fractions are immunoprecipitated with the anti-HA or anti-Ras (Y13-238) antibody and the immunoprecipitates are probed with the anti-HA or anti-Ras (F235) antibody.

RalGDS is present in the cytosol fraction without Ras expression. Ras^{G12V} is located in both the cytosol and membrane fractions and induces the translocation of RalGDS from the cytosol to the membrane fraction. On the other hand, $Ras^{G12V\Delta}$, which lacks the CAAX motif (where A represents an aliphatic amino acid) at the C terminus and is not posttranslationally modified, remains in the cytosol and does not translocate RalGDS to the membrane, although they form a complex. Therefore, membrane localization of Ras through its posttranslational lipid modification is necessary for the translocation of RalGDS from the cytoplasm to the plasma membrane.[18]

Immunocytochemistry

NIH 3T3 cells grown on glass coverslips are microinjected with various plasmids (0.2–0.4 mg/ml) to express Myc–RalGDS, HA–RalGDS, Myc–Ras^{G12V}, Myc–$Rap1^{Q63E}$, and HA–Ral using micromanipulator 5171 and transjector 5246 (Eppendorf-Netheler-Hinz, Hamburg, Germany).[19] The following procedures are performed at room temperature. At 4 hr postmicroinjection, the cells are fixed for 20 min in PBS containing 4% (w/v) paraformaldehyde. After being washed with PBS three times, the cells are

[18] S. Kishida, T. Hinoi, S. Koyama, M. Ikeda, Y. Matsuura, and A. Kikuchi, *J. Biol. Chem.* **271,** 19710 (1996).

[19] K. Matsubara, S. Kishida, Y. Matsuura, H. Katayama, M. Noda, and A. Kikuchi, *Oncogene* **18,** 1303 (1999).

permeabilized for 10 min with PBS containing 0.1% (w/v) Triton X-100 and BSA (2 mg/ml). The cells are washed and incubated for 1 hr with the anti-Myc, anti-Ral, or anti-RalGDS antibodies. After being washed with PBS, they are further incubated for 1 hr with Cy2- or Alexa488-labeled anti-mouse IgG or with Cy3- or Alexa594-labeled anti-rabbit IgG. Coverslips are washed with PBS, mounted on a glass slide, and viewed with a confocal laser-scanning microscope (TCS-NT; Leica-laser-technik, Heidelberg, Germany).

The effector loop of Rap 1 has the same amino acid sequence as that of Ras.[20,21] Rap1 binds to RalGDS with higher affinity than Ras.[22,23] Consistent with previous observations,[24–26] Ras^{G12V} and $Rap1^{Q63E}$, active forms of Ras and Rap1, are observed at the plasma membrane and in the perinuclear region, respectively. Ral is detected at the plasma membrane. RalGDS is found in the cytosol. When RalGDS is expressed with Ras^{G12V}, RalGDS is detected at the plasma membrane along with Ras^{G12V}, whereas when RalGDS is coinjected with $Rap1^{Q63E}$, it is detected in the perinuclear region along with $Rap1^{Q63E}$. Therefore, the subcellular localization of RalGDS is dependent on that of Ras and Rap1.

Direct Interaction of RalGDS with Ras *in Vitro*

Full-Length RalGDS and Ras

Full-length RalGDS and Ras are purified from Sf9 cells.[10] To make the GTPγS- or GDP-bound form of Ras, c-Ras (20 pmol) is incubated for 10 min at 30° in 40 μl of G protein-labeling buffer containing 25 μ*M* GTPγS or GDP. After the incubation, 1 *M* $MgCl_2$ is added at a final concentration of 20 m*M*. The GTPγS- or GDP-bound form of Ras is incubated for 30 min at 4° with RalGDS (20 pmol) in 80 μl of reaction buffer containing 25 μ*M* GTPγS or GDP. The anti-Ras antibody (Y13-238) is then added and this mixture is immunoprecipitated. The precipitates are probed with the anti-RalGDS antibody. RalGDS is coprecipitated with the GTPγS-bound form of Ras but not with the GDP-bound form.[10]

[20] M. Noda, *Biochim. Biophys. Acta* **1155,** 97 (1993).

[21] Y. Takai, K. Kaibuchi, A. Kikuchi, and M. Kawata, *Int. Rev. Cytol.* **133,** 187 (1992).

[22] M. Ikeda, S. Koyama, M. Okazaki, K. Dohi, and A. Kikuchi, *FEBS Lett.* **375,** 37 (1995).

[23] C. Herrmann, G. Horn, M. Spaargaren, and A. Wittinghofer, *J. Biol. Chem.* **271,** 6794 (1996).

[24] F. Beranger, B. Goud, A. Tavitian, and J. de Gunzburg, *Proc. Natl. Acad. Sci. U.S.A.* **88,** 1606 (1991).

[25] D. R. Lowy and B. M. Willumsen, *Annu. Rev. Biochem.* **62,** 851 (1993).

[26] V. Pizon, M. Desjardins, C. Bucci, R. G. Parton, and M. Zerial, *J. Cell. Sci.* **107,** 1661 (1994).

Ras-Interacting Domain of RalGDS and Ras

RIDs of RalGDS and RGL [RalGDS-(764–864) and RGL-(632–734)] are purified from *Escherichia coli* as maltose-binding protein (MBP) fusion proteins.[18] They are referred to as MBP–RIDs here. Ras is purified from *E. coli* or Sf9 cells as a glutathione *S*-transferase (GST) fusion protein. To make immobilized RIDs on amylose resin, MBP–RID (200 μg of proteins) is rocked for 2 hr at 4° with 200 μl of amylose resin in 450 μl of buffer A containing 200 m*M* NaCl and BSA (1 mg/ml). The resin is precipitated by centrifugation and washed with 20 m*M* Tris–HCl (pH 7.5) three times. One microliter of the resin binds 2 pmol of MBP–RID. To make the [^{35}S]GTPγS-bound form of GST–Ras, GST–Ras (100 pmol) is incubated for 10 min at 30° in 50 μl of G protein-labeling buffer containing 10 μM [^{35}S]GTPγS (8000–10,000 cpm/pmol), 0.3% (w/v) CHAPS, and 1 m*M* DMPC. After the incubation, 1 μl of 1 *M* $MgCl_2$ is added. The [^{35}S]GTPγS-bound form of Ras is incubated for 30 min at 4° with immobilized MBP–RID (20 pmol) in 100 μl of the reaction buffer containing 100 μM GTP, 0.3% (w/v) CHAPS, and 0.5 m*M* DMPC. Immobilized MBP–RID is precipitated by centrifugation and washed once with 20 m*M* Tris–HCl (pH 7.5), twice with 100 m*M* Tris–HCl (pH 7.5)–0.5 *M* LiCl, and once with 10 m*M* Tris–HCl (pH 7.5). The remaining radioactivities are measured. During the procedures at 4° of this assay, [^{35}S]GTPγS is not dissociated from Ras. The [^{35}S]GTPγS-bound form of Ras is precipitated with MBP–RID in a dose-dependent manner (Fig. 2).[18,22]

Activation of Ral by Ras through RalGDS

GDP/GTP Exchange Assay of Ral in Intact Cells

COS cells expressing HA–Ral, Myc–Ras^{G12V}, Myc–Ras$^{G12V\Delta}$, Myc–Rapl^{Q63E}, and Myc–RalGDS are serum starved and metabolically labeled with $^{32}P_i$ (0.1 mCi/ml) in phosphate-free RPMI 1640 for 12 hr.[27] After the cells are lysed, HA–Ral is immunoprecipitated with the anti-HA antibody. After extensive washing of the immunoprecipitates, bound guanine nucleotides are eluted in 20 μl of elution buffer [20 m*M* Tris–HCl (pH 7.5), 20 m*M* EDTA, 2% (w/v) SDS, 1 m*M* GDP, and 1 m*M* GTP] for 5 min at 65°, and spotted on a PEI-cellulose TLC plate and developed with 1 *M* KH_2PO_4 (pH 3.5).[28] The plates are dried and analyzed with a Fuji (Tokyo, Japan)

[27] S. Kishida, S. Koyama, K. Matsubara, M. Kishida, Y. Matsuura, and A. Kikuchi, *Oncogene* **15,** 2899 (1997).

[28] T. Satoh, M. Endo, M. Nakafuku, S. Nakamura, and Y. Kaziro, *Proc. Natl. Acad. Sci. U.S.A.* **87,** 5993 (1990).

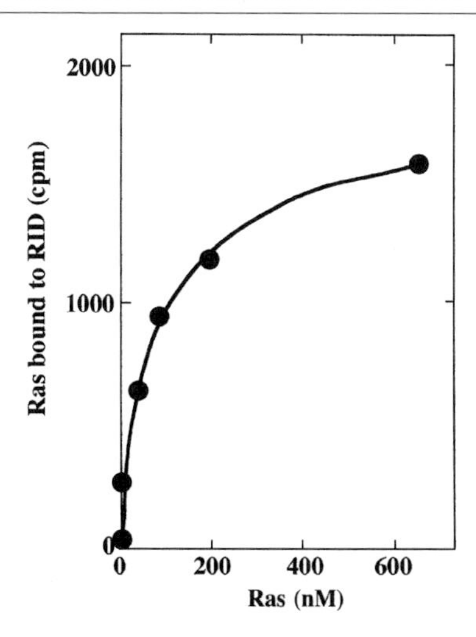

FIG. 2. The binding of Ras to RID. The indicated concentrations of the [^{35}S]GTPγS-bound form of Ras are incubated with 20 pmol of immobilized RID. After the beads have been washed, the remaining radioactivities are measured.

BAS 2000 image analyzer. The relative molar ratio of GTP and GDP is corrected for the number of phosphates per mole of guanosine, and the amount of the GTP bound form of Ral is expressed as a percentage of the total amount of Ral. When the cells are stimulated with EGF, they are pretreated with 50 μM Na_3VO_4 for 30 min.

Expression of RalGDS increases the molar ratio of the GTP-bound form of Ral to total Ral from 5 to 10% (Fig. 3). Although expression of Ras^{G12V} alone does not affect the molar ratio of the GTP-bound form of Ral, coexpression of Ras^{G12V} and RalGDS increases the ratio to 15%. $Ras^{G12V\Delta}$ coexpressed with RalGDS does not activate Ral. Although $Rap1^{Q63E}$ forms a complex with RalGDS, it does not activate Ral. EGF activates Ral only when c-Ras and RalGDS are coexpressed.

Activation of Ral on Reconstituted Liposomes

The posttranslationally modified forms of GST–Ras, GST–RalB, GST–Rap1, and GST–CDC42 are purified from the membrane fraction of Sf9 cells with *n*-OG instead of CHAPS.[27] The membrane extracts are prepared by the use of 1.17% (40 m*M*) *n*-OG in buffer A containing protease inhibitors and glutathione–Sepharose 4B column chromatography is carried out

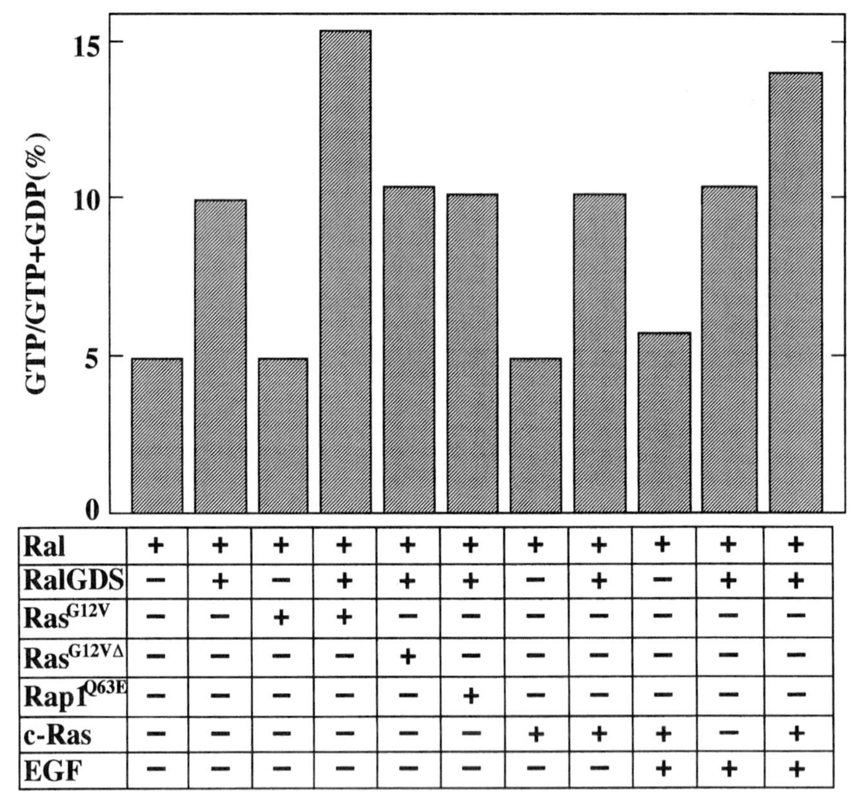

FIG. 3. Ras-dependent Ral activation through RalGDS in intact cells. After COS cells are transfected with various plasmids as indicated, they are metabolically labeled with $^{32}P_i$. Where indicated, the cells are stimulated with EGF (100 ng/ml) for 3 min. The lysates are immunoprecipitated with the anti-HA antibody. The labeled nucleotides are separated by thin-layer chromatography, and GTP and GDP spots are quantified.

in the presence of 0.88% (30 m*M*) *n*-OG in the same buffer. GST–RalGDS and GST–RalGDS-(1–633) are purified from *E. coli.*[18,29]

Liposomes are prepared by sonication of dried lipids containing PC, PS, PE, and PI in buffer A. Sonication is carried out on ice 15 times (45 sec each time), with a 15-sec interval between bursts. To prepare the posttranslationally modified [^{3}H]GDP-bound form of Ral, the modified form of Ral (15 pmol) is incubated with 5 μM [^{3}H]GDP (12,000–15,000 dpm/pmol) in 20 μl of G protein-labeling buffer containing 15 m*M* *n*-OG

[29] H. Murai, M. Ikeda, S. Kishida, O. Ishida, M. Okazaki-Kishida, Y. Matsuura, and A. Kikuchi, *J. Biol. Chem.* **272,** 10483 (1997).

for 10 min at 30°. The posttranslationally modified [^{3}H]GDP-bound form of Ral (10 pmol) is added to the liposomes with the posttranslationally modified GTPγS- or GDP-bound form of Ras and Rap1, and incubated for 10 min on ice in 400 μl of buffer A containing 1 m*M* *n*-OG (1/25th the critical micellar concentration). The final concentrations of PS, PI, PC, and PE are 600, 60, 280, and 640 μ*M*, respectively. After the incubation, the mixture is overlaid on a discontinuous sucrose density gradient that consists of 1 ml of 0.05% (w/v) CHAPS in buffer A containing 1.2 *M* sucrose and 2.5 ml of the same buffer containing 0.15 *M* sucrose in a 4-ml tube from the bottom in this order, and centrifuged for 2 hr at 4° at 100,000*g*. After the centrifugation, fractions of 200 μl each are collected and the radioactivity of a 20-μl aliquot of each fraction is counted after filtration through the

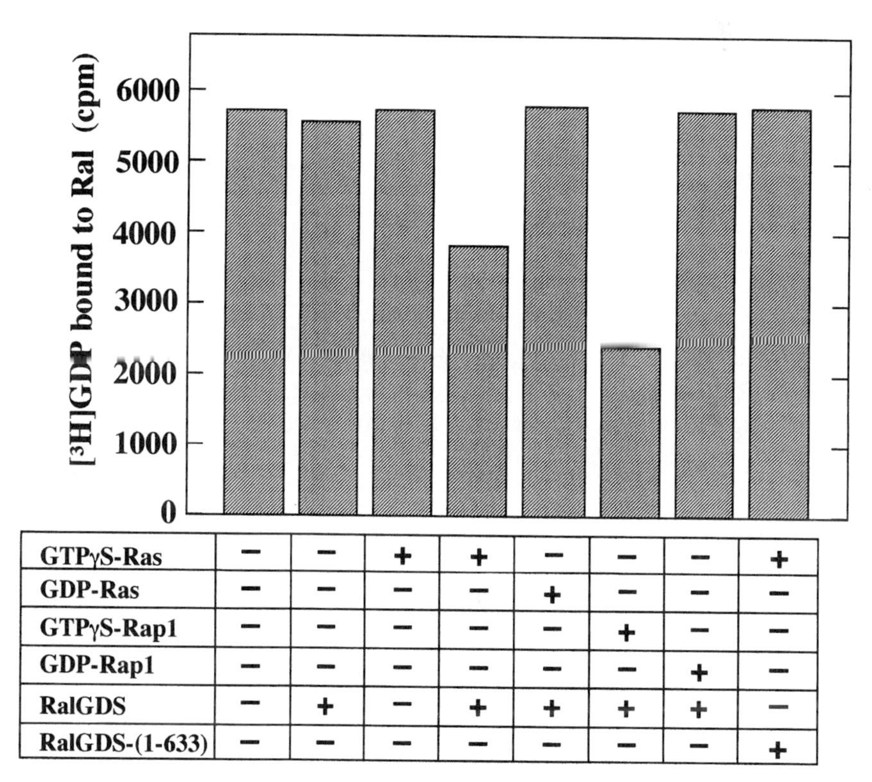

GTPγS-Ras	−	−	+	+	−	−	−	+
GDP-Ras	−	−	−	−	+	−	−	−
GTPγS-Rap1	−	−	−	−	−	+	−	−
GDP-Rap1	−	−	−	−	−	−	+	−
RalGDS	−	+	−	+	+	+	+	−
RalGDS-(1-633)	−	−	−	−	−	−	−	+

FIG. 4. Ras-dependent Ral activation through RalGDS in the liposome reconstitution system. The [^{3}H]GDP-bound form of Ral is incorporated alone or with the GTPγS-bound or GDP-bound form of Ras or Rap1 in the liposomes. The liposomes are incubated alone, with 94 n*M* RalGDS, or with 94 n*M* RalGDS-(1–633) for 30 min at 30°, and the [^{3}H]GDP remaining on Ral is measured.

nitrocellulose filters. Most of the liposomes are recovered at the 0.15–1.2 *M* sucrose interface and approximately 30–40% of the posttranslationally modified form of [^{3}H]GDP-bound Ral is recovered in the liposomes. The amount of Ral recovered in the liposomes is not affected by the presence of other small G proteins. The molar ratio of Ral to other small G proteins is 1:3–4. GST–RalGDS or GST–RalGDS-(1–633) (7.5 pmol each) is incubated for 30 min at 30° with the liposomes that contain the [^{3}H]GDP-bound form of Ral (0.5 pmol) and the GTPγS- or GDP-bound form of Ras or Rap1 (2 pmol) in 80 μl of buffer A containing 200 μM GTP. [^{3}H]GDP that remains on Ral is quantified by rapid filtration on nitrocellulose filters.[1]

RalGDS stimulates the dissociation of [^{3}H]GDP from Ral in the presence of the GTPγS-bound form of Ras or Rap1, but not in the presence of the GDP-bound form of Ras or Rap1 (Fig. 4). RalGDS-(1–633) that lacks RID does not stimulate the GDP/GTP exchange of Ral in the presence of the GTPγS-bound form of Ras, although RalGDS-(1–633) activates Ral in the solution assay system.[27] These reconstitution assays demonstrate that Ras and Rap1 have the ability to activate Ral through RalGDS and that the binding of RalGDS to Ras is necessary for transmitting the signal from Ras to Ral.

Comments

These results demonstrate that RalGDS interacts with the GTP-bound from of Ras and Rap1 in intact cells and *in vitro* and that Ras and Rap1 activate Ral in a liposome reconstitution assay. Ras but not Rap activates Ral through RalGDS in intact cells. The difference in effect of Rap on Ral activation between intact cells and *in vitro* would be due to the distinct subcellular localization of Rap1 and Ral. Once activated, Ras induces the translocation of RalGDS from the cytoplasm to the plasma membrane, where Ral is also present, thereby stimulating the GDP/GTP exchange of Ral.

[10] RAS Interaction with Effector Target RIN1

By YING WANG and JOHN COLICELLI

Introduction

Eukaryotic cells possess a staggering repertoire of signal transduction pathways that mediate responses to extracellular cues. This is accomplished at many levels, each contributing a factor of multiplicity to the possible outcomes. In the case of the mammalian small G protein RAS, the factors influencing cell response include structural divergence among different gene products (H-, K-, and N-RAS), regulated posttranslational modification, subcellular sequestration, multiple avenues for incoming activation signals, and diversity of attenuation regulators (discussed elsewhere in this volume). Critical determinants of signal output also include immediate downstream RAS interaction partners that function to accept the activation message and dispatch it appropriately. Through differences in their availability and biochemical properties, these "effector" proteins provide needed specificity.

RAS effectors are expected to share several biochemical properties, including the preferential binding of activated (GTP-bound) over inactivated (GDP-bound) RAS. They are also predicted to show similar, but not necessarily identical, binding configurations on the RAS "effector-binding domain." To a large extent, the RAF protein kinases have served as the paradigmatic effectors of RAS. RAF proteins have been characterized extensively and their role in propagating RAS signals has been studied in detail. Other signaling proteins that demonstrate the RAS-binding properties discussed above include phosphatidylinositol 3-kinase (PI3K), RALGDS (guanine nucleotide dissociation stimulator), AF6 (ALL-1 fusion partner from chromosome 6), and RIN1 (RAS interaction/intereference 1) (reviewed in Refs. 1–3).

The human RIN1 protein was originally isolated because of its ability to interfere with RAS signaling in a yeast model system.[4] Subsequent studies demonstrated that RIN1 binds directly to RAS through a GTP-dependent

[1] A. B. Vojtek and C. J. Der, *J. Biol. Chem.* **273,** 19925 (1998).

[2] A. Wittinghofer, *Biol. Chem.* **379,** 933 (1998).

[3] M. E. Katz and F. McComick, *Curr. Opin. Genet. Dev.* **7,** 75 (1997).

[4] J. Colicelli, C. Nicolette, C. Birchmeier, L. Rodgers, M. Riggs, and M. Wigler, *Proc. Natl. Acad. Sci. U.S.A.* **88,** 2913 (1991).

0076-6879/00 $35.00

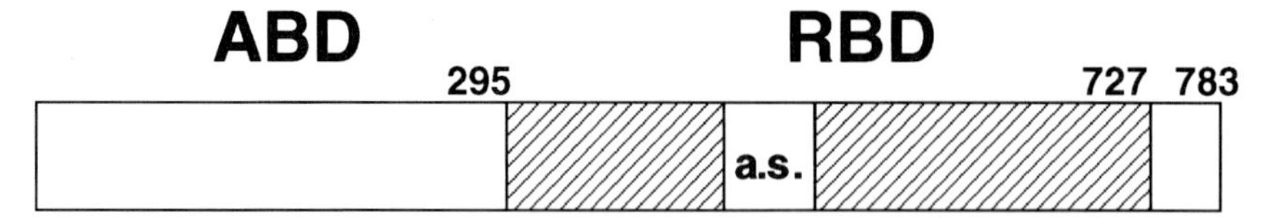

FIG. 1. Linear map of RIN1. The RIN1 protein is depicted with the ABL-binding domain (ABD) and RAS-binding domain (RBD) indicated. Within the RBD is a 62-amino acid sequence (a.s.) that is missing in protein (RIN1Δ) derived from an alternatively spliced message.

interaction that requires an intact effector-binding domain.[5] The RAS-binding domain (RBD) is located in the carboxy-terminal region of RIN1. This same region also binds to 14-3-3 proteins,[6] a property shared with other RAS effectors.[7–9] Interactions with 14-3-3 dimers are believed to modulate a variety of signaling pathways. Separate from the RBD, the amino terminus of RIN1 serves as both a binding partner and substrate for the tyrosine kinase c-ABL and its mutant transforming allele, BCR/ABL.[6,10] The RIN1–BCR/ABL interaction correlates with a demonstrated biological function: RIN1, or the ABL-binding domain (ABD) alone, enhances the cellular transformation and leukemogenesis properties of BCR/ABL.[10] The bivalent interaction properties of RIN1 (Fig. 1), and the role of RAS and ABL in tumorigenesis, are consistent with a signal regulatory function.

It is noteworthy that the tissue expression pattern of RIN1 is significantly more restricted than that of RAF proteins,[6,11,12] suggesting a cell type-specific function for the RIN1 protein. In addition, the RIN1 message is subject to alternative splicing that gives rise to multiple gene products in mammalian cells.[6,13] One form of the protein has an internal deletion, relative to the other (Fig. 1). This shorter variant has reduced affinity for RAS but still shows significant binding.[6] The larger, and most highly expressed, form of the protein has a predicted length of 783 amino acids

[5] L. Han and J. Colicelli, *Mol. Cell. Biol.* **15,** 1318 (1995).

[6] L. Han, D. Wong, A. Dhaka, D. Afar, M. White, W. Xie, H. Herschman, O. Witte, and J. Colicelli, *Proc. Natl. Acad. Sci. U.S.A.* **94,** 4953 (1997).

[7] W. J. Fantl, A. J. Muslin, A. Kikuchi, J. A. Martin, A. M. MacNicol, R. W. Gross, and L. T. Williams, *Nature* (*London*) **37,** 612 (1994).

[8] E. Freed, M. Symons, S. G. Macdonald, F. McCormick, and R. Ruggieri, *Science* **265,** 1713 (1994).

[9] N. Bonnefoy-Bérard, Y. C. Liu, M. von Willebrand, A. Sung, C. Elly, T. Mustelin, H. Yoshida, K. Ishizaka, and A. Altman, *Proc. Natl. Acad. Sci. U.S.A.* **92,** 10142 (1995).

[10] D. Afar, L. Han, J. McLaughlin, S. Wong, A. Dhaka, K. Parmr, N. Rosenberg, O. Witte, and J. Colicelli, *Immunity* **6,** 773–782 (1997).

[11] S. M. Storm, J. L. Cleveland, and U. R. Rapp, *Oncogene* **5,** 345 (1990).

[12] J. E. Lee, T. W. Beck, L. Wojnowski, and U. R. Rapp, *Oncogene* **12,** 1669 (1996).

[13] A. Patel and J. Colicelli, unpublished data (1999).

and a molecular mass of 84 kDa. This species typically migrates in the 90- to 95-kDa range in sodium dodecyl sulfate–polyacrylamide gel electrophoresis (SDS–PAGE). Part of the difference between predicted and observed molecular weight may arise from phosphorylation of RIN1.[6,14,15] Here we focus on the RAS-binding properties of RIN1 and describe in detail several methodologies useful for the analysis of this interaction. Engineered constructs of the RBD portion of RIN1 (residues 294–727), defined from deletion mutation analysis, are used in some experiments. The RBD of RIN1 appears to be significantly larger than the minimum-length binding domains of other RAS effectors. In addition, there is little commonality among RAS effectors at the level of amino acid sequence, although structural studies suggest that some conformational features are shared.[16] Further analysis, including structural studies, will be required to further localize the RAS-binding determinants within the RBD fragment of RIN1.

Methods

Assay of RIN1–RAS Interaction by Two-Hybrid Analysis

The two-hybrid technique[17] provided the first indirect evidence of a physical interaction between human RIN1 and human H-RAS. This method uses the yeast *Saccharomyces cerevisiae* to detect protein–protein interactions through the juxtaposition of DNA-binding and transcription activation domains fused to the proteins being tested. We have published several experiments demonstrating interactions between RIN1 or the RBD and wild-type or mutant forms of human or yeast RAS.[5,6] Both configurations of hybrids have proved useful (RIN1 fused to either a DNA-binding domain or an activation domain used with a complementary fusion of RAS).

The procedures we describe here use both RIN1 and a naturally expressed alternate splice variant, RIN1Δ. In each case, the protein is fused to the GAL4 transcription activation domain. The H-RAS proteins are fused to the LexA DNA-binding protein. The yeast strain used in these assays, L40, carries both selectable (*HIS3*) and screenable (*lacZ*) reporter genes.

Plasmids. The LexA DNA-binding domain fusions of H-RAS, RAS^{V12}, RAS^{A15}, and RAS^{L35R37} constructs have been generously provided by A. Vojtek (University of Michigan, Ann Arbor, MI) and J. Cooper (NCI-Frederick Cancer Research and Development Center, Frederick, MD).

[14] Y. M. Lim and J. Colicelli, unpublished data (1999).

[15] Y. Wang and J. Colicelli, unpublished data (1999).

[16] L. Huang, F. Hofer, G. S. Martin, and S. H. Kim, *Nature Struct. Biol.* **5,** 422 (1998).

[17] S. Fields and O. Song, *Nature* (*London*) **340,** 246 (1989).

They employ the plasmid pBTM116, a 2μ-based yeast expression vector that carries the *TRP1* selectable marker and uses a portion of the *ADH1* promoter to express LexA alone or as part of a fusion protein.

The RIN1 sequence is ligated into the *Sal*I and *Not*I sites of the GAL4 activation domain vector pGAD425, a modified form of pGAD424,[18] using a strategy previously described.[6] The splice variant of RIN1, referred to as RIN1Δ because of a 62-codon deletion,[6] is ligated into pGAD425 by an analogous procedure.

Procedures. Yeast strain L40[19] is used to perform interaction studies. The genotype of this strain is *MAT***a** *his3Δ200 trp1-901 leu2-3,112 ade2 LYS2::lexAop-HIS3 URA3::lexAop-lacZ*. The expression of *HIS3* and *lacZ* coding sequences is driven, respectively, by minimal *HIS3* and *GAL1* promoters fused to multimerized LexA-binding sites. The expression of *HIS3* enables this strain to grow in the absence of histidine, while the expression of β-galactosidase activity from the *lacZ* gene can be detected by color assay using 5-bromo-4-chloro-3-indolyl-β-D-galactosidase (X-Gal). To determine if RIN1 can interact with human H-RAS, L40 is first transformed with pBTM-RAS and pGAD-RIN1 constructs and the cells are then plated on synthetic medium without tryptophan (pBTM116 marker) and leucine (pGAD425 marker) and incubated at 30° for 2–3 days.

To assay histidine prototrophy, individual transformants are picked and streaked to plates lacking tryptophan, leucine, and histidine. Incubation at 30° for 2 days is sufficient to judge reporter gene activation. To assay the activation of the *lacZ* reporter, a master plate of yeast patches derived from selected colonies is replica plated onto a nylon filter. The cells are permeabilized by immersing the filter in liquid nitrogen. The filter is placed in *lacZ* assay buffer (Z buffer) containing X-Gal (0.5 mg/ml) and incubated at 37°. The expression of β-galactosidase, indicated by the development of blue color, is apparent within 1 hr. Using two-hybrid assays, we have demonstrated that RIN1, expressed as a LexA fusion protein, is clearly capable of interacting with H-RAS (Fig. 2). No signals (*HIS3* or *lacZ*) are detected when either partner is replaced by vector alone (data not shown). The activated mutant H-RAS^{V12} also shows strong interaction with RIN1. However, two signaling-defective forms of RAS show no detectable interaction. Loss of binding by the dominant negative mutant, RAS^{A15}, demonstrates the preference of RIN1 for the active conformation of RAS. Simi-

[18] P. L. Bartell, C. T. Chien, R. Sternglanz, and S. Fields, *in* "Cellular Interactions in Development: A Practical Approach" (D. A. Hartley, ed.). Oxford University Press, Oxford, 1993, p. 153.

[19] S. M. Hollenberg, R. Sternglanz, P. F. Chang, and H. Weintraub, *Mol. Cell. Biol.* **15,** 3813 (1995).

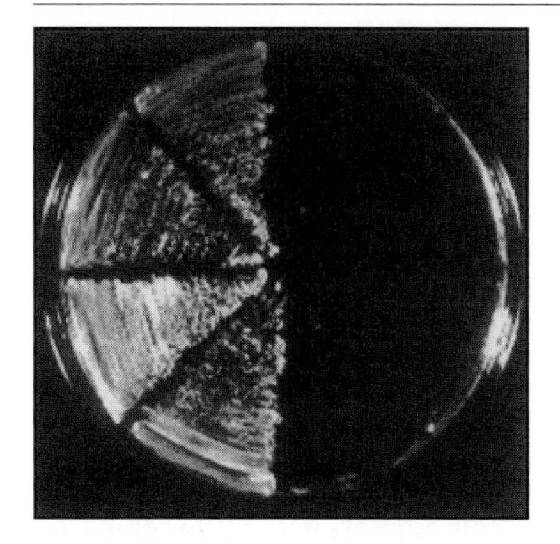

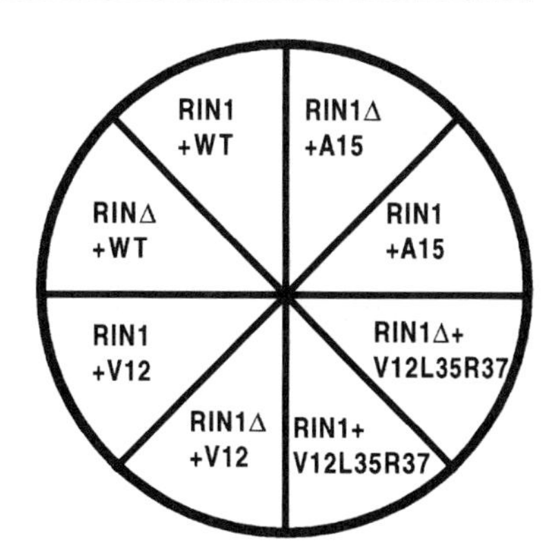

FIG. 2. Interaction of RIN1 with RAS detected by two-hybrid analysis. Yeast cells transformed with plasmids expressing RIN1 or RIN1Δ fused to an activation domain and wild-type or mutant forms of human H-RAS were streaked on synthetic medium without leucine, tryptophan, and histidine.

larly, the lack of signal from the compound effector mutant, RAS^{L35R37}, indicates that, as with other RAS effectors, residues in this domain are critical requirements for binding. An examination of two-hybrid binding to well-characterized RAS effector domain single-position mutants[6] indicates that some effector mutants of RAS still show RIN1 binding, but the mutant binding "signature" is distinct from that of RAF1.

Data from *lacZ* reporter assays of L40 cells expressing RIN1 and RAS two-hybrid constructs have consistently reflected results from the histidine auxotrophic assays.[5] Filter assays using X-Gal, as described above, appear to be more reflective of interaction intensity, however. In addition, assays of *lacZ* activity using liquid medium-cultured cells and the substrate *o*-nirophenyl galactoside (ONPG) can be carried out by a modification[20] of established procedures,[21] and can be performed in a semiquantitative fashion.[22] Interaction strength and characterization studies using purified proteins (discussed below) both confirm and extend findings from the two-hybrid system.

Using similar fusion protein techniques, binding of RIN1 to 14-3-3 proteins has also been demonstrated[6] and deletion analysis indicates that RAS and 14-3-3 binding determinants are colocalized within the RBD (Fig.

[20] B. H. Spain, D. Koo, M. Ramakrishnan, B. Dzudzor, and J. Colicelli, *J. Biol. Chem.* **270,** 25435 (1995).

[21] L. Guarente, *Methods Enzymol.* **101,** 181 (1983).

[22] J. Estojak, R. Brent, and E. A. Golemis, *Mol. Cell. Biol.* **15,** 5820 (1995).

1). The role of 14-3-3 binding in the affinity of RIN1 for RAS is currently under investigation.

Analysis of RAS–RIN1 Interaction by in Vitro Binding Assays

A particularly useful methodology for studying the biochemical interaction of RAS with its effectors is *in vitro* binding. In principle, any coprecipitation method that does not interfere with the protein–protein interaction should be applicable, and this approach has been useful for analysis of other RAS–effector binding studies.[23,24]

We describe several methods for detection of direct binding between RIN1 and RAS. In the methods described below, coprecipitation of RIN1 and RAS is achieved by tagging one of the components with either glutathione *S*-transferase (GST) or a hexahistidine (His_6) sequence. The tagged proteins are precipitated with glutathione–Sepharose or Talon beads, respectively. The RIN1–RAS binding determined by these methods relies on the ability of the RAS-binding domain of RIN1 to form a stable complex with RAS protein. With these assays, we have shown that RIN1 binds tightly to activated RAS in a manner that is resistant to dilution and repeated washes in buffer.

Buffers

Guanine nucleotide loading buffer: 20 m*M* Tris–HCl (pH 7.5), 6 m*M* MgCl, 10 m*M* EDTA, 1 m*M* dithiothreitol (DTT), and 10% (v/v) glycerol

Binding buffer 1: 20 m*M* Tris–HCl (pH 7.4), 25 m*M* NaCl, 6 m*M* $MgCl_2$, 1 m*M* EDTA, 1 m*M* DTT, 0.01% (v/v) Nonidet P-40, 10% (v/v) glycerol, and 1% (w/v) dry milk

Binding buffer 2: 20 m*M* Tris–HCl (pH 7.5), 80 m*M* NaCl, 6 m*M* $MgCl_2$, 1 m*M* EDTA, 1 m*M* DTT, 0.01% (v/v) Nonidet P-40, 10% (v/v) glycerol, and 1% (w/v) dry milk

Washing buffer: 20 m*M* Tris–HCl (pH 7.4), 80 m*M* NaCl, 6 m*M* $MgCl_2$, 1 m*M* EDTA, 1 m*M* DTT, 0.5% (v/v) Nonidet P-40, and 10% (v/v) glycerol

Lysis/IP buffer: 50 m*M* HEPES (pH 7.5), 150 m*M* NaCl, 6 m*M* $MgCl_2$, 100 m*M* NaF, 1 m*M* EDTA, 1% (v/v) Nonidet P-40, and 10% (v/v) glycerol

[23] X. F. Zhang, J. Settleman, J. M. Kyriakis, E. Takeuchi-Suzuki, S. J. Elledge, M. S. Marshall, J. T. Bruder, U. R. Rapp, and J. Avruch, *Nature* (*London*) **364,** 308 (1993).

[24] P. Rodriguez-Viciana, P. H. Warne, R. Dhand, B. Vanhaesebroeck, I. Gout, M. J. Fry, M. D. Waterfield, and J. Downward, *Nature* (*London*) **370,** 527 (1994).

Z buffer: 60 mM $Na_2HPO_4 \cdot 7H_2O$, 35 mM $NaH_2PO_4 \cdot H_2O$, 10 mM KCl, 1 mM $MgSO_4 \cdot 7H_2O$, 38 mM 2-mercaptoethanol (adjust to final pH 7)

1. Determination of RIN1–RAS Binding Using in Vitro-Translated RIN1 Protein

The human GST–H-RAS and maltose-binding protein (MBP)–RAF1 bacterial expression constructs have been generously provided by A. Vojtek and J. Cooper.[25,26] GST–H-RAS fusion protein is purified from BL21 cells. Cell pellets from 400-ml cultures induced with 0.3 mM isopropyl-β-D-thiogalactopyranoside (IPTG) for 3 hr are washed with phosphate-buffered saline (PBS) and disrupted by sonication in 10 ml of PBS plus 1% (v/v) Triton X-100. During and after lysis, all procedures are carried out at 4° or on ice and all buffers contain a standard protease inhibitor cocktail (0.1 mM phenylmethylsulfonyl fluoride [PMSF], leupeptin [2 μg/ml], pepstatin [1 μg/ml]). Cleared lysates are incubated with 400 μl of glutathione beads in a 15-ml tube and rotated at 4° for 1 hr. The beads are washed two times with PBS and two times with 20 mM Tris–HCl (pH 7.5) and 10% (v/v) glycerol. Alternate guanine nucleotide-bound forms of GST–RAS proteins are prepared by incubation with 1.2 mM GDP or guanosine 5′-(3-O-thio)triphosphate (GTPγS) in 400 μl of nucleotide loading buffer for 30 min at 30°. Guanosine-5′-(β,γ-imido)triphosphate (GMPPNP) can be used in place of GTPγS with comparable results (each compound works best when used directly after dissolving). The nucleotide exchange is stopped by addition of $MgCl_2$ to a final concentration of 20 mM and the reaction is chilled at 4°. For RIN1–RAS binding, a fragment of the RIN1 protein including the RBD is expressed from an *in vitro* transcription vector and RIN1 RNA is synthesized and translated with a coupled transcription/translation system (Promega, Madison, WI) in the presence of [^{35}S]methionine. To determine RIN1–RAS binding, 30 μl of a 50% slurry of glutathione beads, precoated with GST–RAS fusion protein and loaded with the appropriate guanine nucleotide, is added to 500 μl of binding buffer 1 containing a 0.6 mM concentration of the appropriate guanine nucleotide and 6 μl of [^{35}S]RIN1 protein. A reaction using beads precoated with GST protein is carried out in parallel as a negative control. The mixture is incubated at 4° on a rotator for 1 hr. The beads are then washed four times with washing buffer. The bound protein is eluted from the beads by boiling with protein sample buffer and then separated by SDS–PAGE. The output of RIN1 binding

[25] A. B. Vojtek, S. M. Hollenberg, and J. A. Cooper, *Cell* **74,** 205 (1993).

[26] J. R. Fabian, A. B. Vojtek, J. A. Cooper, and D. K. Morrison, *Proc. Natl. Acad. Sci. U.S.A.* **91,** 5982 (1994).

is determined by autoradiography and normalization of RAS is assessed by Western blot with RAS-specific monoclonal antibody (Transduction Laboratories, Lexington, KY). Results obtained by this method show that RIN1 binding is dependent on RAS activation and can be blocked by mutations in the RAS effector domain.[5] The same procedure is used to illustrate that RIN1 binding is competitive with RAF1 binding. Increasing amounts of the Ras-binding domain of RAF1 added into the binding reaction lead to a reduction in the amount of RIN1 associated with GST–RAS (Fig. 3). A mutant, binding-incompetent form of RAF1 shows no competitive effect.

2. Determination of RIN–RAS Interaction Using Purified RIN1 and Ras-Binding Domain Proteins

The purification of RAS effector proteins from recombinant sources has proved to be invaluable for studying their biochemical properties and biological functions. Attempts to express RIN1 in bacterial strains, using a variety of vectors, have consistently resulted in poor yields, however. We therefore turned to the baculovirus–Sf21 insect cell system and have demonstrated purification of both full–length RIN1 and RBD. It should be noted that proteins expressed in this eukaryotic system typically maintain their biological activity and immunological reactivity. We purified RIN1 (and RBD) proteins from the baculovirus–Sf21 cell system using the pBAC-PAK-1 His_6-tagged vector (Clontech, Palo Alto, CA) with modifications in the polylinker region. Translation starts from the native initiation codon of RIN1 (within a unique *Nco*I site) and includes six histidine residues at the carboxy terminus. This configuration is specifically chosen on the basis of observations suggesting that alteration of the amino terminus of RIN1 could perturb its function (our unpublished data, 1999). The RBD fragment is cloned in a similar fashion using an engineered *Nco*I site in frame with the vector-provided ATG. As described in the following sections, the yield of RIN1 from this system is approximately 5 mg/liter of culture, with the RBD construct typically producing severalfold more protein. Both proteins expressed in insect cells have proved useful for a variety of biochemical and functional studies.

Generation and Isolation of Recombinant Baculovirus. The full-length human RIN1 cDNA, originally cloned from a U118-MG cell library,[5] is modified by polymerase chain reaction (PCR) to incorporate a *Bgl*II site at the carboxy terminus. This permits the cloning of an *Nco*I-to-*Bgl*II fragment encompassing RIN1 into the QE60 (Qiagen, Valencia, CA) vector and results in a carboxy-terminal extension of six histidine residues. To construct the RBD of RIN1, oligonucleotide-directed mutagenesis is used

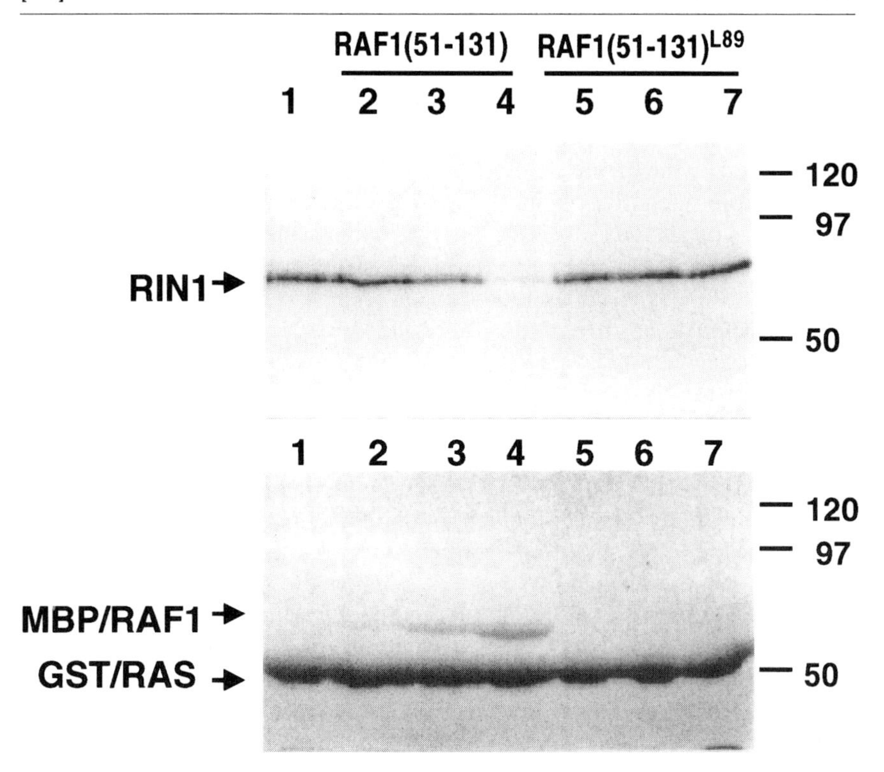

FIG. 3. Binding of *in vitro*-translated RIN1 to RAS: Competition with RAF1. A fragment of ^{35}S-labeled RIN1 (amino acids 294–783) bound to GST–RAS(GTPγS) was analyzed by SDS–PAGE and autoradiography (*top*) and Coomassie blue staining (*bottom*, same gel). Lane 1, RIN1 binding to RAS; lanes 2–4, same reaction in the presence of increasing amounts of wild-type RAF1 (amino acids 51–131 fused to MBP); lanes 5–7, same reaction in the presence of increasing (and equal to corresponding lanes 2–4) amounts of RAF1^{L89} (amino acids 51–131). Molecular mass markers are shown at the right (in kDa). [Derived from work presented in L. Han and J. Colicelli, *Mol. Cell. Biol.* **15,** 1318 (1995).]

to introduce a unique *Bgl*II site at the position of RIN1 corresponding to the carboxy end of RBD (56 residues prior to the natural carboxy terminus[6]). The *Bgl*II site is fused, in frame, to the unique *Bgl*II site of QE60, again resulting in the addition of six histidine residues. RIN1(His_6) can be transferred to a modified form of the baculovirus expression vector pBacPAK-His1 (Clontech) as two fragments (*Nco*I–*Kpn*I and *Kpn*I–*Hin*dIII) to yield pBacRIN1, in which expression is controlled by the polyhedrin promotor. The RBD(His_6) is moved as a *Kpn*I (start of RBD)-to-*Hin*dIII (downstream of stop codon in QE60) fragment into the modified vector to yield pBacRBD.

To generate recombinant baculovirus, 100 ng of pBacRIN1 or pBacRBD is cotransfected with 5 μl of gapped and linearized pBacPAK6 viral DNA (Clontech) into Sf21 (*Spodoptera frugiperda,* fall armyworm ovary) cells. Cotransfection is carried out by lipofection, using the synthetic lipid bacfectin provided by the manufacturer (Clontech). Recombinant virus is isolated from occlusion-negative plaques after two rounds of amplification. To verify the expression of RIN1 or RBD, several small-scale cultures are infected with independent isolates of recombinant virus and analyzed by Western blot, using RIN1-specific antibodies.

Expression and Purification of RIN1 and RAS-Binding Domain Proteins. For large-scale cultures, Sf21 cells seeded at a density of 10^6/ml are infected with 10^7 plaque-forming units (PFU) of recombinant virus. Suspension cultures are grown in Grace medium supplemented with 10% (v/v) fetal bovine serum (HyClone, Logan, UT) for 48 hr at 27° prior to harvesting. The expressed RIN1(His_6) and RBD(His_6) polypeptides are purified with Talon metal affinity resin (Clontech) under nondenaturing conditions. Purified full-length RIN1 sometimes contains a fraction of breakdown products, but this can be minimized by strict maintenance of low-temperature conditions (4°) and protease inhibitors. Aliquots of purified proteins should be snap frozen and stored at −80° for single use. Purified RBD appears far less prone to degradation.

The biological activity and immunological reactivity of purified RIN1 and RBD proteins are examined by *in vitro* binding assay and Western blot.

Binding Assay of RIN1(His_6) and Glutathione S-Transferase–RAS. To test binding to immobilized RIN1 or RBD, the GST–RAS fusion proteins are purified from glutathione–Sepharose beads by a modification of a published protocol.[25] Proteins are eluted with 10 mM reduced glutathione in 50 mM Tris–HCl buffer (pH 8.0). The eluted proteins are dialyzed against 2 liters of guanine nucleotide loading buffer (with at least one buffer change) at 4° overnight. The concentrations of GST and GST–RAS fusion proteins are determined by the Bradford method (Bio-Rad, Hercules, CA) and the purities are examined by SDS–PAGE. To load the guanine nucleotides, the purified GST–RAS fusion proteins are incubated with a 1.2 mM concentration of either GTPγS or GDP at 30° for 30 min and the nucleotide-exchange reaction is stopped by adding $MgCl_2$ to a final concentration of 20 mM. The purified proteins are snap frozen and stored in 40-μl aliquots at −80°.

To study the association of RIN1 with RAS, we use Talon metal affinity resin for the specific precipitation of His_6-tagged RIN1(His_6) or RBD(His_6). In one format, GST or GST–RAS fusion proteins are mixed with resin-bound baculoviral histidine-tagged RBD or RIN1 in 10 bead volumes of binding buffer 2 with a 0.6 mM concentration of the appropriate guanine nucleotide. The incubation is carried out at 4° for 1 hr and the beads are washed four times with 1 ml of cold washing buffer. The bound proteins are released from the beads by boiling for 5 min in SDS-containing protein sample buffer and analyzed by Western blot (Fig. 4) with RAS monoclonal antibody (Transduction Laboratories). Alternatively, GST–RAS is first bound to glutathione beads and loaded with guanine nucleotides as described previously. The binding of GST–RAS to RIN1 polypeptides is determined by elution of GST–RAS with reduced glutathione and the result is analyzed by Western blot with RIN1 antibody.

3. Use of RBD(His$_6$) to Pull Down RAS from Mammalian Cell Extracts

To determine further the specificity of RIN1–RAS binding and to examine the biological activity of RIN1 protein expressed from insect cells, we have employed a modified form of the procedure used to examine binding of ^{35}S-labeled RIN1 to purified RAS from bacterial cells (described above). To determine if the high-affinity binding observed in this *in vitro* assay truly reflects the specificity of RIN1–RAS interaction *in vivo*, we have made use of RBD(His$_6$) to pull down RAS from mammalian cells. RBD(His$_6$) is purified as described above. The clarified insect cell lysates containing RBD are aliquoted and stored at −80°. Prior to use, aliquots are thawed and RBD protein is further purified by the Talon affinity purification procedure

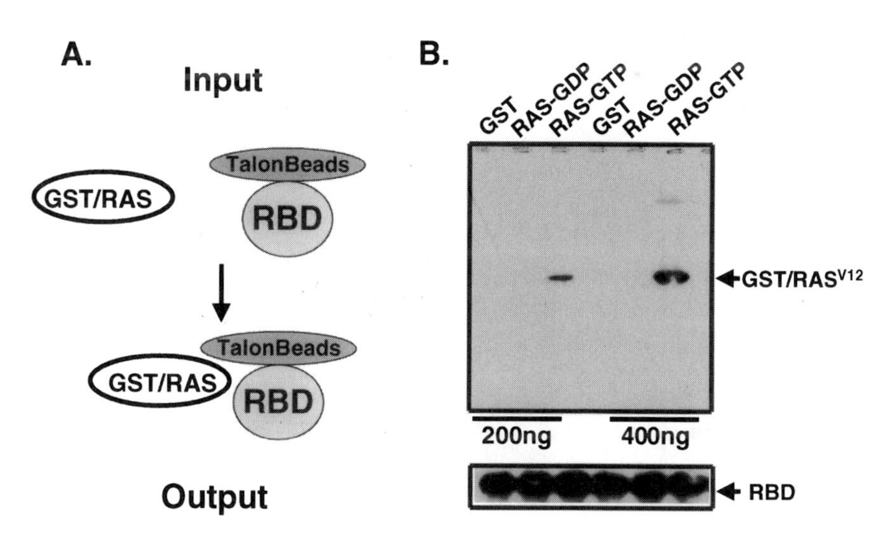

FIG. 4. Binding of baculovirus-produced RBD (RIN1 fragment) with RAS. (A) Experimental scheme shows resin-immobilized RBD(His$_6$) interacting with GST–RAS^{V12}(GTPγS) from a bacterial expression system. (B) Immunoblot of bound proteins stripped from beads. The concentration of RAS protein added to each binding reaction is indicated.

(Clontech). To express RAS proteins, 293T cells are transfected (calcium phosphate method) with pSRα-based retroviral constructs of H-RAS^{V12} and three effector binding domain mutants (generously provided by M. White, University of Texas Southwest Medical Center, Dallas, TX) that also harbor the V12 mutation. The transfected 293T cells are briefly washed with PBS and disrupted with cell lysis buffer. The cleared lysates are incubated with RBD–resin in lysis/IP buffer adjusted to 10 mM $MgCl_2$ at 4° for 1 hr. Resins are then washed four times with the same buffer containing 5 mM $MgCl_2$ and the amount of bound RAS protein is determined by Western blot with RAS monoclonal antibody (Transduction Laboratories). Figure 5 shows that RBD(His_6) specifically and preferentially interacts with RAS^{V12} and RAS^{V12G37}. In contrast, the other effector domain-binding mutants, RAS^{V12S35} and RAS^{V12C40}, have either partially or completely lost binding affinity for RBD. This result is consistent with our previous observation from two-hybrid assays[6] and demonstrates that RIN1 proteins expressed from insect cells retain binding specificity. In addition, the use of RAS effector domain mutants should prove instrumental for examining the role of RIN1 in distinct RAS-controlled signal transduction pathways.

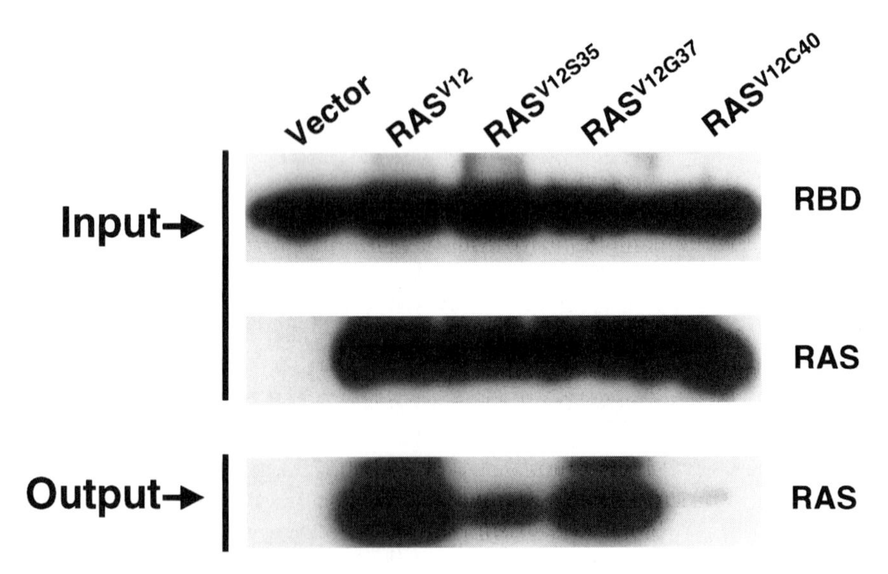

Fig. 5. Efficient pull-down of RAS from mammalian cells, using RBD. Similar experimental scheme as shown in Fig. 4A, except that the source of RAS is extracts from transfected 293T cells. The amount of RBD on the beads and the amount of RAS in the extract are indicated in the top two immunoblot panels (Input). The relative amounts of mutant RAS proteins are indicated in the bottom immunoblot panel (Output).

Coimmunoprecipitation of RIN1 and RAS from Mammalian Cells

In a previous study, we employed coimmunoprecipitation to detect RIN1–RAS complexes in mammalian cells that express both constitutively active RAS and RIN1 proteins.[6] We showed that RIN1 specifically binds to activated mammalian RAS in NIH 3T3 cells. Notably, the *in vivo* association of RIN1 with RAS was not detectable when using a RAS antibody that is directed at a portion of the effector domain. This result, similar to findings for RAF,[27] is consistent with a free effector domain requirement for RIN1 binding.

Conclusions

Both genetic and biochemical approaches can be used to analyze the physical interaction of RIN1 and RAS. Results from various methodologies indicate that RIN1 binds to the effector domain and preferentially recognizes the activated conformation of RAS. The effector mutant binding profile of RIN1, together with its unique tissue expression pattern, distinguishes it from other known RAS effectors. Although no catalytic activity has been ascribed to RIN1, the presence of both ABL- and RAS-binding domains suggests a role in coordinating the transmission of signals through these pathways. The ability to isolate purified RIN1 should facilitate the further characterization of specific binding properties. Of particular interest will be the examination of how interaction affinities are regulated through posttranslational modifications that may also be cell type specific.

[27] F. Finney and D. Herrera, *Methods Enzymol.* **255,** 310 (1995).

[11] Ras and Rap1 Interaction with AF-6 Effector Target

By BENJAMIN BOETTNER, CHRISTIAN HERRMANN, and LINDA VAN AELST

Introduction

The Rap types of small GTPases are members of the Ras superfamily and are the molecules that show the most identity with the oncogenic Ras proteins. Whereas the interaction of activated Ras proteins with their downstream effectors Raf, Ral guanine nucleotide dissociation stimulator (RalGDS), and phosphatidylinositol 3-kinase (PI3K) led to a fairly defined

0076-6879/00 $35.00

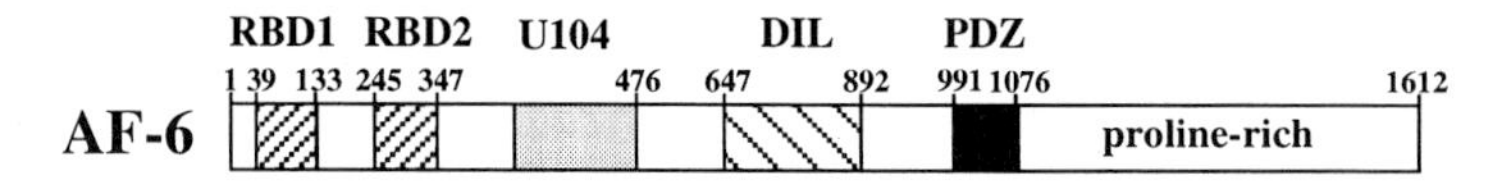

FIG. 1. Schematic representation of AF-6.

picture, the role of Rap1 is as yet poorly understood.[1] Besides its ability to bind to Raf and RalGDS, two-hybrid and *in vitro* experiments suggest that still another molecule, namely AF-6 ALL1-fused gene on chromosome 6, might serve as a relevant target. Both two-hybrid and kinetic studies suggest a strength of interaction that exceeds the one exerted by Ras–AF-6 complexes, which originally led to the identification of AF-6.[2,3] AF-6 has also been described as a fusion partner for ALL1 in acute lymphoblastic leukemias.[4] The AF-6 protein contains a combination of interesting homology regions.[5] At its very NH_2 terminus reside two putative Ras/Rap 1-binding motifs, followed by U104 and DIL motifs, domains that are found in the head portions of microtubule- and actin-based motor proteins, respectively. Located further to the COOH terminus is a PDZ domain succeeded by proline-rich clusters, which may function as docking sites for other molecules (Fig. 1).

In this chapter, we outline the methods that allowed us to investigate the physical interaction between Ras/Rap1 and AF-6, namely two-hybrid analyses, kinetic and thermodynamic studies, and *in vivo* studies utilizing retrovirally engineered cell lines.

AF-6 resides in cell–cell adhesion complexes and could provide the molecular link between the activity of Ras/Rap1 proteins and their effects on intercellular adhesion.

Use of Yeast Two-Hybrid System to Evaluate Ras/Rap1 Interaction with AF-6

Principle

One approach we took to examine the interaction between Ras/Rap1 and AF-6 consists of a two-hybrid interaction trap assay. This system is a

[1] J. L. Bos, *EMBO J.* **17,** 6776 (1998).

[2] L. Van Aelst, M. A. White, and M. H. Wigler, *Cold Spring Harbor Symp. Quant. Biol.* **59,** 181 (1994).

[3] M. Kuriyama, N. Harada, S. Kuroda, T. Yamamoto, M. Nakafuku, A. Iwamatsu, D. Yamamoto, R. Prasad, C. Croce, E. Canaani, and K. Kaibuchi, *J. Biol. Chem.* **12,** 607 (1996).

[4] R. Prasad, Y. Gu, H. Alder, T. Nakamura, O. Canaani, H. Saito, K. Huebner, R. P. Gale, P. C. Nowell, K. Kuriyama, Y. Miyazaki, C. M. Groce, and E. Canaani, *Cancer Res.* **53,** 5624 (1993).

[5] C. P. Ponting and D. R. Benjamin, *Trends Biochem. Sci.* **21,** 422 (1996).

genetic method that allows the determination of physical complexes between two proteins within yeast cells.[6] These proteins are expressed as hybrid proteins, one fused to a DNA-binding domain and the other fused to a transcription-activating domain. If the two proteins associate, a functional transcription factor is reconstituted and a reporter gene is transcribed. Several versions of the two-hybrid system exist; they commonly involve DNA-binding domains derived from Gal4 (GBD) or LexA (LBD) and activation domains from Gal4 (GAD) or VP16 transcriptional activators.[6,7] Numerous yeast strains with different reporter genes for both systems have been constructed, as well as variations of the original two-hybrid system, including a reverse one- and two-hybrid system, and a three-hybrid system. The more recently constructed yeast strains, with multiple reporter genes harboring independent promoters, offer a major advantage because they increase specificity, thus limiting false positives. A detailed description can be found in Vidal and Legrain[8] and in Brent and Finley.[9] The choice of using either the Gal4- or LexA-based system will be dependent on the nature of the protein of interest.[10] In our studies directed to assess the interaction between Ras/Rap1 and AF-6, we found that both Gal4- and LexA-based systems can be used. The use of the LexA-based system is presented here. The application of the two-hybrid system allowed us to assess whether the binding of AF-6 requires Ras/Rap1 to be in a GTP-bound state, to map the minimum domain of AF-6 required for interaction, and to compare binding profiles between Ras, Rap1, AF-6, and other Ras/Rap1 targets.

Materials

Yeast Strain. The yeast strain L40 (*Mat***a** *his3Δ200 trp1-901 leu2-3,112 ade2*) containing *HIS3* and *lacZ* as reporter genes is used for the LexA-based system.[7]

Media. YPD medium contains 10 g of yeast extract, 20 g of Bacto-Peptone (Difco, Detroit, MI), 2% (w/v) glucose, and 20 g of agar for plates, per liter. DO-Leu-Trp and DO-Leu-Trp-His media contain Bacto-Yeast nitrogen base without amino acids (0.67%, w/v), dropout mix (0.2%, w/v), and Bacto-Agar (2%, w/v). Dropout mix is a combination of all essential amino acids minus the appropriate supplement.[10]

[6] C. T. Chien, P. L. Bartel, R. Sternglanz, and S. Fields, *Proc. Natl. Acad. Sci. U.S.A.* **88,** 9578 (1991).

[7] A. B. Vojtek, S. M. Hollenberg, and J. A. Cooper, *Cell* **74,** 205 (1993).

[8] M. Vidal and P. Legrain, *Nucleic Acids Res.* **27,** 919 (1999).

[9] R. Brent and R. L. Finley, Jr., *Annu. Rev. Genet.* **31,** 663 (1997).

[10] L. Van Aelst, *Methods Mol. Biol.* **84,** 201 (1998).

Solutions

Lithium acetate/TE (0.1 *M*, pH 7.5): Combine 0.1 *M* lithium acetate (Sigma, St. Louis, MO), 10 m*M* Tris–HCl (pH 7.5), and 1 m*M* EDTA

Polyethylene glycol (PEG) 3300, 40% (w/v) in 0.1 *M* lithium acetate

Z buffer: Combine $Na_2HPO_4 \cdot 7H_2O$ (16.1 g/liter), $NaH_2PO_4 \cdot H_2O$ (5.5 g/liter), KCl (0.75 g/liter), $MgSO_4 \cdot 7H_2O$ (0.246 g/liter); adjust to pH 7.0 and autoclave.

X-Gal stock solution: Dissolve 20 mg of 5-bromo-4-chloro-3-indolyl-β-D-galactopyranoside (X-Gal) (Boehringer Mannheim, Indianapolis, IN) in 1 ml of *N,N*-dimethylformamide. Store at $-20°$ in the dark.

Z buffer/X-Gal solution: To 100 ml of Z buffer add 270 μl of 2-mercaptoethanol and 1 ml of X-Gal stock solution

Plasmids and Constructs. For the construction of pGAD AF-6N, pGAD AF-6-RBD1, and pGAD AF-6-RBD2: AF-6N (amino acids 1–368) is polymerase chain reaction (PCR) amplified with oligonucleotides AF6N-5′ *Sal*I (5′-GGGACGTCGACTCTCGGCGGGCGGCCGTGACGAG-3′) and AF6N-3′*Xho*I (5′-CGGCAGCTCGAGCTATCTCTCCTTTCCCTTGG-GTGT-3′) and inserted into *Xho*I-digested pGAD1318 plasmid, which is a derivative of pGADGH.[10] AF-6-RBD1 (amino acids 1–140) is amplified with AF6N-5′ *Sal*I and AF6RBD1-3′ *Xho*I (5′-CCGCCGCTCGAGCTAC-TTAGGAGGAATGGC-3′) as 5′ and 3′ oligonucleotides, respectively, and treated as described above. AF-6-RBD2 (amino acids 181–368) is amplified with AF6RBD2-5′BHI (5′-GGGCCGGATCCGCCATTCCTC-CTAAG-3′) and AF6N-3′*Xho*I oligonucleotides and inserted into *Bam*HI–*Xho*I-digested pGAD1318 vector. For the construction of pGAD PI3Kδ-RBD, PI3Kδ-RBD is PCR amplified using 5′ oligonucleotide PI3KδRBD5′ *Bam*HI (5′-CGGCGCGGATCCATGGCCAAGATGTGCCAATTCT-GC-3′) and 3′ oligonucleotide PI3KδRBD3′ *Sal*I (5′-GCCGACGTCGAC-CTAGTTGCTCTGCTCATCCCG-3′), and digested with *Bam*HI and *Sal*I restriction endonucleases prior to ligation into *Bam*HI–*Xho*I-cut pGAD1318. pGAD1318 RalGDS-RBD is obtained in a yeast two-hybrid screen, using LBD Ha-RasV12 as bait and a Jurkat cDNA library cloned in pGAD1318. The plasmids pGADGH cRafN, LBD RasV12, and LBD RasN17 have been previously described.[7,11] For the LBD Rap 1E63 and Rap1N17 constructs, the cDNAs are PCR amplified with pZip-EE-Rap1E63 (obtained from B. Knudsen, Cornell University, Ithaca, NY) and pGTB9 Rap1N17 (from D. Broek, USC, Los Angeles, CA) as templates, and as primers 5′-ATTTATGGATCCTCTAGAATGCGTGAGTA-CAAGCTA-3′ and 5′-CTGACTCTCGAGCTAGAGCAGCAGACAT-

[11] L. Van Aelst, M. Barr, S. Marcus, A. Polverino, and M. Wigler, *Proc. Natl. Acad. Sci. U.S.A.* **90,** 6213 (1993).

GATTT-3′ are used. The products are digested with *Bam*HI and *Xho*I and subcloned into a *Bam*HI–*Sal*I-digested pLexVJ10 plasmid.[10]

Protocols and Results

Analysis of Ras/Rap1–AF-6 Interactions, Using Liquid Assay for β–Galactosidase

As shown in Fig. 1, AF-6 harbors two predicted Ras-binding domains at its N terminus. To determine whether both domains are able to interact with Ras and Rap1, the RBD1, RBD2, and AF-6N domains of AF-6 are fused to the GAL4-activation domain of *GAL4*, whereas Ras and Rap1 mutants are fused to the LexA DNA-binding domain. In analogy to the well-characterized Ras–Raf interaction, we have also tested whether AF-6 requires Ras and Rap1 to be in the GTP-bound state. To this end we use constitutively active mutant forms (RasV12 and Rap1E63) and dominant negative mutant forms (RasN17 and Rap1N17) of Ras and Rap1. The Ras and Rap1 LDB fusion constructs are transformed together with either pGAD AF-6-RBD1, pGAD AF-6-RBD2, or pGAD AF-6N into the yeast strain L40, according to the protocol described below. Before performing the liquid β–galactosidase assay, we always first perform growth selection on selective medium (DO-Leu-Trp-His) and β–galactosidase filter assays (see below). The liquid culture assay allows us to compare and to quantify the strength of interactions.

Yeast Transformation. The protocol described below is a modification of the method of Ito *et al.*[12] and can be applied for both LexA- and Gal4-based systems.

1. Inoculate a single yeast colony into 10 ml of YPD and grow overnight at 30° with shaking. Transfer the overnight preculture into 100 ml of YPD and grow the yeast culture further at 30° with shaking (230 rpm) until an OD_{600} of 0.5–0.8 is reached.
2. Harvest the cells by centrifugation at 1500*g* for 5 min at room temperature and wash them in 25–50 ml of 0.1 *M* lithium acetate in TE.
3. Resuspend the washed cells in 1 ml of 0.1 *M* lithium acetate in TE and incubate for 1 hr at 30° with shaking at 230 rpm. The yeast cells are now competent for transformation. One milliliter of cells allows for 10 transformations.
4. Add 100 μl of competent cells for each transformation into a 1.5-ml microcentrifuge tube.

[12] H. Ito, Y. Fukuda, K. Murata, and A. Kimura, *J. Bacteriol.* **153,** 163 (1983).

5. Add the plasmid DNAs (approximately 0.5 to 2 μg) together with 100 μg of sheared, denaturated salmon sperm DNA to the competent yeast cells and subsequently add 600 μl of sterile PEG–lithium acetate solution. Mix well by inversion.

6. Incubate at 30° for 30–60 min (shaking is not required) and heat shock for 15–30 min in a 42° water bath.

7. Pellet the cells by centrifugation for 30 sec in a microcentrifuge, remove the supernatants, and resuspend the cells in 100 μl of sterile TE.

8. Plate the yeast cells on DO-Leu-Trp medium and incubate at 30° until colonies appear.

Liquid Culture Assay for β–Galactosidase. The liquid culture assay for β–galactosidase provides quantitative data for the Ras/Rap1–AF-6 interaction, allowing us to compare the strength of interactions. The assay described below quantifies the β–galactosidase enzymatic activity by measuring the generation of the yellow compound *o*-nitrophenol (ONP) from the colorless substrate *o*-nitrophenyl galactoside (ONPG).

1. Inoculate single colonies from the yeast transformants in 1 ml of selective medium (DO-Leu-Trp) and grow them overnight. The next day, dilute the cells 5- to 10-fold in 5 ml of fresh DO-Leu-Trp medium and incubate them further until an OD_{600} of approximately 0.8 to 1 is reached. Record the OD_{600} for 1-ml samples of each culture.

2. Transfer 1-ml aliquots (in triplicate) to 12 × 75-mm polypropylene tubes and pellet the cells by centrifugation. Add 1 ml of Z buffer to the cells. Include a control with Z buffer alone.

3. Add 50 μl of $CHCl_3$ and 50 μl of 0.1% (w/v) sodium dodecyl sulfate (SDS) to the tubes and vortex vigorously for 10 sec to resuspend the cells.

4. Prewarm the samples to 30° for 5 min and then add 0.2 ml of ONPG solution to each tube. Vortex and incubate the reactions at 30°, until color develops (between 10 min and 6 hr). Stop the reaction by adding 0.5 ml of Na_2CO_3 (1 *M*) followed by brief vortexing. Centrifuge the samples for 10 min (3500 rpm at room temperature) and remove 1 ml of each sample to a disposable cuvette. Measure the OD at 420 nm against the blank.

5. Calculate the β–galactosidase activity by using the following equation: Activity (in Miller units) = $1000[(OD_{420} - OD_{blank})]/(tVOD_{600})$, where t is time of incubation, V is volume (ml) of initial cells aliquoted, and OD_{600} is cell density of the culture.

The results obtained in the β–galactosidase liquid culture assay are shown in Table I. We noticed that AF-6N and the first domain (RBD1), but not the second domain (RBD2), were able to bind to Ras and Rap1. Furthermore, while both activated mutant forms of Ras and Rap1 interact

TABLE I
INTERACTION BETWEEN Ras/Rap1 AND AF-6

LBD fusion	β-Gal activity of GAD-fused AF-6 domains (Miller units)[a]		
	AF-6N	AF-6–RBD1	AF-6–RBD2
RasV12	99 ± 1.6	210 ± 1.8	0.7 ± 1.4
RasN17	0.6 ± 1.4	0.9 ± 1.4	0.6 ± 1.1
RapE63	120 ± 1.7	350 ± 1.9	1.2 ± 1.1
RapN17	0.7 ± 1.7	0.8 ± 1.9	0.9 ± 1.1
Lamin	0.7 ± 1.3	0.8 ± 1.1	0.7 ± 1.5

[a] Data representative of a typical liquid β-galactosidase assay are shown. β-Gal was assayed with *o*-nitrophenyl-β-galactosidase as described in text. The values represent means ±SD of triplicate determination. (Reproduced from Boettner *et al.*[24] with permission of publisher.)

with AF-6, none of the dominant negative mutant forms show binding activity toward AF-6. This suggests that AF-6 binds Ras and Rap1 in their GTP-bound state. In addition, the data in Table I further indicate that the strength of interaction between Rap1 and AF-6 exceeds that exerted by Ras–AF-6 complexes. The findings that the affinity of interaction between Rap1 and AF-6 is greater than that between Ras and AF-6 are consistent with the kinetic studies described below.

Comparison of Interactions between AF-6, Raf, RalGDS, Phosphatidylinositol 3-Kinase δ, and Ras/Rap1, Using Histidine Prototrophy Assay and Filter Assays for β–Galactosidase Activity

We have made use of the two-hybrid system to see how AF-6 compares with c-Raf, RalGDS, and PI3Kδ in terms of its ability to bind to Ras and Rap1. LBD constructs expressing AF-6-RBD1, c-RafN, RalGDS-RBD, and PI3Kδ-RBD are cotransformed with LBD RasV12 and LBD Rap1E63, respectively, in the yeast strain L40 as described above. The transformants are subjected to histidine prototrophy and β–galactosidase filter assays for assessment of their respective interactions. In the first assay, the growth selection marker *HIS3* is used as a reporter, whereas in the latter transcriptional activity of the bacterial *lacZ* reporter is utilized.

To assay histidine prototrophy, pick individual transformants and spread as small patches on DO-Leu-Trp plates. After 2 days, the grown patches are replica plated first onto one or two DO-Leu-Trp plates to preclean the excess of yeast material and subsequently onto plates that in addition to leucine and tryptophan, also lack histidine (DO-Leu-Trp-His). Precleaning is important to avoid background growth.

To assay for activation of the *lacZ* reporter construct, the grown yeast patches (see above) are replica plated onto a Whatman (Clifton, NJ) No. 50 filter paper placed onto a DO-Leu-Trp plate and grown overnight. The filter with the yeast is placed in a container with liquid nitrogen for about 30 sec to lyse the cells, and then transferred (yeast cells facing up) in a petri dish containing a Whatman filter No. 3 presoaked in Z buffer/X-Gal solution (~2.5 ml of Z buffer/X-Gal solution per petri dish). Incubate at 30° and check periodically for the appearance of blue colonies.

As shown in Fig. 2, RasV12 is able to interact with all targets, the strongest interaction being with c-Raf. Rap1E63 shows strong interaction with AF-6 and RalGDS, but no association is observed with PI3Kδ and only weak to no interaction is observed with c-Raf. This indicates that both GTPases, Ras and Rap1, use effector molecules that are only partially identical and that they exhibit differential binding profiles toward the targets listed above.

Thermodynamic and Kinetic Characterization of Interaction between Ras/Rap1 and Ras-Binding Domain of AF-6

To study protein interactions and their biochemical characterization, large quantities of protein (in the range of milligrams) are required. Al-

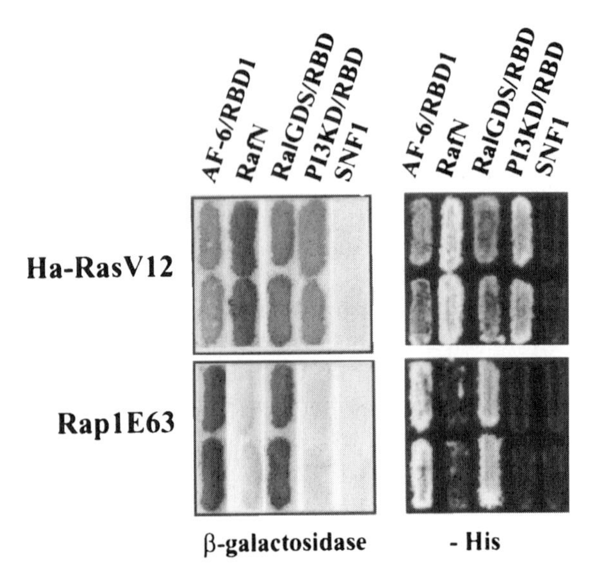

FIG. 2. Analysis of interactions between Ras/Rap1 and their effectors, using histidine prototrophy and filter assay for β-galactosidase. The LexA two-hybrid tester strain, L40, was transformed with plasmids expressing Ras and Rap1 mutants fused to LBD, and effectors of Ras and Rap1 fused to GAD. Transformants were assayed for β-galactosidase expression (*left*) and for their ability to grow on His$^-$ plates. (Reproduced from Boettner *et al.*[24] with permission of publisher.)

though high yields of Ras proteins can be obtained with prokaryotic expression systems, this is usually not the case for their full-length effector proteins. Therefore, most biochemical studies are restricted to protein fragments that are soluble and can be prepared in large amounts. The Ras-binding domain (RBD) has been identified for many effectors such as Raf kinase, RalGDS, PI3K, and AF-6. The RBDs of the first three effectors are similar in size, comprising 80–90 amino acids. Despite the lack of sequence homology their three-dimensional structures are highly related, namely, they all show the ubiquitin fold.[13–15] In contrast, the RBD1 of AF-6 appears to be larger. The above-described two-hybrid studies, as well as biochemical studies,[16] indicated that the N-terminal part comprising the first 141 amino acids is a stable domain competent for strong binding to Ras and Rap1. This fragment was used for thermodynamic and kinetic investigations.

Methods to Quantitate Ras/Rap1 and AF-6 Interactions

Fluorescence Titration

Fluorescence is a property of many biological macromolecules that is widely used in interaction studies.[17] Two different approaches allowing fluorescence measurements are commonly employed. Either intrinsic fluorescence contributed by, for example, tryptophan residues in the protein, or extrinsic fluorescence, using chemically attached fluorescent label, may be used for monitoring binding to ligands or other proteins. Although Ras does not contain tryptophan residues, the mutant Y32W, located in switch I (effector region), shows a small decrease in fluorescence when the bound GTP is hydrolyzed to GDP.[18] Small changes in fluorescence intensity of RasY32W are observed with RalGDS–RBD (decreased fluorescence[19]) and Raf–RBD (increased fluorescence[20]). This effect is larger at lower temperatures and can in principle be used for titration experiments. However, these experiments are feasible only for RalGDS–RBD, because it contains no tryptophan residues. The high background fluorescence caused

[13] N. Nassar, G. Horn, C. Herrmann, A. Scherer, F. McCormick, and A. Wittinghofer, *Nature (London)* **375,** 554 (1995).

[14] L. Huang, F. Hofer, G. S. Martin, and S.-H. Kim, *Nat. Struct. Biol.* **5,** 422 (1998).

[15] E. H. Walker, O. Perisic, C. Ried, L. Stephens, and R. L. Williams, *Nature (London)* **402,** 313 (1999).

[16] T. Linnemann, M. Geyer, B. K. Jaitner, C. Block, H. R. Kalbitzer, A. Wittinghofer, and C. Herrmann, *J. Biol. Chem.* **274,** 13556 (1999).

[17] L. Brand and M. L. Johnson (eds.), *Methods Enzymol.* **278** (1997).

[18] K. Yamasaki, M. Shirouzu, Y. Muto, J. Fujita-Yoshigaki, H. Koide, Y. Ito, G. Kawai, S. Hattori, S. Yokoyama, S. Nishimura, and T. Miyazawa, *Biochemistry* **33,** 65 (1994).

[19] C. Herrmann, G. Horn, M. Spaargaren, and A. Wittinghofer, *J. Biol. Chem.* **271,** 6794 (1996).

[20] J. R. Sydor, M. Engelhardt, A. Wittinghofer, R. S. Goody, and C. Herrmann, *Biochemistry* **37,** 14292 (1998).

$$\begin{array}{ccc} \text{Ras}\cdot\text{mant-GppNHp} + \text{AF-6} + \text{GppNHp} & \overset{K_d}{\rightleftharpoons} & \text{Ras}\cdot\text{mant-GppNHp}\cdot\text{AF-6} + \text{GppNHp} \\ \downarrow k_{-1} & & \downarrow k_{-2} \\ \text{Ras}\cdot\text{GppNHp} + \text{AF-6} + \text{mant-GppNHp} & & \text{Ras}\cdot\text{GppNHp}\cdot\text{AF-6} + \text{mant-GppNHp} \end{array}$$

SCHEME 1.

by the tryptophan residues present in Raf–RBD (one) and in AF-6–RBD1 (two) make a direct titration impossible. Nonetheless, as we discuss below, kinetic measurements using this RasY32W mutant, together with the stopped-flow technique, allow a detailed analysis of Ras/AF-6–RBD1 complex formation.

Another method to characterize the biochemistry of Ras-like GTPases makes use of the fluorescent 2′,3′-*N*-methylanthraniloyl (mant) group attached to the ribose moiety of the nucleotide bound by the GTPases.[21,22] Binding of mant-GDP or mant-GTP to Ras or Rap1 results in a more than 2-fold increase in fluorescence. A small decrease in fluorescence intensity is observed when Raf–RBD or AF-6–RBD1 is bound.[20] As seen with the Y32W mutant, this effect diminishes with increasing temperature. However, in this case, a direct titration of Ras/Rap1–mant-GppNHp with AF-6–RBD1 is feasible with a highly accurate fluorescence detection system.[16] A broader application of the mant label is described in the following sections.

Inhibition of Guanine Nucleotide Dissociation: GDI Assay

As mentioned above, the dissociation of mant-nucleotides from Ras-like GTPases results in a large decrease in fluorescence intensity. It has been further observed that binding of effectors to the Ras proteins inhibits the dissociation of the bound nucleotide.[23] These features have resulted in the development of a method that allows quantification of all Ras GTPase–effector interactions. To block the GTPase activity of Ras or Rap1, the following experiments all use the nonhydrolyzable GTP analog GppNHp (guanyl-5′-yl imidodiphosphate). For example, the Ras–AF-6 interaction is described below. When Ras–mant-GppNHp is incubated together with a large excess of nonlabeled GppNHp, the latter quantitatively displaces the mant-nucleotide (Scheme 1). Because of the large excess of GppNHp, virtually no rebinding of mant-GppNHp occurs, and thus this part of the reaction is ignored in Scheme 1. Equation (1) describes the dependence of the observed dissociation rate constant of mant-GppNHp (k_{obs}) on the AF-6 concentration. The rapid equilibration of the Ras–AF-6 complex is

[21] H. Rensland, A. Lautwein, A. Wittinghofer, and R. S. Goody, *Biochemistry* **30,** 11181 (1991).
[22] C. Lenzen, R. H. Cool, and A. Wittinghofer, *Methods Enzymol.* **225,** 95 (1995).
[23] C. Herrmann, G. A. Martin, and A. Wittinghofer, *J. Biol. Chem.* **270,** 2901 (1995).

illustrated by kinetic studies (see below) and is a prerequisite for Eq. (1) to hold. The rate of nucleotide dissociation is measured in the presence of different AF-6 concentrations, and this essentially corresponds to a titration experiment with k_{obs} as a readout. The K_d, describing the affinity of AF-6 for Ras, is then retrieved by fitting the values of k_{obs} obtained at different AF-6 concentrations to Eq. (1).

$$k_{obs} = k_{-1} - (k_{-1} - k_{-2})\{(\mathrm{Ras}_0 + \mathrm{AF\text{-}6}_0 + K_d) - [(\mathrm{Ras}_0 + \mathrm{AF\text{-}6}_0 + K_d)^2 - 4\mathrm{Ras}_0\mathrm{AF\text{-}6}_0]^{1/2}\}/(2\mathrm{Ras}_0) \quad (1)$$

where Ras_0 and $\mathrm{AF\text{-}6}_0$ denote total concentrations of the Ras protein and AF-6–RBD1, respectively, and K_d is the equilibrium dissociation constant of the complex.

Nucleotide Loading. The synthesis of the fluorescent nonhydrolyzable GTP analog, mant-GppNHp, has been described.[22] To exchange the GTPase-bound GDP for the nonhydrolyzable GTP analog, the Ras protein (20 mg/ml) is incubated for 1 hr at 20° with alkaline phosphatase (2 U/mg) and a 2-fold excess of mant-GppNHp in the presence of 200 m*M* ammonium sulfate. The alkaline phosphatase hydrolyzes GDP, thereby quantitatively loading mant-GppNHp (or GppNHp) onto Ras. In the case of Rap1, 10 m*M* EDTA is also included. Furthermore, excess nucleotide and salt are removed by gel filtration and the mant-GppNHp-bound Ras protein is thereby transferred into the desired buffer.

GDI Assay. The GDI assay is performed by thermostatting 50 n*M* Ras · mant-GppNHp and varying concentrations of AF-6–RBD1. This is done in fluorescence cuvettes at 37°. The buffer contains 5 m*M* $MgCl_2$, 20 m*M* Tris (pH 7.5), and NaCl, to set the desired ionic strength. To obtain reliable results, 8 to 12 different AF-6–RBD1 concentrations ranging below, near, and up to 5 times the K_d value should be tested. The dissociation (displacement) of the mant-nucleotide is initiated by the addition of 100 μM GppNHp, and the fluorescence is excited at 360 nm and monitored at 450 nm. An exponential decay curve is fitted to the fluorescence time trace, yielding k_{obs}. These data are plotted versus the effector concentration, and the K_d value is obtained from the fit according to Eq. (1).

The GDI assay has been applied to many different Ras proteins, their mutant variants, and different effectors such as Raf, RalGDS, and AF-6.[16,19,23] For AF-6–RBD1, a K_d value of 0.25 μM was reported for Rap1A, which binds 12 times more strongly than Ras (K_d of 3 μM[16]). These data are consistent with the two-hybrid results described above.

Stopped-Flow Technique

The dynamics of Ras/Rap1 and AF-6 interactions have been investigated by means of stopped flow. In this technique, two solutions containing

Rap1 · mant-GppNHp and AF-6–RBD1, for example, are rapidly mixed (within 1 msec) and the time trace of the fluorescence change due to association of the proteins is recorded. In contrast to equilibrium fluorescence titration, the stopped-flow technique can be applied even when small changes in fluorescence intensity occur (<5%). In a stopped-flow experiment, this small fluorescence change is sufficient for monitoring the binding, whereas in titration experiments, it is difficult to take many readings for the accurate determination of the K_d value. Stopped-flow experiments yield dissociation and association rate constants (k_{off} and k_{on}) and from their ratio the K_d value can be calculated.

As in the titration experiment, the fluorescence is excited at 360 nm. However, all light emitted above 400 nm is detected with the use of a cutoff filter. The concentration of AF-6–RBD1 should be at least 5-fold in excess of Rap1 · mant-GppNHp, to fulfill pseudo first-order conditions and to allow a single-exponential fit to the recorded trace. From this fit, the obtained k_{obs} is plotted versus the concentration of AF-6–RBD1. According to Eq. (2), the slope of the fitted straight line corresponds to k_{on} and the intercept yields k_{off}.

$$k_{obs} = k_{on}[\text{AF-6–RBD1}] + k_{off} \tag{2}$$

In many cases, as with Rap1/AF-6–RBD1, k_{off} is small and therefore not reliably obtained by this extrapolation. An accurate k_{off} value can, however, be obtained by a displacement experiment. In this case, the Rap1 · mant-GppNHp · AF-6–RBD1 complex is placed in one syringe of the stopped-flow apparatus and is mixed with a large excess (>20-fold) of nonlabeled Rap1 · GppNHp from another syringe. Like in the GDI assay, the time trace is fitted by an exponential curve, in this case yielding k_{off}, the dissociation rate constant of AF-6–RBD1 and Rap1. This method has been used to characterize the Ras and Rap1 binding kinetics with AF-6–RBD1,[16] which are included in Table II. A more than 10% change in fluorescence obtained with saturating amounts of AF6–RBD1 allows comfortable detection of the fluorescence transience.

As a complement to the use of the mant-labeled fluorophore, we also used the intrinsic fluorescence of RasY32W in our stopped-flow assay. The experiments are carried out according to the evaluation strategy described above; however, in this case the fluorescence excitation is set at 290 nm and detection is through a 320-nm cutoff filter. In Fig. 3A, a typical fluorescence trace is shown, demonstrating a small fluorescence change. This is mainly because RasY32W is not sensitive to effector binding,[20] and because of the large background of AF-6–RBD1, owing to the presence of two tryptophans. To reduce the background, we use a concentration of AF-6–RBD1 no more than 5-fold in excess of RasY32W · GppNHp. Furthermore, we have chosen a low temperature, at which the change in fluores-

TABLE II
RESULTS OBTAINED BY STOPPED-FLOW EXPERIMENTS[a]

Complex	$k_{on}(\mu M^{-1}\ sec^{-1})$	$k_{off}(sec^{-1})^b$	$k_{off}(sec^{-1})^c$	$K_d(\mu M)^d$
RasY32W · GppNHp[e]	19	13	11	0.58
Ras · mant-GppNHp[f]	6.4	20.8	15.3	2.4
Rap1 · mant-GppNHp[f]	11.9	—	2.6	0.22

[a] T. Linnemann, M. Geyer, B. K. Jaitner, C. Block, H. R. Kalbitzer, A. Wittinghofer, and C. Herrmann, *J. Biol. Chem.* **274,** 13556 (1999).
[b] Obtained from the intercept of the linear fit (see text).
[c] Obtained from displacement experiment.
[d] Calculated from $k_{off}{}^c/k_{on}$.
[e] 20 m*M* Tris (pH 7.5), 5 m*M* $MgCl_2$.
[f] 20 m*M* Tris (pH 7.5), 5 m*M* $MgCl_2$, 100 m*M* NaCl.

cence is larger. In addition, NaCl is avoided in the buffer in order to have tighter binding. The observed rate constants are plotted in Fig. 4 versus the concentration of AF-6–RBD1, and the slope of the fitted straight line yields $k_{on} = 19\ \mu M^{-1}\ sec^{-1}$. However, the intercept corresponding to $k_{off} = 13\ sec^{-1}$ has an uncertainty of at least 50%. Therefore, we also performed displacement experiments, such as the typical trace shown in Fig. 3B. These experiments result in a much more reliable value of $k_{off} = 11\ sec^{-1}$.

This example emphasizes the strength of the stopped-flow technique. In addition to the kinetic constants, the K_d value can also be obtained using $K_d = k_{off}/k_{on}$ with only a 1.5% change in fluorescence intensity. This small change in fluorescence would not allow determination of the affinity by titration experiments. In Table II, the results obtained from the RasY32W mutant and mant labeling are compared. The results of the two systems agree well with each other, despite the different labels and the different salt concentrations used. Also, the K_d values obtained by stopped flow (Ras, 2.4 μM; Rap, 0.22 μM) correlate well with the K_d values derived from the GDI assay (Ras, 3 μM; Rap1, 0.25 μM).[16] Furthermore, as observed for the interaction of Ras and the effector Raf,[20] the interaction of Ras with AF-6–RBD1 is highly dynamic, with the half-life of the complex being 60 msec (at 10°).

Establishment of Stable Ras- and Rap1-Expressing Cell Lines, Using Retroviral Gene Transfer to Study *in Vivo* Interactions between Ha-Ras/Rap1 and AF-6

Background and General Principles

The generation of stable transfectants has become a common practice to investigate the functions of specific GTPases. Most cell lines expressing

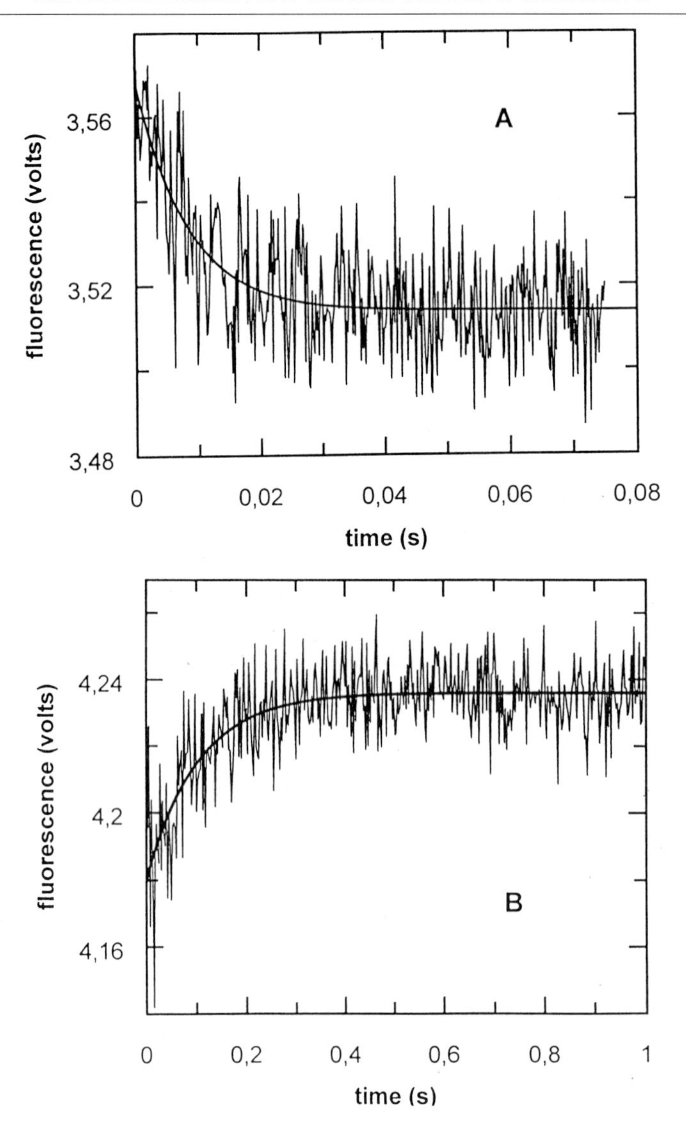

FIG. 3. Stopped-flow experiments. Fluorescence excitation was at 290 nm, and detection was through a 320 nm-cutoff filter. The buffer used consisted of 20 m*M* Tris (pH 7.5)–5 m*M* $MgCl_2$, and experiments were carried out at 10°. The graphs show the experimental trace and the exponential fit. (A) Association of 1 μM RasY32W · GppNHp with 5 μM AF-6–RBD1; (B) dissociation of 2 μM RasY32W · GppNHp · AF-6–RBD1 by displacement with 40 μM Ras · GppNHp.

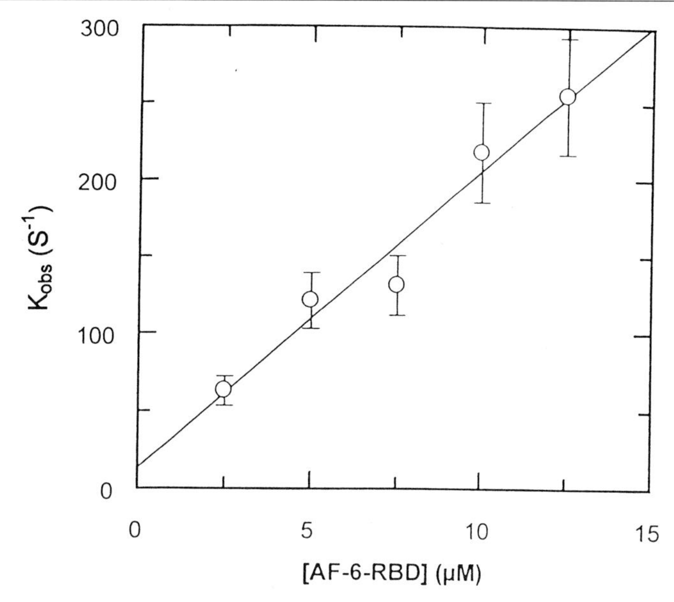

FIG. 4. Binding kinetics of RasY32W/AF-6–RBD1. k_{obs} values obtained in stopped-flow experiments, as in Fig. 3A, are plotted versus AF-6–RBD1 concentrations. According to Eq. (2), the fitted straight line yields k_{on} and k_{off} (see Table II).

dominant active or negative mutant forms of GTPases have been generated by plasmid transfection and subsequent selection for plasmid-encoded marker genes. A more refined method for the stable introduction of Ras-type cDNAs into various acceptor cell lines is viral transduction. The efficiency of gene transfer into cell lines that are relatively intransigent to conventional transfection techniques may be considerably improved by this method. A second benefit offered by this procedure is that transduced host cells tend to lose their inserted sequences to a much lesser extent than sequences that have been conventionally transfected, when kept in culture over a long period of time. The use of packaging cell lines allows us to produce viral particles that are infectious, but cannot replicate once they have entered the host cell. Retroviruses recognize specific receptors to enter their host cells. Ecotropic receptors are present on cells of mouse and rat origin. Amphotropic receptors are present on rodent cells, as well as cells of many other species, including humans. It is possible to use an ecotropic virus in nonrodent cell lines by building the ecotropic receptor into the host cell line. Retroviruses, upon invasion, reverse transcribe virion RNA to generate linear, double-stranded DNA that integrates into the host cell genome.

To address questions of localization and interaction of the AF-6 protein in cells expressing constitutively active forms of Ras and Rap1, cell lines are generated by using retroviral vectors.[24] LinXA cells [an amphotropic packaging line provided by G. Hannon (Cold Spring Harbor Laboratories, Cold Spring Harbor, NY)] are transfected with a pBABE[25] or pWZL vector[26] containing either the RasV12 or RapE63 insert. The virus obtained is then used to infect Madin–Darby canine kidney (MDCK) cells (nontransformed dog epithelial cells). These cell lines consist of a population of transduced cells, thus averaging out effects of retroviral integration. A detailed protocol of retroviral transfection and infection in MDCK cells is given below.

Materials

Retroviral Packaging Cell Line. LinXA (for host cell line bearing the amphotropic receptor) and LinXE (for host cell line bearing the ecotropic receptor) packaging cell lines are a gift from G. Hannon (Cold Spring Harbor Laboratories). Alternatively, the Bosc 23 packaging cell line (for host cell line bearing the ecotropic receptor) or Bing Cak 8 (for host cell line bearing the amphotropic receptor) may be obtained from the American Type Culture Collection (ATCC, Manassas, VA) with permission from Rockefeller University (New York, NY).

Media and Solutions for Retroviral Transfection and Infection. The medium used for LinX packaging cells is Dulbecco's modified Eagle's medium (DMEM; GIBCO-BRL, Gaithersburg, MD) containing 10% (v/v) fetal bovine serum (FBS; HyClone). The MDCK cells are grown in DMEM containing 10% (v/v) FBS, 1% (v/v) penicillin–streptomycin, and 20 m*M* HEPES (GIBCO-BRL). Solutions required are 2.5 *M* $CaCl_2$–0.01 *M* HEPES (pH 5.5) with NaOH (sterile filter, aliquot, and store at $-20°$) and 2× BBS: 50 m*M* *N,N*-bis-(2-hydroxyethyl)-2-aminoethanesulfonic acid (BES), 280 m*M* NaCl, 1.5 m*M* Na_2HPO_4, pH ~7.00 (sterile filter, aliquot, and store at $-20°$). Make batches between 0.05 pH units below and above 7.00 and test which works the best. Polybrene (Sigma, St. Louis, MO) is used at a stock concentration of 8 mg/ml.

Retroviral Transfection and Infection of Host Cell Line

1. Plate out 6.5×10^5 LinX packaging cells per well of a six-well plate. Allow the cells to settle and begin to adhere before returning them

[24] B. Boettner, E. Govek, J. Gross, and L. Van Aelst, *Proc. Natl. Acad. Sci. U.S.A.*, **97,** 9064 (2000).

[25] J. P. Morgenstern and H. Land, *Nucleic Acids Res.* **18,** 3587 (1990).

[26] G. J. Hannon, P. Sun, A. Carnero, L. Y. Xie, R. Maestro, D. S. Conklin, and D. Beach, *Science* **283,** 1129 (1999).

to the incubator (approximately 10 min). Incubate overnight at 37°, 5% CO_2.

2. When the cells have reached approximately 70% confluency, change the medium gently (cells tend to lift easily) and incubate between 1 and 4 hr at 37°C, 5% CO_2.

3. Aliquot the DNA into a sterile 1.5-ml Eppendorf tube. It is best to do an initial titration using a β-galactosidase (β-Gal) construct as a readout for transfection and infection efficiency in order to determine the optimal amount of DNA for each host cell type. However, 6 μg appears to work well in most cases. Dilute DNA to a total of 225 μl with sterile water. To this tube add 25 μl of 2.5 *M* $CaCl_2$–0.01 *M* HEPES (pH 5.5).

4. Next, bubble this mixture with a pasteur or 1-ml disposable plastic pipette, using a mechanical pipette aid. At the same time, add 250 μl of 2× BBS dropwise. Add 500 μl (total) to one well of a six-well plate dropwise around the plate. Incubate for 12–16 hr at 37°, 5% CO_2. (Two to 4% CO_2 is actually optimal at this step, but 5% works as well.)

5. Change the medium and incubate for 60 hr at 32°, 5% CO_2.

6. Remove the virus-laden medium from well and filter through a 0.45-μm pore size syringe filter. Add to one well of host cells and subsequently add Polybrene to a final concentration of 8 μg/ml. Host cells should be plated at a concentration that will give confluency, but not overcrowding, within 2 to 3 days.

7. Spin the cells for 1 hr at room temperature at 1700 rpm in a Beckman (Fullerton, CA) tabletop centrifuge.

8. Additional rounds of infection may give better infection efficiencies, depending on the packaging line and host cell type. Between one and six rounds may be necessary, as suggested by the ATCC, when using the Bosc cell packaging line from Rockefeller University. However, in most cases, one round appears to be sufficient with the LinX packaging cell line. Incubate the cells overnight at 32°, 5% CO_2.

9. On the next day, change the medium and incubate the cells at 37°, 5% CO_2 until the cells become confluent.

10. Transfer the cells to a 10-cm plate and incubate overnight at 37°, 5% CO_2. Add an appropriate amount of antibiotic after the transfer, but not before 48 hr after the medium was last changed. Selection may be started at the time of transfer to a large plate; however, it is best to make sure that the cell density will support viability after selection has begun. This will be dependent on host cell type and can be checked ahead of time using a β-Gal construct. Titration of appropriate antibiotics should also be done ahead of time. Generally, if puromycin is used as the selection drug, 3 days should be adequate to select out cells containing plasmid. For the generation of MDCK RasV12 and RapE63 stable cell lines, we use 10 μg/ml for selection and 5 μg/ml for maintenance. For the analysis of

RasV12- and RapE63-expressing clones, we have made use of anti-Ras and anti-Rap1 monoclonal antibodies obtained from Transduction Laboratories (Lexington, KY).

We have observed that expression of Rap1 does not disturb cell–cell adhesion, whereas cell–cell adhesion complexes in RasV12-expressing clones are disturbed. Furthermore, coimmunoprecipitation experiments using the above-described cell lines have been performed to show *in vivo* association between Ras/Rap1 and AF-6.[24]

Conclusions

In this chapter, we have presented a spectrum of investigative approaches that served to demonstrate specific protein interactions between the Ras/Rap1 GTPases and their potential effector molecule AF-6 *in vivo* and *in vitro*. The two-hybrid analysis and the investigation of the kinetic and thermodynamic properties of Ras–AF-6 and Rap1–AF-6 complexes led to the same overall outcome, namely, that Rap1 appears to form the tightest complex with AF-6–RBD1. These data suggest a role for AF-6 as an effector in Rap1-mediated signaling. However, we cannot exclude its involvement in Ras-induced activities. Further investigations of the function of Ras/Rap1 and AF-6 in different biological contexts will shed more light on these interactions.

Acknowledgments

We thank Eve-Ellen Govek, Arndt Schmidt, and Mingming Zhao for assistance and critical reading of this manuscript. We also thank Greg Hannon for assistance with setting up the retroviral transfection and infection assays. Our research was supported by grants from the National Institutes of Health and Department of Defense. L.V.A. is a recipient of the V Foundation and the Sidney Kimmel Foundation for Cancer Research. B.B. is a fellow of the Gesellschaft der Naturforscher der Leopoldina.

Section II

Screening Analyses

[12] Analysis of Protein Kinase Specificity by Peptide Libraries and Prediction of *in Vivo* Substrates

By ZHOU SONGYANG

Introduction

Protein phosphorylation is crucial to regulating a variety of intracellular biological activities. Therefore defining the specificities of protein kinases is critical to our understanding of signal transduction. There are about 130 protein kinases predicted in yeast.[1] In the more complex multicellular organisms such as *Caenorhabditis elegans* and humans, between 400 and 1000 protein kinases are estimated to exist. Determining which proteins are the *in vivo* substrates of the kinases has long been the bottleneck for elucidating signal transduction pathways. Methods to quickly study and predict kinase substrates specificities are therefore necessary for the genome-wide study of kinases.

Protein kinases are generally divided into three categories based on their abilities to phosphorylate serine/threonine, tyrosine, or both residues. The specificity of a protein kinase is governed by several factors including its intracellular localization and interaction with its substrates. The most important factor is the specificity of the kinase domain. This chapter focuses on using oriented peptide libraries to study the specificities of protein kinases. In the following sections we outline peptide library design strategies, detail the experimental techniques for peptide library synthesis and selection, and discuss how the data obtained are analyzed and used to predict *in vivo* substrates for protein kinases.

Principle of Using Degenerate Peptide Libraries for Studying Protein Kinases

We have developed an oriented peptide library technique similar to the strategy utilized for SH2 domains to rapidly determine optimal motifs for protein kinases.[2,3] Briefly, a particular kinase is mixed with a soluble mixture of peptides identical in length and orientation but differing in amino acid

[1] T. Hunter and G. D. Plowman, *Trends Biochem. Sci.* **22,** 18 (1997).

[2] Z. Songyang, S. E. Shoelson, M. Chaudhuri, G. Gish, T. Pawson, W. G. Haser, F. King, T. Roberts, S. Ratnofsky, R. J. Lechleider, *et al., Cell* **72,** 767 (1993).

[3] Z. Songyang, S. Blechner, N. Hoagland, M. F. Hoekstra, H. Piwnica-Worms, and L. C. Cantley, *Curr. Biol.* **4,** 973 (1994).

0076-6879/00 $35.00

composition except for a single phosphorylatable amino acid fixed in the center. Phosphorylated peptides, which generally represent only a small fraction, are quantitatively separated from the bulk of nonphosphorylated peptides. The mixture is then sequenced to determine the abundance of amino acids at each degenerate position surrounding the phosphorylation site as compared with the same position in the starting mixture. The results should indicate which amino acids are preferred at a given position.

Three-Step Peptide Library Screening

As illustrated in Fig. 1, if little is known about the specificity of a kinase, a three-step screening strategy is employed. The kinase is first subjected to an initial peptide library pool screening that helps to determine the type of libraries for further use (see the next section). Next, completely degenerate libraries (primary libraries) are used to identify the key residues for substrate recognition. In primary libraries, the phosphorylatable amino acids (Tyr, Ser, or Thr) are fixed and surrounded by random sequences to

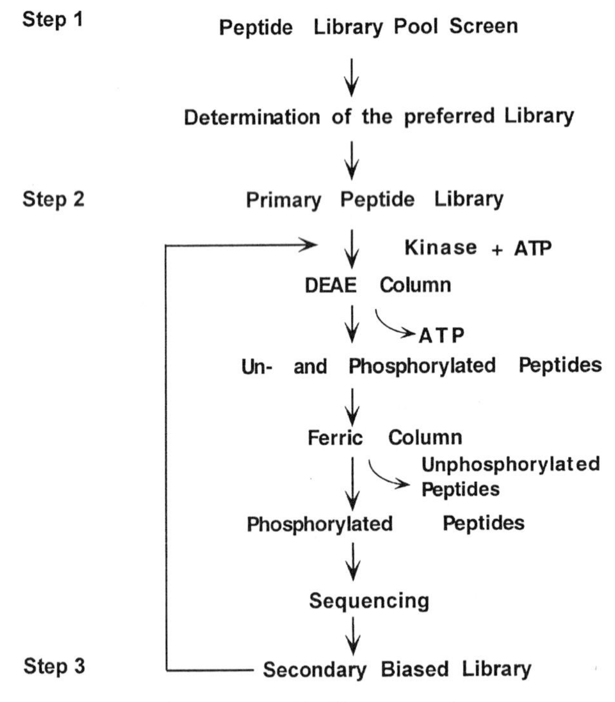

FIG. 1. Three-step peptide library screening strategy.

orient the libraries. Because the amount of peptides that can be efficiently phosphorylated by the kinase of interest is often underrepresented in primary libraries, the data from these libraries may contain considerable background value. In the third step, secondary libraries that are much less degenerate are constructed on the basis of the results from the primary library screens. The background should now be considerably minimized, because key residues required for phosphorylation that have been identified in the primary library screens can be fixed as well. Screening is then repeated using secondary libraries to define kinase specificities.

Peptide Library Pool Screening of Protein Kinases

As mentioned above, initial screening of a relatively unknown kinase involves a step we call peptide library pool screening. When nothing is known about its substrates, the protein kinase will be used to phosphorylate all the peptide libraries available in the laboratory (Table I). The phosphorylated kinase mixtures will then be spotted on phosphocellulose paper (p81 paper; Whatman, Clifton, NJ). The library that gives the highest count will be used as a guide for further studies. Table I is a sample list of the libraries currently available in the laboratory.

TABLE I
PEPTIDE LIBRARIES FOR STUDYING PROTEIN KINASES[a]

Library	Sequence
	For serine/threonine kinases
SPL	Met-Ala-Xaa-Pro-Xaa-Ser-Pro-Xaa-Xaa-Xaa-Xaa-Xaa-Ala-Lys-Lys-Lys
SD	Met-Ala-Xxx-Xxx-Xxx-Xxx-Ser-Xxx-Xxx-Xxx-Xxx-Ala-Lys-Lys-Lys
SD2	Met-Ala-Xaa-Xaa-Xaa-Xaa-Ser-Xaa-Xaa-Xaa-Xaa-Ala-Lys-Lys-Lys
SI	Met-Ala-Xaa-Xaa-Xaa-Xaa-Ser-Ile-Xaa-Xaa-Xaa-Ala-Lys-Lys-Lys
SP	Met-Ala-Xaa-Xaa-Xaa-Xaa-Ser-Pro-Xaa-Xaa-Xaa-Ala-Lys-Lys-Lys
RS	Met-Ala-Xaa-Xaa-Xaa-Xaa-Arg-Xaa-Xaa-Ser-Xaa-Xaa-Xaa-Xaa-Xaa-Ala-Lys-Lys-Lys
RSF	Met-Ala-Xaa-Xaa-Xaa-Xaa-Arg-Xaa-Xaa-Ser-Phe/Ile-Xaa-Xaa-Xaa-Xaa-Ala-Lys-Lys-Lys
	For tyrosine kinases
YD	Met-Ala-Xxx-Xxx-Xxx-Xxx-Tyr-Xxx-Xxx-Xxx-Xxx-Ala-Lys-Lys-Lys
YD2	Met-Ala-Xaa-Xaa-Xaa-Xaa-Tyr-Xaa-Xaa-Xaa-Xaa-Ala-Lys-Lys-Lys
E6YD	Ala-E-E-E-E-Tyr-Xaa-Xaa-Xaa-Xaa-Xaa-Xaa-Ala-Lys-Lys-Lys-Lys

[a] Xxx indicates all amino acids except Trp, Cys, Tyr, Ser, or Thr. Xaa indicates all amino acids except Trp and Cys.

Peptide Library Design

General Strategy

Proper orientation of the library is easily achieved for protein kinases because the phosphorylatable amino acids (Tyr, Ser, and Thr) can be used to orient the library. For example, the primary library SD (Met-Ala-Xxx-Xxx-Xxx-Xxx-Ser-Xxx-Xxx-Xxx-Xxx-Ala-Lys-Lys-Lys, a Ser degenerate library) is constructed for serine/threonine kinases. To ensure that the only potential site of phosphorylation is the Ser at residue 7, phosphorylatable amino acids are omitted at all degenerate positions. Xxx hence indicates all amino acids except Trp, Cys, Tyr, Ser, or Thr.[3] Each degenerate position is thus fixed relative to the Ser and the phosphorylated peptides are no longer out of phase when sequenced.

For primary libraries, phosphorylatable amino acids (Tyr, Ser, and Thr) are usually avoided in the degenerate positions. However, for secondary libraries in which residues in addition to the phosphorylatable amino acid can be fixed, Tyr, Ser, and Thr can be included in the degenerate positions because phosphorylation of these residues at any degenerate position is negligible. For the SP library (Met-Ala-Xaa-Xaa-Xaa-Xaa-Ser-Pro-Xaa-Xaa-Xaa-Ala-Lys-Lys-Lys),[4] the chance of Ser and Pro being adjacent to one another at the degenerate positions is quite small ($<5 \times 18^{-2}$, i.e., ~1.5%). Thus, peptides phosphorylated at positions other than the fixed position would not interfere with the sequencing of oriented peptides. To avoid problems with oxidation, Cys is omitted for all peptide libraries. The Cys residue can be added and studied after the optimal peptide substrates have been determined.

Other Considerations

The number of degenerate positions should not exceed 15, although the length of peptide libraries can vary. A library of 15 degenerate positions has already 20^{15} different molecules, which translates into ~100 mg for a 2-kDa peptide. We started with eight degenerate positions because four residues N terminal and C terminal to the phosphorylation site would include the region most likely to be involved in catalytic recognition. If 18 different amino acids are present in any 1 of the 8 degenerate positions

[4] Z. Songyang, K. P. Lu, Y. T. Kwon, L. H. Tsai, O. Filhol, C. Cochet, D. A. Brickey, T. R. Soderling, C. Bartleson, D. J. Graves, A. J. DeMaggio, M. F. Hoekstra, J. Blenis, T. Hunter, and L. C. Cantley, *Mol. Cell. Biol.* **16,** 6486 (1996).

(for primary libraries), the total theoretical degeneracy of this library would be 18^8, or >10 billion.

Including a short leading sequence in front of the degenerate positions is generally beneficial. In the case of the SD library (Met-Ala-Xxx-Xxx-Xxx-Xxx-Ser-Xxx-Xxx-Xxx-Xxx-Ala-Lys-Lys-Lys), the Met-Ala sequence at the N terminus of the peptide libraries not only allows quantification, but also provides two amino acids to verify that peptides from this mixture are being sequenced. Similarly, the Ala at residue 12 makes it possible to quantify and estimate the amount of peptides lost during sequencing. The poly(Lys) tail prevents washout during sequencing, as well as improves the solubility of the mixture (no solubility problems occur at neutral pH and 5-mg/ml concentration). Importantly, the poly(Lys) sequence allows the peptides to stick to phosphocellulose paper (p81 paper), thus facilitating the separation of [γ-^{32}P]ATP (see Peptide Separation Using DEAE Column, below).

Procedures Using Oriented Peptide Libraries

Peptide Library Synthesis

Synthesis of degenerate peptide libraries is accomplished according to the standard benzotriazolyl *N*-oxytrisdimethylaminophosphonium hexafluorophosphate/1-hydroxybenzotriazole (BOP/HOBt) coupling protocols, using a peptide biosynthesizer (ABI 431A; Perkin Elmer Biosystems, Foster City, CA).

Synthesis of Degenerate Peptide Libraries

1. Add equal moles of different 9-fluorenylmethyloxycarbonyl (Fmoc)-blocked amino acids (except for Cys) simultaneously at 5-fold excess of the coupling resin at the degenerate positions. The ratio of input Fmoc-blocked amino acids should be adjusted on different synthesizers to achieve even distribution of degenerate amino acids.
2. Deprotect and then cleave the peptides off resins with trifluoroacetic acid (TFA).
3. Lyophilize the supernatants from step 2 to obtain crude peptide library powder mixtures.
4. Sequence 1–2 μg of the peptide libraries on an ABI peptide sequencer (Perkin-Elmer Biosystems) to confirm that all amino acids are present at similar amounts (within a factor of 3) at all degenerate positions.
5. Check the peptide library by mass spectrometry. A good library is indicated by the bell-shaped molecular weight distribution.

Protein Kinases

Kinase Preparation

Protein kinases are most commonly obtained through overexpression as recombinant proteins in either bacteria or eukaryotic cells. Using baculoviruses for expression in insect cells [Sf9 (*Spodoptera frugiperda* ovary) cells] is more desirable for most kinases and often yields active enzymes. Expressed protein kinases can be purified by conventional liquid chromatography (e.g., FPLC, fast protein liquid chromatography), affinity chromatography, or immunoprecipitation.[5] The amount of enzymes required to phosphorylate enough peptides for sequencing depends on the specific activity of individual kinases. In general, microgram quantities of kinases are used in our experiments.

Kinase Assay

Standard Kinase Reactions

1. Add the protein kinase (soluble or immobilized) to 300 μl of solution containing 300 μg to 1 mg of degenerate peptide mixture, 100 m*M* ATP, with a trace amount (~6×10^5 cpm) of [γ-^{32}P]ATP (3000 mCi/mmol) in kinase buffer [50 m*M* Tris (pH 7.4), 1 m*M* dithiothreitol, and 10 m*M* $MgCl_2$ (for Ser/Thr kinases) or 10 m*M* $MnCl_2$ (for Tyr kinases)]. Include a mock reaction in which no kinase is added as a control.
2. Incubate the mixture at 25–30° for 2 hr to phosphorylate roughly 1% of the peptide mixture.
3. Spot 10 μl of the reaction mixture on p81 paper. Wash three times with 100 ml of 75 m*M* phosphoric acid. If the counts are >1000 cpm, it indicates that enough phosphorylated peptides are present.
4. Terminate the reaction by adding acetic acid to a final concentration of 15% (v/v).

Phosphopeptide Separation

One of the most critical steps in using peptide libraries to study the substrate specificity of protein kinases is to isolate the phosphorylated peptides (<1% of the total peptide mixtures) from the large amount of unphosphorylated peptides. A ferric chelation column is used to achieve efficient separation, because the ferric ion can specifically bind the phosphate moiety.[6,7] However, the high concentration of ATP present in the

[5] T. Hunter and B. M. Sefton (eds.), *Methods Enzymol.* **200** (1990).

[6] G. Muszynska, G. Dobrowolska, A. Medin, P. Ekman, and J. O. Porath, *J. Chromatogr.* **604,** 19 (1992).

[7] G. Muszynska, L. Andersson, and J. Porath, *Biochemistry* **25,** 6850 (1986).

kinase reaction mixture may block the binding of most of the phosphate to the ferric chelation column. It is therefore necessary to first remove the free ATP from the peptide mixture.

Peptide Separation Using DEAE Column

A DEAE anion-exchange column can be used to quickly remove ATP, because (1) all peptides in the library contain a polylysine tail that is positively charged in 30% (v/v) acetic acid,[8] and (2) ATP becomes negatively charged at pH < 5.0.

SEPARATION OF PEPTIDES FROM ATP USING DEAE COLUMN

1. Wash 1–1.5 ml of DEAE-Sephacel beads twice with 10 ml of 30% (v/v) acetic acid. Remove the supernatants after centrifugation at 1000*g* for 2 min at room temperature.
2. Pack the beads in a 15-ml disposable column with 10 ml of 30% (v/v) acetic acid and allow the acetic acid to drip through by gravity. The final volume of the beads should be about 1 ml.
3. Load the peptide supernatants (from step 3 of Kinase Assay, above) carefully onto the top of the beads and allow all the liquids to run into the column.
4. Elute the column with 30% (v/v) acetic acid. Discard the first 600 μl of flowthrough and collect the next 1 ml.
5. Lyophilize the collected fraction on a Speed-Vac evaporator (Savant, Hicksville, NY).

The use of acetic acid helps to solubilize peptides as well as denature protein kinases and the contamination proteins. Under the above-described conditions, peptide mixtures are in the void volume because of their poly-(Lys) tail, while ATP and denatured protein kinases are retained on the column. We have determined that after the first 600-μl void volume, the next 1 ml contains phosphorylated as well as nonphosphorylated peptides that are free of [γ-^{32}P]ATP. Because this peptide fraction is free of [γ-^{32}P]ATP, its radioactivity provides an initial estimate of the fraction of the total peptide mixture that has been phosphorylated.

Ferric Chelation Column

A ferric chelation column (IDA beads; Pierce, Rockford, IL) is used for separation of phosphopeptides. To avoid loss of a subfraction of phosphopeptides during the quantitative removal of the nonphosphorylated peptides from the mixture, we have modified the loading and elution condi-

[8] B. E. Kemp, E. Benjamini, and E. G. Krebs, *Proc. Natl. Acad. Sci. U.S.A.* **73,** 1038 (1976).

tions from published procedures (used to separate tryptic phosphopeptides of phosphorylated proteins from the bulk of nonphosphorylated tryptic peptides).[6,7]

Phosphopeptide Separation on Ferric Chelation Column

1. Charge a 0.3-ml column of iminodiacetic acid (IDA)-coupled agarose beads (Pierce) with 2 ml of 20 m*M* ferric chloride at 0.5 ml/min.
2. Wash with 3 ml of distilled H_2O at 1 ml/min.
3. Wash with 3 ml of buffer C (500 m*M* NH_4HCO_3, pH 8.0) at 1 ml/min.
4. Wash again with 3 ml of distilled H_2O.
5. Equilibrate with 3 ml of buffer A [50 m*M* 2-(*N*-morpholino)ethanesulfonic acid (MES), 1 *M* NaCl, pH 5.5].
6. Dissolve the dried sample of peptide/phosphopeptide mixture in 200 μl of buffer A and load carefully onto the ferric column.
7. Elute the column with 3 ml of buffer A followed by 3 ml of distilled H_2O at 0.2 ml/min.
8. Elute the phosphopeptides with 3 ml of buffer C.
9. Elute the Fe^{3+} with 100 m*M* EDTA (pH 8.0).
10. Collect the buffer C eluate, which contains phosphopeptide, and lyophilize several times to get rid of most of the ammonium bicarbonate salt.
11. Resuspend the phosphopeptide mixture in 40 μl of distilled H_2O and adjust to neutral pH.
12. Spot and sequence 20 μl of the phosphopeptide mixture on an ABI peptide sequencer (Perkin-Elmer Biosystems).

The ferric chelation column is efficient in separating the phosphorylated from unphosphorylated peptides. However, a small percentage (~0.1%) of the degenerate unphosphorylated peptides rich in acidic amino acids are also copurified because they can bind weakly to the ferric column. This can be problematic for peptide libraries in which acidic residues (Asp and Glu) are fixed. To lower the background, acidic residues should be avoided on the fixed positions of peptide libraries. If they must be included in the fixed positions, the pH value of buffer A should be increased to 6.5 and the volume of IDA beads should be reduced.

Enough phosphopeptides (e.g., up to 1% input peptide library) should be purified for sequencing so that the phosphorylated peptides are in much excess of the contaminated unphosphorylated peptides. In a typical reaction, 1 mg of a peptide library (15-mer) is added. When 1% of the peptide mixture is phosphorylated, the total quantity of phosphopeptides is approximately 5 nmol. Roughly 1–2 nmol of the phosphopeptide mixture is typically added to the sequencer. This means that in a cycle in which all 18 residues are equally abundant, the yield of each amino acid is (1 nmol)(1/18) = 55 pmol.

Data Analysis

To determine the substrate preference of a kinase, it is necessary to obtain sequence data of three mixtures: the original peptide library mixture, the purified phosphopeptide mixture, and the purified background mixture from the mock experiment (see Kinase Assay, above). Sequencing of these phosphopeptide mixtures reveals the abundance of amino acids at each degenerate position. In theory, at a given sequencing cycle, the abundance of each amino acid in the phosphopeptide mixture could be divided by its abundance in the starting mixture to determine the optimal peptide motif for a protein kinase. Variations in the abundance of amino acids at a particular position in the starting mixture or variations in amino acid yield from the sequencer can thus be canceled out. If the kinase is insensitive to any amino acid at a given degenerate position, the relative abundance of all amino acids in the phosphopeptide mixture for this cycle will be the same as in the starting mixture.

If much less (e.g., 0.5%) of the total mixture is phosphorylated, a correction should be made for the ~0.1% of the nonphosphorylated peptides. The contamination with nonphosphorylated peptides can be estimated from the amount of Ser (or Thr and Tyr) that can be sequenced in the fixed phosphorylatable amino acid cycle (Ser, Thr, or Tyr). This is made possible because phosphorylated residues cannot be detected by the sequencer. In control reactions in which mock phosphorylation has been carried out with the peptides, the fractions of phosphopeptides are usually rich in Asp and Glu at every degenerate cycle, presumably because of their interaction with the Fe^{3+}. The abundance of each amino acid at each cycle from this control is subtracted to correct for the background.

To calculate the relative preference of amino acids at each degenerate position, the corrected data are compared with the starting mixture to generate ratios of amino acid abundance. The sum of the abundance of each amino acid at a given cycle is normalized to the number of amino acids present (e.g., 15 or 18). Each amino acid at a particular position thus has a value of 1 in the absence of selectivity. Even though any value greater than 1 should theoretically indicate a preference for that particular amino acid, values higher than 1.5 are generally more reliable because of the complexity of the data and subsequent calculation. Graphic plots showing enrichment values of amino acids at all degenerate positions can be generated after the calculation. The entire process is summarized below.

Data Analysis to Determine Substrate Specificity

1. Normalize the amount of each amino acid at the degenerate positions: we normalize the total amount of amino acids (in picomoles) at a given degenerate position to that at the first degenerate position.
   a. $P(ij)$ indicates amount for amino acid j at position i for the kinase experiment.
   b. $P_n(ij)$ indicates normalized $P(ij)$.
   c. $P_n(ij) = P(ij)$ [Sum($P1$)/Sum(Pi)].
2. Normalize sequences of the control experiment and original peptide library as 1.
   a. $C_n(ij)$ indicates normalized amount for amino acid j at position i in the control experiment.
   b. $R_n(ij)$ indicates normalized amount for amino acid j at position i in the original peptide library.
3. Subtract the control experiment values from those of the kinase experiment: $P_n(ij) - KC_n(ij)$. K can be calculated by the relative amount of fixed Ser or Tyr, $K = P(\text{Ser})/C(\text{Ser})$.
4. Calculate the relative abundance: $A(ij) = [P_n(ij) - KC_n(ij)]/R_n(ij)$.
5. Normalize to the total number of amino acids included at the degenerate position. If 18, then $A_n(ij) = A(ij)$ [18/Sum(Aij)]. $A_n(ij)$ represents the enrichment value for amino acid j at position i.

Prediction of *in Vivo* Substrates

Using the method described above, a consensus sequence motif should be reached for the kinase of interest. The consensus sequence can now be used to search the databases for candidate substrates that contain this motif. This section shows how multiple substrates can be predicted, using the protein Ser/Thr kinase Akt as an example. To identify the peptide substrate specificity of Akt, purified Akt kinase is used to phosphorylate an SD2 library (Met-Ala-Xaa-Xaa-Xaa-Xaa-Ser-Xaa-Xaa-Xaa-Xaa-Ala-Lys-Lys-Lys). Sequencing and comparison of the phosphorylated peptide mixtures to the original library mixtures reveal that Akt prefers Arg three residues (position −3) N terminal to the phosphorylation site. Secondary screening using an Arg-Ser-fixed RS random peptide library (Met-Ala-Xaa-Xaa-Xaa-Arg-Xaa-Xaa-Ser-Xaa-Xaa-Xaa-Xaa-Ala-Lys-Lys-Lys) shows that Akt also prefers Arg five residues (position −5) N terminal to the Ser residue and hydrophobic residues at positions +1 and +5 (Fig. 2).

The Akt substrate consensus motif identified by the peptide libraries described above has been subsequently used to search protein databases, using the Prowl Protein-info program (Rockfeller University, New York, NY). The search has yielded 43 potential targets that contain Ser sites for

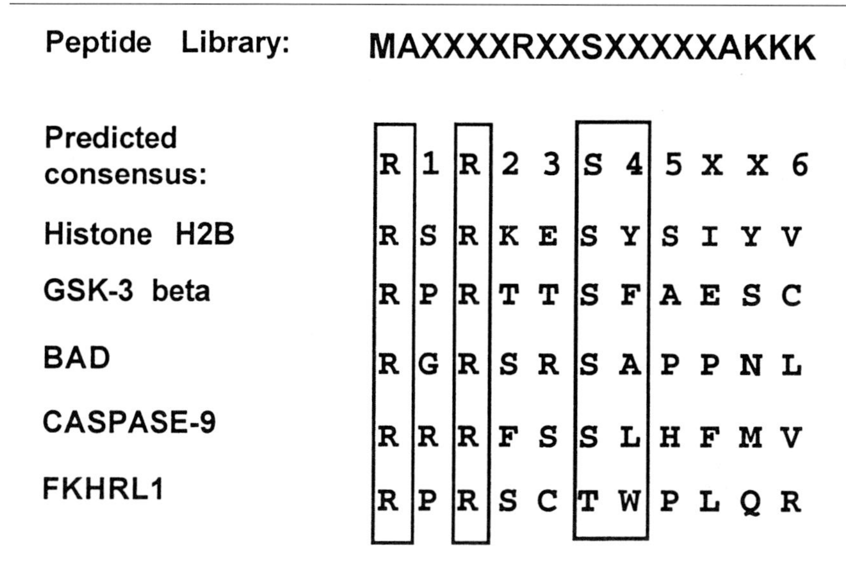

FIG. 2. Comparison of Akt peptide substrate specificity identified with peptide libraries and known substrates. Nonhydrophobic amino acids are preferred at positions 1, 2, 3, and 5; hydrophobic amino acids are preferred at positions 4 and 6.

the Akt kinase in human cells (Table II). Importantly, most of the known *in vitro* substrates of Akt, i.e., GSK-3 (substrate 11), Bad (substrate 41), PFK-2 (substrate 16), and caspase-9 (substrate 31), are among the predicted targets.[9–12] Other interesting candidates include a cell receptor (substrate 6), transcription factors (substrates 2, 26, 30, and 32), and RNA splicing factors (substrates 9, 13, and 27). Whether they can be phosphorylated by Akt may be tested individually *in vitro*. It is possible that some of these predicted proteins are *in vivo* targets of Akt.

Predicting Kinase Substrates and Developing Inhibitors of Protein Kinases

The oriented peptide library method is extremely helpful in predicting the optimal substrates and *in vivo* targets of various protein kinases. It

[9] J. Deprez, D. Vertommen, D. R. Alessi, L. Hue, and M. H. Rider, *J. Biol. Chem.* **272,** 17269 (1997).

[10] S. R. Datta, H. Dudek, X. Tao, S. Masters, H. Fu, Y. Gotoh, and M. E. Greenberg, *Cell* **91,** 231 (1997).

[11] M. H. Cardone, N. Roy, H. R. Stennicke, G. S. Salvesen, T. F. Franke, E. Stanbridge, S. Frisch, and J. C. Reed, *Science* **282,** 1318 (1998).

[12] J. Downward, *Curr. Opin. Cell Biol.* **10,** 262 (1998).

TABLE II
POTENTIAL Akt SUBSTRATES PREDICTED BY DATABASE SEARCH[a]

Substrate	Sequence	Residues
1. CGMP-inhibited cAMP phosphodiesterase	PAPVRRDRSTSIKLQEA	460–466
2. MPOU homeobox protein	EPSKKRKRRTSFTPQAI	235–241
3. PML-3	IRGAVRSRSRSLRGSSH	703–709
4. Myotonic dystrophy kinase	VWRPPRSRPRSLNPRTV	596–602
5. Rb-binding protein 1 (RBBP-1)	AKKTNRGRRSSLPVTED	142–148
6. Similar to human TRAMP protein	MAFRRRTKSYPLFSQ	4–10
7. BS-84	STEGSRSRSRSLDIQPS	19–25
8. KIAA0042	VSSLSRRRSRSLMKNRR	1187–1193
9. Splicing factor, arginine/serine-rich 7	SPSRSRSRSRSISRPRS	197–203
10. Bullous pemphigoid antigen 2	QVYAGRRRRRSIAVKP	1522–1528
11. **GSK-3 β**	MSGRPRTTSFAESCK	4–10
12. Fusion protein	IRGAVRSRSRSLRGSSH	645–651
13. Splicing factor SRP20	RRSFSRSRSRSLSRDRR	129–129
	SLSRDRRRERSLSRERN	133–139
14. Gene X104 protein	DLSRDRSRGRSLERGLD	195–201
	ARTRDRSRGRSLERGLD	215–221
	DRDRDRSRGRSIDQDYE	239–245
15. Calcium-sensing receptor	RSNVSRKRSSSLGGSTG	896–902
16. **6-Phosphofructo-2-kinase**	SSRLQRRRGSSIPQFTN	29–35
17. Inhibitor (aa 1–79)	LCFEGRKRQTSILIQKS	65–71
18. Titin	YDFYYRPRRRSLGDISD	3308–3314
19. Tuberin	EKDSFRARSTSLNERPK	934–940
20. Tuberous sclerosis protein 2	EKDSFRARSTSLNERPK	934–940
21. T cell receptor α chain	DGLEERGRFSSFLSRSK	74–80
23. T lymphoma invasion and metastasis-inducing protein 1 (TIAM1 protein)	SASKRRSRFSSLWGLDT	690–696
24. Glioma pathogenesis-related protein	WPIYPRNRYTSLFLIVN	230–236
25. Platelet CGI-PDE	PAPVRRDRSTSIKLQEA	460–466
26. Transcriptional corepressor SMRT	ERDRDREREKSILTSTT	786–792
27. Splicing factor SRP55	SRRSSRSRSRSISKSRS	36–42
30. Nuclear protein SKIP (SNW1 protein)	KARSQRSRQTSLVSSRR	28–34
31. **Caspase-9 precursor**	DCEKLRRRFSSLHFMVE	191–197
32. Forkhead-related transcription factor 4 (FREAC-4)	VPAQRRRRRRSYAGEDE	53–59
33. Similar to *C. elegans* protein	WSPVMRARKSSFNVSDV	602–608
34. Retina-derived POU-domain factor-1	EPSKKRKRRTSFTPQAL	565–571
35. Similar to rat RHOGAP	DKAKKRHRNRSFLKHLE	123–129
36. FMI protein	VSQAPRGRGTSLNFAEF	269–275
37. Sperm membrane protein	STEGSRSRSRSLDIQPS	11–17
38. T3 receptor cofactor-1, TRAC-1	ERDRDREREKSILTSTT	106–112
39. Topoisomerase-III	EKTVFRARFSSITDTDI	149–155
40. Myosin phosphatase target subunit 1	STTEVRERRRSYLTPVR	663–669
41. **BCL-X/BCL-2-binding protein bad**	GAVEIRSRHSSYPAGTE	70–76
42. IL-1 receptor-associated kinase-2; IRAK-2	ACLCLRRRNTSLQEVCG	483–489
43. Profilaggrin	ENKENRKRPSSLERRNN	122–128

[a] Boldface entries indicate known substrates.

should aid greatly in studying substrates of protein kinases in an organism whose genome has been completely sequenced. In particular, this technique can rapidly identify the optimal peptide substrates of a kinase without any prior knowledge of the kinase. Using existing programs such as BLAST, Fasta, or Findpatterns (GCG, Madison, WI), these optimal peptide sequences can be used to search protein databases. Alternatively, new search programs can be written by using the peptide library selection profile as matrixes. Any proteins that match are likely *in vivo* substrates of the kinase studied. Meanwhile, the sequence of a protein can be scanned for potential phosphorylation sites for any given protein kinase. Both approaches will provide a shortcut in studying signaling pathways regulated by protein kinases.

The predicted optimal peptide substrates can facilitate the designing of inhibitors for protein kinases. First, peptide or peptide mimic inhibitors can be made on the basis of the optimal peptides. Second, the optimal peptides can be used to screen chemical libraries for potential inhibitors. Finally, structural analysis of kinases in complex with their optimal peptides would provide a basis for modeling and designing drugs that specifically intervene with protein kinase-mediated signaling.

Acknowledgment

I thank Dr. Thomas Franke for providing Akt DNA.

[13] Peptide Library Screening for Determination of SH2 or Phosphotyrosine-Binding Domain Sequences

By ZHOU SONGYANG and DAN LIU

Introduction

Tyrosine Phosphorylation-Dependent Interactions

Protein tyrosine kinases (PTKs) play a crucial role in cellular proliferation and differentiation. Signaling cascades initiated by PTKs are largely transduced via protein modules (or protein domains) that interact with phosphotyrosine-containing sequences.[1] These protein modules have been found in many signaling proteins. One good example is the Src homology

[1] T. Pawson and J. D. Scott, *Science* **278,** 2075 (1997).

0076-6879/00 $35.00

domains (SH1, −2, and −3) that have been found in a variety of signaling molecules.[1,2] Therefore understanding these phosphotyrosine interaction domains should help greatly in our quest for dissecting and delineating the signaling networks that govern cellular responses. Major progress has been made in identifying and decoding the specificities of phosphotyrosine-interacting domains, i.e., SH2 and phosphotyrosine-binding (PTB/PID) domains.[3,4] This chapter focuses on using the oriented peptide library technique to determine the specificities of these two domains, especially the PTB domains.

From SH2 to Phosphotyrosine-Binding Domains

SH2 domains recognize the phosphotyrosine moiety. They were originally identified on the basis of homologies between the Src PTK and several other signaling proteins.[5] Tyrosine phosphorylation by PTKs creates binding sites for SH2 domains. Proteins containing the SH2 domains can thus be recruited to the site of phosphorylation and signaling complexes can form in a phosphotyrosine-dependent manner. We determined that the primary sequences adjacent to the phosphotyrosine were required for specific SH2 domain recognition, and showed that residues C terminal to the phosphotyrosine were critical for high affinity and specificity in SH2 domain binding.[6–9]

While the specificities of various SH2 domains were being mapped with oriented peptide libraries, studies from several laboratories were indicating that residues N terminal of the phosphotyrosine were also critical for specific binding to certain SH2-containing proteins such as SHC.[2] The SH2 domain of SHC was thought to mediate SHC binding to the tyrosine-phosphorylated NPXY motifs on several signal molecules.[10–12] Still more puzzling was the

[2] T. Pawson and J. Schlessinger, *Curr. Biol.* **3,** 434 (1993).

[3] T. Pawson, *Nature* (*London*) **373,** 573 (1995).

[4] S. E. Shoelson, *Curr. Opin. Chem. Biol.* **1,** 227 (1997).

[5] T. Pawson, *Oncogene* **3,** 491 (1988).

[6] Z. Songyang, S. E. Shoelson, M. Chaudhuri, G. Gish, T. Pawson, W. G. Haser, F. King, T. Roberts, S. Ratnofsky, R. J. Lechleider, *et al., Cell* **72,** 767 (1993).

[7] Z. Songyang, S. E. Shoelson, J. McGlade, P. Olivier, T. Pawson, X. R. Bustelo, M. Barbacid, H. Sabe, H. Hanafusa, T. Yi, *et al., Mol. Cell. Biol.* **14,** 2777 (1994).

[8] L. C. Cantley, K. R. Auger, C. Carpenter, B. Duckworth, A. Graziani, R. Kapeller, and S. Soltoff, *Cell* **64,** 281 (1991).

[9] W. J. Fantl, J. A. Escobedo, G. A. Martin, C. W. Turck, M. del Rosario, F. McCormick, and L. T. Williams, *Cell* **69,** 413 (1992).

[10] S. A. Prigent and W. J. Gullick, *EMBO J.* **13,** 2831 (1994).

[11] K. S. Campbell, E. Ogris, B. Burke, W. Su, K. R. Auger, B. J. Druker, B. S. Schaffhausen, T. M. Roberts, and D. C. Pallas, *Proc. Natl. Acad. Sci. U.S.A.* **91,** 6344 (1994).

[12] R. M. Stephens, D. M. Loeb, T. D. Copeland, T. Pawson, L. A. Greene, and D. R. Kaplan, *Neuron* **12,** 691 (1994).

fact that many of the mapped *in vivo* SHC binding sites were in poor agreement with the optimal motif we had predicted for the SH2 domain of SHC, using the original peptide library.[7] It turns out that there exists a second type of phosphotyrosine-binding domain in the N terminus of SHC. It was named the PTB domain.[13,14] In light of those results, we went on to compare the specificities of the SH2 and PTB domains, using different oriented peptide libraries.

A number of PTB domains have since been discovered in different signaling proteins, some of which are listed in Fig. 1.

Peptide Library Design

Double-Degenerate Phosphopeptide Library

The original phosphopeptide libraries used to study the SH2 domains varied only amino acids C terminal of the phosphotyrosine and cannot be used to examine specificity N terminal to the phosphotyrosine. We therefore designed a double-degenerate phosphopeptide library, DDL (Gly-Ala-X-X-X-pTyr-X-X-X-Lys-Lys-Lys, where X indicates all amino acids except Cys and Trp).[6,15] In this library, the three residues both N and C terminal to the phosphotyrosine are random. Cys and Trp were omitted to avoid oxidation and sequencing problems. This library has a theoretical degeneracy of 18^6 (>32 million); its complexity is about three orders of magnitude higher than the C-terminal degenerate library described earlier.

In theory, it is possible to make a peptide library as degenerate as possible. However, the more degenerate a library is, the less sensitive sequencing will be. The N-terminal Met-Ala sequence is chosen as a leader because of their reliability in amino acid coupling and sequencing. The poly(Lys) tail C terminal to the degenerate residues is added to ensure high solubility of the peptides and their firm attachment to hydrophilic sequencing membranes.

Secondary Structure-Constrained Phosphopeptide Library

The specificities of SHC and IRS-1 PTB domains indicate that PTB domains prefer to bind peptides that form β turns N terminal to the tyro-

[13] W. M. Kavanaugh and L. Williams, *Science* **266,** 1862 (1994).

[14] P. Blaikie, D. Immanuel, J. Wu, N. Li, V. Yajnik, and B. Margolis, *J. Biol. Chem.* **269,** 32031 (1994).

[15] Z. Songyang, B. Margolis, M. Chaudhuri, S. E. Shoelson, and L. C. Cantley, *J. Biol. Chem.* **270,** 14863 (1995).

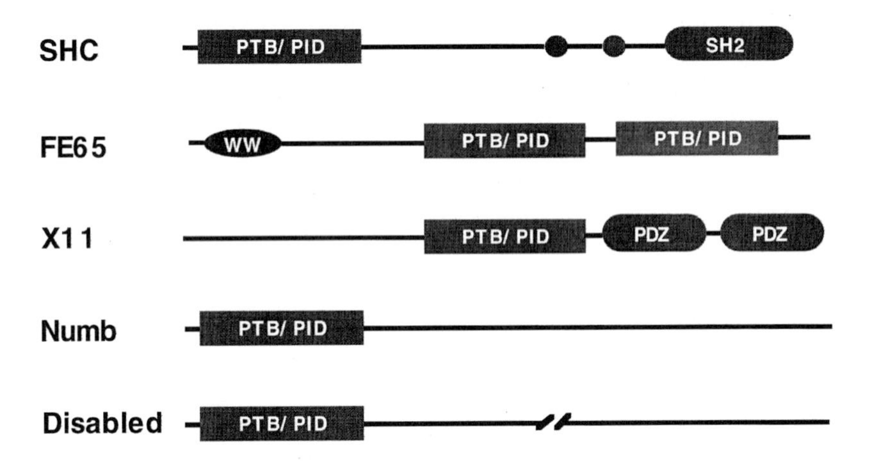

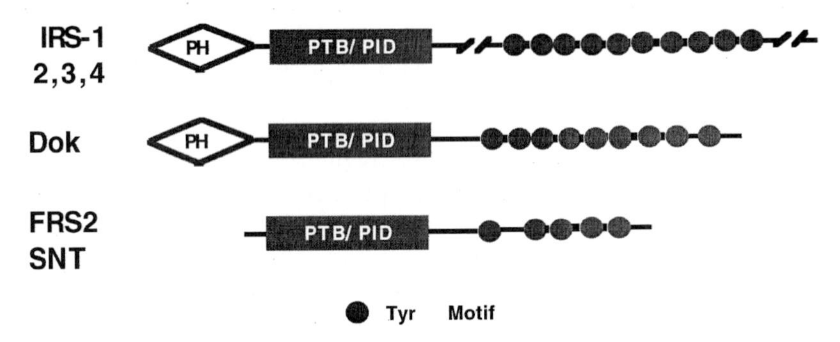

FIG. 1. A comparison of proteins that contain PTB/PID domains.

sine.[16,17] To mimic such conformation, we have synthesized a secondary structure-constrained library, PTBL (Met-Ala-X-X-X-Asn-X-X-pTyr-X-Ala-Lys-Lys-Lys, where X indicates all amino acids except Cys and Trp).[18]

[16] T. Trub, W. E. Choi, G. Wolf, E. Ottinger, Y. Chen, M. Weiss, and S. E. Shoelson, *J. Biol. Chem.* **270,** 18205 (1995).

[17] G. Wolf, T. Trub, E. Ottinger, L. Groninga, A. Lynch, M. F. White, M. Miyazaki, J. Lee, and S. E. Shoelson, *J. Biol. Chem.* **270,** 27407 (1995).

[18] J. P. O'Bryan, C. B. Martin, Z. Songyang, L. C. Cantley, and C. J. Der, *J. Biol. Chem.* **271,** 11787 (1996).

Asn is fixed at the −3 position because it is frequently found in turns or loop structures as inferred from known structures of hundreds of proteins. In addition, the NXXY sequence is known to prefer to adopt the β-turn conformation.[19,20] This library will therefore increase the probability of isolating peptides that are recognized by PTB domains from the random peptide mixtures. Furthermore, it will allow us to examine binding specificities of PTB domains for positions further N terminal of the phosphotyrosine of their ligands.

Tyrosine-Degenerate Peptide Library

A surprising finding about PTB domains is that not all PTB domains recognize phosphotyrosine-containing sequences. For instance, PTB domains found in FE65, X11, Numb, and Dab interact with sequences that contain aromatic residues in a non-phosphorylation-dependent manner.[21–23] To study the specificities of non-phosphotyrosine-binding PTB domains, we constructed the YD2 library (Met-Ala-X-X-X-X-Tyr-X-X-X-X-Ala-Lys-Lys-Lys, where X is all amino acids except Cys and Trp).[22] In this case, the peptide library is oriented via the fixed Tyr residue, which is frequently recognized by PTB domains.

Methods

Peptide Library Synthesis

All peptide libraries are synthesized according to the N^{α}-Fmoc-based strategy and conducted on a Milligen/Biosearch 9600 synthesizer (PE Biosystems, Foster City, CA), using standard BOP/HOBt coupling protocols. Similar to the C-terminal degenerate phosphopeptide library,[6] the double-degenerate peptide library DDL is synthesized by a split-and-pool approach. However, for the YD2 and PTBL libraries, amino acids are no longer coupled by the split-synthesis approach at the degenerate positions. Instead, the resins for 18 different N^{α}-Fmoc-amino acids are mixed at 5-fold excess and coupled simultaneously. The quality of the peptide libraries synthesized in this way is similar to that generated by split synthesis as

[19] W.-J. Chen, J. L. Goldstein, and M. S. Brown, *J. Biol. Chem.* **265,** 3116 (1990).

[20] A. Bansal and L. M. Gierasch, *Cell* **67,** 1195 (1991).

[21] J. P. Borg, J. Ooi, E. Levy, and B. Margolis, *Mol. Cell. Biol.* **16,** 6229 (1996).

[22] S. C. Li, Z. Songyang, S. J. Vincent, C. Zwahlen, S. Wiley, L. Cantley, L. E. Kay, J. Forman-Kay, and T. Pawson, *Proc. Natl. Acad. Sci. U.S.A.* **94,** 7204 (1997).

[23] B. W. Howell, L. M. Lanier, R. Frank, F. B. Gertler, and J. A. Cooper, *Mol. Cell. Biol.* **19,** 5179 (1999).

judged by peptide sequencing and mass spectrometry. This coupling process may bias toward amino acids that are easy to couple. For example, sequences such as Val-Val-Val will be underrepresented in the mixture. However, such bias is unlikely to affect the experimental readouts because such sequences are rare and may be substituted by related sequences such as Val-Z-Val, Z-Val-Val, and Val-Val-Z (where Z is any amino acid except Val, Cys, and Trp). Furthermore, when both mixed and split-coupled peptide libraries are used to study SH2 domain specificities, the results obtained are identical.

Affinity Purification of Phosphopeptides Using GST–SH2 Domains

Fusion Protein Purification. Methods for preparation of glutathione *S*-transferase (GST)–SH2 and GST–PTB domain fusion proteins are essentially the same as previously described.[24]

1. Dilute overnight cultures of *Escherichia coli* transformed with various pGEX plasmids in Terrific broth (TB) or LB with ampicillin (50 μg/ml) and shake at 37°. When the OD_{600} reading reaches ~0.8 (about 2–3 hr), add isopropyl-β-D-thiogalactopyranoside (IPTG) to a final concentration of 0.2 m*M* to induce expression for 4 to 8 hr at 37°.
2. Collect and lyse the bacteria on ice for 20 min in lysis buffer [3 ml/g bacteria, 50 m*M* Tris (pH 8.0), 1 m*M* EDTA, 100 m*M* NaCl] with lysozyme (1 mg/ml), 1 m*M* dithiothreitol (DTT), and 100 μ*M* phenylmethylsulfonyl fluoride (PMSF).
3. Add sodium deoxycholic acid (1-mg/ml final concentration) and shake the lysate at 25° for 30 min.
4. Add 100 μg of crude DNase I to the lysates to clear the chromosome DNA for 1 hr at 25°. Alternatively, sonicate the lysate four times (10 sec each) at 50% power. Then centrifuge the lysate at 5000 rpm for 15 min.
5. Collect and incubate the supernatant with glutathione beads at 4° for 1–2 hr.
6. Wash the beads three times with phosphate-buffered saline (PBS: 150 m*M* NaCl, 3 m*M* KCl, 10 m*M* Na_2HPO_4, 2 m*M* KH_2PO_4, pH 7.2) containing 0.5% Nonidet P-40 (v/v) (NP-40) and 1 m*M* DTT, and once with PBS.

Now the beads are ready for affinity purification of peptides.

Affinity Purification and Sequencing

1. Pack 300 μl of fusion protein beads (100–300 μg of protein containing the domain of interest, and also of a mock control; see step 6 from the

[24] Z. Songyang and L. C. Cantley, *Methods Enzymol.* **254,** 523 (1995).

previous section) in a 3-ml syringe that is used as an affinity column. *Note:* We have doubled the volume of GST fusion beads used as compared with our previously published protocols. Under this new condition, the amount of peptides selected by the GST–SH2 and PTB domains can be significantly increased. This adjustment may be especially important for fusion proteins that are of low yield.

2. Wash the beads twice with 2 ml of PBS. Resuspend the degenerate peptides (1 mg) in 100 μl of PBS and load it onto the top of the column. Stop the column flow and allow the column to stand at room temperature for 10 min.

3. Wash the column quickly three times with 2 ml of ice-cold PBS that contains 0.5% (v/v) NP-40 and once with 2 ml of ice-cold PBS in the cold room. *Note:* NP-40 helps to block nonspecific binding. To avoid peptide dissociation, pressure is added to accelerate the wash so that the entire washing time takes less than 1 min. Quick washing is critical because phosphopeptide and SH2 domain association exhibits a fast off-rate (~10 sec).[25]

4. Load 300 μl of 30% (v/v) acetic acid solution onto the column at room temperature to elute the peptides. Collect the flowthrough, dry it down in a Speed-Vacevaporator (Savant, Hicksville, NY), and resuspend it in 40 μl of doubly distilled H_2O. *Note:* We have previously used a sodium phenylphosphate solution to elute the bound peptides, because phenylphosphate can compete with phosphotyrosine. However, we found that acetic acid could also effectively disassociate the bound peptides from the domains. This replacement may be particularly beneficial in the case of certain PTB domains that do not bind phosphotyrosine-containing sequences (see below).

5. Centrifuge the mixture at 10,000*g* for 10 min at room temperature to recover the peptides. (At this point, peptides are soluble whereas GST fusion proteins are denatured and insoluble.)

6. Sequence these purified peptide mixtures on an Applied Biosystems 477A protein sequencer (Perkin-Elmer Biosystems, Foster City, CA).

Data Interpretation and Analysis

Because a mixture of peptides rather than individual peptides has been affinity isolated, the selectivity can be determined only by comparing the purified mixture with the original crude mixture and calculating the relative enrichment value (see below).[6]

At a given cycle (e.g., cycle Z, which corresponds to position Z), if a

[25] S. Felder, M. Zhon, P. Hu, J. Urena, A. Ullrich, M. Chaudhuri, M. White, S. E. Shoelson, and J. Schlessinger, *Mol. Cell. Biol.* **13,** 1449 (1993).

particular amino acid (e.g., amino acid *B*) is responsible for high-affinity binding, peptides containing that amino acid at position *Z* will be prevalent in the mixture. Theoretically, a comparison of the sequence of this mixture with that of the crude material should show the greatest ratio for amino acid *B* at cycle *Z*. On the other hand, if the domain of interest has no preference for any amino acids at a degenerate cycle, every amino acid should have a similar ratio. If the sum of ratios is normalized to the total number of degenerate amino acids, amino acids that are not selected by the domain will have a value of 1 or less.

Two changes have been made to the calculation, because of practical considerations. First, the total amount of picomoles at a degenerate cycle should be normalized to that of the first degenerate cycle. This is because the rate of washout during sequencing differs from day to day, and from sequencer to sequencer. Second, the amount of each amino acid eluted from GST–domain beads should be divided by that of the control GST column at the same cycle instead of the original peptide mixture. This step should eliminate any variations resulting from the affinity columns and the sequencing. To obtain the enrichment value, the background level (background binding to GST) is subtracted and the data normalized again.

Using data obtained this way, the order of preference for every amino acid at a given degenerate position can be easily determined. A larger number indicates a stronger selection. In addition, a comparison of enrichment values between the different degenerate positions will indicate which position is more selective.

Insights into SH2 and Phosphotyrosine Binding Domain Binding Specificities

SH2 Domain Preference for Sequences C Terminal of Phosphotyrosine

We have used the double-degenerate peptide library (GAXXX-pYXXXKKK) to study six SH2 domains (p85, Crk, Nck, SHC, GRB2, and Csk).[15] This library has predicted the same optimal motifs as the original library, which varied only amino acids C terminal of the phosphotyrosine. These SH2 domains recognize short pY-containing peptide motifs and three to six specific residues immediately C terminal to the pY. They have no apparent preference for amino acids N terminal to the phosphotyrosine. Interestingly, the Crk, Nck, and SHC SH2 domains showed a weak selection for Tyr at the pY−1 position. The p85, Crk, and SHC domains also exhibited a weak selection for acidic amino acids at the pY−2 and −3 positions.

Our studies using the double-degenerate peptide library confirmed that SH2 domains prefer sequences primarily C terminal to the phosphotyrosine.

In fact, SH2 domains can be divided into two major groups based on structural information and known consensus motifs of more than 30 SH2 domains.[6,7] One group, including those of Src family kinases, Crk, and Nck, contains aromatic residues at their βD5 position and prefer to bind phosphopeptide sequences that have acidic/hydrophilic residues at the pY+1 and hydrophobic residues at the pY+3 positions. The other group, such as those of phosphatidylinositol 3-kinase (PI-3 kinase) and phospholipase Cγ (PLC-γ), on the other hand, contain Cys/Ile/Val residues at their βD5 position and prefer sequences that have hydrophobic residues at both the pY+1 and pY+3 positions. The GRB2 SH2 domain may be unique in that it strongly favors Asn at the pY+2 position.

Specificity of Phosphotyrosine-Binding Domain

Phosphotyrosine-Binding Domain Preference for Sequences N Terminal to Tyrosine. The SHC PTB domain shows an entirely different selection when tested with the double-degenerate library. It prefers peptides with Asn three residues N terminal to the phosphotyrosine. The optimal motif selected was Asn-Pro-X-pTyr-Phe-X-Arg (NPXY) (Table I). In contrast to SH2 domains, the PTB domain selectivity appears to be dominated by contact with residues N terminal of the phosphotyrosine moiety.

The finding that the PTB domain of SHC preferentially recognizes sequences N terminal to the phosphotyrosine has provided a new insight into protein–protein interactions mediated by tyrosine phosphorylation.

TABLE I
RECOGNITION SPECIFICITIES OF DIFFERENT PTB DOMAINS

PTB domain	Binding specificity	Phosphotyrosine dependent	Analyzed by peptide libraries	Predicted turn conformation
SHC-A	(L/F)XNPX**pY**F	Yes	DDL, PTBL	Yes
SHC-C	(Y/F/L)XNP(S/T)**pY**F	Yes	PTBL	Yes
IRS-1	(I/L)X(I/F)XXNPX**pY**	Yes	PTBL	Yes
SHB	DDX**pY**	Yes	DDL	Yes
Numb	YIGP**pY**	Partial	YD	Yes
	YIGP**Y**			
FRS2/SNT	IMENPQ**pY**f	Yes	—	Yes
	FGFR1 residue 401–434	?		
X11	QNGYENPT**Y**KFF	No	—	Yes
FE65	QNGYENPT**Y**KFF	No	—	Yes
Dab	QNGYENPT**Y**KFF	No	—	Yes

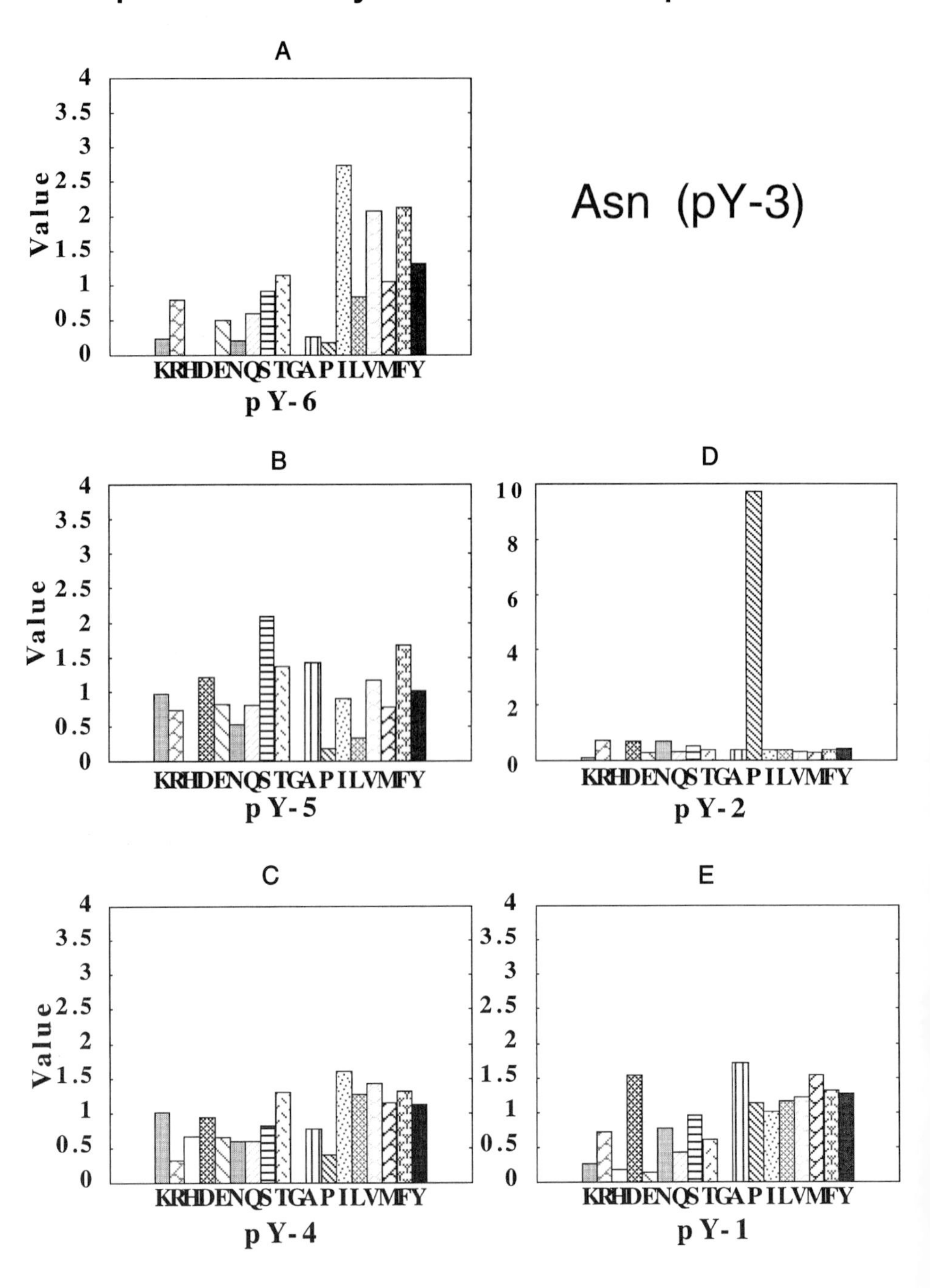
Peptide library: MAXXXNXXpYXAKKK
Asn (pY-3)
A
B
C
D
E
Value
4
3.5
3
2.5
2
1.5
1
0.5
0
10
8
6
4
2
KRHDENQSTGAPILVMFY
pY-6
pY-5
pY-4
pY-2
pY-1

The NPXY sequence is found in regions of tyrosine-phosphorylated TrkA, ErbB3, and polyoma middle t antigen to which SHC binds.[10–12] Therefore SHC binding to these proteins is likely to be mediated by its PTB domain rather than the SH2 domain. Consistent with our peptide library prediction that Asn at pY−3 and Pro at pY−2 are important for binding to the PTB domain, mutation of these two residues in polyoma middle t antigen and ErbB3 inhibited *in vivo* SHC association.[10,11] Some of these known sites for SHC binding also have phenylalanine or hydrophobic amino acids at the pY+1 position, in agreement with the weak selection for these residues in our peptide library results.

The fact that the SHC PTB domain prefers the NPXY motif suggested that PTB domains recognize turn structures and their binding is probably conformation dependent. We therefore constructed the PTBL library (Met-Ala-X-X-X-Asn-X-X-pTyr-X-Ala-Lys-Lys-Lys), which adopts a more rigid conformation. The unique feature that distinguishes this library from all previous libraries is that an Asn residue is fixed, which forces the peptides to adopt a β-turn conformation.

Using this fixed library, we examined the specificities of a number of PTB domains including those of SHC-A, SHC-C,[18] and IRS-1 (Fig. 2). Although the IRS-1 PTB domain exhibits extremely low sequence homology (15% identical) to the SHC PTB domain, these two domains have a similar specificity as determined by the PTBL library. Both SHC and IRS-1 PTB domains prefer the phosphorylated NPXY sequence. In addition, these PTB domains strongly select peptides with hydrophobic amino acids at positions 5 to 8 residues N terminal to the phosphotyrosine. Such preference is entirely consistent with both nuclear magnetic resonance (NMR) and crystal structure data of SHC and IRS-1 PTB domains. For instance, hydrophobic pockets were found on these PTB domains that could coordinate the pY−5 (for the SHC PTB) and pY−6 to −8 residues (for the IRS-1 PTB domain).[17]

Specificities of Phosphotyrosine-Binding Domains That Recognize Non-Phosphotyrosine-Containing Sequences. The original PTB/PID domain nomenclature may be somewhat misleading because some PTB domains can also bind unphosphorylated sequences. For example, PTB domains of Dab and FE65 can both associate with the β-amyloid precursor protein in a phosphorylation-independent manner.[21,23] Interestingly, the site on the am-

---

FIG. 2. Ligand-binding specificity of the IRS-1 PTB domain. The PTBL library was used to examine the binding preference of the IRS-1 PTB domain. (A–E) Selection of different amino acids at positions −6 to −1 N terminal to the phosphotyrosine. A value larger than 1.5 indicates significant preference.

yloid precursor protein that directly interacts with these PTB domains is the NPTY sequence. Synthetic peptides made from these sequences bind with high affinity to the PTB domains of FE65, X11, and Dab.[21,23] Replacement of the Tyr residue with phosphotyrosine did not affect such interactions, indicating that phosphorylated tyrosine is not always required for PTB domain-mediated interactions. Even more intriguing are the results from studies of the Numb PTB domain. Our peptide library screen using the YD library has shown that the Numb PTB domain prefers the GPYL motif. At the same time, the Numb PTB domain can also bind the phosphorylated GPpYL peptide with 10-fold higher affinity. Thus, a PTB domain may be capable of binding both phosphorylated and unphosphorylated sequences.

Phosphotyrosine-Binding Domains Recognizing Turn–Loop Structures. The difference in recognition specificities between SH2 and PTB domains may be a manifestation of the structural differences between these two domains. The NPXY motif, originally discovered to code for protein internalization, can form a tight turn structure as determined by NMR studies.[19,20] On the basis of early studies of PTB domains of SHC and IRS-1, the PTB domain was thought to recognize phosphopeptide motifs in which the phosphotyrosine was preceded by β-turn-forming amino acids (often NXXpY).[16–18] We have confirmed this observation with studies of several other PTB domains including those of Shb[26] and Numb.[22] Our data suggest that PTB domains probably recognize turn/loop structures in general (Table I).

Turn-forming residues may not contribute to direct binding of ligands; the specificity of PTB domain recognition therefore is determined by hydrophobic amino acids five to eight residues N terminal to the tyrosine.[22,27–29] As discussed above, several PTB domains have been found to associate with their targets in a non-phosphotyrosine-dependent manner. These results do not contradict with the idea that PTB domains may have evolved to bind turn/loop structural motifs. All the PTB domains identified to date that mediate phosphorylation-independent interactions recognize sequences containing Asn and Gly (Table I). These residues are predicted to form β or helix turns. It would be interesting to determine what effect

[26] M. L. Lupher, Jr., Z. Songyang, S. E. Shoelson, L. C. Cantley, and H. Band, *J. Biol. Chem.* **272,** 33140 (1997).

[27] M. M. Zhou, K. S. Ravichandran, E. F. Olejniczak, A. M. Petros, R. P. Meadows, M. Sattler, J. E. Harlan, W. S. Wade, S. J. Burakoff, and S. W. Fesik, *Nature* (*London*) **378,** 584 (1995).

[28] M. J. Eck, S. Dhe-Paganon, T. Trub, R. T. Nolte, and S. E. Shoelson, *Cell* **85,** 695 (1996).

[29] Z. Zhang, C. H. Lee, V. Mandiyan, J. P. Borg, B. Margolis, J. Schlessinger, and J. Kuriyan, *EMBO J.* **16,** 6141 (1997).

phosphorylation may have on the interaction between these PTB domains and their targets.

Acknowledgment

We thank Dr. Steve Shoelson for kindly providing the IRS-1 PTB domain.

[14] Expression Cloning of Farnesylated Proteins

By DOUGLAS A. ANDRES

Introduction

The posttranslational modification of proteins by the covalent attachment of farnesyl and geranylgeranyl groups to cysteine residues at or near the C terminus via a thioether bond is now well established in mammalian cells.[1] Among these proteins are the low molecular mass Ras-related GTP-binding proteins, the gamma (γ) subunits of heterotrimeric G proteins, and nuclear lamins. These proteins are involved in cellular signaling, regulation of cell growth, intracellular vesicle transport, and organization of the cytoskeleton. Farnesylation and geranylgeranylation of cysteinyl residues within the C termini of these proteins have been shown to promote both protein–protein and protein–membrane interactions.[7] Isoprenylation, and, in some cases, the subsequent palmitoylation, provide a mechanism for the membrane association of these proteins, which lack a transmembrane domain, and appear to be a prerequisite for their *in vivo* activity.[3,4]

Three distinct protein prenyltransferases catalyzing these modifications have been identified.[5,6] Two geranylgeranyltransferases (GGTases) have been characterized, and are known to modify distinct protein substrates. The C*aa*X GGTase (also known as GGTase-1) geranygeranylates proteins that end in a C*aa*L/F sequence, where C is cysteine, *a* is usually an aliphatic amino acid, and the C-terminal amino acid group is leucine (L) or phenylalanine (F). Rab GGTase (also known as GGTase-2) catalyzes the attachment

[1] P. J. Casey and M. C. Seabra, *J. Biol. Chem.* **271,** 5289 (1996).
[2] J. A. Glomset and C. C. Farnsworth, *Annu. Rev. Cell Biol.* **10,** 181 (1994).
[3] K. Kato, A. D. Cox, M. M. Hisaka, S. M. Graham, J. E. Buss, and C. J. Der, *Proc. Natl. Acad. Sci. U.S.A.* **89,** 6403 (1992).
[4] J. F. Hancock, A. I. Magee, J. E. Childs, and C. J. Marshall, *Cell* **57,** 1167 (1989).
[5] F. L. Zhang and P. J. Casey, *Annu. Rev. Biochem.* **65,** 241 (1996).
[6] S. Clarke, *Annu. Rev. Biochem.* **61,** 355 (1992).

0076-6879/00 $35.00

of two geranylgeranyl groups to paired C-terminal cysteines in most members of the Rab family of GTP-binding proteins. These proteins terminate in CC, CXC, or CCXX motifs, where X is a small hydrophobic amino acid. A final set of regulatory proteins is modified by protein farnesyltransferase (FTase). All known farnesylated proteins terminate in a tetrapeptide C*aa*X box, wherein C is cysteine, *a* is an aliphatic amino acid, and X has been shown to be a C-terminal methionine, serine, glutamine, cysteine, or alanine.[7]

The identification of novel prenylated proteins has most often involved the labeling of cultured cells with [^{3}H]mevalonolactone, which is enzymatically converted to [^{3}H]farnesyl pyrophosphate (FPP) and [^{3}H]geranylgeranyl pyrophosphate (GGPP) prior to being incorporated into protein. Cellular proteins can then be analyzed by gel electrophoresis and autoradiography. The labeling procedure is improved by the inclusion of hydroxymethylglutaryl (HMG)-CoA reductase inhibitors, which prevent dilution of the [^{3}H]mevalonolactone by blocking mevalonate synthesis within the cell.[8,9] However, labeling is still inefficient, because of slow [^{3}H]mevalonate uptake, and is therefore limited to the detection of abundant proteins.[10] A second limitation of [^{3}H]mevalonate labeling is that it does not distinguish between farnesylated and geranylgeranylated proteins. The discovery that mammalian cells can utilize the free allylic isoprenoids, geranylgeraniol (GG-OH) and farnesol (F-OH), for the isoprenylation of cellular proteins provides a method for making this distinction.[11–13] Cellular radiolabeling utilizing free prenyl alcohols overcomes a number of limitations imposed by mevalonate labeling; F-OH and GG-OH are efficiently utilized in a wide range of mammalian cell lines without the inclusion of HMG-CoA reductase inhibitors, and each prenyl alcohol labels a distinct subset of the cellular prenylated proteins, which allows farnesylated or geranylgeranylated proteins to be specifically radiolabeled. This improved method, however, is still limited to the analysis of abundant proteins and requires extensive protein purification before the amino acid sequence of any *in vivo*-radiola-

[7] W. J. Chen, D. A. Andres, J. L. Goldstein, D. W. Russell, and M. S. Brown, *Cell* **66,** 327 (1991).

[8] R. A. Schmidt, C. J. Schneider, and J. A. Glomset, *J. Biol. Chem.* **259,** 10175 (1984).

[9] J. L. Goldstein and M. S. Brown, *Nature (London)* **343,** 425 (1990).

[10] J. Faust and M. Krieger, *J. Biol. Chem.* **262,** 1996 (1987).

[11] D. C. Crick, D. A. Andres, and C. J. Waechter, *Biochem. Biophys. Res. Commun.* **237,** 483 (1997).

[12] D. A. Andres, D. C. Crick, B. S. Finlin, and C. J. Waechter, *Methods Mol. Biol.* **116,** 107 (1999).

[13] D. C. Crick, C. J. Waechter, and D. A. Andres, *Biochem. Biophys. Res. Commun.* **205,** 955 (1994).

beled protein can be obtained. In this chapter we outline an enzymatic screening protocol to identify novel farnesylated proteins from bacterial expression libraries. This method allows the systematic screening of large cDNA libraries in a manner that is quicker and simpler than the conventional approach of purifying radiolabeled proteins, obtaining protein sequence, and cloning by degenerate oligonucleotides.

Methods

Plating and Screening Library

Bacterial expression libraries were originally developed as a system to clone genes using antibody probes. In these methods, bacteria are infected with a phage, such as λgt11, that contains cDNA fragments fused to β-galactosidase. After induction with isopropyl-β-D-thiogalactopyranoside (IPTG), the phage protein is produced and is present in plaques that can be transferred to nitrocellulose filters. These filters are then probed with antibody, DNA, or protein probes.[14] However, we have found that a plasmid expression system, rather than bacteriophage λ vector-based expression, is necessary to achieve the protein expression levels required for successful prenylation library screening (B. Finlin and D. Andres, unpublished observations, 1999). Maximal protein expression is necessary because of the low specific activity of the radioisotopes, the low energy of tritium beta(β) particles, and the relatively low proportion of proteins capable of being labeled once bound to filters.[15,16]

One of the most important aspects of bacterial expression cloning is the quality of the library. Many commercial libraries contain a relatively small number of independent cDNA inserts (fewer than 1×10^6). Unless the library is directionally cloned, only one in six of these clones expresses a protein in the correct reading frame and orientation. These factors, combined with the fact that commercial libraries are often subject to serial amplification, can make it difficult to clone a gene of interest. Thus, it is usually best to start with a library that contains the maximal number of independent clones (greater than 3×10^6) and that has not been repeatedly amplified. In addition, it may be necessary to screen several million colonies owing to the low percentage of colonies that express properly orientated

[14] J. Sambrook, E. F. Fritsch, and T. Maniatis, "Molecular Cloning: A Laboratory Manual." Cold Spring Harbor Laboratory Press, Cold Spring Harbor, New York, 1989.

[15] D. A. Andres, H. Shao, D. C. Crick, and B. S. Finlin, *Arch. Biochem. Biophys.* **346,** 113 (1997).

[16] H. Shao, K. Kadono-Okuda, B. S. Finlin, and D. A. Andres, *Arch. Biochem. Biophys.* **371,** 207 (1999).

fusion proteins. One final factor that should be considered is the construction of the library. Our cloning efforts have focused on oligo(dT)-primed directional libraries to bias clone representation to the C termini.

Preparation of Colony Protein Filters

In this protocol we describe the screening of a primary library constructed to produce glutathione *S*-transferase (GST) fusion proteins.[15] The methods for construction of a library using this expression vector have been described elsewhere.[15] However, this protocol can be used with any high-level bacterial plasmid expression system.

1. Plate between 0.5×10^4 and 1×10^4 bacterial transformants of an appropriate cDNA expression library on individually labeled 82-mm-diameter Magna-lift nitrocellulose filters (Micron Separations, Inc., Westboro, MA), overlaid on LB plates containing carbenicillin (carb, 100 μg/ml) and kanamycin (kan, 50 μg/ml) (for a pGEX-KG N/K expression library[15] or appropriate antibiotic for the library of choice). Allow the filter to become thoroughly wet before applying the bacteria, in a small volume (up to 0.2 ml for an 82-mm filter), to the center of each filter. Using a sterile glass spreader, disperse the fluid evenly over the surface of the filter. Leave a border 2–3 mm wide at the edge of the filter free of bacteria. Incubate the plates at 37° until small colonies (0.1 to 0.2 mm in diameter) appear (about 8–12 hr). The colonies should be well formed but not touching. It is important that the colonies remain small in order to avoid smearing.

2. Overlay each original filter with a dry Nytran$^+$ filter (Schleicher & Schuell, Keene, NH) to generate a replica filter. To make a replica, place the original filter face up on a dry Whatman (Clifton, NJ) No. 1 filter, overlay with a labeled Nytran$^+$ filter (labeled sides together), and press firmly with a flat object to ensure uniform transfer. Orient the two filters by making a series of alignment marks with a needle through both filters before separating so that appropriate colonies can be later identified and recovered. Peel the filters apart and return the master filter to a fresh LB (plus carb/kan) plate. Incubate the plate at 37° to allow regeneration of the colonies (1–4 hr). After recovery, master filters are placed on Whatman No. 1 filters saturated with LB containing 10% (v/v) glycerol in sealed petri dishes, stored at −80°, and later thawed to recover putative positive clones.

Lay the replica filters on a fresh LB (carb/kan) plate and incubate the plates at 37° until colonies appear (4–6 hr). Library protein expression is induced by placing the replica filters on fresh LB (carb/kan) plates containing 1 m*M* IPTG for 6 hr at 37°.

3. Bacterial colonies are lysed and the liberated fusion proteins are bound to the replica filter by incubating the filters (colonies face up) sequen-

tially on Whatman filter paper soaked in the following solutions (remove excess solution from the filters to limit colony blurring): solution A [150 m*M* NaCl, 100 m*M* Tris, (pH 8.0), 5 m*M* $MgCl_2$, DNase I (2 μg/ml), lysozyme (50 μg/ml)] for 20 min, solution B [150 m*M* NaCl, 100 m*M* NaOH, 0.1% (w/v) sodium dodecyl sulfate (SDS)] for 5 min, and solution C [150 m*M* NaCl, 100 m*M* Tris-HCl, (pH 6.5)] for 5 min. Filters are air dried and can be stored at room temperature for up to 10 days before proceeding to the prenylation assays.

4. It is helpful to prepare a positive control for incubation with the library filters. GST–Ras fusion proteins expressed in *Escherichia coli* and engineered to contain consensus prenylation motifs (CVLS or CVIM for FTase and GGTase, respectively) are useful controls to test experimental conditions and establish detection limits. GST alone or GST–Ras protein engineered with an SVLS C terminus is a convenient negative control. These bacteria can be plated and processed as described above, or purified fusion proteins can be directly applied to a Nytran$^+$ filter. These control filters are then added to the library screening.

Prenyl Labeling of Filters

Recombinant prenyltransferase enzymes are expressed with baculovirus and partially purified as described.[17] Methods for obtaining high levels of FTase and GGTase I activity from insect cell culture, and the biochemical analysis of these proteins, are described elsewhere.[17,18] In this protocol, we describe expression screening with FTase. However, the following protocol can be readily adapted to GGTase-I expression screens.

1. To prepare protein filters for enzymatic screening with [^{3}H]FPP or [^{3}H]GGPP, air-dried filters must first be wetted and thoroughly blocked. Filters are placed in blocking buffer [50 m*M* Tris (pH 8.0), 150 m*M* NaCl, bovine serum albumin (BSA, 20 mg/ml), and 0.05% (v/v) Tween 20] and incubated for 1–2 hr with gentle shaking at 25° (10 ml/filter, four buffer changes). To reduce nonspecific background labeling, it is necessary to remove excess bacterial debris from the filters. Gently scrape the bacterial debris from the surface of the filters, using Kimwipes soaked in blocking buffer. Use a 100-mm crystallizing dish for blocking and incubation and a 170-mm crystallizing dish for washing.

[17] D. C. Crick, J. Suders, C. M. Kluthe, D. A. Andres, and C. J. Waechter, *J. Neurochem.* **65,** 1365 (1995).

[18] S. A. Armstrong, V. C. Hannah, J. L. Goldstein, and M. S. Brown, *J. Biol. Chem.* **270,** 7864 (1995).

2. To begin farnesyl radiolabeling, place the filters in blocking buffer [50 mM Tris (pH 8.0), 150 mM NaCl, bovine serum albumin (20 mg/ml), and 0.05% (v/v) Tween 20]. Incubate the filters for 1 hr with gentle shaking at 25° (10 ml/filter, four buffer changes). All the remaining steps are carried out at 30° with vigorous shaking (to ensure that filters remain wetted). It is important that the filters be transferred one at a time from one solution to the next to prevent the filters from sticking together.

3. Wash the filters four times in TBS [50 mM Tris, (pH 8.0), 150 mM NaCl] for 5 min (10 ml per 82-mm filter).

4. Wash the filters once with prenylation buffer [50 mM HEPES, (pH 7.6), 20 mM $MgCl_2$, 5 mM dithiothreitol (DTT), and 5 μM $ZnCl_2$] for 10 min (10 ml/filter). To label the library filters, add fresh prenylation reaction mixture [prenylation buffer containing 0.27 μM [^{3}H]farnesyl pyrophosphate (30–50 Ci/mmol; American Radiolabeled Chemicals, St. Louis, MO) and 15–50 ng of recombinant FTase per milliliter], 1–2 ml/filter. Incubate for 1 hr with continuous shaking.

5. Terminate the reaction by washing the filters with TBS (10 ml/filter). Perform 6–10 washes, 5 min each.

6. A series of additional washes is necessary to completely remove unincorporated isoprenoid lipids. Perform three washes (5 min each, 10 ml/filter) with 50% (v/v) ethanol. Repeat these wash steps with ethanol–HCl (9:1, v/v), followed by three additional 5-min washes with 100% ethanol (10 ml/filter).

7. The filters are then air dried, dipped in Amplify (Amersham, Arlington Heights, IL) fluorographic reagent, placed on plastic backing, dried for 1 hr at 50°, and exposed with an intensifying screen to Kodak (Rochester, NY) X-Omat AR film at −70° for 7–45 days. Bacterial colonies from areas of the filters where positive signals were obtained are recovered and rescreened by dilution cloning as described previously[15] to isolate individual bacterial colonies expressing prenylated fusion proteins.

Colony Purification

The random probability of finding the highly degenerate farnesylation consensus sequences (C*aa*X, followed by a stop codon) appearing in any reading frame (some of which might specify artifactual prenylation) is approximately 1–3 × 10^{-4}. Indeed, we have isolated a number of these artifactual sequences during our screens.[15,16] Therefore, it is recommended that the following secondary screening procedure be used to optimize the chance that an individual clone is a bona fide, cDNA-encoded protein.

1. From a single 82-mm filter we usually pick 2–10 potential positives. These positives appear as small spots on the film and it is difficult to

determine false and true positives by inspection, on the basis of size or intensity (Fig. 1). Positive colonies should be picked from the thawed master filter, streaked, and rescreened. Usually only 50–60% of these primary positives will test positive on secondary screening.

To pick the positives, the film is aligned with the replica filters and the orientation marks are transferred to the film. These marks are then used to align the film with the original master filter. Positive colonies are then picked and regrown in LB (carb/kan) medium.

2. Putative positives are then rescreened as described above, using an 82-mm filter. It is important to have 100–500 colonies per filter to ensure that no positive colony is missed and to allow the isolation of a pure colony. Once pure bacteria stocks are obtained, the expression plasmids can be isolated and sequenced by conventional methods.[14]

3. A third screen is generally performed prior to sequence analysis. This involves performing an *in vitro* prenylation reaction and SDS–polyacrylamide gel electrophoresis (PAGE) of the affinity-purified GST fusion protein isolated from a culture of positive *E. coli*. The bacteria are

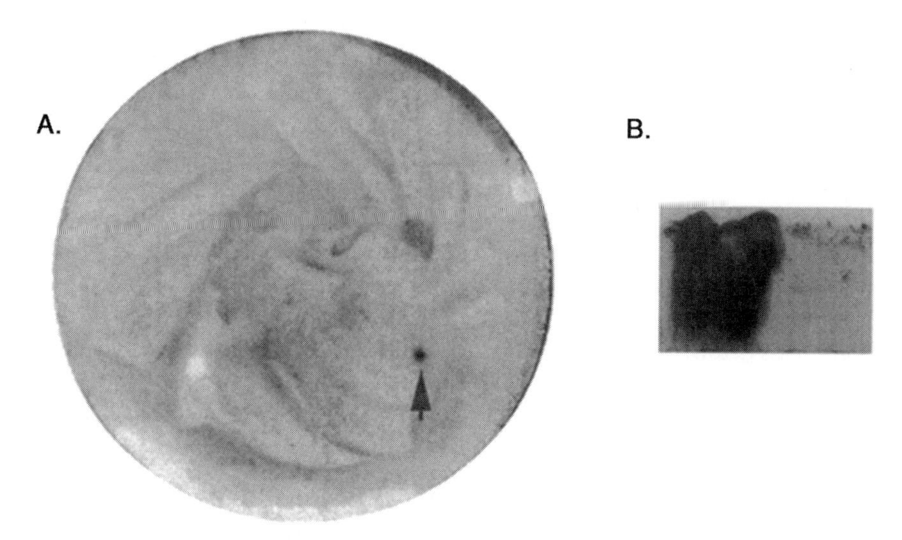

FIG. 1. Expected results from primary and secondary screens. (A) Primary screen: an autoradiograph of a primary screen exposed for 25 days shows one potential positive (arrow). Both this colony and another colony were picked, streaked to a new filter, and subjected to a secondary screen. (B) Secondary screen: this shows the secondary screen for both primary colonies exposed for 14 days. A shorter exposure time is possible because the colonies are larger with fewer plaques per plate. In this case, the bacteria were grown as a dense patch for secondary screening. Nonetheless, each colony is different and exposure times can vary greatly. In our experience dark symmetrical spots are rarely true positives; however, the original colony expressed an authentic farnesylated protein.

grown and fusion proteins are isolated as described previously.[15] To assay for farnesyltransferase substrates the affinity-purified fusion protein is incubated in a final volume of 50 μl of buffer FTC [FT buffer containing 50 μM $ZnCl_2$, 3 mM $MgCl_2$, 0.6 μM [^{3}H]farnesyl pyrophosphate (49,500 dpm/pmol; New England Nuclear, Boston, MA), and 10 ng of recombinant farnesyltransferase] for 30 min at 37°. The reaction is terminated by the addition of SDS sample buffer and analyzed for the presence of farnesylated proteins by electrophoresis on 10% SDS–polyacrylamide gels. Gels are treated with Amplify (Amersham), dried, and exposed with an intensifying screen to Kodak X-Omat AR film at −70° for 2–10 days. The molecular mass of the fusion protein does not necessarily correspond to the size of the mature protein, because the entire coding region may not be present in the cDNA fusion. In our libraries, the vector-encoded GST partner comprises 27 kDa of the fusion protein molecular mass. The molecular mass of GST fusion proteins can be rapidly determined by immunoblotting extracts of IPTG-induced bacterial cultures with anti-GST antisera as described previously.[15] We have found that a large molecular mass (>50-kDa GST fusion) of the fusion is the best indicator of the isolation of bona fide prenylated cDNA-encoded proteins.

Results

Using these techniques we have successfully screened human and rat retina and total mouse embryo expression libraries. With the advances in the genome initiative, particularly the availability of large numbers of expressed sequence tags (ESTs), expression cloning of novel prenylated proteins may not be the most effective approach in identifying these proteins. Instead, the biochemical characterization of ESTs containing putative consensus C-terminal prenylation motifs may prove more satisfactory. However, the extension of this expression cloning technique to additional organ systems and particularly to organisms that are currently not included in the genome initiative should allow the isolation of additional novel isoprenylated proteins.

[15] Expression Cloning to Identify Monomeric GTP-Binding Proteins by GTP Overlay

By DOUGLAS A. ANDRES

Introduction

The Ras-related GTP-binding proteins constitute a large family of regulatory molecules that have been implicated in the control of a wide range of cellular processes, including cell growth and differentiation, intracellular vesicular trafficking, nuclear transport, and cytoskeletal reorganization. To date, six subfamilies have been identified: Ras, Rho, Rab, Ran, ARF (ADP-ribosylation factor), and Rem/Rem2/Rad/Gem/Kir.[1–3] These subfamilies are defined largely by primary sequence relationships but also by their regulation of common cellular functions. All GTPases of the Ras superfamily contain five highly conserved amino acid motifs involved in guanine nucleotide binding and hydrolysis.[1,4] These primary sequence motifs have been conserved throughout evolution and define a conserved structure whose importance has been confirmed through exhaustive mutational analysis.[5] Therefore, the sequence of all small GTPases share approximately 20–30% amino acid identity, whereas the sequence identity is considerably higher within subfamilies.

To date, more than 70 small Ras-related GTP binding proteins have been identified, and additional members of the family continue to be discovered.[2] The majority of these proteins have been isolated by screening genomic or cDNA libraries at low stringency with oligonucleotide mixes corresponding to the conserved guanine nucleotide-binding regions or to short peptide sequences conserved within individual Ras-related subfamilies. Although these strategies have proved successful, their use in the identification of novel mammalian small GTP-binding proteins, or of Ras-related proteins in plants and lower organisms, may not be as effective, particularly if these novel GTPases are distant in sequence from previously identified Ras family members.

An alternative approach to identifying low molecular weight GTP-

[1] H. R. Bourne, D. A. Sanders, and F. McCormick, *Nature* (*London*) **348,** 125 (1990).

[2] M. Zerial and L. A. Huber, "Guidebook to the Small GTPases." Oxford University Press, Oxford, 1995.

[3] B. S. Finlin and D. A. Andres, *J Biol. Chem.* **272,** 21982 (1997).

[4] A. Valencia, P. Chardin, A. Wittinghofer, and C. Sander, *Biochemistry* **30,** 4637 (1991).

[5] H. R. Bourne, D. A. Sanders, and F. McCormick, *Nature* (*London*) **349,** 117 (1991).

0076-6879/00 $35.00

binding proteins is provided by resolving proteins by sodium dodecyl sulfate–polyacrylamide gel electrophoresis (SDS–PAGE) and probing with [α-^{32}P]GTP after transfer onto nitrocellulose membrane.[6,7] This ligand-blotting method has been used to generate two-dimensional gel electrophoresis maps of many previously identified Ras-related GTPases in mammalian cells. However, this technique cannot be generally applied to the identification of novel small G proteins because of their low cellular abundance.

To expand the available methods for identifying Ras-related GTP-binding proteins, we developed an expression cloning method. The approach relies on the proved ability of Ras-related proteins to bind GTP after denaturation and transfer to nitrocellulose membrane.[8,9] In this chapter we outline the use of this method for the identification of low molecular weight GTP-binding proteins from bacterial expression libraries. This method allows the systematic screening of large cDNA libraries in a manner that is simpler and more sensitive than purifying proteins identified by classic two-dimensional electrophoresis techniques.

Methods

Plating and Screening Library

Because the expression cloning procedure is based on the proved ability of Ras-related GTP-binding proteins to bind GTP tightly, the limitations of the GTP overlay method lie in the representation of the library and the limit of sensitivity as determined by the expression level of the individual GTP-binding proteins. We initially attempted to screen λ expression libraries.[10] However, we failed to detect a significant number of initial positive plaques that were capable of binding GTP. In addition, it proved difficult to routinely detect these primary plaques in secondary assays, suggesting that we were working near the limits of sensitivity afforded by this assay. LacZ fusion proteins expressed in *Escherichia coli* bacteriophage λ vectors are synthesized at up to 100 pg/plaque.[11] Because protein expression levels are generally higher in plasmid libraries, and we had proved the feasibility of detecting plasmid-encoded isoprenylated proteins by expression clon-

[6] L. A. Huber, O. Ullrich, Y. Takai, A. Lutcke, P. Dupree, V. Olkkonen, H. Virta, M. J. de Hoop, K. Alexandrov, M. Peter, M. Zerial, and K. Simons, *Proc. Natl. Acad. Sci. U.S.A.* **91,** 7874 (1994).

[7] E. G. Lapetina and B. R. Reep, *Proc. Natl. Acad. Sci. U.S.A.* **84,** 2261 (1987).

[8] P. S. Gromov, P. Madsen, N. Tomerup, and J. E. Celis, *FEBS Lett.* **377,** 221 (1995).

[9] Y. Nagano, R. Matsuno, and Y. Sasaki, *Anal. Biochem.* **211,** 197 (1993).

[10] K. Kadono-Okuda and D. A. Andres, *Anal. Biochem.* **254,** 187 (1997).

[11] J. Sambrook, E. F. Fritsch, and T. Maniatis, "Molecular Cloning: A Laboratory Manual." Cold Spring Harbor Laboratory Press, Cold Spring Harbor, New York, 1989.

ing,[12] we adapted the nucleotide overlay methods to plasmid-based expression libraries.

One of the most important aspects of any cloning experiment is the quality of the library. Many commercial libraries contain a relatively small number of independent cDNA inserts (fewer than 1×10^6). Unless the library is directionally cloned, only one in six of these clones expresses a protein in the correct reading frame and orientation. These factors, combined with the fact that commercial libraries are often subject to serial amplification, can make it difficult to clone a gene of interest, particularly from a rare mRNA. Thus, it is best to begin this process with a library that contains the maximal number of independent clones (greater than 3×10^6) and that has not been repeatedly amplified. In addition, it may be necessary to screen several million colonies owing to the low percentage of colonies that express properly orientated fusion proteins. One final factor that should be considered is the construction of the library. Our cloning efforts have focused on oligo(dT)-primed directional libraries with cDNA inserts of >1.5 kb to bias clone representation to cDNAs large enough to encode full-length Ras-like GTP-binding proteins.

Small GTP-binding proteins have been shown to be readily detected by ligand blotting after separation by SDS–PAGE and transfer to nitrocellulose filters.[2,6,7] The limitation to the application of direct GTP ligand-binding assays to the screening of bacterial expression libraries has been the presence in *E. coli* of endogenous nucleotide-binding proteins, such as bacterial elongation and initiation factors, which cause high levels of nonspecific GTP binding.[9] It was shown that SDS treatment of nitrocellulose filters containing bacterial cell lysates effectively eliminated the GTP-binding activity of endogenous bacterial G proteins without affecting the GTP-binding activity of recombinant Ras-related proteins.[8] This suggested a simple method for screening plasmid expression libraries.

Preparation of Colony Protein Filters

In this protocol we describe the screening of a human retinal expression library in which cDNA inserts are under the inducible control of the *lac* promoter.[10] However, this protocol can be used with any high-level bacterial plasmid expression system.

1. Plate between 0.5×10^5 and 2×10^5 bacterial transformants of an appropriate cDNA expression library on individually labeled 132-mm-diameter Magna-lift nitrocellulose filters (Micron Separations, Inc., West-

[12] D. A. Andres, H. Shao, D. C. Crick, and B. S. Finlin, *Arch. Biochem. Biophys.* **346,** 113 (1997).

boro, MA), overlaid on LB plates containing carbenicillin (carb, 100 μg/ml) (for the Soares human adult retinal cDNA library (N2b4HR)[10] or an appropriate antibiotic for the library of choice. Allow the filter to become thoroughly wet before applying the bacteria, in a small volume (up to 0.6 ml for a 132-mm filter), to the center of each filter. Using a sterile glass spreader, disperse the fluid evenly over the surface of the filter. Leave at the edge of the filter a border 2–3 mm wide that is free of bacteria. Incubate the plates at 37° until small colonies (0.1 to 0.2 mm in diameter) appear (about 8–12 hr). The colonies should be well formed but not touching. It is important that the colonies remain small in order to avoid smearing during replica plating.

2. Overlay each original filter with a dry Nytran$^+$ filter (Schleicher & Schuell, Keene, NH) to generate a replica filter. To make a replica, place the original filter face up on a dry Whatman (Clifton, NJ) No. 1 filter, overlay with a labeled nitrocellulose filter (labeled sides together), and press firmly with a flat object to ensure uniform transfer. Orient the two filters by making a series of alignment marks with a needle through both filters before separating then, so that individual colonies can be later identified and recovered. Peel the filters apart and return the master filter to a fresh LB (carb) plate. An additional replica is prepared from the master in an identical manner. Key the second replica to the existing holes in the master filter. Incubate the master plate at 37° to allow regeneration of the colonies (1–4 hr). After recovery, master filters are stored at 4° on fresh plates (carb), and later used to recover putative positive clones.

Lay the replica filters on fresh LB (carb) plates and incubate the plates at 37° to allow colonies to recover (1–3 hr). Library protein expression is induced by placing the replica filters on fresh LB (carb) plates containing 1 m*M* isopropyl-β-D-thiogalactopyranoside (IPTG) for 6 hr at 37°.

3. Bacterial colonies are lysed and the liberated fusion proteins are bound to the replica filter by exposing the filters (colonies face up) to chloroform vapor for 15 min.[13] Filters are air dried and can be stored for up to 10 days at 4° before proceeding to GTP overlay assays.

4. It is helpful to prepare a positive control for incubation with the library filters. GST–Ras fusion proteins expressed in *E. coli* are useful controls with which to test experimental conditions and establish detection limits.[10] GST alone is a convenient negative control. These bacteria can be plated and processed as described above for a library filter. These control filters are then added to the library screening.

[13] D. M. Helfman, J. R. Feramisco, J. C. Fiddes, G. P. Thomas, and S. H. Hughes, *Proc. Natl. Acad. Sci. U.S.A.* **80,** 31 (1983).

Probing Filters by GTP Overlay

1. Filters are placed face up on 3MM paper (Whatman, Clifton, NJ) soaked in solution A [2% (w/v) SDS, 0.3% (v/v) Tween 20] at 50° for 15 min and then washed at room temperature for 90 min in buffer B [25 m*M* Tris–190 m*M* glycine (pH 8.3), 20% (v/v) methanol, three buffer changes, 10 ml/filter]. To reduce nonspecific background labeling, it is necessary to remove excess bacterial debris from the filters. Gently scrape the bacterial debris from the surface of the filters, using Kimwipes soaked in buffer B. Use a 170-mm crystallizing dish for the incubation and washing of filters. Unless noted, it is important that the filters be transferred one at a time from one solution to the next to prevent the filters from sticking together.

2. Incubate the filters in buffer C [50 m*M* Tris-HCl (pH 7.4), 0.3% (v/v) Tween 20, and 10 μM $MgCl_2$] for 30 min with gentle shaking at room temperature (10 ml/filter, two buffer changes). All the remaining steps are carried out at room temperature with vigorous shaking (to ensure that filters remain wetted).

3. To begin the GTP overlay assay add fresh buffer D [buffer C containing 100 m*M* dithiothreitol (DTT), 100 μM ATP, and 1 μCi of [α-^{32}P]GTP/ml] at 2–3 ml/filter. We probe 10 filters at a time in ~25 ml of buffer. Incubate for 1 hr with continuous shaking. For best results the [α-^{32}P]GTP must be fresh (aged less than 1 half-life). In addition, the concentration of free $MgCl_2$ in both the binding and wash buffers can be adjusted in an attempt to optimize GTP-binding conditions for individual GTPases. We have found that 10 μM $MgCl_2$ is the minimal concentration that allows efficient GTP binding (using H-Ras as the binding target). However, magnesium concentrations as high as 1 m*M* have yielded excellent results.

4. Remove the probe solution and rinse the filters once with buffer E (buffer C containing 10 m*M* $MgCl_2$) without changing containers, to remove the bulk of the unbound radioactivity. The high concentration of magnesium is designed to stabilize the GTP-bound state of Ras-related GTPases.

5. Wash the filters in buffer E three additional times for 10 min at room temperature (10 ml/filter).

6. Air dry the filters and expose to Kodak (Rochester, NY) X-Omat AR film for 16–48 hr at −70° with intensifying screens. It is important to check that the overlay assay has given a strong signal with the positive control filter.

Colony Purification

1. From a single 132-mm filter we usually pick 10–50 potential positives (we have screened only normalized libraries and would expect many fewer

colonies when screening a primary library). These positives appear as small spots on the film and it is difficult to determine false and true positives by inspection, on the basis of size or intensity.[10] Therefore, only colonies that give positive signals from both filters are selected for further analysis. Bacterial colonies from areas of the master plate where positive signals were obtained are recovered and rescreened by dilution cloning as described previously[10] to isolate individual bacterial colonies expressing GTP-binding proteins. Usually 80–90% of these primary positives will test positive on secondary screening (see Fig. 1).

To pick the positives, the autoradiograph is aligned with the replica filters and the orientation marks are transferred to the film. These marks are then used to align the film with the original master filter. Positive colonies are then picked and regrown in LB (carb). If a large number of primary positives is identified, they can be stored as frozen glycerol stocks prior to secondary screening.

2. Putative positives are then rescreened as described above, using an 82-mm filter. It is important to have 100–400 colonies per filter to ensure that no positive colony is missed and to allow the isolation of a pure colony. Once pure bacteria stocks are obtained, the expression plasmids can be isolated and sequenced by conventional methods.[11]

3. We have performed a third screen prior to sequence analysis when using an amplified (normalized) library to eliminate the analysis of redundant clones. To avoid the repetitive isolation of identical clones from the

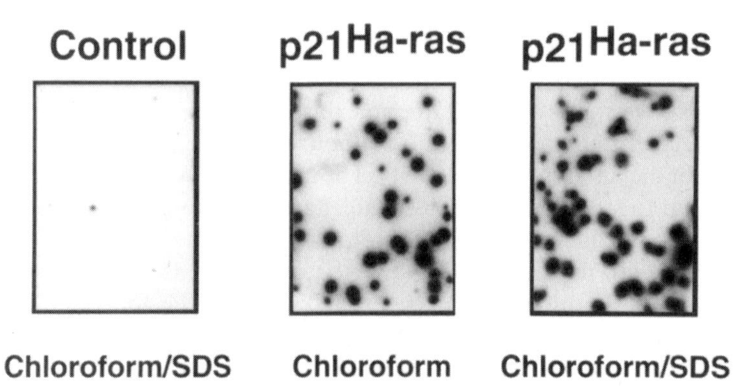

FIG. 1. Ras-related GTP-binding proteins can be detected in bacterial colony blots. An equal number of IPTG-induced bacterial colonies containing either plasmid-encoded Ha-Ras or empty plasmid (control) were transferred to nitrocellulose filters and protein blots were generated by exposure to chloroform vapor. Blots were treated with SDS (*right* and *left*) or left untreated (*center*), renatured with 0.3% (v/v) Tween 20, and subjected to [α-^{32}P]GTP ligand blot analysis. The filters were exposed to Kodak X-Omat film with an intensifying screen at $-70°$ for 12 hr.

library, only a fraction (10–20) of the initial positive colonies are regrown and subjected to a second round of ligand blotting. The cDNA clones from these purified colonies are sequenced to assess their identities. Once the identity of the individual cDNA clones has been established, gene-specific oligonucleotide primers are synthesized for each novel cDNA. The remaining uncharacterized primary positive bacterial stocks are then analyzed by polymerase chain reaction (PCR), using these gene-specific primers.[10] PCR is a sensitive and rapid method for identifying and subtracting identical cDNA clones from the remaining primary positives and greatly enhances the effectiveness of the screen.

Results

We have used this method to screen human retinal (see Fig. 2) and mouse embryo cDNA expression libraries. To date, all colonies that were positive in secondary screening have contained a Ras-related small GTP-binding protein cDNA clone and all these cDNAs have encoded full-length coding regions. DNA sequence analysis of the isolated cDNAs has so far identified a series of Ras-related GTP-binding proteins.[10] These include the Ras subfamily members G25K and Rap1B, the Rab subfamily members Rab1A and Rab5B, the ARF subfamily proteins ARF1 (ADP-ribosylation factor 1), and ARL3 (ADP-ribosylation factor-like protein 3), the Rho subfamily protein Rac1, and a novel Ras-related GTP. These clones represent at least one member from each of the major Ras subfamilies and serve to validate the expression cloning strategy and underscore the fact that the only property required for identification in this screen is the ability to specifically bind GTP after a cycle of protein denaturation. The fact that all the clones are full length is an added benefit of this experimental ap-

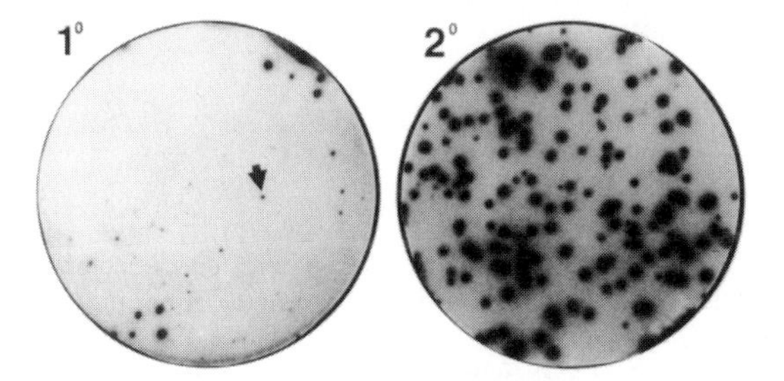

FIG. 2. Isolation of GTP-binding proteins from a bacterial expression cDNA library. A human retinal cDNA library plated at 1×10^4 colonies per filter was screened by [α-^{32}P]GTP ligand blot analysis (*left*). The indicated single colony (arrow) was selected and rescreened by the same method to allow the isolation of a single bacterial colony (*right*).

proach and results from the need for each cDNA to encode a coding region large enough to maintain the conserved three-dimensional structure required for efficient guanine nucleotide binding.[1,4,5] In addition, we have converted a λ Uni-ZAP XR fetal retina library to a plasmid expression library, using helper phage rescue, and rescreened it by GTP ligand blotting with excellent results. Thus, this assay can be applied to any commercial λ cDNA library that can be efficiently converted to plasmid, greatly extending the number of cDNA libraries that may be screened.

Knowledge of the number and function of Ras-related GTP-binding proteins in plants and parasitic organisms is quite limited when compared with the extensive knowledge about the same proteins in animals and yeast. To date, almost all the small GTP-binding proteins that have been cloned from these hosts have been isolated by conventional methods on the basis of sequence similarity or by the functional complementation of yeast mutants.[14,15] It is surprising that no Ras proteins have been cloned, although many researchers have used oligonucleotide probes based on Ras sequence in their searches. Indeed, several downstream proteins known to be regulated by Ras action, including Raf and mitogen-activated protein kinase (MAPK) homologs, have been identified in plants.[16,17] This strongly suggests that plant Ras family members remain to be identified. Because the method described here relies solely on guanine nucleotide-binding specificity, it may allow the isolation of cDNAs encoding novel Ras-related GTP-binding proteins from plants and other organisms.

In conclusion, we have developed an expression cloning strategy that allows the rapid isolation of full-length cDNAs encoding small GTP-binding proteins that may escape detection by conventional cloning methods. Identification of novel GTP-binding proteins as well as a complete cataloging of all Ras-related GTPases will provide new insight into the diverse cellular pathways regulated by these proteins and will be critical to understanding the regulation and complicated biological interactions among members of the Ras protein family.

[14] I. Moore, T. Diefenthal, V. Zarsky, J. Schell, and K. Palme, *Proc. Natl. Acad. Sci. U.S.A.* **94,** 762 (1997).

[15] L. M. Chen, Y. Chern, S. J. Ong, and J. H. Tai, *J. Biol. Chem.* **269,** 17297 (1994).

[16] C. Wilson, N. Eller, A. Gartner, O. Vicente, and E. Heberle-Bors, *Plant Mol. Biol.* **23,** 543 (1993).

[17] J. J. Kieber, M. Rothenberg, G. Roman, K. A. Feldmann, and J. R. Ecker, *Cell* **72,** 427 (1993).

[16] Retrovirus cDNA Expression Library Screening for Oncogenes

By GWENDOLYN M. MAHON and IAN P. WHITEHEAD

Although conventional, plasmid-based, gene transfer assays have been used to identify protooncogenes, only a handful of transforming sequences have been detected in this manner. The limited success of these efforts reflects technical limitations of the systems used, rather than exhaustion of the pool of cDNAs with oncogenic potential. The most severe limitations associated with the use of plasmid-based libraries for expression cloning (low efficiencies of transfection and expression) have now been overcome by the development of retrovirus-based cDNA transfer systems.[1–4] In these systems, cDNA expression libraries are constructed in retroviral plasmids, and then converted into libraries of infectious retroviral particles. Four major advantages are obtained through the use of retroviral library transfer: (1) the ability to screen large numbers of cDNA clones on an equivalent number of recipient cells, (2) the relatively high levels of expression obtained with retrovirally transferred cDNAs, (3) the potential to use cell lines that have been inaccessible to expression cloning because of low transfection efficiencies, and (4) the development of highly efficient recovery mechanisms for the proviral inserts.

We have described a retrovirus-based cDNA expression system and its successful application to the identification of novel oncogenes.[2] This system, which is described here in detail, permits the stable transfer and expression of large numbers of cDNA clones into equivalent numbers of recipient cells. This allows for the efficient screening of complex cDNA libraries, and facilitates the identification of transforming sequences that are present at low frequency within the cDNA population. Although we are describing methods that identify transforming cDNAs on the basis of their focus-forming activity,[5] this expression system can be readily adapted for use in alternative screens for growth transformation.

[1] T. Kitamura, M. Onishi, S. Kinoshita, A. Shibuya, A. Miyajima, and G. P. Nolan, *Proc. Natl. Acad. Sci. U.S.A.* **92,** 9146 (1995).

[2] I. Whitehead, H. Kirk, and R. Kay, *Mol. Cell. Biol.* **15,** 704 (1995).

[3] T. Tsukamoto, T. Huang, R. C. Guzman, X. Chen, R. V. Pascual, T. Kitamura, and S. Nandi, *Biochem. Biophys. Res. Commun.* **265,** 7 (1999).

[4] G. J. Hannon, P. Sun, A. Carnero, L. Y. Xie, R. Maestro, D. S. Conklin, and D. Beach, *Science* **283,** 1129 (1999).

[5] G. J. Clark, A. D. Cox, S. M. Graham, and C. J. Der, *Methods Enzymol.* **255,** 395 (1995).

0076-6879/00 $35.00

Construction of Retrovirus-Based cDNA Expression Libraries

Retroviral Vectors

We have described the construction of a retroviral vector (pCTV3B) that has been designed specifically for use in cDNA expression cloning (Fig. 1).[2] The vector has been made as compact as possible (~5560 bp) to maximize the stability of cDNAs, and all splice sites have been removed so that they cannot activate cryptic splice sites that may be present within the cDNA inserts. In addition to viral sequences, the viral transcripts that are derived from pCTV3B contain an inserted cDNA, a bacterial selectable

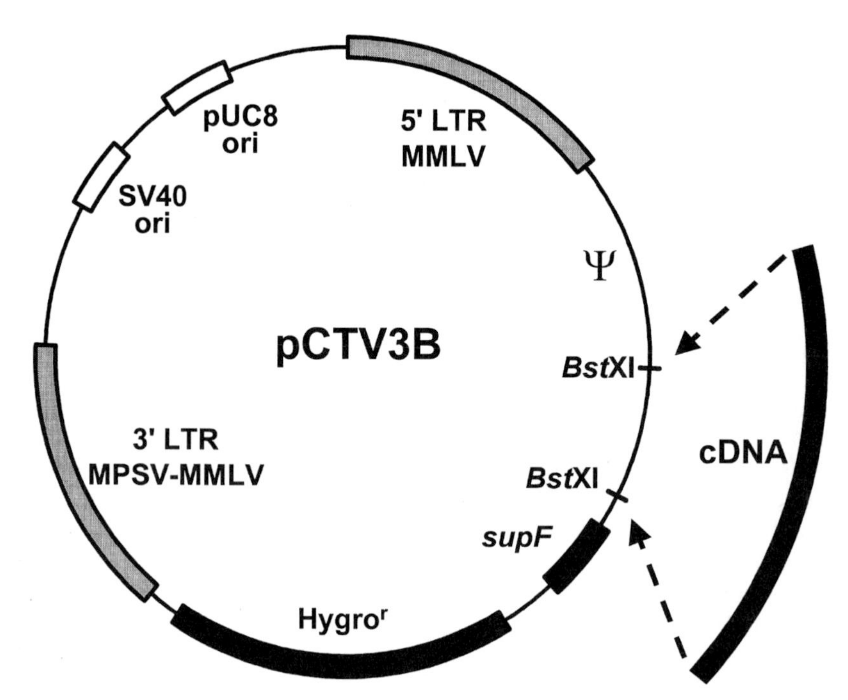

FIG. 1. Structure of the pCTV3B retrovirus-based expression cloning vector. pCTV3B consists of (1) a 5′ Mo-MuLV (MMLV) LTR with an extended *gag* region (Ψ) that lacks the normal initiation codon, (2) a cDNA cloning site consisting of two *Bst*XI sites separated by a 400-bp stuffer fragment, (3) a *supF* gene that provides a bacterial selectable marker, (4) the hygromycin phosphohydrolase gene, which provides a mammalian selectable marker (Hygro^r), (5) a composite MPSV/MoMuLV 3′ LTR, (6) the replication origin from pUC8 to facilitate propagation in bacterial cells, and (7) the simian virus 40 (SV40) origin of replication. After viral transmission, the region between (and including) the 5′ and 3′ LTRs becomes stably integrated into the host genome as a proviral insert. Both the 5′ and 3′ LTRs of the provirus are derived from the composite MPSV/Mo-MuLV 3′ LTR.

marker (*supF*),[6] and a marker that permits selection in mammalian cells (Hygro^r).[7] An extended *gag* region ensures efficient packaging of this viral transcript when it is expressed in an appropriate packaging cell line.[8]

The pCTV retroviral vectors utilize the *supF*–p3 selection system for *Escherichia coli* that was developed by B. Seed.[9] The *supF* gene encodes a tRNA molecule that can rescue amber mutations within antibiotic resistance genes, and thus provides a compact (~220 bp) bacterial selectable marker for the retroviral vectors. The p3 plasmid is a 50-kbp, single-copy, stably propagated plasmid that encodes a wild-type kanamycin resistance (Kan^r) gene as well as amber mutant ampicillin and tetracycline resistance genes. *Escherichia coli* strains that contain p3 (e.g., MC1061/p3[10]) can be used to select for plasmids that are carrying *supF*.

Isolation of Poly(A) mRNA

The successful construction of a complex library (>10^6 clones) generally requires 3–5 μg of good quality poly(A) mRNA. Such quantities can be readily obtained from tumor-derived, mammalian cell lines. Total RNA is first isolated with Trizol reagent (GIBCO-BRL, Gaithersburg, MD) according to the manufacturer instructions. We collect lysates from at least 2×10^8 cells (e.g., 20 confluent 100-mm dishes), which will yield in excess of 1 mg of total RNA. The RNA is then subjected to two rounds of purification on oligo(dT)-cellulose columns to obtain 10–50 μg of purified poly(A) mRNA.[11] Because we generally use random priming during cDNA synthesis, it is important to carry out the second round of purification to remove as much ribosomal RNA as possible.

Preparation of cDNAs

When constructing cDNA libraries that are to be screened for oncogenes, we find it advantageous to use random priming and bidirectional cloning. The majority of oncogenes that have been described, to date, are activated by truncation, and random priming enriches the library for truncated cDNAs. Bidirectional cloning allows us to simultaneously screen for oncogenes and tumor suppressors. The protocol that we are describing generates blunt-ended cDNAs that can be fused to *Bst*XI linkers, and then

[6] B. Seed, *Nucleic Acids Res.* **11,** 2427 (1983).

[7] H. U. Bernard, G. Krammer, and W. G. Rowekamp, *Exp. Cell. Res.* **158,** 237 (1985).

[8] M. A. Bender, T. D. Palmer, R. E. Gelinas, and A. D. Miller, *J. Virol.* **61,** 1639 (1987).

[9] B. Seed, *Nature (London)* **329,** 840 (1987).

[10] B. Seed and A. Aruffo, *Proc. Natl. Acad. Sci. U.S.A.* **84,** 3365 (1987).

[11] T. Maniatis, E. F. Fritsch, and J. Sambrook, "Molecular Cloning: A Laboratory Manual." Cold Spring Harbor Laboratory Press, Cold Spring Harbor, New York, 1989.

cloned into the retroviral pCTV3B vector. We use SuperScript II reverse transcriptase (GIBCO-BRL) for first-strand synthesis. Second-strand synthesis is carried out by nick translation using *E. coli* DNA polymerase I, *E. coli* RNase H, and *E. coli* DNA ligase (all from GIBCO-BRL).

All glassware and solutions used for first-strand synthesis must be RNase free. For each reaction it is preferable to work with at least 3–5 μg of good quality mRNA that has been resuspended in RNase-free Milli-Q (Millipore, Danvers, MA) water. This should generate sufficient quantities of cDNA to allow size fractionation of the libraries. In a sterile 1.5-ml tube combine 4 μl of a 50-ng/μl random hexamer mix 5 μg of mRNA, and RNase-free water to a final volume of 21 μl. Incubate at 70° for 10 min, and then chill on ice. Add 8 μl of 5× first-strand buffer (GIBCO-BRL), 4 μl of 0.1 *M* dithiothreitol (DTT), and 2 μl of a 10 m*M* dNTP mix. Mix the contents gently by vortexing, and incubate for 2 min at 37°. Add 5 μl of SuperScript II (200 U/μl), mix gently, and continue the incubation (37°) for 1 hr. Place the mixture on ice and proceed with second-strand synthesis.

To the first-strand reaction mixture add the following, in order (on ice): 71 μl of distilled H_2O, 30 μl of 5× second-strand buffer (GIBCO-BRL), 3 μl 10 m*M* dNTP mix, 1 μl (10 U) of *E. coli* DNA ligase, 4 μl (40 U) of *E. coli* DNA polymerase I, and 1 μl (2 U) of *E. coli* RNase H. Mix by gently vortexing, and incubate at 16° for 2 hr. Then, add 2 μl (10 U) of T4 DNA polymerase and incubate at 16° for a further 5 min. Place on ice and stop the reaction with 10 μl of 0.5 *M* EDTA. Extract with 150 μl of phenol–chloroform, and precipitate with 70 μl of 7.5 *M* ammonium acetate and 500 μl of ice-cold, 100% ethanol. Centrifuge at 14,000*g* for 20 min at 4°, wash the pellet with ice-cold 70% (v/v) ethanol, and resuspend in 20 μl of distilled H_2O. Load an aliquot (2 μl) on a 1.2% (w/v) agarose gel to check the quality and size range of the cDNAs.

Linkering of cDNAs

The *Bst*XI linkering system that has been adapted for use with the pCTV vectors has been described previously.[9] The *Bst*XI linkers substantially decrease the number of transformants that do not contain an insert, and reduce the possibility of having two cDNAs ligated into the same vector. These adapters have also been designed to introduce stop codons in all three reading frames at the 3′ end of the cDNA, thus preventing translational readthrough into the vector sequences. Oligonucleotides should be synthesized with 5′-phosphates or should be kinased prior to annealing (13-mer, 5′-TCA GTT ACT CAG G; 17-mer, 5′-CCT GAG TAA CTG ACA CA). To anneal oligonucleotide pairs, combine equimolar amounts of each oligonucleotide, heat to 60°, and then allow to slowly cool to room temperature. Ligation reactions consist of 1–5 μg of cDNA, 0.5 μg of linkers, 4 μl of 5× ligation buffer, 2 U of T4 DNA ligase, and water to 20 μl. Incubate the ligation reactions at 16° for 4–8 hr.

Size Fractionation of cDNA

To avoid the selection against large cDNAs that often occurs during library construction, we have found it beneficial to separate our linkered cDNAs into several size classes prior to ligation into the retroviral vector. This facilitates the separate transformation of larger cDNAs and affords the opportunity to adjust their representation in the final libraries. Load the cDNA linkering reaction onto a 1.2% (w/v) agarose gel (containing ethidium bromide) with flanking size markers. Run the cDNA a sufficient distance into the gel to obtain reasonable marker separation while still minimizing the size of the gel slice. Slice off the flanking marker lanes, expose them to UV, and mark off the appropriate size ranges. Do not expose the cDNA lane to UV. We normally isolate fractions with size ranges of 1000 to 2500 bp and 2500+ bp. Realign the marker lanes with the cDNA lane and excise the appropriately sized gel segments. Purify the cDNA by the GeneClean procedure (Bio 101, La Jolla, CA). This provides reasonable yields of cDNA while eliminating unligated adapters.

Ligation of Linkered cDNA

The cloning site of the pCTV3B retroviral vector consists of two *Bst*XI sites separated by a 400-bp stuffer fragment (Fig. 1).[11] Complete digestion of the vector with *Bst*XI removes the stuffer and generates two noncomplementary ends that can be ligated to the *Bst*XI-linkered cDNAs. The retroviral vector is digested to completion with *Bst*XI, dephosphorylated, and purified on a 1.2% (w/v) agarose gel followed by electroelution and precipitation. As with the cDNA, it is important not to expose the vector DNA to UV at any time. Ligation reactions consist of 20 ng of vector, an equimolar amount of cDNA, 1 μl of 10× ligation buffer [250 m*M* Tris-HCl (pH 7.8), 50 m*M* $MgCl_2$, 10 m*M* DTT, 10 m*M* ATP], H_2O to 10 μl total volume, and 0.5 U of T4 Ligase. This ligation reaction includes a minimal ionic strength ligation buffer that is required specifically for electroporation. Incubate the ligation at room temperature for 3 hr, and then add 10 μl of H_2O and heat kill at 72° for 20 min.

Generating Retroviral Plasmid Libraries

Libraries of retroviral plasmids are generated by electrotransformation of MC1061/p310 *E. coli* [we use a Bio-Rad (Hercules, CA) Gene Pulser

II].[10] The transformation mixes are plated in soft agar, which substantially reduces the variation in colony size and allows large numbers of colonies to be plated (up to 750,000) on single 15-cm plates. Because the size of the library is determined, in part, by the success of the electroporation, we use only electrocompetent MC1061/p3 cells that have a competence greater than 5×10^9.

Thaw electrocompetent MC1061/p3 *E. coli* on ice until completely melted. Add up to 1.5 μl of the ligation mix to a prechilled tube, and then add and gently mix 40 μl of electrocompetent cells. Allow to sit for 1 min on ice and then transfer to a prechilled 1-mm cuvette. Electroporate the cells (1600 V, 25 μF, and 200 Ω) and then immediately add 1 ml of YT broth (with 20 m*M* glucose) to the cuvette.[11] Transfer the cells to a new tube, incubate in a 37° water bath for 10 min, and then shake at 37° for an additional 110 min. Bring the volume of the cell solution up to 5 ml with more YT and warm to 37° in a water bath. Put 5 ml of melted 1.2% (w/v) agar–YT containing ampicillin (100 μg/ml) and tetracycline (15 μg/ml) in a second tube and incubate in a 44° water bath. Once both tubes are temperature equilibrated, combine the two tubes, mix by inversion, and quickly pour on the surface of a 15-cm plate of 1.2% (w/v) agar–YT [ampicillin (50 μg/ml)–tetracycline (7.5 μg/ml)]. Let the soft agar set at room temperature for 15 min and then incubate at 37° for 20 hr. The library size can now be determined by counting the number of colonies on the plates.

Recovering Plasmid Library from Soft Agar

Once the plasmid library has been grown in soft agar, we recommend that the cells be harvested, and the DNA isolated, within 24 hr. Plates can be stored at 4° prior to harvesting. Gently wash the surface of the plate with YT to remove most of the large surface colonies. Scrape off the soft agar into 50-ml Falcon tubes, mash it well, and then transfer it to a 50-ml syringe. Add 10 ml of YT to the slurry and run it three times through an 18-gauge needle and then three times through a 22-gauge needle. Mix in 10 ml of Sephadex G-25 (medium; Sigma, St. Louis, MO) that has been autoclaved in YT (G-25/YT). Prepare spin columns (one per plate) from 20-ml syringe barrels, plugged with a wad of glass wool and resting in 50-ml Falcon tubes. Add a 10-ml bed of G-25/YT to each column and then overlay with the library/G-25/YT mix. Centrifuge the columns at 500 rpm in a swinging bucket rotor for 5 min at room temperature, and then increase to 1000 rpm for an additional 10 min. Centrifuge the eluted medium at 4000*g* for 10 min at room temperature to recover the cells, and then resuspend the pellet in 10 volumes of YT. To prepare a glycerol stock, add 0.5 ml of the cells to 0.5 ml of YT–40% (v/v) glycerol, freeze in a dry ice–ethanol bath for 10 min, and then store at −70°. To prepare a library of plasmid DNA, dilute the remaining cells to an A_{600} of 0.1 in YT, and then grow to an A_{600} of 1.5. DNA can now be prepared by alkaline lysis,[12] followed by purification through ethidium bromide–cesium chloride gradients.[11]

Converting Plasmid Libraries to Viral Libraries

Packaging cells are mammalian cell lines that stably express all the viral components that are necessary to recognize and package viral mRNAs. Such cell lines can facilitate the simultaneous and proportional conversion of a plasmid library into a library of retroviral particles. When pooled plasmid DNAs from the cDNA library are introduced into these cell lines, viral mRNAs (including the inserted cDNA) are transcribed, packaged into infectious retroviral particles, and released into the medium. These particles can then be harvested and either stored or used for screening. Packaging cell lines vary in the viral titers that can be obtained. We have found that the high titers ($>10^6$/ml) that can be routinely obtained with the BOSC23 cell line are more than sufficient to ensure the efficient screening of highly complex libraries.[13]

Pooled plasmid DNAs from the cDNA libraries are introduced into the BOSC23 cells by calcium phosphate transfection. Unlike most adherent cell lines, BOSC23 cells do not form even monolayers and will begin clumping before confluence is reached. To ensure a high efficiency of transfection, it is important that this clumping be minimized. To achieve this, we begin splitting the cells 1:1, 2–3 days prior to transfection. If this is repeated on two or three consecutive days, it is possible to obtain high-density plates of well-spread cells. These cells can be readily trypsinized, and can be accurately counted and plated.

Plate 2×10^6 BOSC23 cells per plate (60 mm), 24 hr prior to transfection. Optimal transfection density is about 80% confluence. Be precise—relatively small variations in cell density, either high or low, can sometimes have dramatic effects on transfection efficiencies and/or cell survival. Prior to transfection, change the medium to 4 ml of Dulbecco's modified Eagle's medium (DMEM)–10% (v/v) fetal bovine serum (FBS) containing 25 μM choloroquine. Transfect by adding 10 μg of DNA to 500 μl of HEPES-buffered saline (pH 7.05),[11] and then add (while vortexing) 50 μl of 1.25 M $CaCl_2$. Immediately add this solution to the cells. At 10 hr, replace the medium with 4 ml of DMEM–10% (v/v) FBS. It is important to remove

[12] H. C. Birnboim and J. Doly, *Nucleic Acids Res.* **7,** 1513 (1979).

[13] W. S. Pear, G. P. Nolan, M. L. Scott, and D. Baltimore, *Proc. Natl. Acad. Sci. U.S.A.* **90,** 8392 (1993).

the chloroquine in this timely fashion to prevent cell killing. We change the volume of the medium to 2.5 ml (fresh medium) at 36 hr posttransfection and then collect the medium 24 hr later. The virus-containing medium should then be filtered (45-μm pore size filter) to remove cells. The virus can now be frozen at $-80°$, without any significant loss of titer, or used immediately for infections.

Screening Retrovirus-Based Expression Libraries for Oncogenes

Infection of Recipient Cell Lines

We generally use the focus formation assay as the basis of our screens for transforming cDNAs.[5] These screens are performed in adherent, murine cell lines such as NIH 3T3 (fibroblasts) or C127 (epithelial), which are readily infectable by ecotropic virus, and which are susceptible to single-hit transformation. If screens are to be conducted with human cell lines as recipients it would be necessary to use amphotropic packaging cell lines to generate infectious virus.

Plate 2×10^5 cells/plate (100 mm), 24 hr prior to infection. Immediately prior to infection, dilute the viral soup 1:1 in DMEM that contains 10% calf serum (CS) and Polybrene (16 μg/ml; final concentration, 8 μg/ml). Aspirate the medium from the cells and replace with the medium containing virus and Polybrene. Allow the infection to proceed for 3–5 hr and then replace the medium with fresh DMEM–10% (v/v) CS. Change the medium every 2–3 days for up to 21 days and identify foci that are formed. We do not perform these screens under selective conditions because the infection frequency is usually high, and selection often disrupts the integrity of the monolayer, thus generating spontaneous foci. Individual foci are then scraped from the surface of the plate, transferred to individual plates (35 mm), and amplified clonally. Clonal amplification is done in the presence of hygromycin (200 μg/ml) to ensure the presence of a proviral insert. To determine viral titer, we remove a small aliquot of the viral soup, perform serial dilutions, and then infect recipient cells, under selection, to determine the number of resistant colonies that can be generated.

Polymerase Chain Reaction Amplification of Proviral cDNA Inserts

A polymerase chain reaction (PCR)-mediated DNA amplification protocol has been developed to facilitate the recovery of proviral cDNA inserts (Fig. 2).[2] High molecular weight DNA, prepared from the transformed cells, is PCR amplified with a set of vector primers that flank the cDNA

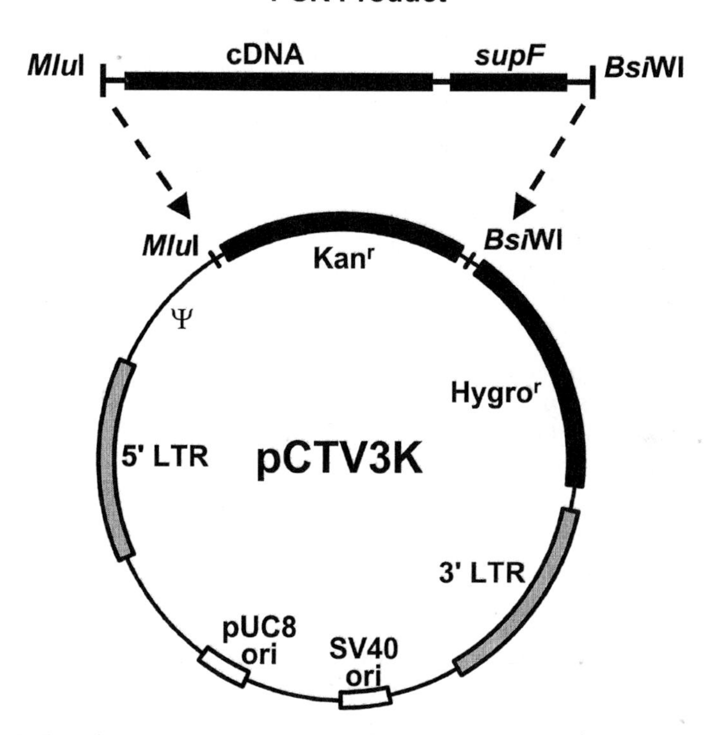

FIG. 2. Procedure for recovering transforming cDNAs from proviral inserts. Genomic DNA is isolated from transformed cell clones and then used as a template for PCR amplification, using a set of vector primers that flank the cDNA insert and the linked *supF* gene. The amplified fragment is cut with *Mlu*I and *Bsi*WI and cloned into complementary sites within the pCTV3K vector. Recombinants are isolated on the basis of the acquisition of the suppressor tRNA activity (*supF*).

insert and the adjacent *supF* gene. *supF* encodes suppressor tRNA activity and can be used as a selectable marker in bacterial strains that contain the p3 plasmid. PCR products that contain the *supF* sequences are then cloned into a retroviral vector that lacks a bacterial selectable marker and selected on the basis of their acquired suppressor tRNA activity. The PCR process utilizes *Pfu* DNA polymerase (Stratagene, La Jolla, CA), which, unlike more common versions of *Taq*, has a proofreading function associated with it. Thus, errors introduced into the DNA sequences by the polymerase are kept to a minimum.

Add 600 μl of lysis buffer [1 m*M* EDTA, 1% (w/v) SDS, and proteinase K (100 μg/ml) added directly before use] to a 35-mm well of confluent cells and incubate at 37° for 90 min. Transfer the lysate to a 1.5-ml tube and extract twice with an equal volume of phenol–chloroform. Precipitate the DNA twice with 1/10 volume of 10 *M* ammonium acetate and 2 volumes of 95% (v/v) ethanol. Wash thoroughly with 70% (v/v) ethanol after each precipitation. Dry the pellet and resuspend in 150 μl of distilled H_2O. Prepare PCRs containing 100 ng of genomic DNA template, 100 ng of each primer, 2.5 μl of dNTP mix at 2 m*M* each, 2.5 μl of *Pfu* polymerase 10× buffer, 2.5 U of *Pfu* DNA polymerase, and water to 25 μl. We use the following primers for amplifying pCTV3-derived proviruses: pCTV-5′ CCT CAC TCC TTC TCT AGC TC and pCTV-3′-TCG AAT CAA GCT TAT CGA TAC G. PCR cycles are 95° for 60 sec, 50° for 30 sec, 68° for 6 min; for 30 cycles. At this point we run 5 μl of the PCR reaction mix on a 1.2% (w/v) polyacrylamide gel to identify any bands that have been amplified (often there is more than one). We then do a Southern blot with a probe specific for the *supF* gene. This allows us to distinguish bands that are derived from bona fide proviral inserts, and allows us to identify amplified products that are present in subvisual amounts.

Cloning Polymerase Chain Reaction Products

pCTV3K is a specialized retroviral vector that has been developed for use in cloning PCR-amplified proviral inserts (Fig. 2).[2] The vector is derived from pCTV3B and contains a Kan^r marker in place of *supF*. Ligations are performed by replacing the Kan^r marker of pCTV3K with the amplified transforming cDNA and its linked *supF* gene. Because pCTV3K does not contain any ampicillin or tetracycline resistance, recombinants can be recovered in p3 hosts with no background derived from religated vector.

Bring the volume of the PCR to 180 μl with water and then add 20 μl of 10 *M* ammonium acetate. Extract once with 200 μl of phenol–chloroform and then precipitate with 2 volumes 95% (v/v) ethanol. Centrifuge for 30 min at low temperature (15°) and rinse the pellet with 70% (v/v) ethanol. Dissolve the pellet in water, and digest with 10 U each of *Mlu*I (37°) and *Bsi*WI (55°). Run the digests on a 1.2% (w/v) agarose gel and purify the proviral fragments (determined by the Southern blot) by electroelution and ethanol precipitation. Combine 20 ng of the PCR fragment with 40 ng of pCTV3K that has been digested with *Mlu*I and *Bsi*WI, and ligate for 1 hr at room temperature. Use the ligation mix to transform chemically competent ($>5 \times 10^6$) MC1061/p3. Plasmids can be prepared by an alkaline lysis procedure,[12] followed by RNase digestion. The plasmid DNA that is recovered is a fully reconstituted retroviral vector that can be used for functional testing of the transforming cDNA.

Concluding Remarks

It is our experience that these screens most often fail because of a lack of diligence in the cDNA recovery process. Because of the high efficiencies associated with retroviral infections, it is not uncommon to obtain multiple proviral inserts in a single transformed cell clone. Although the smaller, more abundant PCR products are easiest to recover, it is often the larger, subvisual bands that are transforming. Such bands, if identified by Southern blot, can be readily cloned into pCTV3K. On occasion, it is not possible to recover a transforming cDNA from a transformed clone. This will occur if transformation is due to a spontaneous event, insertional mutagenesis, or the combined expression of two or more cDNAs. Because the system has not been designed to characterize such transforming events, we have generally found it prudent to set these clones aside. If the screens are performed correctly, and carefully, a large proportion of transformed cell clones will yield transforming cDNAs.

[17] Identification of Ras-Regulated Genes by Representational Difference Analysis

By JANIEL M. SHIELDS, CHANNING J. DER, and SCOTT POWERS

Introduction

ras genes are among the most frequently mutated oncogenes found in human cancers.[1,2] Since its discovery in 1982 as a transforming oncogene, there have been numerous studies to understand the function of Ras and the pathways utilized by Ras to mediate its actions. To date, most of the studies have focused on the identification of the Ras effector proteins and analyses of the various signaling pathways activated. In contrast, while great strides have been achieved in understanding signaling from the plasma membrane to the nucleus, relatively little is known about which genes are the targets of these signals. Ras signaling pathways lead to the activation of a variety of transcription factors including the Ets family (Elk-1, Ets1, Ets2), Jun, ATF-2 (activating transcription factor 2), Fos, NF-κB, and SRF (serum response factor).[3] Consequently, Ras can stimulate transcription from promoters that contain AP-1 (Fos : Jun, Jun : Jun, Jun : ATF-2 dimers),

[1] M. Barbacid, *Annu. Rev. Biochem.* **56,** 779 (1987).
[2] J. L. Bos, *Cancer Res.* **49,** 4682 (1989).
[3] R. Treisman, *Curr. Opin. Cell Biol.* **8,** 205 (1996).

0076-6879/00 $35.00

NF-κB, serum response elements (Elk-1 and SRF), and ets (Elk-1, Ets) DNA-binding motifs. However, there is no clear picture as to which genes are then affected by the activation of these transcription factors in response to Ras activation. In addition, little is known about the expression of genes that may be repressed (such as tumor suppressors) as a consequence of Ras activation. In this chapter, we describe the use of one method, representational difference analysis (RDA), to identify those genes aberrantly expressed by Ras activation. RDA has been applied successfully to identify genes whose expression is deregulated by a variety of oncoproteins including the Ewing's sarcoma EWS/FLI1 fusion protein, Fos, Myc, and ErbB3.[4–7]

Background

A majority of the studies of Ras-mediated signal transduction have been done with fibroblast cell lines, in particular NIH 3T3 mouse fibroblasts. However, most *ras* mutations occur in cancers derived from epithelial or hematopoietic cells. Although both Ras and Raf [a downstream effector of Ras that activates the mitogen-activated protein (MAP) kinase kinase/extracellular signal-regulated kinase (MEK/ERK) pathway], can fully transform NIH 3T3 cells, only activated Ras can transform rat intestinal epithelial (RIE-1) cells.[8] This suggests that in contrast to fibroblasts, where Raf-dependent signaling alone is sufficient to cause transformation, in RIE-1 cells, Raf-independent pathways play a critical role in mediating transformation. Therefore, these cells provide a powerful cell system for defining Raf-independent signaling events critical for Ras transformation. To define those genes important for the maintenance of transformation in a biologically relevant cell system, we chose to identify by RDA those genes whose expression is aberrantly regulated by Ras, through Raf-independent pathways, using two RIE-1 cell lines stably expressing either activated K-Ras4B(12V) or Raf-1 (Raf-22W).

Representational Difference Analysis

Representational difference analysis is a method that was originally developed to find small differences between the sequences of two genomic

[4] A. Arvand, H. Bastians, S. M. Welford, A. D. Thompson, J. V. Ruderman, and C. T. Denny, *Oncogene* **17,** 2039 (1998).

[5] A. V. Bakin and T. Curran, *Science* **283,** 387 (1999).

[6] B. C. Lewis, H. Shim, Q. Li, C. S. Wu, L. A. Lee, A. Maity, and C. V. Dang, *Mol. Cell. Biol.* **17,** 4967 (1997).

[7] C. F. Edman, S. A. Prigent, A. Schipper, and J. R. Feramisco, *Biochem. J.* **323,** 113 (1997).

[8] S. M. Oldham, G. J. Clark, L. M. Gangarosa, R. J. Coffey, Jr., and C. J. Der, *Proc. Natl. Acad. Sci. U.S.A.* **93,** 6924 (1996).

DNA populations.[9] The method builds on earlier subtractive hybridization methods and couples this to polymerase chain reaction (PCR) amplification. Subsequently, RDA was adapted to compare differences between two cDNA populations,[10] which is the method described in this chapter. The method is powerful in that it eliminates those cDNAs found in both populations and amplifies sequences differentially expressed in only one population. This is the main advantage of RDA over differential display, which amplifies all the mRNA present in two populations, requiring comparisons of the products on gels to identify bands present in one population but not the other. This results in a high frequency of false positives for differential display. In contrast, a major strength of RDA is that it has a low background of false positives. In addition, it is fairly fast (once cDNA is made, the method can be completed in about 2 weeks), technically not difficult, and allows for the isolation of genes that are either up- or downregulated. Other than a PCR machine, the method utilizes standard laboratory equipment and reagents. In addition, RDA is a relatively inexpensive method when compared with chip array technology, which may not be affordable for many laboratories. Chip array technology also requires the investigator to preselect which genes to analyze. One weakness of the RDA method, however, is that the average size of the cloned products is about 260 bp. To isolate full-length clones for use in functional studies, the clone of known genes can be obtained from published sources, or a full-length cDNA library can be generated and screened with the 260-bp fragment as a probe, or reverse transcriptase (RT)-PCR can be used if the sequence is available in a database.

How Representational Difference Analysis Works

Figure 1 is a schematic illustrating the hybridization and amplification portion of the method once the tester and driver cDNAs have been generated. The tester cDNA is the experimental population that is under question; in this case cDNA from RIE-1 cells stably expressing activated K-Ras4B(12V). The driver cDNA is the population used to "drive" the removal of sequences found in both populations; in this case, from RIE-1 cells stably expressing activated Raf-1 (Raf-22W). In brief, specific adaptors (which serve as priming sites for PCR and restriction enzyme-cutting sites) are added to the ends of the tester cDNA, which is then mixed with excess driver cDNA, denatured, and annealed. Three types of hybrids can form: tester/tester, tester/driver, and driver/driver. After PCR amplification with

[9] N. Lisitsyn, N. Lisitsyn, and M. Wigler, *Science* **259,** 946 (1993).

[10] M. Hubank and D. G. Schatz, *Nucleic Acids Res.* **22,** 5640 (1994).

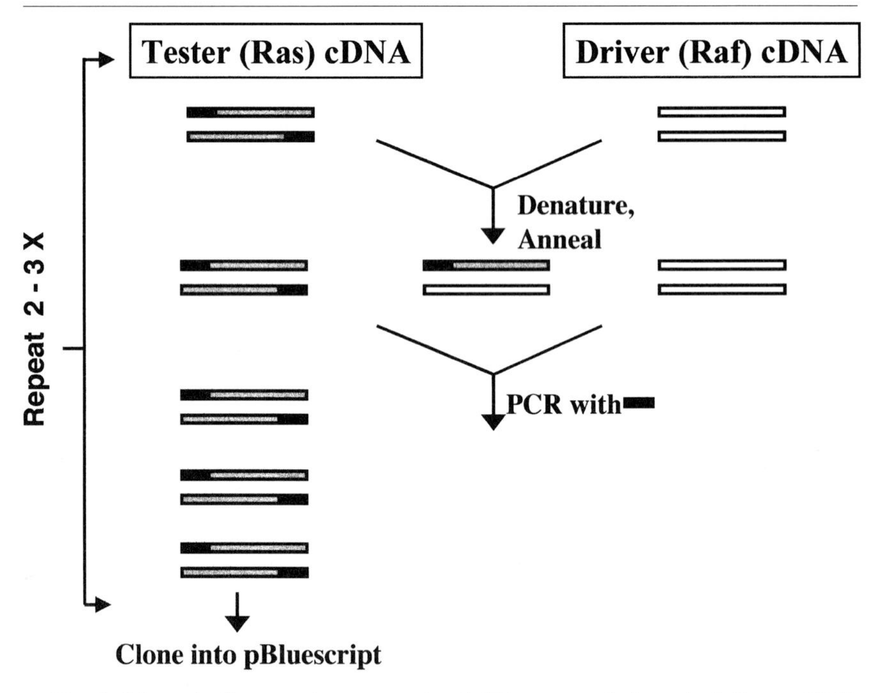

FIG. 1. Schematic diagram of representational difference analysis method. Tester cDNA from RIE/K-Ras cells (*left*) is annealed to excess driver cDNA from RIE/Raf cells (*right*), PCR amplified, and cloned into a suitable vector (e.g., pBluescript; Stratagene, La Jolla, CA).

an oligonucleotide primer that recognizes the adaptor sequence, driver/driver hybrids are not amplified because the driver cDNAs do not contain the priming site, tester/driver hybrids are amplified only from the single tester strand resulting in linear amplification, and tester/tester hybrids are amplified exponentially. The amplified products are then digested with mung bean nuclease to remove all single-stranded PCR products. The resulting difference products are then used to repeat the subtraction and amplification procedure with a greater amount of driver cDNA than that used in the first round of subtraction, to sequentially reduce the cloning of false positives and amplify the truly differentially expressed genes. After the final round of subtraction and amplification (usually no more than two or three rounds), the final PCR products are digested with a restriction endonuclease that cleaves off the adaptor ends to allow for subcloning into a suitable vector.

Cloning of Genes Upregulated by Ras

To clone genes upregulated by Ras, but not Raf, in RIE-1 cells, cDNA from RIE-1 cells stably expressing K-Ras(12V) was used as tester cDNA and subtracted with driver cDNA from RIE-1 cells stably expressing Raf(22W) as described in Fig. 1. As diagrammed in Fig. 2, those genes (B and D) upregulated by Ras through Raf-independent pathways would not be expressed in the Raf-expressing cells, and consequently would be cloned from the Ras tester cDNA population.

Cloning of Genes Downregulated by Ras

To clone genes downregulated by Ras, but not Raf, in RIE-1 cells, the two populations of cDNA are simply switched. That is, in this case, cDNA from Raf-expressing RIE-1 cells is used as the tester cDNA whereas cDNA from K-Ras-expressing RIE-1 cells is used as the driver cDNA. The idea here is that genes downregulated by Ras through Raf-independent pathways would not be downregulated in the Raf-expressing cells. As diagrammed in Fig. 3, genes (G and I) identified from the Raf tester cDNA are cloned because they are downregulated in the Ras-expressing cells.

Cell Culture, mRNA Isolation, and cDNA Generation

Because the method is PCR based, it is important to minimize in the two cell populations gene expression differences caused by, e.g., cell density,

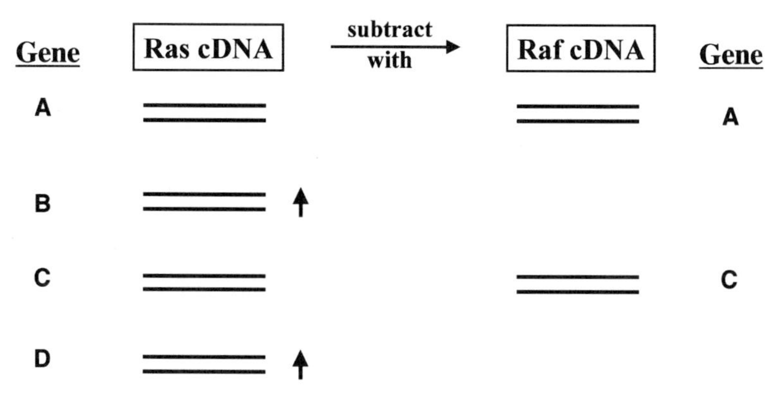

FIG. 2. Schematic of how genes upregulated by Ras are cloned. Ras cDNA is subtracted with excess Raf cDNA. Genes expressed in both cells are subtracted out (genes A and C) while genes upregulated by Ras, but not Raf (genes B and D), are cloned from the Ras cDNA.

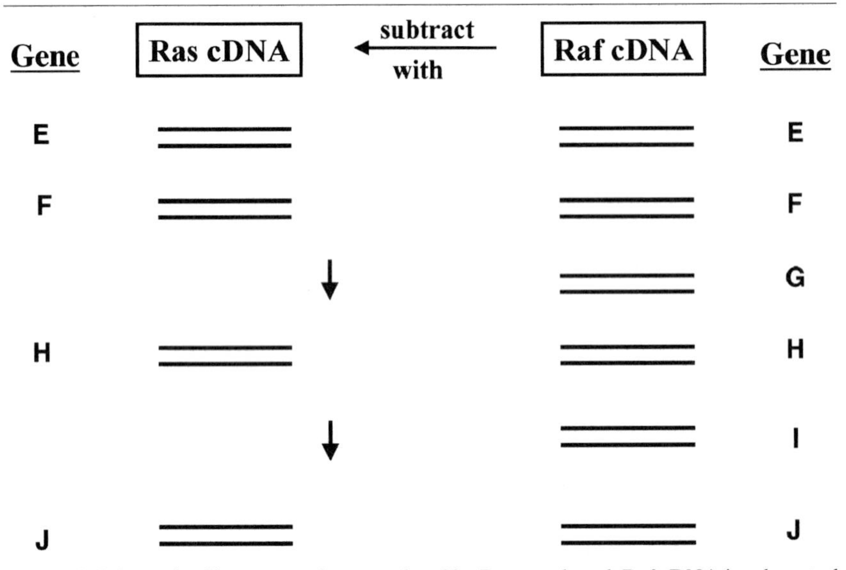

FIG. 3. Schematic of how genes downregulated by Ras are cloned. Raf cDNA is subtracted with excess Ras cDNA. Genes expressed in both cells are subtracted out (genes E, F, H, and J) while genes downregulated by Ras, but not Raf (genes G and I), are cloned from the Raf cDNA.

passage number, and time of feeding before harvesting RNA. It is best, if possible, to culture and harvest the cells at the same time to minimize these effects. Total RNA can be isolated with guanidine thiocyanate[11] and mRNA can be purified with oligo(dT)[12] by standard methods or by commercially available kits such as mRNA Direct (Qiagen, Valencia, CA). Purified mRNA should first be checked for purity by Northern blot analysis to verify that the RNA is not degraded. We use the Superscript Choice (GIBCO, Grand Island, NY) kit for cDNA synthesis essentially as described by the manufacturer.

Representational Difference Analysis Protocol

The method used here is based on that of Hubank and Schatz.[10] Oligonucleotides needed are as follows: R-Bgl-12, 5′-GATCTGCGGTGA-3′; R-Bgl-24, 5′-AGCACTCTCCAGCCTCTCACCGCA-3′; J-Bgl-12, 5′-GATCTGTTCATG-3′; J-Bgl-24, 5′-ACCGACGTCGACTATCCAT-

[11] J. J. Chirgwin, A. E. Przbyla, R. J. MacDonald, and W. J. Rutter, *Biochemistry* **18,** 5294 (1979).

[12] H. Aviv and P. Leder, *Proc. Natl. Acad. Sci. U.S.A.* **69,** 1408 (1972).

GAACA-3′; N-Bgl-12, 5′-GATCTTCCCTCG-3′; and N-Bgl-24, 5′-AGG-CAACTGTGCTATCCGAGGGAA-3′.

Preparation of Tester and Driver cDNA

Both cDNA populations are first amplified by PCR and then digested with the restriction enzyme *Dpn*II, a 4-base pair (bp) cutter, to generate amplifiable "representations" of the original cDNA. It should be noted that a small percentage of cDNAs will not contain two *Dpn*II sites and so will not be amplifiable. If this is an issue a parallel RDA can be done with another 4 bp-cutting restriction enzyme such as *Bfa*I and the oligonucleotides modified accordingly. Digest 2 μg of each cDNA (Ras and Raf) with 20 U of *Dpn*II (final volume, 100 μl), extract with 100 μl of phenol–chloroform (1:1, v/v), and precipitate with 2 μg of glycogen carrier (Boehringer Mannheim, Indianapolis, IN), 50 μl of 10 *M* ammonium acetate, and 650 μl of ethanol. Place on ice for 20 min, spin in a microcentrifuge at 4° for 15 min, wash the pellet with 70% (v/v) ethanol, dry the pellet, and resuspend in 20 μl of TE [10 m*M* Tris (pH 7.5), 1 m*M* EDTA]. Mix 12 μl (1.2 μg) of *Dpn*II-cut cDNA with 4 μl of desalted R-Bgl-24 (2 mg/ml), 4 μl of desalted R-Bgl-12 (1 mg/ml), 6 μl of 10× T4 DNA ligase buffer [500 m*M* Tris (pH 7.8), 100 m*M* $MgCl_2$, 100 m*M* dithiothreitol (DTT), 10 m*M* ATP], and 31 μl of water. Place in a PCR machine at 50° for 1 min and then cool to 10° at 1°/min. T4 DNA ligase, 3 μl (400 U/μl; New England BioLabs, Beverly, MA), is added and incubated at 14° overnight. Dilute the ligations with 140 μl of TE. To generate 0.5–1.0 mg of each representation, prepare 20 to 30 PCR amplification reactions of each as follows: mix 138 μl of water, 40 μl of 5× PCR buffer [335 m*M* Tris (pH 8.8), 20 m*M* $MgCl_2$, 80 m*M* $(NH_4)_2SO_4$, and bovine serum albumin (BSA, 166 μg/ml)], 17 μl of dNTPs (4 m*M* each), 2 μl of R-Bgl-24 (1 mg/ml), and 2 μl of diluted ligation. Place in the PCR machine at 72° for 3 min, add 1 μl (5 U) of *Taq* DNA polymerase, and overlay with mineral oil. Incubate for 5 min at 72° and perform PCR at 1 min at 95°, 3 min at 72° for 18 to 20 cycles, and 10 min at 72°. Per 800 μl of PCR product, extract with 800 μl of phenol–chloroform, and precipitate with 75 μl of 3 *M* sodium acetate (pH 5.3) and 800 μl of 2-propanol. Leave on ice for 20 min, spin at 4° for 15 min in a microcentrifuge, wash the pellet with 70% (v/v) ethanol, and resuspend in TE at 0.5 mg/ml. Digest 300 μg of each representation as follows: mix 600 μl of representation DNA, 140 μl of 10× *Dpn*II buffer, 75 μl of *Dpn*II (10 U/μl), and 585 μl of water at 37° for 4 hr. Extract with 1400 μl of phenol–chloroform and precipitate with 140 μl of 3 *M* sodium acetate (pH 5.3) and 1400 μl of 2-propanol on ice for 20 min. Spin for 15 min at 4° in a microcentrifuge, wash the pellet with 70% (v/v) ethanol, and resuspend at 0.5 mg/ml. The Raf cDNA prepared this way is the digested Raf driver DNA.

To prepare Ras tester DNA, size fractionate 40 μl (20 μg) of *Dpn*II-digested Ras representation over a TAE–1.2% (w/v) agarose gel to remove the adaptors and cut out the band containing digested DNA. Purify the DNA in gel slice, using a Qiaex gel extraction kit (Qiagen). This is the digested Ras tester DNA. Ligate J oligonucleotides to Ras tester DNA as follows: mix 2 μg of Ras tester DNA, 6 μl of 10× T4 ligase buffer, 4 μl of J-Bgl-24 (2 mg/ml), and 4 μl of J-Bgl-12 (1 mg/ml) to a final volume of 57 μl with water. Anneal in the PCR machine at 50° for 1 min and then cool to 10° at 1°/min. Add 3 μl of T4 DNA ligase and ligate overnight at 14°. Dilute the Ras/tester/J DNA to 10 ng/μl with TE.

Subtractive Hybridization

Mix 80 μl (40 μg) of Raf/driver and 40 μl (0.4 μg) of Ras/tester/J DNA and extract with phenol–chloroform and precipitate with 30 μl of 10 *M* ammonium acetate and 380 μl of ethanol on dry ice for 10 min. Microcentrifuge for 15 min at 4°, wash with 70% (v/v) ethanol, dry, and resuspend with 4 μl of EE × 3 buffer [30 m*M* hydroxyethyl piperazine propanesulfonic acid (EPPS, pH 8.0), 3 m*M* EDTA]. Incubate at 37° for 5 min, vortex, spin, and overlay with mineral oil. Denature for 5 min in a 98° water bath, place in the PCR machine at 67°, add 1 μl of 5 *M* NaCl, and incubate for 20 hr.

Generation of First Difference Product (DP1)

Remove the mineral oil from off the hybridized DNA and dilute with 8 μl of TE followed with 25 μl of TE and finally with 362 μl of TE. Set up four PCRs, each containing 120 μl of water, 40 μl of 5× PCR buffer, 17 μl of dNTPs, and 20 μl of hybridization mix. Place in the PCR machine at 72° for 3 min, add 1 μl of *Taq* DNA polymerase, incubate at 72° for 5 min, add 2 μl of J-Bgl-24 (1 mg/ml), and do 10 PCR cycles at 95° for 1 min, 70° for 3 min, and 10 min at 72° and then cool to 25°. Combine all four tubes, extract with phenol–chloroform, and precipitate with 2 μg of glycogen carrier, 75 μl of 3 *M* sodium acetate, and 800 μl of 2-propanol on ice for 20 min. Spin for 15 min at 4°, wash with 70% (v/v) ethanol, and resuspend in 40 μl of 0.2× TE. Digest single-stranded DNA as follows: mix 20 μl of DNA, 4 μl of 10× mung bean nuclease (MBN) buffer, 14 μl of water, and 2 μl of MBN at 30° for 35 min. Add 160 μl of 50 m*M* Tris, pH 8.9, and incubate at 98° for 5 min and then place on ice. Set up four PCRs on ice, each containing 120 μl of water, 40 μl of 5× PCR buffer, 17 μl of dNTPs, 2 μl of J-Bgl-24 (1 mg/ml), and, adding this last, 20 μl of

MBN-treated DNA. Perform PCR at 95° for 1 min, cool to 80°, and add 1 μl of *Taq* DNA polymerase. Do 18 cycles at 95° for 1 min, 70° for 3 min, and 72° for 10 min. Combine the four tubes, extract with phenol–chloroform, precipitate as described above, and resuspend the final pellet in TE at 0.5 μg/μl.

To change the adaptors, digest 40 μl (20 μg) with 150 U of *Dpn*II in a volume of 300 μl, extract with phenol–chloroform, and precipitate with 33 μl of 3 *M* sodium acetate and 800 μl of ethanol at −20° for 20 min. Spin for 15 min at 4°, wash with 70% (v/v) ethanol, and resuspend the pellet with 40 μl of TE (0.5 μg/μl). Dilute 1 μl with 9 μl of water. Mix 4 μl of diluted DNA, 6 μl of 10× T4 DNA ligase buffer, 4 μl of N-Bgl-24 (2 mg/ml), 4 μl of N-Bgl-12 (1 mg/ml), and 39 μl of water. Place in the PCR machine at 50° for 1 min, cool to 10° at 1°/min, add 3 μl of T4 ligase, and incubate at 14° overnight.

Generation of Second Difference Product (DP2)

Dilute ligation with 100 μl of TE to 1.25 ng/μl. Add 40 μl (50 ng) of N-ligated second difference product (DP2) with 80 μl of Raf driver DNA and proceed with the subtraction and amplification steps described above, using the J oligonucleotides with ligations at 14° and amplifications at 70°.

Generation of Third Difference Product (DP3)

Dilute J-ligated DP2 to 1 ng/μl. Mix 10 μl with 990 μl of TE, yielding 10 pg/μl. Hybridize 100 pg (10 μl) of J-ligated DP2 and 40 μg (80 μl) of Raf driver DNA. Perform the above-described subtraction and amplification for 22 cycles.

Comparison of Difference Products

To verify that the subtractions worked, a small aliquot (0.5 μg) of each difference product and the original representation should be run in a 1.4% (w/v) agarose gel. As shown in Fig. 4, a sequential enrichment of specifically sized bands should be observed. In addition, the bands of the final difference product from the Ras minus Raf subtraction should be different in size from those of the final difference product of the Raf minus Ras subtraction (compare lanes 5 and 6, Fig. 4). Note that there is little difference between the band patterns of DP3 and DP4, indicating that DP3 would probably be a better choice for final cloning to prevent loss of some genes that may have occurred in the generation of DP4.

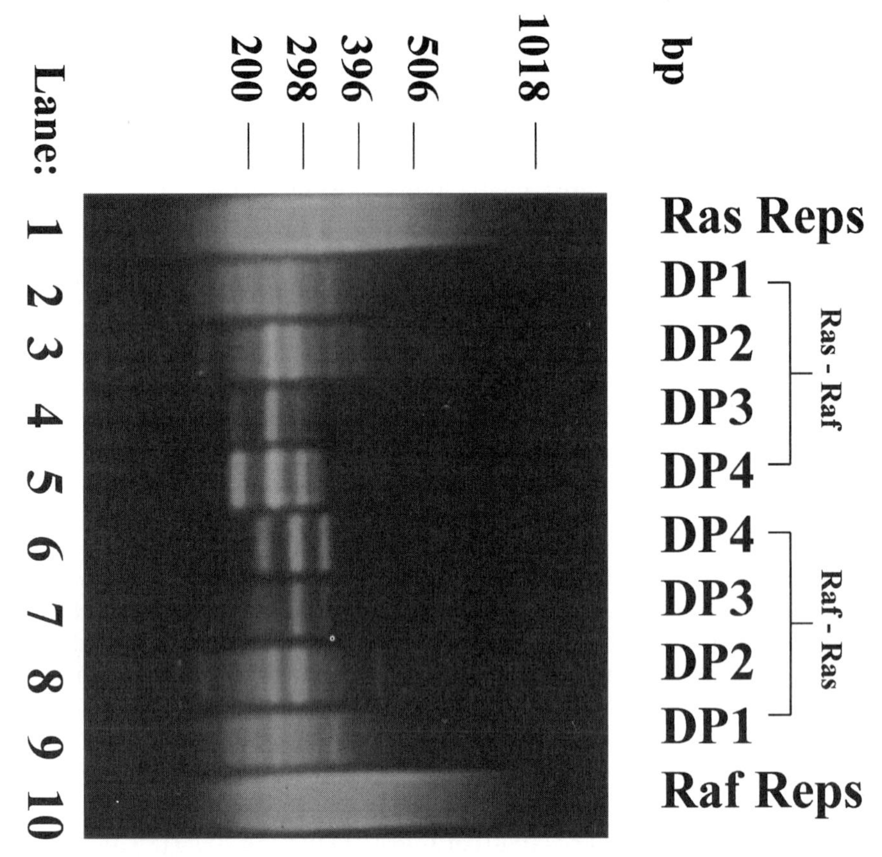

FIG. 4. Comparison of RDA difference products generated from the subtractions. Shown is 0.5 μg of DNA from the Ras representation (lane 1) followed by difference products 1–4 from the Ras minus Raf subtraction (lanes 2–5) followed by difference products 1–4 from the Raf minus Ras subtraction (lanes 6–9) and the Raf representation (lane 10). Note the difference in band sizes in the final difference products (DP4) of the Ras minus Raf (lane 6) compared with the Raf minus Ras (lane 7) subtractions. This is a good indication that the subtractions worked and that the bands represent differentially expressed genes.

Cloning Final Difference Product

Digest DP3 with *Dpn*II, size fractionate on a 1.4% (w/v) agarose gel, isolate DNA bands, and ligate into the *Bam*HI site of a suitable cloning vector (e.g., pBluescript).

Gene Identification and Differential Expression Verification

Sequence the inserts of the clones and compare them with the GenBank database. Verify that cloned genes are truly differentially expressed by

Northern blot analysis of total RNA from untransformed RIE-1 cells, and from Ras- and Raf-expressing cells, as shown in Fig. 5.

Screening Cloned Genes

Probably the most time-consuming part of the analysis is the screening process to determine which genes to choose for further characterization. To confirm that the aberrant expression patterns observed for the clones is not cell line specific (hence RIE-1 specific), Northern blot analyses of other cell lines harboring Ras mutations should be examined. For this purpose, our analyses have included the human colon adenocarcinoma

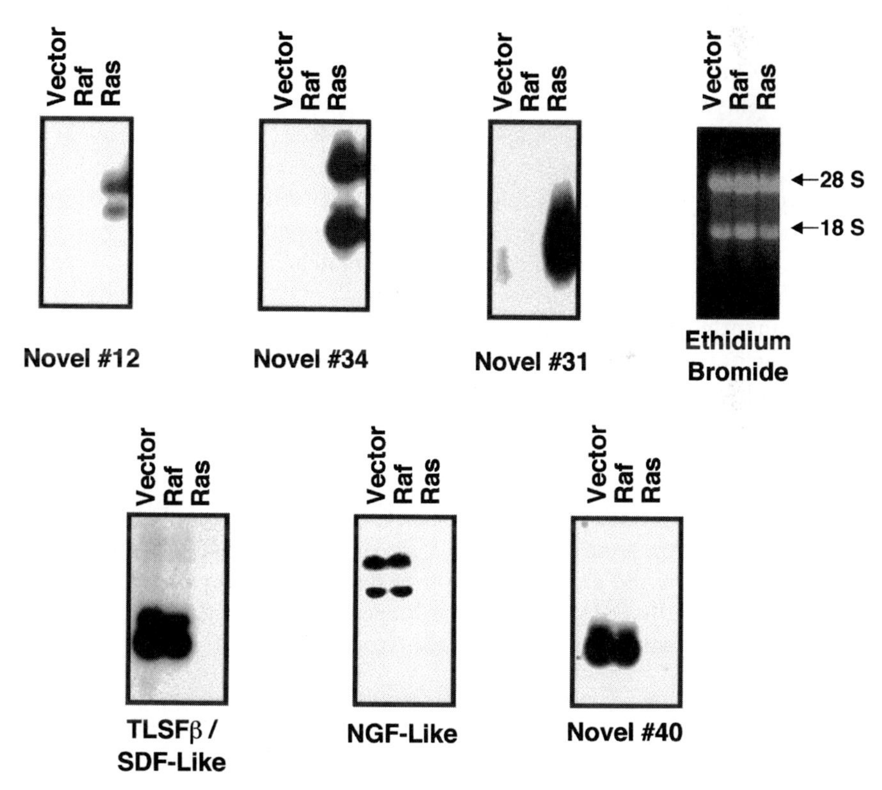

FIG. 5. Northern blot analysis of genes upregulated and downregulated by Ras, but not Raf, in RIE-1 cells. Shown is 25 μg of total RNA from RIE-1 cells stably expressing empty vector, Raf, or Ras and probed with ^{32}P-labeled DNA from each gene shown. Numbers represent the clone number of novel genes not found in GenBank. Clones 12, 34, and 31 were upregulated by Ras whereas TLSF/SDF-like, NGF-like, and clone 40 were downregulated by Ras.

(DLD-1)[13] and the human fibrosarcoma (HT 1080)[14] cell lines, which carry *ras* mutations, along with their derivative cell lines (DKS-8 and MCH 603c8, respectively), in which the mutant *ras* allele has been deleted, as controls (data not shown). If the aberrant expression patterns seen in the RIE-1 cells are also observed in other cell lines expressing activated Ras, but not in their respective wild-type *ras* cell lines, it may be concluded that the aberrant expression is associated with Ras transformation in a variety of cell types.

Biological Significance Assays

As a further means to determine which genes are important for Ras transformation of RIE-1 cells, those genes upregulated by the Ras-expressing RIE-1 cells can be forcibly expressed in the Raf-expressing cells and their ability to transform the cells determined. Likewise, to determine which genes are important in inhibiting transformation, those genes downregulated by Ras can be forcibly expressed in the Ras-expressing RIE-1 cells and their ability to reverse transformation assessed. However, a caveat of this approach is that forced expression of any one gene may not be sufficient to drive or prevent transformation. Therefore, another approach to determine biological significance involves comparing expression of the genes in human tumor samples with expression in matched surrounding normal tissue by Northern blot analysis. This approach may also reveal genes that are frequently aberrantly expressed in tumors by a variety of mechanisms including, but not limited to, Ras transformation. Hence, this might identify genes that may provide important diagnostic markers for human cancer.

Summary

In conclusion, RDA provides a fast, technically simple, and inexpensive way to characterize genes aberrantly expressed due to Ras transformation. The identification and characterization of these genes may provide insight not only into the mechanism by which Ras causes transformation, but also may identify novel targets for rational drug design and development of anticancer drugs.

[13] S. Shirasawa, M. Furuse, N. Yokoyama, and T. Sasazuki, *Science* **260,** 85 (1993).

[14] R. Plattner, M. J. Anderson, K. Y. Sato, C. L. Pasching, C. J. Der, and E. J. Stanbridge, *Proc. Natl. Acad. Sci. U.S.A.* **93,** 6665 (1996).

[18] Differential Display Analysis of Gene Expression Altered by *ras* Oncogene

By HAKRYUL JO, YONG-JIG CHO, HONG ZHANG, and PENG LIANG

Introduction

The coordinated regulation of gene expression is a key cellular function that specifies cell characteristics as well as controls normal physiological processes of the organism. Deregulation of this gene expression leads to a variety of abnormal conditions such as cancer. *ras* is among the oncogenes most frequently mutated in various types of human cancers. The mutated Ras protein constitutively elicits multiple mitogenic signals to the nucleus to alter the expression of target genes that are involved in a broad range of normal cellular functions. Thus, the identification of these genes may provide an important tool toward the understanding of the pathogenic processes. As a first step to understand at the molecular level the profound effect elicited by oncogenic Ras protein, we have looked for its target genes in murine model cell lines, using the differential display method.[1] Our initial screening has isolated a number of genes either up- or downregulated by oncogenic *ras* activation.[2,3] Although the functional analyses of the role of these genes in Ras-mediated cell transformation will be more challenging and time consuming, differential display has become an efficient tool that has helped us move to the next step. Here, we focus primarily on the technical aspects of the differential display method and experimental designs that are critical for the identification of relevant genes.

The development of cancer is a multistep process in which a wide range of normal cellular functions is affected.[4] Conventional studies toward the understanding of tumorigenic mechanisms have been focused on the identification of oncogenes and tumor suppressor genes that are mutated in human cancers. About 100 oncogenes and a dozen tumor suppressor genes have been isolated from various kinds of cancers.[5] Some of these genes are involved in early stages of tumorigenesis and the others have been shown

[1] P. Liang and A. B. Pardee, *Science* **257,** 967 (1992).

[2] P. Liang, L. Averboukh, W. Zhu, and A. B. Pardee, *Proc. Natl. Acad. Sci. U.S.A.* **91,** 12515 (1994).

[3] H. Jo, H. Zhang, R. Zhang, and P. Liang, *Methods* **16,** 365 (1998).

[4] T. Hunter, *Cell* **64,** 249 (1991).

[5] R. Sager, *Proc. Natl. Acad. Sci. U.S.A.* **94,** 952 (1997).

0076-6879/00 $35.00

to be associated with later progression.[6] The identification of mutations in these genes may provide diagnostic information as to the risk factor for an individual to develop certain types of cancers. However, cancer phenotypes result from the alteration of gene expression and no simple relationship between mutated genes and cancer phenotypes is found. This observation suggests that the identification of genes deregulated in cancers is equally important, and those genes may provide important insights leading to a better understanding of tumorigenesis. This idea is further supported by the fact that most cancer genes are either directly (such as *p53, myc*, and WT-1) or indirectly (such as *ras, erbB2*, and APC) involved in the regulation of downstream gene expression.

The *ras* oncogene is one of the most commonly mutated in human cancers (>50% of colon cancer and >90% of pancreatic cancer), and thus has been the area of intensive biochemical and genetic studies.[7] These studies have revealed downstream signaling events and components that relay *ras*-induced mitogenic signals to the ultimate transcription factors that regulate the expression of genes largely involved in cell growth and differentiation.[8] The oncogenic form of Ras protein constitutively elicits downstream signaling such that the normal regulations of gene expression are impaired and alterations of a subset of gene expression ensue, which eventually lead to disruption of normal cellular functions. Consistent with this idea, it has been shown that downstream transcription factors are essential for *ras*-mediated cell transformation.[9–11] However, compared with our knowledge of *ras* signaling events, the target genes that are responsible for the various phenotypic changes, such as cell transformation, resulting from *ras* activation are largely unknown. Thus identification of the genes altered in the process of *ras*-mediated cell transformation would provide important clues toward the understanding of the molecular mechanism underlying this process.

Differential display has been developed to identify genes differentially expressed in a given biological system,[1] and has been widely applied to a

[6] K. W. Kinzler and B. Vogelstein, *Cell* **87,** 159 (1996).

[7] J. B. Gibbs, A. Oliff, and N. E. Kohl, *Cell* **77,** 175 (1994).

[8] S. A. Moodie, B. M. Willumsen, M. J. Weberand, and A. Wolfman, *Science* **260,** 1658 (1993).

[9] S. Langer, D. Bortner, M. Roussel, C. Sherr, and M. Ostrowski, *Mol. Cell. Biol.* **12,** 5355 (1992).

[10] R. Johnson, B. Spiegelman, D. Hanahan, and R. Wisdom, *Mol. Cell. Biol.* **16,** 4504 (1996).

[11] T. S. Finco, J. K. Westwick, J. L. Norris, A. A. Beg, C. J. Der, and A. S. Baldwin, *J. Biol. Chem.* **272,** 24113 (1997).

variety of biological problems resulting from altered gene expression.[12,13] Here we provide a step-by-step description of the method, which was used to systematically identify genes altered in *ras*-mediated cell transformation.

Experimental Designs

Although differential display has been widely used to identify differentially expressed genes, careful experimental designs could enhance the probability of identifying more relevant genes for a given biological problem under investigation.[14] Because the ultimate goal of this study is to dissect the key pathways employed by oncogenic *ras* during cell transformation, it is desirable to use simple model systems in which to look for genes directly controlled by *ras* oncogenes. This requirement brought us to two important considerations in designing experimental systems and selecting the candidate genes. First, the genetic heterogeneity of the systems to be compared should be minimal. The Ras signaling pathway shares downstream effector components with other intracellular signaling pathways. As a result, the differences in genetic backgrounds would make the functional assessments of the candidate genes difficult and thus the understanding of underlying mechanism more complicated. For instance, identification of the genes altered in normal and tumor tissues would certainly provide physiologically more relevant information, but it hardly allows identification of the cause of these gene alterations. And also, if the gene is previously uncharacterized, it is still necessary to go back to the experimentally accessible system to assess the biochemical contribution of this gene product in the tumorigenic process. Second, the altered expression profile of the gene must be reverted when the initiating event is removed, and the experimental systems to which such manipulation is applicable would further narrow down the candidate genes from the initial screening.

With these considerations in mind, we have employed differential display to identify *ras* oncogene-regulated gene expression under three different conditions.

1. We have compared mRNA expression between Rat-1, an immortalized rat embryo fibroblast cell line, and its H-*ras*-transformed derivatives.[2] The Ras effect is easily discernible by morphological changes such as the loss of contact inhibition in culture, and tumorigenicity in nude mice.

[12] P. Liang and A. B. Pardee, *Curr. Opin. Immunol.* **7,** 274 (1995).
[13] P. Liang and A. B. Pardee, *Methods Mol. Biol.* **85,** 3 (1997).
[14] P. Liang (ed.), *Methods* **16** (1998).

2. To eliminate *ras*-independent gene expression between cell lines, we have also compared gene expression in an inducible oncogenic H-*ras* cell line, Rat-1 : iRas,[15] with temporal activation of *ras* expression by isopropyl-β-D-thiogalactopyranoside (IPTG).

3. To further confirm or dissect the pathways under which the genes of interest are regulated, we have used mitogen-activated protein (MAP) kinase kinase inhibitor PD98059 to selectively inhibit oncogenic *ras*-mediated cell transformation.[16] mRNA expression patterns between the normal Rat-1 cell line and its transformed derivative, Rat-1(ras), grown without or with the inhibitor, were compared.

Materials and Methods

Materials

Reverse transcriptase (RT) buffer (5×): 125 m*M* Tris-HCl (pH 8.3), 188 m*M* KCl, 7.5 m*M* $MgCl_2$, and 25 m*M* dithiothreitol (DTT)
Moloney murine leukemia virus (Mo-MuLV) reverse transcriptase (100 U/ml)
dNTPs (250 μM)
5′-AAGCTTTTTTTTTTTG-3′ (2 μM)
5′-AAGCTTTTTTTTTTTA-3′ (2 μM)
5′-AAGCTTTTTTTTTTTC-3′ (2 μM)
Arbitrary 13-mers (2 μM)
Polymerase chain reaction (PCR) buffer (10×)
dNTP stock (25 μM)
Glycogen (10 mg/ml)
Distilled H_2O
Loading dye
AmpliTaq DNA polymerase (5 units/ml), Perkin-Elmer (Norwalk, CT)
[α-^{35}S]dATP (>1000 Ci/mmol), or [α-^{33}P]dATP (>2000 Ci/mmol)
RNase-free DNase I (10 U/ml)
Thermocycler
DNA-sequencing apparatus
Farnesyltransferase inhibitor (FTI; Merck, Rahway, NJ)
PD98059 (CalBiochem, La Jolla, CA)

[15] S. A. McCarthy, M. L. Samuels, C. A. Pritchard, J. A. Abraham, and M. McMahon, *Genes Dev.* **9,** 1953 (1995).
[16] R. Zhang, Z. Tan, and P. Liang, *J. Biol. Chem.* **275,** 24436 (2000).

Although individual components for differential display may be purchased separately from various suppliers, most of them can be obtained in kit forms from GenHunter (Nashville, TN).

DNase I Treatment of Total RNA

Purification of polyadenylated RNAs is neither necessary nor helpful for differential display. The major pitfalls in using polyadenylated mRNAs are the frequent contamination of the oligo(dT) primers, which gives high background smearing in the display, and the difficulty in assessing the integrity of the mRNA templates. Total cellular RNAs can be easily purified by a one-step acid–phenol extraction method using RNApure reagent (GenHunter). However, whatever methods are used for total RNA purification, trace amounts of chromosomal DNA contamination in the RNA sample could be amplified along with mRNAs, thereby complicating the pattern of displayed bands. Therefore removal of all contaminating chromosomal DNA from RNA samples is essential before carrying out differential display.

Incubate 10 to 100 μg of total cellular RNA with 10 units of DNase I (RNase free) in 10 m*M* Tris-HCl (pH 8.3), 50 m*M* KCl, 1.5 m*M* $MgCl_2$ for 30 min at 37°. Inactivate DNase I by adding an equal volume of phenol–chloroform (3:1, v/v) to the sample. Mix by vortexing and leave the sample on ice for 10 min. Centrifuge the sample for 5 min at 4° in an Eppendorf centrifuge. Save the supernatant, and ethanol precipitate the RNA by adding 3 volumes of ethanol in the presence of 0.3 *M* sodium acetate and incubate at −80° for 30 min. Pellet the RNA by centrifugation at 4° for 10 min. Rinse the RNA pellet with 0.5 ml of 70% (v/v) ethanol [made with diethyl pyrocarbonate (DEPC)-treated H_2O] and redissolve the RNA in 20 ml of DEPC-treated H_2O. Measure the RNA concentration at OD_{260} with a spectrophotometer by diluting 1 μl of the RNA sample in 1 ml of H_2O. Check the integrity of the RNA samples before and after cleaning with DNase I by running 1–3 μg of each RNA on a formaldehyde–7% (v/v) agarose gel. Store the RNA sample at a concentration higher then 1 mg/ml at −80° before using for differential display.

Reverse Transcription of mRNA

Set up three reverse transcription reactions for each RNA sample in three microcentrifuge tubes (0.5-ml size), each containing one of the three different anchored oligo(dT) primers as follows.

For 20-μl final volume:

Component	Volume (μl)
Distilled H_2O	9.4
RT buffer (5×)	4
dNTP stock (250 μ*M*)	1.6
Total RNA (DNA free)	2 (0.1 μg/μl, freshly diluted)
$AAGCT_{11}M$ (2 μ*M*)	2 (M can be either G, A, or C)
Total:	19.0

Program the thermocycler to 65°, 5 min; 37°, 60 min; 75°, 5 min; to 4°. One μl, of Mo-MuLV reverse transcriptase is added to each tube 10 min later at 37° and mixed well quickly by finger tipping. Continue the incubation and at the end of the reverse transcription reaction, spin each tube briefly to collect condensation. Set the tubes on ice for PCR or store at −80° for later use.

Polymerase Chain Reaction

Set up PCRs at room temperature as follows.

For a 20-μl final volume for each primer set combination:

Component	Volume (μl)
Distilled H_2O	10
PCR buffer (10×)	2
dNTP stock (25 μ*M*)	1.6
Arbitrary 13-mer (2 μ*M*)	2
$AAGCT_{11}M$ (2 μ*M*)	2
RT mix from RT step above	2
$AAGCT_{11}M$	2
[α-^{33}P]dATP (2000 Ci/mmol)	0.2
AmpliTaq (5 U/ml) (Perkin-Elmer)	0.2
Total:	20

Make core mixes whenever possible to avoid pipetting errors (e.g., aliquot RT mix and arbitrary primer individually). Otherwise it would be difficult to pipette 0.2 μl of AmpliTaq. Mix well by pipetting up and down. Add 25 μl of mineral oil if needed. Perform PCR at 94° for 30 sec, 40° for 2 min, and 72° for 30 sec for 40 cycles; 72° for 5 min; and to 4°. (For the Perkin-Elmer 9600 thermocycler it is recommended that the denaturation temperature be shortened to 15 sec and the rest of parameters be kept the same.)

Denaturing Polyacrylamide Gel Electrophoresis

Prepare a denaturing polyacrylamide (6%, w/v) gel in TBE buffer. Let it polymerize for at least 2 hr before use. Prerun the gel for 30 min. It is crucial that the urea in the wells be completely flushed out right before loading the samples. For best resolution, flush every four to six wells each time during sample loading while trying not to disturb the samples that have already been loaded.

Mix 3.5 μl of each sample with 2 μl of loading dye and incubate at 80° for 2 min immediately before loading onto a polyacrylamide (6%, w/v) DNA-sequencing gel. Electrophorese for about 3.5 hr at 60-W constant power (with voltage not to exceed 1700 V) until the xylene dye (the slower moving dye) reaches the bottom. Turn off the power supply and blot the gel onto a piece of 3 MM paper (Whatman, Clifton, NJ). Cover the gel with a plastic wrap and dry at 80° for 1 hr (do not fix the gel with methanol–acetic acid). Orient the autoradiogram and dried gel with radioactive ink or needle punches before exposing to X-ray film.

Reamplification of cDNA Bands

After developing the film (overnight to 72 hr of exposure), orient the autoradiogram with the gel. Locate bands of interest either by marking with a clean pencil from underneath the film or by cutting through the film with a razor blade. The other way that has been found to work well is to punch through the film with a needle at the four corners of each band of interest. (Handle the dried gel with gloves and save it between two sheets of clean paper.) Cut out the located band with a clean razor blade. Soak the gel slice along with the 3M paper in 100 μl of distilled H_2O for 10 min. Boil the tube with the cap tightly closed (e.g., with Parafilm) for 15 min. Spin at maximum speed in a Eppendorf centrifuge at room temperature for 2 min to collect condensation and pellet the gel and paper debris. Transfer the supernatant to a new microcentrifuge tube. Add 10 μl of 3 *M* sodium acetate, 5 μl of glycogen (10 mg/ml), and 450 μl of 100% ethanol. Let sit for 30 min on dry ice or in a −80° freezer. Spin for 10 min at 4° to pellet DNA. Remove the supernatant and rinse the pellet with 200 μl of ice-cold 85% (v/v) ethanol (DNA is lost if less concentrated ethanol is used). Spin briefly and remove the residual ethanol. Dissolve the pellet in 10 μl of PCR-grade H_2O and use 4 μl for reamplification. Save the rest at −20° in case of mishaps. Reamplification should be done with the same primer set and PCR conditions except that the dNTP concentrations are at 20 μ*M* (use 250 μ*M* dNTP stock) instead of 2–4 μ*M* and no isotopes are added. A 40-μl reaction is recommended.

A 40-μl final volume for each primer set combination:

Component	Volume (μl)
Distilled H_2O	20.4
PCR buffer (10×)	4
dNTP stock (250 μ*M*)	3.2
Arbitrary 13-mer (2 μ*M*)	4
$AAGCT_{11}M$ (2 μ*M*)	4
cDNA template from RT	4
AmpliTaq (5 U//ml)	0.4
Total:	40

Run 10–15 μl of the PCR sample on a 1.5% (w/v) agarose gel to check that the size of reamplified PCR products is consistent with their size on the denaturing polyacrylamide gel. Subclone the PCR products into vectors such as the PCR-TRAP cloning system (GenHunter). Alternatively, PCR samples can be run in duplicate to perform reverse Northern blots to screen differentially expressed cDNAs before cloning.

Confirmation of Differentially Expressed cDNAs

An efficient method for screening cDNAs has been described previously.[17] After cloning into the PCR-TRAP vector followed by bacterial transformation, colony PCR is performed to identify transformants that contain cDNA of the expected size. At least three of these PCR products are analyzed for a cDNA band by a reverse Northern dot-blot method. The cDNA probes are prepared from 10–50 μg of each of the two RNA samples by reverse transcription in a 50-μl reaction, using a ReversePrime kit (GenHunter), on a PCR thermocycler. Samples are heated for 5 min at 65° followed by a 10-min incubation at 37°, and then 1000 units of Mo-MuLV reverse transcriptase is added followed by an additional 1 hr of incubation. After the reverse transcription, a Quick Spin column (Boehringer Mannheim, Indianapolis, IN) is used to remove the unincorporated ^{32}P. Equal counts ($5–10 \times 10^6$ cpm) of the cDNA probes from Rat-1 and Rat-1(ras) are heat denatured and used to probe the duplicate blots. The blots are washed twice at room temperature for 15 min with 1× saline–sodium citrate (SSC) and 0.1% (w/v) sodium dodecyl sulfate (SDS), and

[17] H. Zhang, R. Zhang, and P. Liang, *Nucleic Acids Res.* **24,** 2454 (1996).

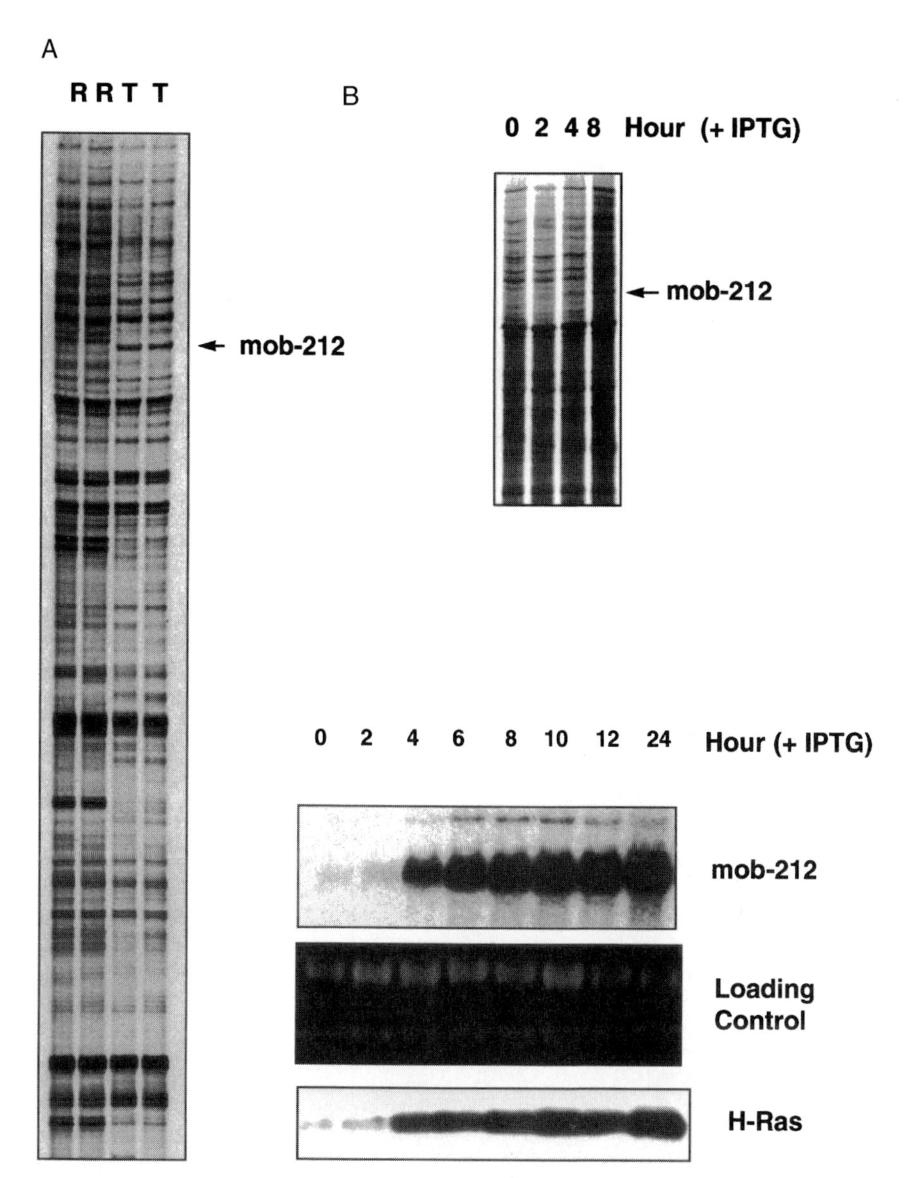

FIG. 1. Isolation of oncogenic *ras* target genes by differential display. (A) Total RNAs from rat embryo fibroblast Rat-1 and an oncogenic H-*ras*-transformed derivative, T101-4, were compared by differential display in duplicate. The arrow indicate a message activated in the transformed cells. Note both up- and several downregulated messages were reproducibly detected in this given differential display PCR. (B) Total RNAs from the Rat-1 : iRas cell line before and after IPTG induction of the oncogenic H-*ras* were compared by differential display. The arrow indicates that the same message seen in (A) is induced after 4 hr. *Bottom:* Northern blot confirmation of the *ras*-dependent expression of the gene. Western blot analysis indicates that the oncogenic H-Ras was induced 4 hr after the addition of IPTG.

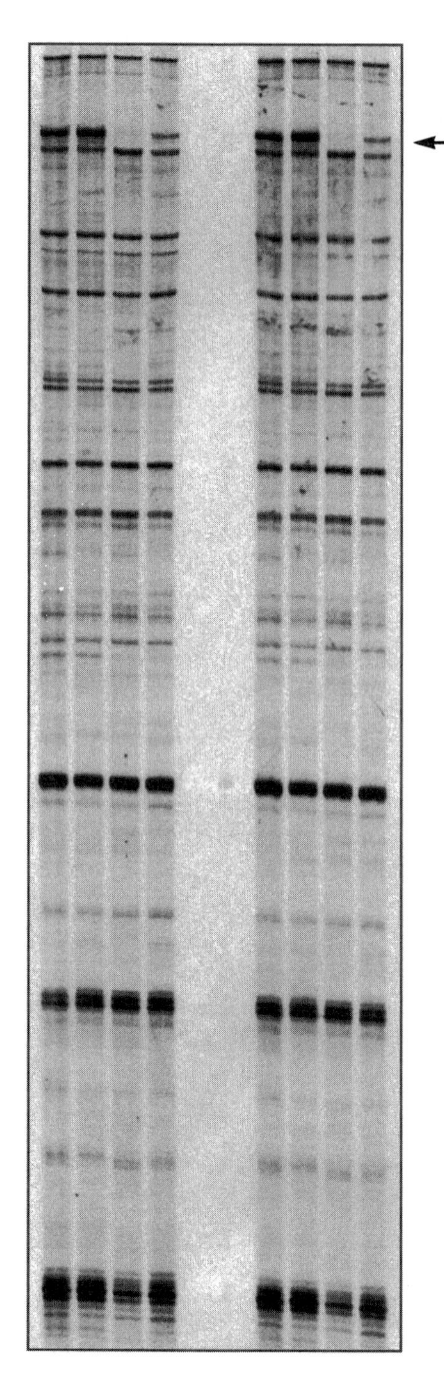
Rat1
Rat1(ras)
Rat1
Rat1(ras)
- + - +
- + - +
PD98059

then once at 60° for 15 min with 0.25× SSC and 0.1% (w/v) SDS. The membranes are exposed overnight at −80° with an intensifying screen.

Experimental Results and Discussions

Reproducibility: Key to Cutting Down Rate of False Positives

To identify genes that are involved in the key steps in *ras*-induced cell transformation, differential display was carried out to compare RNA samples from the Rat-1 cell line and its derivative transformed by H-*ras*. The reactions were prepared in duplicate at both RT and PCR steps to ensure reproducibility of the comparison. Figure 1A shows a representative result of such analysis. It should be noted that one of the greatest advantages of differential display over other methodologies is the simultaneous detection of both up- and downregulated gene expression.

Expression Kinetics of ras Target Genes

The other major advantage of differential display is its ability to simultaneously compare more than two RNA samples. This allows criteria to be set more strictly so that more relevant genes may be identified. For example, the comparison of multiple RNA samples could help to narrow down disease-specific rather than individual specific genes, or genes regulated in a temporal fashion by a drug or an oncogene such as *ras*, before embarking on time-consuming functional studies. With this in mind, the same differential display reaction performed in Figure 1A was carried out with RNA isolated from another cell line, Rat-1 : iRas, that contains H-*ras* under IPTG-inducible promoter control. The temporal induction of the gene upregulated by *ras* as shown in Fig. 1A was confirmed here by both differential display and Northern blot analysis. The downregulated messages revealed by Fig. 1A, however, appeared to be cell line specific, as none of them were affected by the induction of oncogenic H-*ras* (data not shown). Therefore the use of an inducible system could help eliminate heterogeneity in the genetic background of different specimens being compared.

FIG. 2. Identification of oncogenic H-Ras-activated genes through the MAP kinase pathway. Both Rat-1 and Rat-1(ras) cells were grown in the absence or presence of a 5 μM concentration of the MAP kinase inhibitor PD98059 for 24 hr. The total RNAs were compared by differential display to reveal genes of interest. The arrow denotes a message turned off by the *ras* oncogene in Rat-1(ras) cells, but reactivated by PD98059.

Further Dissection of Ras Signaling Pathways under Which ras Target Genes Are Regulated

Ras sits near the top of the signal transduction pathway, just downstream of the growth factor receptor. One of the major pathways downstream of Ras is the MAP kinase pathway, which can be selectively blocked by inhibitors such as PD98059, developed by Parke Davis (Morris Plains, NJ).[16] To identify genes regulated by the *ras* oncogenes through the MAP kinase pathway, the normal Rat-1 cell line and its derivative transformed by H-*ras*, Rat-1(ras), are grown in the absence or presence of PD98058 for 24 hr and their RNAs compared by differential display in duplicate. The representative result is shown in Fig. 2, revealing a gene whose expression is turned off by activated Ras through the MAP kinase pathway, as PD98059 could reactivate its expression.

Summary

The goal of the signal transduction pathways, such as those controlled by Ras, is in large part to ensure highly stringent regulation of the target genes in the nucleus, which are collectively responsible for the signal output, or phenotypes, of the cell. Understanding of the Ras effect ultimately requires the identification of these downstream target genes. Reverse genetic approaches would trace back the pathways by which they are regulated by Ras. While newer methods such as DNA microarray are emerging, differential display has allowed the identification of a greater number of differentially expressed genes than have been cloned by all the other methods combined, based on Medline search. Much of this success has been attributed to its simplicity (RT-PCR and DNA-sequencing gel) and versatility (compare more than two RNAs for both up- and downregulated genes). It has become obvious that finding the genes by either differential display or DNA microarray is only the first step toward the understanding of biological problems under investigation. It is hoped that finding the right genes through careful experimental designs, such as outlined here, will narrow down the number of relevant genes and increase the odds for solving the puzzles of nature, such as *ras*.

[19] cDNA Array Analyses of K-Ras-Induced Gene Transcription

By Gaston G. Habets, Marc Knepper, Jaina Sumortin, Yun-Jung Choi, Takehiko Sasazuki, Senji Shirasawa, and Gideon Bollag

Introduction

Among all oncogene-encoded proteins, the Ras proteins are probably the best studied with regard to their biochemical mechanism and how the mutations that are found in human tumors affect their activity. Also, the signaling processes in which Ras plays a crucial role have been well studied. Through intense study in numerous models it has become evident that the Ras proteins control signaling at crucial positions in pathways that influence DNA synthesis, apoptosis, cell differentiation, cell adhesion, and cytoskeletal processes. The best characterized pathway leads through the Raf–MEK–ERK kinase cascade [MEK, mitogen-activated protein (ERK) kinase; ERK, extracellular signal-regulated kinase] to activation of nuclear transcription factors.[1,2] In addition, Ras activates various other effectors that can contribute to transformation.[3] Ras effectors such as phosphatidylinositol 3-kinase (PI3-kinase), and Ral guanine nucleotide dissociation stimulator (Ral GDS) each control separate pathways that are likely to influence gene transcription on their own.[4]

To characterize the quantitative and qualitative differences in gene expression during tumor transformation, we chose to conditionally reconstitute the expression of an activated K-*ras* allele in the K-RasD13 knockout cell clones DKO-4 and Hke-3.[5] These cells are derived from the DLD1 and HCT116 colon carcinoma cell lines, respectively. To control the expression of exogenous activated genes encoding K-RasD12 and K-RasV12, we used the ecdysone-inducible vector system (Invitrogen, Carlsbad, CA). In this system, the expression of the exogenous gene is under the control of ecdysone, an insect steroid that is inert in mammalian cells.[6]

[1] C. J. Marshall, *Cell* **80,** 179 (1995).
[2] C. S. Hill and R. Treisman, *Cell* **80,** 199 (1995).
[3] M. A. White, C. Nicolette, A. Minden, A. Polverino, L. Van Aelst, M. Karin, and M. H. Wigler, *Cell* **80,** 533 (1995).
[4] T. Hunter, *Cell* **88,** 333 (1997).
[5] S. Shirasawa. M. Furuse, N. Yokoyama, and T. Sasazuki, *Science* **260,** 85 (1993).
[6] D. No, T. P. Yao, and R. M. Evans, *Proc. Natl. Acad. Sci. U.S.A.* **93,** 3346 (1996).

0076-6879/00 $35.00

Generation of Ecdysone Receptor-Expressing Cells

Ecdysone-responsive K-Ras-expressing cells are generated as follows. First, cells are selected that expressed a functional ecdysone receptor (zeocin selection; Invitrogen); subsequently these cell lines are transfected with the activated *ras* genes under the control of an inducible promoter (hygromycin B selection; Roche, Indianapolis, IN). The details of these selections are provided below.

The effective concentration of zeocin (Invitrogen) and hygromycin B (Roche) is determined for all cell lines by plating 5×10^5 cells per well in six-well dishes (Corning, Acton, MA) in complete medium [2 ml/well: Dulbecco's modified Eagle's (DME)-high glucose medium, 10% (v/v) fetal calf serum (FCS), penicillin (200 U/ml), streptomycin (200 U/ml), and 2 m*M* L-glutamine; all from Irvine Scientific (Santa Ana, CA)]. Cells are plated in increasing concentrations (0.1–1.0 mg/ml) of zeocin or hygromycin and the effective concentration is estimated after 5 days by microscopy. For all the clones the effective selection concentration is 0.6 mg/ml for both zeocin and hygromycin B. Both the DKO-4 and Hke-3 cell clones are resistant to G418 (G418-sulfate, 0.6 mg/ml; GIBCO, Grand Island, NY) as a result of the construct by which the activated K-*ras* allele is inactivated.[5]

First, DKO-4 and Hke-3 cells are generated that express both subunits of the ecdysone receptor. For this, the cells are transfected with the pVgRxR plasmid (retinoid X receptor; Invitrogen). Cells are plated at 5×10^5 cells per well in six-well dishes and transfected with the plasmid by means of the SuperFect reagent (Qiagen, Valencia, CA). Forty-eight hours after transfection the cells are trypsinized and replated in 10-cm dishes at 1:5, 1:10, 1:20, and 1:40 dilutions in complete medium with zeocin (0.6 mg/ml). Medium is refreshed twice weekly and after 14 days single clones are isolated with cloning rings and expanded in 24-well dishes.

The zeocin-resistant clones are analyzed for the presence of a functional ecdysone receptor by transient transfections with a plasmid containing the β-galactosidase gene (*lacZ*) under the control of an ecdysone-responsive promoter (pIND/*lacZ;* Invitrogen). Transient transfections with this plasmid are done in duplicate in six-well plates with SuperFect. Forty-eight hours after transfection, one well of each transfection is induced with a 10 μM final concentration of ponasterone A (Invitrogen), an analog of the insect steroid ecdysone. The duplicate wells are mock treated and used as uninduced controls. Ponasterone A is stored at $-20°$ and a 1 m*M* stock solution is made fresh from the desiccated powder in ethanol, after which the appropriate dilution is made in complete medium. Cells are induced for 48 hr, and then washed once with PBS$^-$ [phosphate-buffered saline (PBS) without Ca^{2+} or Mg^{2+}] and fixed for 10 min at 4° in PBS$^-$, 0.2%

(v/v) glutaraldehyde, 2% (v/v) formaldehyde. The fixed cells were washed twice with PBS^- and stained for 1–2 hr at 37° in staining mix [5 m*M* potassium ferrocyanide, 5 m*M* potassium ferricyanide, 5-bromo-4-chloro-3-indolyl-β-D-galactopyranoside (X-Gal, 1 mg/ml), 2 m*M* $MgCl_2$]. Clones are selected on the basis of the increased number and intensity of blue-staining cells on induction with the ecdysone analog. Quantitation of inducibility is also analyzed by cotransfection of the pIND/*lacZ* plasmid with a plasmid encoding the luciferase gene under the control of the constitutive cytomegalovirus (CMV) promoter pCMVluciferase (pCMVluc). For this, the ecdysone receptor transfectants are seeded in duplicate in 12-well culture plates and transfected with the pINDlacZ and pCMVluc plasmids in a 5:1 weight ratio with SuperFect. Twenty-four hours after transfection, ponasterone A is added to a final of 10 μ*M* to one of the duplicate wells. Quantitation is performed with the Tropix (Bedford, MA) Dual Lite chemiluminescent reporter gene assay system. This assay allows the combined detection of luciferase and β-galactosidase activity. Inducibility varies from 1- to 5-fold (Table I, *lacZ* induction). Alternatively, cells are tested for induced expression of the secreted placental alkaline phosphatase (SEAP) gene product. For this, the cells are transiently transfected with a plasmid in which the SEAP coding region (pSEAP2-Basic; Clontech, Palo Alto, CA) is cloned under the control of the ecdysone-inducible promoter (pIND; Invitrogen). After transfection and induction, a fraction of the supernatant is assayed for SEAP activity with the Phospha-light chemiluminescent reporter assay (Tropix). The luminescence is detected in a ML2250 microtiter plate luminometer (Dynatech, Vienna, VA) (Table I).

Characterization of Ecdysone Receptor-Expressing DKO-4 and Hke-3 Clones

We have verified that the ecdysone receptor-expressing clones of DKO-4 and Hke-3 cells retain their untransformed phenotype by confirming their lack of growth in soft agar and tumor formation on subcutaneous injection in nude mice. The parental transformed DLD1 and HCT116 cells are included as positive controls (Table I). For the soft agar assay, cells are seeded at 2000, 4000, or 8000 cells per well of 6-well plates in 0.4% (w/v) agar containing complete medium. A 3% (w/v) Bacto-agar (Difco, Detroit, MI) stock solution is melted in a microwave and kept at 56° in a water bath. The required complete medium is preheated for no longer than 15 min in the water bath at 56° and the plates are prepared by adding 1 ml of full medium containing 0.6% (w/v) Bacto-agar as the bottom layer. For the top agar, 3.6 ml of complete medium is preheated as described above and 0.6 ml of 3% (w/v) Bacto-agar is added and mixed. This mixture is

TABLE I
CELL LINE CHARACTERISTICS

Cell line	Clone	β-Gal stain[a]	*lacZ*[b]	SEAP[c]	K-Ras expression	K-Ras induction[d]	Soft agar growth[e]	Tumor formation[f]	Latency[g]
DLD1	Parent	None	1.03	nd	wt/K-RasD13		+++/+++	5/5	22
DKO-4	Parent	None	1.02	0.97	wt		−/−	1/5	73
	DKOc8	+	1.23	1.14	wt		−/−	nd	nd
	DKOc10	−/+	1.09	1.33	wt		−/−	0/5	90
	DKOc10A6	nd	nd	nd	wt/K-RasV12	+	−/−	nd	nd
	DKOc10A8	nd	nd	nd	wt/K-RasV12	+	(+)/(+)	nd	nd
	DKOc10A9	nd	nd	nd	wt/K-RasD12	++	−/−	nd	nd
	DKOc10B4	nd	nd	nd	wt/K-RasD12	+++	−/−	nd	nd
	DKOc10B8	nd	nd	nd	wt/K-RasD12	+++	−/−	nd	nd
	DKOc10B9	nd	nd	nd	wt/K-RasD12	++	−/−	nd	nd
	DKOc10B12	nd	nd	nd	wt/K-RasD12	+++	−/−	nd	nd
	DKOc10C1	nd	nd	nd	wt/K-RasV12	+	−/−	nd	nd
	DKOc10C2	nd	nd	nd	wt/K-RasV12	+	−/−	nd	nd
	DKOc10C8	nd	nd	nd	wt/K-RasD12	+	−/−	nd	nd
	DKOc10C9	nd	nd	nd	wt/K-RasD12	(+)	−/−	nd	nd
	DKOc10C10	nd	nd	nd	wt/K-RasD12	(+)	−/−	nd	nd
	DKOc10D2	nd	nd	nd	wt/K-RasV12	+	−/−	nd	nd
	DKOc11	+	1.26	0.7	wt		−	nd	nd
	DKOc16	++	1.92	1.67	wt		(+)/(+)	1/5	64
HCT116	Parent	None	nd	nd	wt/K-RasD13		+++/+++	4/5	36
Hke-3	Parent	None	nd	nd	wt		(+)/(+)	4/5	50
	Hkec2	++	1.75	4.55	wt		−	nd	nd
	Hkec8	−/+	0.9	1.03	wt		−	2/5	50
	Hkec18	−/+	1.05	5.55	wt		+	nd	nd
	Hkec31	+++	5.45	3.37	wt		−	0/5	90

Abbreviation: nd, Not determined.

[a] Relative number of cells staining blue after stimulation with ponasterone A.

[b] Ratio of β-galactosidase activity determined by Tropix Dual Lite chemiluminescent reporter with or without stimulation with ponasterone A.

[c] Ponasterone A-stimulated SEAP activity determined by Tropix Phospha-light chemiluminescent reporter assay.

[d] Relative levels of ponasterone A-induced exogenous K-Ras expression as determined by immunostaining of Western blots.

[e] Relative number of cell colonies in soft agar medium in the presence (0.1, 0.3, 1.0, and 3.0 μM) or absence of ponasterone A. +++, more than 200 colonies; (+), fewer than 20 colonies; −, 5 or fewer.

[f] Number of mice that developed tumors of 1 ml in size.

[g] Latency depicts the number of days at which the first mouse in each group developed a tumor volume of 1 ml.

allowed to cool to 40° before being added to the cells (150 μl) along with the required concentration of ponasterone A (150 μl). The cells are mixed by pipetting up and down twice in a 5-ml pipette and 1 ml is plated per well. After solidification of the soft agar the cells are incubated at 37° and fed with 0.25 ml of full medium twice weekly. Ponasterone A is added once per week. After 2 weeks the colonies of cells are stained by adding 1 ml of staining solution per well. Staining solution is made up by heating 0.6 ml of warm neutral red to 37° for 30 min and then adding 33.5 ml of preheated medium (56°) and 1.3 ml of 3% (w/v) Bacto-agar. The wells are stained at room temperature for at least 2 hr to overnight and then scored.

The tumorigenic capacity of each obtained cell clone is determined by injecting subcutaneously five athymic (*nu/nu*) nude mice (NCr; Taconic Farms, Germantown, NY) per cell clone with 1×10^6 cells in a 100-μl volume (1×10^7/ml PBS) in each flank. Mice are monitored daily, and animal weight and tumor size are measured twice weekly. Mice are killed when one of their tumors reaches a volume of 1200 μl (Table I).

Generation of Ecdysone-Inducible K-Ras Plasmids

The coding regions for K-RasD12 and KRasV12 are amplified by polymerase chain reaction (PCR) from plasmids containing the K-*ras* open reading frame (ORF) with flanking sequences. The primers generate a fragment containing the complete ORF with start and stop codons flanked by a *Bam*HI at the 5′ end and an *Spe*I site at the 3′ end (sense primer, 5′-TAGGATCCATGACTGAATATAAACTT; antisense primer, 5′-GGACTAGTTACATAATTACACACTT). The PCR contains 50 pmol of each primer and 0.5 unit of *Taq* polymerase. The *Bam*HI- and *Spe*I-digested PCR product is ligated in frame with an upstream Myc or hemagglutinin (HA) epitope tag (*Bam*HI–*Xba*I sites) of pcDNA1 (Invitrogen)-derived vectors under the control of the CMV promoter. The 5′ *Bam*HI sequence introduced by the primer disrupts most of the Kozak consensus sequence of K-*ras* and thereby strengthens translation initiation at the start site of the epitope tag. After ligation and transformation, each of the plasmids created is confirmed by sequencing the complete ORF encoding the Myc- or HA-tagged K-Ras proteins. The ORF and flanking polylinker sequences are released from these plasmids by a *Hin*dIII–*Apa*I digestion and ligated into the *Hin*dIII–*Apa*I-digested pIND hygro plasmid (Invitrogen) downstream of the ecdysone-responsive promoter. Plasmids containing the correct insert are analyzed for induction and expression of the K-*ras* genes by transiently transfecting human 293-EcR cells (Invitrogen). These 293-HEK-derived cells express the ecdysone receptor. The cells are stimulated with ponasterone A at 10 μ*M* for 16 hr. After stimulation, the medium is removed and cells from six-well plates are lysed in 100 μl of TG lysis buffer [20 m*M* Tris-HCl (pH 7.5), 137 m*M* NaCl, 1 m*M* EGTA, 1% (v/v) Triton X-100, 10% (v/v) glycerol, 1.5 m*M* $MgCl_2$, 1 m*M* $NaVO_4$, 25 m*M* NaF, and 1 tablet of Complete (Roche) protease inhibitor cocktail per 25 ml]. The lysates are transferred into Eppendorf tubes and cleared by centrifugation at 13,000*g* for 10 min at 4°. Twenty microliters of the cleared lysate is denatured by adding 5 μl of sample buffer [8% (w/v) sodium dodecyl sulfate (SDS), 200 m*M* Tris-HCl (pH 6.8), 20% (v/v) glycerol, 20% (v/v) 2-mercaptoethanol, 20 m*M* bromphenol blue] and heating in a heating block at 95° for 3 min. The samples are separated on SDS–4–20% polyacrylamide minigels (TG 4–20%; Novex, San Diego, CA), and electrotransferred to Immobilon-P membranes (Millipore, Bedford, NY), using semidry Multiphor II unit (Amersham Pharmacia Biotech, Uppsala, Sweden). The expressed proteins are detected on Western blots, using 1 : 10,000 dilutions of monoclonal antibody (MAb) specific for the Myc epitope (clone 9E10, 0.3 μg/ml) or the hemagglutinin (HA) tag (clone 12CA5, 0.2 μg/ml). After incubation with 1 : 10,000 dilutions of the secondary antibody, peroxidase-conjugated goat anti-mouse IgG (Pierce Chemical, Rockford, IL), the immunoreaction is visualized by enhanced chemiluminescence (ECL; Amersham, Arlington Heights, IL) and exposure to film (Hyperfilm; Amersham).

The tagged K-Ras proteins are clearly detectable after 16 hr of induction with 10 μM ponasterone A (Fig. 1A). The expression from the ecdysone-inducible promoter is tightly controlled by the presence of the inducer. Also, phosphorylation of ERK is clearly induced by the expression of the activated Ras proteins and detected by a phospho-p44/p42-specific ERK monoclonal antibody (Thr202/T204 E10, 1 : 1000 dilution; New England BioLabs, Beverly, MA) (Fig. 1B).

Generation of K-Ras-Inducible Cell Clones

All the DKO-4- and Hke-3-generated ecdysone receptor cell clones are transfected with the pINDKRasD12 or pINDKRasV12 plasmids, both of which confer hygromycin resistance to the transfectants. Transfection and isolation of the hygromycin and zeocin double-resistant transfectants are carried out as described above. The colonies are picked and expanded in 24-well dishes. From the 24-well plates, all the clones are split into three wells. One plate is kept as a backup and the remaining duplicate wells are induced with 10 μM ponasterone A, or mock treated, for 48 hr. Expression of the tagged K-Ras proteins is analyzed by Western blots of total cell lysate as described above.

In this manner, eight clones with the K-RasD12 gene (Fig. 2A) and five clones with the KRasV12 gene under the control of the ecdysone-responsive

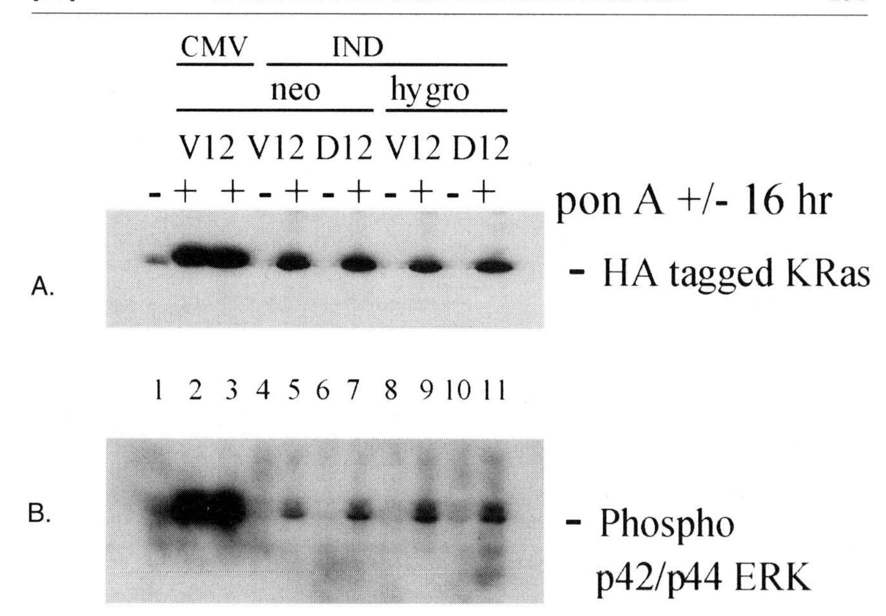

FIG. 1. Expression of activated K-RasD12 and -V12 proteins on transient transfection of 293-HEK cells. (A) Immunodetection of hemagglutinin (HA)-tagged K-Ras proteins. Cells were transfected with control plasmid (lane 1), activated K-*ras* under the control of the constitutive cytomegalovirus-responsive (CMV, lanes 2 and 3) or ecdysone-responsive (IND, lanes 4–11) promoters. K-RasV12-encoding (V12 lanes) and K-RasD12-encoding (D12 lanes) plasmids were used. Cells transfected with plasmids that confer G418 neomycin resistance (neo, lanes 2–7) or hygromycin (hygro, lanes 8–11) are indicated. Lysates in lanes 3, 5, 7, 9, and 11 were derived from cells that were stimulated with 10 μM ponasterone A (pon A) for 16 hr. The position of HA-tagged K-Ras protein is indicated (about 25 kDa). Additional unspecific proteins with higher molecular weight are visible in all lanes and illustrate the equal loading of the samples. (B) Duplicate Western blot of (A) was incubated with a phosphospecific ERK MAb (E10; New England BioLabs).

promoter (Table I) are maintained. None of the clones obtained shows any detectable exogenous Ras protein product in the absence of the ecdysone analog, confirming tight control of the system. At identical concentrations of ponasterone A, the levels of the induced K-Ras protein levels vary among the clones (Fig. 2A). When analyzed with a K-Ras-specific monoclonal antibody (F234; Santa Cruz Biotechnologies, Santa Cruz, CA) the exogenous Ras protein is expressed up to 5-fold over the endogenous wild-type K-Ras protein (Fig. 2B). The levels of induced K-Ras expression are monitored as a function of varying concentrations of ponasterone A (0.1–10 μM) and durations of induction (4–120 hr) (e.g., see Fig. 3). Epitope-tagged K-Ras protein is detectable as early as 8 hr after induction and accumulates up to 24 hr after induction. Titration of ponasterone A shows a clear dose

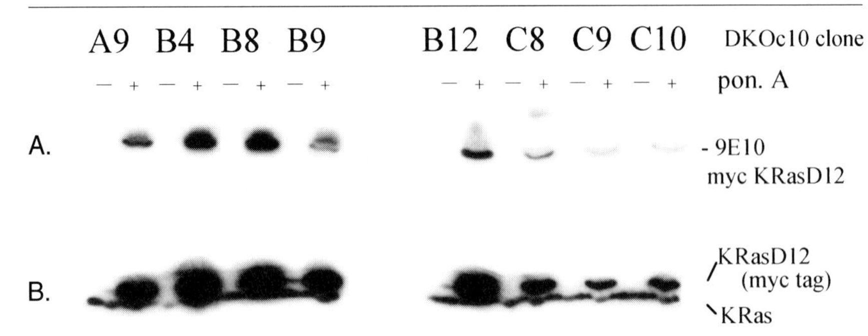

FIG. 2. Isolation of hygromycin-resistant DKOc10 transfectants with inducible expression of K-RasD12. Cells were stimulated (+ lanes) for 48 hr with 10 μM ponasterone (pon A); minus lanes (−) are unstimulated controls. Fifty micrograms of each lysate was loaded in each lane. (A) Eight clones with inducible expression of K-RasD12 were selected to illustrate the varying expression level of the Myc-tagged exogenous K-RasD12 protein. (B) A duplicate immunoblot of (A) was incubated with a K-Ras-specific monoclonal antibody (F234; Santa Cruz Biotechnologies) to illustrate the level of induced expression in the various clones. Both the exogenous Myc-tagged K-RasD12 running at 25 kDa and the endogenous wild-type protein running slightly faster were detected.

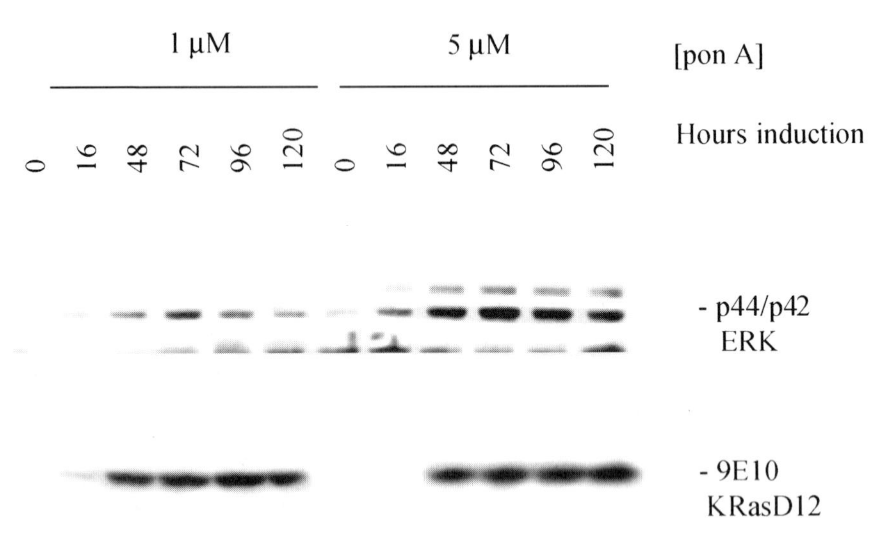

FIG. 3. Kinetics of K-Ras induction and phosphorylation of ERK in the DKOB8 cell clone. Equal amounts of total cell lysate were analyzed by Western blot for expression levels of Myc-tagged K-RasD12 protein (9E10 MAb; bottom) after 0–120 hr of stimulation with 1 or 5 μM ponasterone A (pon A). The upper part of the blot was incubated with a phospho-ERK-specific monoclonal antibody (9-E10 MAb; *top*).

dependency of K-*ras* gene expression, allowing control over the expression of the activated alleles in these cells. All clones are characterized for soft agar growth capacity, changed morphology (Table I and Fig. 4), and cell doubling time (not shown) in the presence of various concentrations of ponasterone A. Despite the clear induced expression of the activated K-RasD12 and K-RasV12 proteins, none of the isolated cell clones showed any change in morphology, or in cell doubling time, or evidence of induced transformation.

Isolation of Cytoplasmic Poly(A)$^+$ RNA Fraction

For most expression analyses we use the inducible Myc-tagged K-RasD12 DKOc10B8 clone (DKOB8) (Figs. 2 and 3). Cells cultured in 150-cm^2 culture plates are induced for 72 hr in the presence of serum and harvested at 75% confluency. Typically, a near-confluent plate is split 1:10 and 24 hr later the cells are stimulated with 3 μM ponasterone A or mock treated. Under these conditions ERK is clearly phosphorylated on induced expression of activated K-Ras. Seventy-two hours after induction the medium is removed and the cytoplasmic fraction is isolated on ice by adding 2.5 ml of ice-cold OCL buffer (10 m*M* Tris-HCl (pH 7.5), 140 m*M* NaCl, 5 m*M* KCl, 1% (v/v) Nonidet P-40, 1 m*M* dithiothreitol (DTT); Qiagen, Valencia, CA). We have reasoned that, by isolating the cytoplasmic fraction of poly(A)$^+$ RNA, we would analyze the more relevant population closer to the actually translated messenger RNAs. Transport of RNA from the nucleus is one of the steps in mRNA regulation. For example, the induction of cyclin D1 is at least partly regulated by increased export of mature

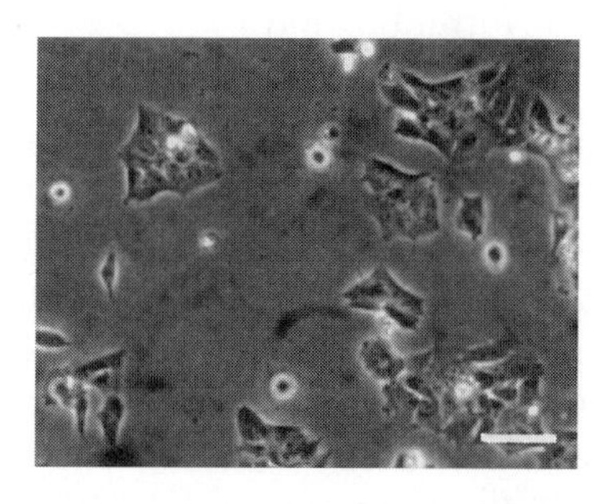

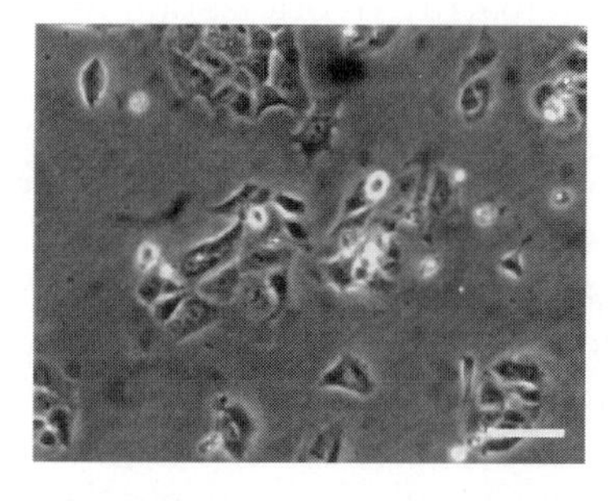

FIG. 4. Phase-contrast microscopic image of the DKOc10B8 with (48 hr, 5 μM pon A) and without induced expression of K-RasD12 (control). No clear difference was observed in the morphology and growth characteristics of the cells. Bar: 30 μm.

transcripts from the nucleus into the cytoplasm, with little change in the total level of transcription.[7] The cell lysate is collected with a disposable cell scraper (Sarstedt, Newtown, NC), transferred into microcentrifuge tubes, and left on ice for 5 min. Nuclei and other cell debris are then pelleted by centrifugation for 2 min in a microcentrifuge at 500*g* and 4°. The supernatants are pooled and transferred directly into 10 volumes of Trizol (GIBCO) to isolate the total RNA according to the manufacturer procedure. The final RNA precipitate is dissolved in H_2O, and a fraction is taken to measure the absorption spectrum between 220 and 320 nm. All RNAs are stored in 70% (v/v) ethanol at −80°. In the absence of salt, no precipitate is formed and the ethanol prevents degradation. When needed, aliquots are dispensed after placing the tubes on ice for 5 min, and the RNA is reprecipitated after adding a 1/30 volume of 2 *M* sodium acetate, pH 5. Microgram amounts of RNA are precipitated in this manner and pelleted by centrifugation at 10,000*g* for 30 min at 4°. After removing the supernatant, the precipitated pellet is carefully washed with 100 μl of 75% (v/v) ethanol and centrifuged as before. The supernatant is removed with a microtiter pipette and the pellet is directly dissolved in the required volume of H_2O. The poly(A)$^+$ RNA fraction is isolated with Oligotex beads (Qiagen). Concentration measurements and storage of the poly(A)$^+$ mRNA fraction are as described above. Typically, 20–50 μg of poly(A)$^+$ mRNA (2–5%) is isolated from 1000 μg of total RNA. Integrity of the isolated RNA is judged by running 15 μg of total RNA on a denaturing formaldehyde gel according to standard procedures. After staining the gel (H_2O–ethidium bromide, 1 μg/ml) the presence of intact 18S and 28S ribosomal RNAs in a 1 : 2 ratio indicates that the RNA was intact.

cDNA Microarray Hybridization and Data Normalization

Labeling of the cDNA probes with the fluorophores, hybridization, and scanning are performed by Incyte Pharmaceuticals (Palo Alto, CA). In our analyses, we have compared two series of cDNA probe hybridizations: (1) RNA isolated from uninduced DKOc10B8 cell clone is compared with samples isolated from cells exposed to 72 hr of induced expression of the exogenous K-RasD12 gene (3 μM ponasterone A), and (2) the DKOc10 ecdysone receptor-positive cell clone is compared with the transformed (parental) DLD1 cell line. All hybridizations are performed in duplicate with independently isolated mRNAs. Hybridizations are performed on arrays 2.0, 3.0, 4.0, and V of the UniGEM series (Incyte). Each of these

[7] D. Rousseau, R. Kaspar, I. Rosenwald, L. Gehrke, and N. Sonenberg, *Proc. Natl. Acad. Sci. U.S.A.* **93,** 1065 (1996).

arrays consists of approximately 10,000 expressed sequence tag (EST) clone inserts arrayed on a glass slide. Clones arrayed on UniGEM V are all sequence confirmed. All uncorrected intensity data of background (B1 and B2) and probe signals (P1 and P2) in both channels are normalized according to the following procedure. For each element, the signals P1 and P2 with the signals over background ratio (P1/B1 and P2/B2) provided by Incyte are used to calculate the background signal in each channel (B1 and B2). The average signal of each channel is normalized to 1 by dividing all P1 and P2 values by the average signal from all well-measured elements in that channel. Well-measured elements are defined as elements that have a greater than 40% hybridization area and an uncorrected signal value that is greater than 2.5-fold over the standard deviation of the background signal.[8] These normalization steps correct for differences in total amounts of probe or labeling between both channels and individual hybridizations. By taking the ratio of the normalized signals for each element, the differences, in the arraying procedure or hybridization between duplicate gene arrays with the same layout can be minimized.

BioMind Database and Data Analyses

We have developed an in-house database in which we provide a standardized annotation for the arrayed genes. The database is gene array platform independent and links the various EST sequences spotted on the cDNA arrays to their individual Unigene cluster (*http://www.ncbi.nlm.nih.gov/UniGene/*), which is used as gene identifier. The zipped Unigene Hs.data file (Build #96) is downloaded from the NCBI data repository *ftp://ncbi.nlm.nih.gov/repository/unigene/* and PERL scripts are used to parse the cluster information into the BioMind data sets. The total of more than 60,000 ESTs that are arrayed on the UniGEM and Stanford arrays[9,10] collapses into about 27,000 individual clusters. For all clones we have retained the array and hybridization specific annotation, such as information on sequence confirmation and quality control.

The reproducible and significant data are selected as follows. The data needed to fullfill the criteria of well measured spots for at least one channel

[8] J. L. DeRisi, V. R. Iyer, and P. O. Brown, *Science* **278,** 680 (1997).

[9] V. R. Iyer, M. B. Eisen, D. T. Ross, G. Schuler, T. Moore, J. C. F. Lee, J. M. Trent, L. M. Staudt, J. Hudson Jr, M. S. Boguski, D. Lashkari, D. Shalon, D. Botstein, and P. O. Brown, *Science* **283,** 83 (1999).

[10] C. M. Perou, S. S. Jeffrey, M. van de Rijn, C. A. Rees, M. B. Eisen, D. T. Ross, A. Pergamenschikov, C. F. Williams, S. X. Zhu, J. C. Lee, D. Lashkari, D. Shalon, P. O. Brown, and D. Botstein, *Proc. Natl. Acad. Sci. U.S.A.* **96,** 9212 (1999).

(see above), and the variation between the duplicate normalized intensities for that spot, should be less than 40% in that channel. This selection of gene expression data is then analyzed, using the gene expression analysis program GeneSpring (Silicon Genetics, Redwood City, CA).

Hybridization and TaqMAN Data; and Results

The induction of K-Ras is detected on both the UniGEM 2 and UniGEM V arrays. On the V array a sequence-confirmed cDNA insert is arrayed that represents the complete K-Ras cDNA (Table II). Hybridizations on this spot measure a 30-fold ratio between the uninduced and induced transcript levels in the DKOc10B8 clone (Table II). This hyperexpression at the transcript level is only partially reflected by the 5-fold increase in K-Ras protein level (Fig 2B). Three individual ESTs representing K-Ras are arrayed on the UniGEM 2. Only one of these ESTs detects in duplicate the 30-fold overexpression of the exogeneous sequence as is detected on the UniGEM V arrays. The other two ESTs overlap only partly or not at all with the K-*ras* ORF (Table II) and therefore detect only partially or not at all the exogenous K-Ras expression. Expression of the wild-type and mutated K-*ras* alleles in the transformed DLD1 cell is comparable to the expression of the single wild-type allele in the DKOc10 cell clone. We have analyzed the total RNA of the same RNA samples used for the hybridizations with the TaqMAN procedure (Perkin-Elmer Biosystems, Foster City, CA) to confirm the overexpression of K-Ras. For this, K-Ras mRNA expression is measured in triplicate by real-time PCR, using the ABI prism 7700 sequence detection system (Perkin-Elmer Biosystems). K-Ras expression is normalized by comparison with glyceraldehyde-3-phosphate dehydrogenase (GAPDH) expression, and samples are quantitated using 5-fold dilutions of total RNA (1–125 ng) from DKOB8 (72-hr induction) cells. All other samples contain 25 ng of total RNA. One-step RT-PCR are started with an initial reverse transcription step at 48° for 30 min, followed by denaturation at 95° for 10 min. The subsequent PCR is repeated 40 times at 95° for 15 sec and 60° for 1 min each. The RT-PCR TaqMAN mixture (25 μl) consists of the indicated amount of total RNA; 1× TaqMAN buffer A; 5.5 mM $MgCl_2$; 0.3 mM dATP, dCTP, dGTP, and dUTP; 0.625 unit of AmpliTaq Gold DNA polymerase; 6.25 units of MultiScribe reverse transcriptase; 5 units of RNase inhibitor; 0.9 μM primers for K-Ras mRNA (5′-TCTTGGATATTCTCGACACAGCA-3′ AND 5′-CCCTCCCCAG-TCCTCATGTA-3′); and 100 nM K-Ras-specific probe (VIC-TGGT-CCCTCATTGCACTGTACTCCTCTTG-TAMRA). All reagents, including the primers and probe for human GAPDH, are obtained from Perkin-Elmer Biosystems.

TABLE II

Ratios of Normalized Intensities Detected for Various Genes

Detection[a]:	UniGEM V	UniGEM 2	UniGEM 2	UniGEM 2	TaqMAN	UniGEM V	UniGEM V	UniGEM V	UniGEM V	UniGEM V	UniGEM V
Gene symbol:	K-RAS2[b]	K-RAS2[c]	K-RAS2[d]	K-RAS2[e]	K-RAS2	MYC	CCND1	JUN	FOS	EGR1	CDKN1A
DKOB8 ind/unind	31.7	48.8	3.3	[illegible].2	nd	0.7	1	2.7	3.9	5.2	3.4
DKOB8 ind/unind	29.1	33	4	[illegible].3	60.1	0.8	0.8	3.1	2.6	8	5.3
DLD1/DKOc10	1.2	1.5	1.2	[illegible].8	nd	1	0.9	2.7	0.5	0.7	1.2
DLD1/DKOc11	1.4	1	0.9	[illegible]1	1.8	0.7	1.2	1.3	1.2	1.4	1.5

[a] UniGEM and TaqMAN are trademarks of Incyte Pharmaceuticals and Perkin-Elmer Biosystems, respectively.

[b] Full-length K-Ras clone on UniGEM V gene array (nucleotides 1–[illegible]775). Exogenous expressed K-RasD12 sequence is only the open reading frame (nucleotides 193–759).

[c] EST on UniGEM 2 covering nucleotides 258–714 of K-Ras cDNA sequence.

[d] EST on UniGEM 2 covering nucleotides 572–1354 of K-Ras cDNA sequence.

[e] EST on UniGEM 2 covering nucleotides 2735–3154 of K-Ras cDNA sequence.

The K-Ras induction ratio is about 2-fold higher for the ponasterone A-stimulated DKOB8 cells when comparing the TaqMAN data with those of the gene array hybridization (Table II). This difference is probably due to the larger dynamic range of the TaqMAN procedure. Obviously, the large induction at the mRNA level is only partially reflected at the protein level. Also in the low range, the TaqMAN analyses seemed more sensitive then the array hybridization. With the TaqMAN procedure, the total expression of the diploid wild-type and mutated K-*ras* alleles in the DLD1 cells is 1.8-fold higher than the remaining wild-type allele in the DKOc10 cells (Table II).

Inactivation of the activated allele in the DLD1 cells has been shown to result in a decreased expression of Myc.[5] This 9-fold reduced expression of Myc in the knockout DKO-4 cells compared with the parental DLD1 cells is not detected in our gene array hybridization analyses (Table II). Similarly, we do not find any differential expression of cyclin D1 (CCND1; Table II) when comparing the transformed DLD1 cells with the untransformed DKOc10, or during the induction of the DKOB8 cells. Cyclin D1 is reported to be upregulated on expression of activated Ras.[7,11] In contrast, the induced expression of the activated K-Ras proteins clearly results in activation of the Raf–MEK–ERK pathway as detected by increased phosphorylation of ERK. Also, the transcript levels of various early response genes such as *fos*, *jun*, and *Egr*1 are increased on induced expression of the activated K-Ras protein. Jun transcript levels are also increased in the transformed DLD1 cells when compared with DKOc10 cells and have been shown to be required for Ras-induced cell transformation.[12] However, increased expression of these protooncogenes is apparently insufficient to establish a transformed phenotype in concert with activated K-Ras expression in the DKOB8 cells.

The observed hyperexpression of the activated K-Ras protein could be one reason why we have failed to induce the transformed phenotype in these cells. In our gene array analyses exogenous levels of the K-Ras transcript are greater than 30-fold over the endogenous wild-type allele, and result in about 5-fold higher K-RasD12 protein levels (Fig. 2B). Furthermore, about 6-fold more K-Ras transcripts are detected with the TaqMAN procedure in the uninduced DKOB8 clone when compared with the parental DKOc10 clone that lacks the plasmid with inducible K-RasD12 (not shown). This apparent leakiness of the ecdysone-responsive promoter is, however, not reflected in detectable amounts of tagged K-Ras protein (Fig.

[11] J. Filmus, A. I. Robles, W. Shi, M. J. Wong, L. L. Colombo, and C. J. Conti, *Oncogene* **9,** 3627 (1994).

[12] R. Johnson, B. Spiegelman, D. Hanahan, and R. Wisdom, *Mol. Cell. Biol.* **16,** 4504 (1996).

2 and 3). High-intensity signaling through Ras and Raf is known to inhibit cell proliferation, probably through increased expression and activation of the cyclin-dependent kinase inhibitor p21ClP1.[13,14] It should be noted, however, that we have tested a large spectrum of K-Ras expression by analyzing all clones, which vary greatly in induction levels (Fig. 2B). Furthermore, all these clones are analyzed for their capacity to grow in soft agar in the presence of 10% serum and a large concentration range of ponasterone A (Table I). Ras-induced inhibition of cell proliferation through activation of p21ClP1 can be prevented by simultaneous activation of the Ras-related Rho GTPase.[15] Rho is efficiently activated by lysophosphatidic acid, which is present in serum.[16] Transcript levels for the cyclin-dependent kinase inhibitor p21CIP1 are, despite the presence of serum, clearly induced on expression of activated Ras in our hybridization experiments (CDKN1A; Table II). It is not clear if the increased expression of p21CIP1 correlates with increased activity in inhibiting cyclin-dependent kinases. No impact on the cell-doubling time is observed with the varying levels of exogenous K-Ras. The increased expression of p21CIP1 has most likely a negative impact on inducing transformation by activated K-Ras.

Still, the differentially expressed genes in these experiments are likely responding to signals from pathways controlled by K-Ras and comparison with the differentially expressed genes in the transformed cells versus the nontransformed DKOc10 clone may reveal known genes and unknown ESTs essential for the transformed phenotype.

Conclusions

We have described a set of experiments designed to identify transcripts that respond to K-Ras activity. These experiments were conducted in cell lines whose morphological transformation had been reverted by deletion of an activated K-*ras* allele. To our surprise, reintroduction of variable levels of activated K-Ras failed to restore morphological transformation to these cell lines. Nonetheless, biochemical analysis of these K-Ras-expressing cells confirms activation of K-Ras pathways and accompanying gene expression. The methods described here should prove useful to unravel

[13] A. Sewing, B. Wiseman, A. C. Lloyd, and H. Land, *Mol. Cell. Biol.* **17,** 5588 (1997).

[14] M. F. Olson, H. F. Paterson, and C. J. Marshall, *Nature* (*London*) **394,** 295 (1998).

[15] D. Woods, D. Parry, H. Cherwinski, E. Bosch, E. Lees, and M. McMahon, *Mol. Cell. Biol.* **17,** 5598 (1997).

[16] F. Imamura, K. Shinkai, M. Mukai, K. Yoshioka, R. Komagome, T. Iwasaki, and H. Akedo, *Int. J. Cancer* **65,** 627 (1996).

K-Ras-dependent transcription. Given the central role of K-Ras in many types of cancer, it is hoped that the identification of K-Ras transcriptional targets will lead to new therapeutic targets, and ultimately to new cancer cures.

[20] Ras Signaling Pathway for Analysis of Protein–Protein Interactions

By AMI ARONHEIM

Introduction

The rapid advance in the various genome projects provides the research community with numerous known and novel genes encoding proteins with no assigned function. It is well accepted that protein–protein interaction plays a major role in all biological processes of bacteria, viruses, plants, and animal cells. Proteins are composed of modular structures that enable the formation of large active units that function in the correct place and time. Understanding the multiple interaction surfaces and partners of a protein of interest is the main goal in molecular biology research today in almost all research arenas and disciplines. Toward this end, multiple and diverse methods have been developed and are being employed, including biochemical, biophysical, and genetic systems.[1] Among the latter, the two-hybrid system is the most commonly used.[2–4] The two-hybrid system is based on the reconstitution of a functional transcription factor via protein–protein interaction employing a transcriptional readout in yeast. Although the two-hybrid system is a simple method and can be easily performed in a nonyeast laboratory, the assay exhibits several limitations and inherent problems.[5,6] Because the two-hybrid system is dependent on a transcriptional readout, it cannot be used with proteins that exhibit intrinsic transcriptional activity. Moreover, the reconstitution of the transcription factor (typically Gal4) should occur in the yeast nucleus. This may not be suitable for proteins of cytoplasmic origin. In addition, it results in problems of toxicity for several

[1] A. R. Mendelsohn and R. Brent, *Science* **284,** 1948 (1999).
[2] S. Fields and O. K. Song, *Nature* (*London*) **340,** 245 (1989).
[3] J. B. Allen, M. W. Walberg, M. C. Edwards, and S. J. Elledge, *Trends Biochem. Sci.* **20,** 511 (1995).
[4] J. Boeke and R. K. Brachmann, *Curr. Biol.* **8,** 561 (1997).
[5] C. Evangelista, D. Lockshon, and S. Fields, *Trends Cell Biol.* **6,** 196 (1996).
[6] K. Hopkin, *J. Natl. Inst. Health Res.* **8,** 27 (1996).

METHODS IN ENZYMOLOGY, VOL. 332
0076-6879/00 $35.00

proteins when expressed in yeast nucleus, such as those encoded by homeo box genes and cell cycle regulators. An additional problem is the repetitive isolation of "positive" interacting partners following a library screening approach using different bait proteins. Some of these proteins are heat shock proteins, chaperones, and ribosomal RNA. Yet while these numerous proteins may pass bait specificity tests, they are considered "false positives" and result in wasted effort and confusion. To overcome these problems, we have developed a protein recruitment system that complements the limitations of the two-hybrid system.

The protein recruitment system developed is based on the Ras signaling pathway in yeast. Ras activity is absolutely required for cell growth in yeast and, therefore, mutations in various components of the pathway result in growth arrest. One such mutant has a mutation in the Ras guanyl nucleotide exchange factor, Cdc25-2. The substitution of Glu-1328 with a lysine residue (E1328K) in Cdc25 renders it inactive at elevated temperatures, e.g., 36°, but allows cell growth at 24°. Therefore, this yeast mutant serves as an attractive strain with which to identify mammalian homologs of Cdc25 in complementation assays. However, the mammalian homolog for Cdc25, hSos, is unable to complement the Cdc25-2 mutation in its native form, unless it is translocated to the plasma membrane.[7] Membrane translocation occurs naturally in mammalian cells following growth factor stimulation via the recruitment of an adapter protein, Grb2, to the autophosphorylated receptor.[8] Alternatively, hSos can be constitutively localized to the plasma membrane by fusion with classic membrane localization signals such as myristoylation or farnesylation sequences.[7] Overexpression of hSos at the plasma membrane accelerates the exchange of GDP for GTP on Ras, thereby resulting in its activation. This mechanism has been shown to function both in mammalian and yeast cells.[7] On the basis of this observation, that hSos can complement the Cdc25-2 mutation only when it is localized to the plasma membrane, we have developed a protein recruitment system in which hSos membrane localization occurs through protein–protein interaction. In principle, a protein of interest is fused to hSos, to either the amino or carboxy terminus (the bait), whereas a protein partner is fused to membrane localization sequences (the prey). When protein–protein interaction occurs between the bait and the prey proteins it is expected to result in hSos translocation and thereby Ras activation. This system, designated the Sos recruitment system (SRS), is efficient in detection of protein–protein interaction between known proteins as well as for

[7] A. Aronheim, D. Engelberg, N. Li, N. Al-Alawi, J. Schlessinger, and M. Karin, *Cell* **78,** 949 (1994).

[8] L. Buday and J. Downward, *Cell* **73,** 611 (1993).

screening cDNA expression libraries fused to a v-Src membrane localization signal.[9–11] The fact that the expression of the prey protein is designed to be under the control of an inducible promoter (*GAL1* promoter[12]) greatly facilitates the analysis of clones isolated from a library screening. In addition to isolation of bait cDNAs encoding prey proteins that specifically interact with the bait proteins, the Ras proteins were found to efficiently complement the Cdc25-2 mutations by virtue of bypassing the requirement for a functional exchange factor. This inherent problem of the system could be partially overcome by expression of the mammalian GTPase-activating protein simultaneously.[13] An improved protein recruitment system has been developed that overcomes several problems of the SRS.[14] This system, designated the Ras recruitment system (RRS), is based on a similar approach; however, it has an absolute requirement that the Ras protein be localized to the inner leaflet of the plasma membrane for its function.[15] This membrane requirement for Ras function made it an attractive target for anticancer drugs[16] for malignancies that involve Ras function. Similar to the SRS, the bait cDNA is fused to an activated Ras that lacks its farnesylation sequence. The RRS exhibits several advantages over the SRS, including effector size, lower rate of self-activation, and reduced number of predicted "false positives," and it provides all the advantages of the SRS over the two-hybrid approach. The RRS, similar to the SRS, can be used to test known and novel protein–protein interactions.[14,17] In addition, the RRS can be used to study protein–protein interactions directly in mammalian cells.[18] In mammalian cells, Ras activation results in induction of the mitogen-activated protein kinase (MAPK) cascade, leading to increased transcription of Ras-responsive genes. Therefore, Ras activity can be monitored at multiple stages along the MAPK cascade or simply by using different reporter genes placed under the control of Ras-responsive promoter elements. This provides a possibility to quantitate the strength of protein–

[9] A. Aronheim, E. Zandi, H. Hennemann, S. Elledge, and M. Karin. *Mol. Cell. Biol.* **17,** 3094 (1997).

[10] X. Yu, L. C. Wu, A. M. Bowcock, A. Aronheim, and R. Baer, *J. Biol. Chem.* **273,** 25388 (1998).

[11] J. Andreev, J. P. Simon, D. D. Sabatini, J. Kam, G. Plowman, P. A. Randazzo, and J. Schlessinger, *Mol. Cell. Biol.* **19,** 2338 (1999).

[12] R. W. West, S. Chen, H. Putz, G. Butler, and M. Banerjee, *Genes Dev.* **1,** 1118 (1987).

[13] A. Aronheim, *Nucleic Acids Res.* **25,** 3373 (1997).

[14] Y. C. Broder, S. Katz, and A. Aronheim, *Curr. Biol.* **8,** 1121 (1998).

[15] J. F. Hancock, A. I. Magee, J. Childs, and C. J. Marshall, *Cell* **57,** 1167 (1989).

[16] J. Travis, *Science* **260,** 1877 (1993).

[17] A. Aronheim, Y. C. Broder, A. Cohen, A. Fritsch, B. Belisle, and A. Abo, *Curr. Biol.* **8,** 1125 (1998).

[18] M. Maroun and A. Aronheim, *Nucleic Acids Res.* **27,** e4 (1999).

protein interactions, initially identified in yeast, using Ras-responsive reporter gene assays. This chapter describes in detail the technical aspects of the study of protein–protein interactions, using the SRS and RRS.

Materials and Solutions

Yeast Media

Ynb galactose medium (500 ml)

Yeast nitrogen without amino acids (Difco, Detroit, MI)	0.85 g
Ammonium sulfate	2.5 g
Galactose (Sigma, St. Louis, MO)	15 g
D-Raffinose (Sigma)	10 g
Glycerol	10 g
Bacto-agar (Difco)	20 g

Ynb glucose medium (500 ml)

Yeast nitrogen without amino acids	0.85 g
Ammonium sulfate	2.5 g
Glucose (Sigma)	10 g
Bacto-agar	20 g

The following amino acids (Sigma) should be added to a final concentration of 50 μg/ml, excluding the amino acids that are encoded by the transfected plasmid: leucine, uracil, tryptophan, methionine, lysine, adenine, and histidine.

YPD medium

Yeast extract (Difco)	1% (w/v)
Bacto-peptone (Difco)	2% (w/v)
Glucose	2% (w/v)
Bacto-agar	4% (w/v)

Yeast Solutions

LISORB: In TE (10 m*M* Tris, pH 8.0; 1 m*M* EDTA), combine
- Lithium acetate, 100 m*M*
- Sorbitol, 1 *M*

LIPEG: In TE, combine
- Polyethylene glycol (PEG) 3350, 40% (w/v)
- Lithium acetate, 100 m*M*

Salmon sperm DNA

Salmon sperm (Sigma) stock solution is prepared at 10 mg/ml and sonicated for 10 min. Prior to transfection, salmon sperm is boiled for 10 min and cooled on ice for 5 min.

Whenever a high efficiency of transformation is necessary, such as for library screening, YeastMaker carrier DNA from Clontech (Palo Alto, CA) is used.

STET
- Sucrose, 8% (w/v)
- Tris (pH 8.0), 50 m*M*
- EDTA, 50 m*M*
- Triton X-100, 5% (v/v)

Mammalian Cell Transformation Solutions

HBS×2: Combine
- HEPES, 50 m*M*
- NaCl, 280 m*M*
- Na_2HPO_4, 1.5 m*M*
- and adjust the solution to pH 7.1 with NaOH

Chloramphenicol acetyltransferase (CAT) reaction mixture: per reaction, combine

Tris (pH 8.0), 1 *M*	40 μl
n-Butyryl coenzyme A, 5 mg/ml	3 μl
[^{14}C]Chloramphenicol	1 μl
Doubly distilled water	6 μl

Materials

Biodegradable counting scintillant (BCS; Amersham, Arlington Heights, IL)
n-Butyryl coenzyme A (Sigma)
[^{14}C]Chloramphenicol (New England Nuclear, Boston, MA)
Dimethyl sulfoxide (DMSO; Sigma)
Glass beads, 425–600 μm (Sigma)
HEPES (Sigma)
Luciferase assay reagent (Promega, Madison, WI)
Tetramethylpentadecane (Sigma)

Methods

General Yeast Transformation Protocol

Because the SRS and RRS are based on a temperature-sensitive yeast strain they are highly subject to problems of contamination by regular yeast strains. Therefore, the following two controls should be included in every transfection.

1. Plating ~10^6 cells (100 μl of culture ready for transfection) into a YPD plate incubated directly at 36°. This control tests the *cdc25-2* culture and provides an estimation of the rate of revertants. Typically 10–20 colonies exhibit growth at 36°.
2. A control transfection tube should include both pYes2 and pADNS expression plasmids. After the transfection procedure the control transformants are plated on Ynb glucose (−Leu −Ura) plate and incubated directly at 36°. This control gives an estimate regarding revertants and possible contamination accumulated during the transfection procedure. No colony is expected to grow on this plate.

Transformants incubated at 24° appear 3–4 days after transfection.

1. Place a single *cdc25-2* yeast colony into 200 ml of YPD medium. Grow the cells overnight at 24° to 2–10 × 10^6 cells/ml.
2. Pellet the cells for 5 min at 2500 rpm and resuspend the pellet in 20 ml of LISORB. After two washes with LISORB, resuspend the cells with LISORB to 2–5 × 10^8 cells/ml and rotate the cells for 30 min at 24°.
3. For each transfection add 10 μl of preboiled sheared salmon sperm DNA (20 mg/ml) and 2–3 μg of each plasmid DNA.
4. Mix the DNA by vortexing and add 200 μl of the prewashed yeast cells. Vortex the cell–DNA mixture briefly and add 1.2 ml of LIPEG; mix well.
5. Incubate for 30 min at room temperature with constant rotation.
6. To increase the transfection efficiency, add 100 μl of DMSO and mix well before heat shocking for 10 min at 42°.
7. Spin the transfection mixture for 1 min and discard the supernatant.
8. Respin for 30 sec and completely remove the remaining PEG with a 200-μl pipette tip. Resuspend the pellet in 150 μl of 1 *M* sorbitol and plate the cells on appropriate medium. Incubate the plates at 24° in a humidified incubator for 4 days.
9. After colonies have appeared, plate single colonies with a grid plate on appropriate Ynb glucose plates; grow at 24° for an additional 2 days before further replica plating onto Ynb galactose, Ynb glucose, and YPD plates incubated at 36°.

Bait Test. Prior to library screening, test the growth abilities of the bait designed in *cdc25-2* cells coexpressed with a number of nonrelevant preys and, more importantly, with a known interacting prey, if available. Colonies expressing the bait in different combinations are streaked on the appropriate plate and subsequently replica plated into a set of duplicate plates containing either YPD medium, glucose and galactose medium. One set is incubated at 24°, to test the toxicity of the bait and the prey, whereas the

Transfected plasmids: 1. ADNS(LEU)-Ras(61)ΔF-Bait/ 5'Sos-Bait
2. Yes(TRP)-mGAP
3. Yes(URA)-M-cDNA

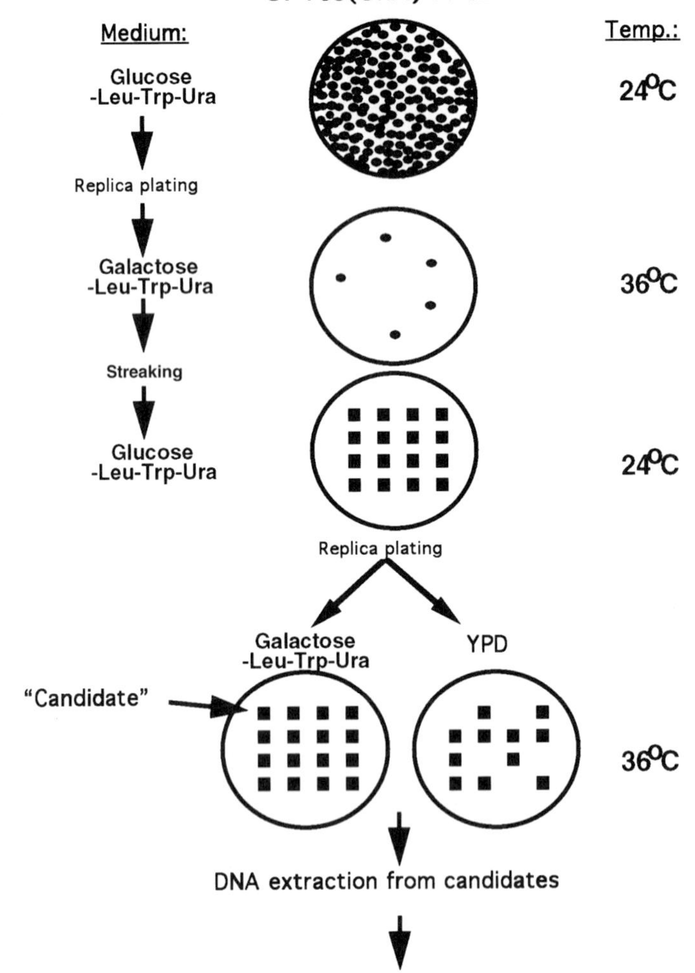

Reintroduction with either specific bait (Sp.) or non specific bait (Non-Sp.)

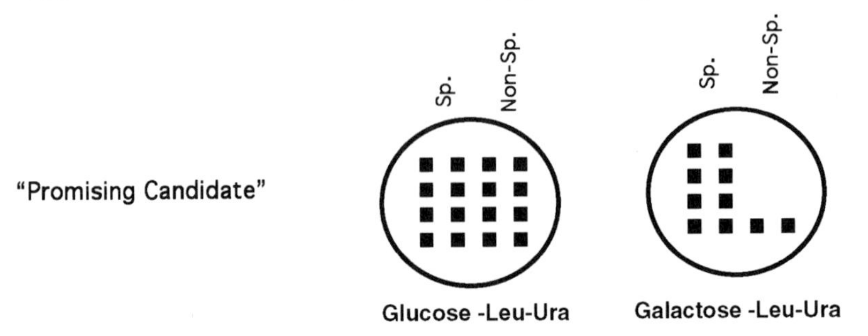

other set is incubated at 36° to test the growth of the bait-expressing cells dependent on the presence of the appropriate prey.

Library Screening

Library screening (see the flow chart in Fig. 1) with the SRS/RRS requires that special libraries be used. The cDNA library, fused to the v-Src myristoylation signal, is routinely inserted by *Eco*RI–*Xho*I digestion in the pYes2(URA)-derived expression vector. To reduce the isolation of mammalian Ras false positives, the plasmid encoding mGAP (mammalian GTPase-activating protein) is coexpressed with the bait and the cDNA library expression plasmids.[13] The mGAP protein is expressed under the control of the *GAL1* promoter using the pYes2(TRP)-based expression vector. Efficient elimination of mammalian Ras false positives requires the expression of mGAP by a multicopy expression plasmid, because a single-copy plasmid encoding mGAP is unable to eliminate all the mRas false positives when expressed in cdc25-2 cells (A. Aronheim, unpublished results, 1999).

To obtain high transformation efficiency, the bait and mGAP expression plasmids are first introduced into the *cdc25-2* yeast strain. Transformants are isolated and used to inoculate a 3-ml liquid culture for overnight growth at 24°. The culture is subsequently transferred to 200 ml of liquid culture for an additional overnight incubation at 24°. The culture is than pelleted and transferred to 200 ml of YPD medium for a recovery period of 3–5 hr. Subsequently, the cells are used to transform 20 tubes, each with 3 μg of library plasmid, resulting in 5000–10,000 transformants on each 10-cm plate. After 5–7 days at 24°, plates are used for replica plating to Ynb galactose plates (lacking leucine, uracil, and tryptophan) medium incubated for 3–4 days at 36°. Colonies that exhibit growth are selected and placed on an appropriate Ynb glucose plate containing the appropriate amino acids and bases, using a grid, and incubated at 24° for 2 days. These clones are tested for their ability to grow at 36°, depending on the presence of

FIG. 1. Library screening flow chart. Cdc25-2 cells are transfected with three expression plasmids: (1) the bait plasmid fused to either activated Ras devoid of its farnesylation sequence Ras(61)ΔF or truncated hSos (5′Sos). The bait is expressed under the control of the alcohol dehydrogenase promoter in an ADNS-based plasmid, which provides complementation of leucine auxotrophy (Leu); (2) the mammalian GTPase expression plasmid designed to be under the control of the *GAL1* promoter in the Yes2-derived expression plasmid (InVitrogen) that complements tryptophan auxotrophy (Trp); (3) the prey expression plasmid fused to the v-Src myristoylation sequence designed to be under the control of the *GAL1* promoter in a Yes2-derived expression plasmid (InVitrogen) that complements the uracyl auxotrophy (Ura). The cells are treated as described in text.

galactose in the medium. The growth is compared with the growth obtained on a YPD–glucose plate. Those clones that show preferential growth when grown on galactose medium are considered candidates. To test the specificity of the library plasmid, plasmid DNA is extracted from candidate clones and is used to cotransform *cdc25-2* cells with either the specific bait or a nonrelevant bait. Candidate clones that exhibit bait-specific growth are further analyzed.

DNA Plasmid Isolation from Yeast

In principle, yeast candidate clones contain three different DNA plasmids (bait, prey, and GAP expression plasmids). To reduce the isolation of nonrelevant plasmids galactose-dependent clones are grown overnight at 24° in 3 ml of glucose–Ynb medium lacking uracil, which is supplemented by the library plasmid. Cells are pelleted at 2000 rpm for 5 min and washed once with 1 ml of doubly distilled water.

Resuspend the pellet in 100 μl of STET. Add 0.2 g of 0.45-mm glass beads and vortex vigorously for 5 min. After addition of another 100 μl, vortex briefly and boil for 3 min (punch a hole in the lid). Cool on ice and spin in a microcentrifuge for 10 min at 4°. Transfer 100 μl of the supernatant to 50 μl of 7.5 *M* ammonium acetate, incubate at −20° for 1 hr, and centrifuge for 10 min at 4°. Transfer 100 μl of the supernatant to 200 μl of ice-cold ethanol, mix well, and recover the DNA by centrifugation for 10 min at 4°. Wash with 150 μl of 70% (v/v) ethanol and resuspend the pellet in 24 μl of doubly distilled water. To increase the isolation of the library plasmid, the DNA mixture is digested with *Not*I (a rare "eight-cutter" restriction enzyme), which linearizes the bait and GAP plasmids but has no recognition site within the library-derived expression plasmid. After 1 hr of digestion, DNA is extracted with phenol–chloroform and recovered by ethanol precipitation, using 2 μl of tRNA (10 mg/ml) as carrier. The DNA is dissolved in 10 μl of doubly distilled water. Commercially available plasmid libraries (Stratagene, La Jolla, CA) are designed to provide chloramphenicol resistance and therefore the isolation of the library plasmid is highly facilitated.

Highly competent bacteria are used to transform 1 μl of the isolated plasmid DNA. Bacteria are plated on Luria–Bertani (LB) medium plus ampicillin (Amp, 100 μg/ml) and single colonies are selected for the preparation of single DNA minipreparations to be further analyzed by digestion with *Eco*RI–*Xho*I restriction enzymes for identification of the cDNA inserts. Individual library plasmids are used to retransform *cdc25-2* yeast cells with either the specific bait or a nonspecific bait. Library plasmids that allow cell growth only in the presence of the specific bait are further analyzed.

293-HEK Transfection

After identification and verification of a DNA plasmid that provides efficient yeast growth at the restrictive temperature only in the presence of the specific bait, it is possible to test the interaction directly in mammalian cells. Toward this end, 293-human embryonic kidney (HEK) cells (300,000) are plated on 60-mm plates 1 day before transfection. Twelve micrograms of PEG-prepared DNA plasmid is used for each transfection. The DNA mixture contains 3 μg of each of the following plasmids: polyoma enhancer-CAT reporter gene, 4×AP-1-luciferase reporter gene, pcDNA (InVitrogen, Carlsbad, CA)-derived bait expression plasmid, and prey expression plasmid. In control transfections, in which either the bait or prey expression plasmid is omitted, pcDNA empty empression vector is used to adjust the DNA content. The volume of DNA mixture is adjusted to 450 μl with sterile doubly distilled water and 50 μl of 2.5 *M* $CaCl_2$ is added. The DNA–$CaCl_2$ mixture is slowly added to a sterile tube containing 500 μl of 2× Hanks' buffered saline (HBS) by air bubbling and incubated for 15 min at room temperature. Five hundred microliters of the transfection mixture is added to the cells. After 5 hr, the medium is replaced with fresh medium. Cells are harvested 40 hr after the addition of DNA to the cells. Cells are collected by resuspending in 1 ml of phosphate-buffered saline followed by centrifugation. The cell pellet is resuspended in 100 μl of 100 m*M* potassium phosphate buffer, pH 7.8, containing 1 m*M* DTT. Cell extract is prepared by three cycles of freezing (liquid nitrogen) and thawing (37°) followed by 5 min of centrifugation at 4°. The supernatant is transferred to new tubes and used for further analysis.

Luciferase Reporter Assay

The luciferase assay is performed with 10–25 μl of cell extract, using the luciferase assay system (Promega) according to the manufacturer instructions. Measurements are made with a TD-20/20 luminometer (Turner Designs, Sunnyvale, CA).

Chloramphenicol Acetyltransferase Reporter Assay

The CAT assay is performed with 10–25 μl of cell extract adjusted to 50 μl with doubly distilled water. Subsequently, 50 μl of CAT reaction mixture is added and incubated at 37° for 1 hr. The enzymatic reaction is stopped by addition of 200 μl of *N,N,N′,N′*-tetramethyl-*p*-phenylenediamine (TMPD)/xylene (2 : 1, v/v), vortexing for 1 min, followed by centrifugation at 16,000 rpm for 2 min. 100 μl of the acetylated upper phase is transferred into scintillation tubes containing 1 ml of BCS (biodegradable counting scintillant). The percentage of acetylated chloramphenicol is determined with a conventional β counter.

Perspective

The protein recruitment systems we have developed serve as attractive assays for the identification and characterization of protein–protein interactions. The RRS and SRS features overcome the limitations and problems that arose while using the conventional two-hybrid system and so far serve as the alternatives of choice for identification of protein–protein interactions when the two-hybrid system fails.

Acknowledgments

This research was supported by the Israel Science Foundation founded by the Israel Academy of Sciences and Humanities-Charles H. Revson Foundation. A. Aronheim is a recipient of an academic lectureship from Samuel and Miriam Wein.

[21] Isolation of Effector-Selective Ras Mutants by Yeast Two-Hybrid Screening

By KIRAN J. KAUR and MICHAEL A. WHITE

Introduction

It is becoming increasingly apparent from the work of many laboratories that the pleiotropic effect of Ras activation on cells is mediated through multiple effector pathways. A growing roster of proteins can interact directly with activated Ras, and can potentially function as Ras effectors in cells.[1,2]

A method to dissect the involvement of Ras-binding proteins in Ras function, which is broadly applicable to any protein with multiple binding partners, is to generate mutants of Ras that are defective for different target interactions. Examination of the phenotypes induced on expression of these mutants in cells can lead to information about which target interactions are required to generate these phenotypes. We describe here the procedures developed, using the yeast two-hybrid system, to identify muta-

[1] M. E. Katz and F. McCormick, *Curr. Opin. Cell Biol.* **7,** 75 (1997).
[2] C. J. Marshall, *Curr. Opin. Cell Biol.* **8,** 197 (1996).

0076-6879/00 $35.00

tions in Ras that selectively uncouple the association of Ras with various Ras-binding proteins.

Materials

Plasmids

pBTM116[3] is a LexA DNA-binding domain fusion vector containing the *TRP1* gene, which serves as a selectable marker in yeast, the complete LexA coding sequence upstream of a polylinker, and a β-lactamase gene for selection in *Escherichia coli*. This vector is used for expression of the mutant Ras library. pGADGH,[4] pGADGE, and pGAD424 contain the GAL4 activation domain sequences upstream of a multiple cloning site. The plasmids also contain the *LEU2* gene for selection in yeast and the *bla* gene for selection in *E. coli*. The target plasmids we use that display good Ras interaction in the yeast two-hybrid system are pGADGH-Raf,[5] pGADGH-Rin1,[6] pGAD424-AF6,[7] pGADGE-RalGDS,[8] and pGADGH-Nore1.[9]

Yeast Strains and Media

Ras mutants are expressed in L40[3] [*MAT* **a** *HIS3-200 trp1-901 leu2-3,112 ade2 LYS2::(lexAop)4-his3 URA3::(lexAop)8-lacZ Gal4*]. Ras targets are expressed in AMR70 [*MATα HIS3 lys2 trp1 leu2 URA3::(lexAop)8-lacZGAL4*]. *HIS3* and *lacZ* serve as reporters for interacting proteins.

YPD is a complete nutritive medium containing 20 g of peptone, 10 g of yeast extract, and 20 g of glucose per liter.[10] Twenty percent agar is added for solid medium. CSM is a defined growth medium containing 1.7 g of yeast nitrogen base without amino acids and ammonium sulfate, 5 g of ammonium sulfate, 20 g of glucose, and all essential amino acids.[10] Selective medium is prepared from mixes lacking one or more amino acid. CSM-W, CSM-L, CSM-WL, and CSM-WLH solid media is required.

[3] A. B. Vojtek, S. M. Hollenberg, and J. A. Cooper, *Cell (Cambridge, Mass.)* **74,** 205 (1993).
[4] G. J. Hannon, D. Demetrick, and D. Beach, *Genes Dev.* **7,** 2378 (1993).
[5] L. Van Aelst, M. Barr, S. Marcus, A. Polverino, and M. Wigler, *Proc. Natl. Acad. Sci. U.S.A.* **90,** 6213 (1993).
[6] L. Han and L. Collicelli, *Mol. Cell Biol.* **15,** 1318 (1995).
[7] L. Van Aelst, M. A. White, and M. Wigler, *Cold Spring Harbor Symp. Quant. Biol.* **59,** 181 (1994).
[8] M. A. White, T. Vale, J. H. Camonis, E. Schaefer, and M. H. Wigler, *J. Biol. Chem.* **271,** 16439 (1996).
[9] D. Vavvas, A. Li, J. Avruch, and X.-F. Zhang, *J. Biol. Chem.* **273,** 5439 (1998).
[10] F. Sherman, *Methods Enzymol.* **194,** 3 (1991).

Methods

Production of Library of Randomly Mutated ras Genes

To isolate Ras mutants defective for specific target interactions, a library of randomly mutated *ras* genes is generated that is expressed in yeast as fusions to the LexA DNA-binding protein. We have found the polymerase chain reaction (PCR)-based mutagenesis protocol described by Zhou *et al.*[11] to be a simple and efficient method for producing mutant libraries. The full-length Ha-Ras or Ras family member cDNA is randomly mutagenized by PCR amplification with error-prone *Taq* DNA polymerase, using standard PCR conditions and a dilute template concentration. The following two primers are used to generate Ha-Ras cDNA products that can be ligated as in-frame fusions with the LexA protein in pBTM116: a forward primer with a *Bam*HI restriction site (5′-CCGCAGGATCCCAGAGCTT-CACCATTGAA-3′) and a reverse primer with a *Sal*I site (5′-CCACAGG-TCGACGAAATTCGCCCGGAATT-3′). PCR conditions in a volume of 200 μl are as follows: 2 fmol of DNA template, 10 m*M* Tris-HCl (pH 8.3), 50 m*M* KCl, 1.5 m*M* $MgCl_2$, 50 μM dNTPs, 30 pmol each of forward and reverse primers, and 5 units of *Taq* DNA polymerase. The cycle profile is 1 min at 95°, 2 min at 58°, and 30 sec at 72° for 30 cycles. The product is purified, digested with *Bam*HI and *Sal*I, and ligated into pBTM116 also digested with *Bam*HI and *Sal*I.

The nucleotide misincorporation frequency of *Taq* DNA polymerase is approximately 10^{-5} errors per nucleotide synthesized.[11] Given a 600-base pair (bp) template, the above-described protocol results in a single-base substitution in approximately 35% of the product. Therefore, the minimal complexity of the final library should be 1.5×10^4 *E. coli* clones (3-fold oversampling) in order to represent all possible mutations. It is important to note that the method described above will not produce mutations that require alteration of more than one nucleotide in any particular codon.

Display of Mutant Ras Library in Yeast

The library of Ras mutants is transformed into a haploid yeast strain containing reporter genes under the control of LexA-dependent promoters. We routinely use the L40 strain, which contains both *HIS3* and *lacZ* reporter genes. The yeast transformants are plated on CSM-W at a density that will result in the growth of approximately 1000 W^+ colonies per 15-cm plate. Each colony presumably represents one clone from the Ras library, and sufficient colonies are plated to represent the entire library (~15,000 colo-

[11] Y. Zhou, A. Zhang and R. H. Ebright, *Nucleic Acids Res.* **19,** 6053 (1991).

nies). Displaying the library this way will allow all Ras variants present in the library to be simultaneously tested for interaction with multiple Ras targets (as described below).

1. Inoculate 250 ml of YPD with a single fresh colony of L40 and incubate overnight at 30° with shaking.

2. The culture should be grown to a cell density of approximately 1×10^7 cells/ml. Pellet the yeast at 3000 rpm for 5 min. Wash two times with 10 ml of sterile H_2O.

3. Resuspend the pellet in 0.1 *M* lithium acetate and incubate at 30° for 1 hr with shaking.

4. Pellet the cells at 3000 rpm for 5 min and resuspend in 2 ml of 0.1 *M* lithium acetate.

5. Add 10 μg of Ras mutant library DNA (quick-boiled or alkaline lysis-quality DNA is sufficient) and 20 ml of 40% (w/v) polyethylene glycol (MW 3500), and incubate for 1 hr at 30°.

6. Heat shock at 42° for 20 min.

7. Pellet the cells for 10 min at 5000 rpm. Wash once in 10 ml of YPD.

8. Resuspend the pellet in 2 ml of YPD. Add glycerol to 15% (v/v) and freeze 200-μl aliquots at $-80°$.

9. Thaw one aliquot and plate 1/100, 1/10, and 1/2 of the volume on CSM-W selective medium and incubate at 30° for 3 days.

10. On the basis of the number of colonies appearing on the CSM-W plates, calculate the total number of transformants present in the transformation mix. Typically 1×10^5 W^+ colonies are recovered. Plate the appropriate volume of the remaining frozen aliquots to result in approximately 1000 W^+ colonies per 15-cm CSM-W plate. For a Ras mutant library, approximately 15 plates will be required for enough colonies to saturate the screen. Once colonies have appeared, plates can be stored for several weeks at 4°.

Display of Ras Targets in Yeast

Each of the Ras targets to be used in the screen is introduced into AMR70 or an equivalent yeast strain that is *mat*α and contains the appropriate genetic markers. The choice of full-length versus truncated versions of Ras targets can affect the outcome of the screens, as multiple domains of a protein can be involved in modulating the specificity of interactions.[12] We generally use only full-length versions of Ras targets in order to avoid this complication. The choice of targets is obviously limited to those that

[12] J. K. Drugan, R. Khosravi-Far, M. A. White, C. J. Der, Y.-S. Sung, Y.-W. Hwang, and S. L. Campbell, *J. Biol. Chem.* **271,** 233 (1996).

have been demonstrated to display an interaction with Ras in the yeast two-hybrid system. Mammalian proteins that can associate with Ras in the yeast two-hybrid system include all three Raf family members,[3] the RalGDS (Ral guanine nucleotide dissociation stimulator), Rgl, and Rlf members of the Ral GEF (guanine exchange factor) family,[13] Rin1,[6] AF6,[7] Nore1,[9] and the p110 subunit of phosphatidylinositol 3-kinase δ (PI3K δ).[14]

1. Plasmids containing the *Leu2* gene, and expressing Ras targets fused to a *trans*-acting transcriptional activation domain (i.e., VP16 or pGAD variants), are individually introduced into AMR70 by small-scale yeast transformation.
2. L^+ colonies derived from the yeast transformations are used to inoculate 10-ml CSM-L liquid cultures. Grow the cultures to saturation (1–2 days at 30°).
3. Evenly spread 200 μl of each culture onto 15-cm CSM-L agar plates, and incubate overnight to produce dense monolayer cultures. Each monolayer culture contains enough cells for three separate replica-plate mating assays. Therefore, preparation of five plates for each target provides sufficient material for the Ras library screen.

Replica-Plate Mating Assay

Ras variants with selective target interactions are isolated by testing the interaction of each library clone with each Ras target. This is accomplished *en masse* through introduction of Ras targets by mating (Fig. 1).

1. Using a 15-cm replica-plate cylinder and sterile velvets, make one replica of each library plate (CSM-W with colonies expressing LexA–Ras fusions in L40) on YPD for each Ras target to be tested. Typically up to six replicas can be produced from one imprinted velvet. It is critical that the colonies not be smeared when replica plated. Retain the original CSM-W library plates, and allow the colonies to regrow. These will be used later to recover pBTM116-ras plasmids expressing mutants with selective target interactions.
2. Overlay the AMR70 tester strains (CSM-L with lawns of AMR70 expressing the Ras targets) by replica plating onto the YPD plates produced above. Incubate the YPD plates overnight at 30° to allow zygotes to form.
3. Replica plate the YPD plates onto 15-cm CSM-WL plates to select for diploid colonies. Incubate at 30° for 1–2 days.

[13] R. M. Wolthuis and J. L. Bos, *Curr. Opin. Genet. Dev.* **9,** 112 (1999).

[14] D. Chantry, A. Vojtek, A. Kashishian, D. Holtzmann, C. Wood, P. W. Gray, J. A. Cooper, and M. F. Hoekstra, *J. Biol. Chem.* **272,** 19236 (1997).

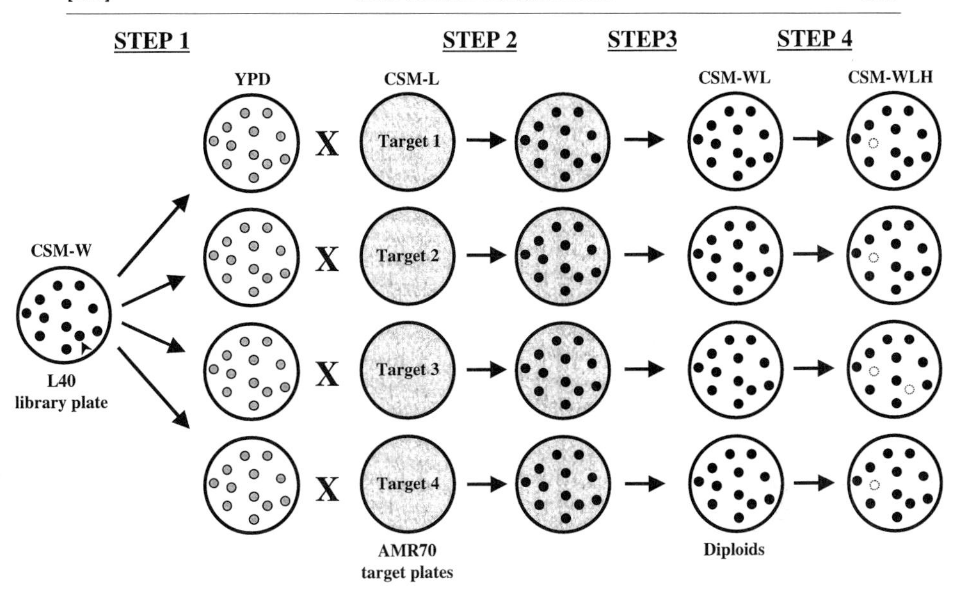

Fig. 1. Plate mating assay to screen for Ras mutants with selective target interactions. Shown is a schematic representation of the steps described in the protocol. An example of a "yeast colony" expressing a Ras mutant that fails to interact with target 3 but retains interaction with all other tested targets is indicated by the arrowhead.

4. Replica plate the resulting diploid colonies onto CSM-WLH and CSM-WL overlaid with filter paper. Score the interactions by observing growth on CSM-WLH and/or β-galactosidase activity. The orientation of the colonies on the plates allows comparison of the interactions of each library colony with each Ras target. The majority of colonies will express wild-type Ras or Ras variants with benign mutations, and will score positive for all tested targets. Colonies that score negative with all targets likely express unstable or truncated proteins, or Ras variants with other uninformative alterations.

Plasmid Rescue

The pBTM116-ras plasmids, expressing mutants with selective target interactions, are recovered from the haploid L40 colonies on the CSM-W plates.

1. Locate the appropriate L40 haploid colonies by their orientation on the plates, and use to inoculate 5 ml of CSM-W. Incubate overnight at 30°.
2. Pellet the cells at 3000 rpm for 5 min at room temperature.

TABLE I
INTERACTION OF RAS VARIANTS WITH RAS TARGETS IN YEAST TWO-HYBRID SYSTEM

Variant	Raf1	RalGDS	Rin1	Nore1	AF6
Ras12V	+++	+++	+++	+++	+++
Ras12V,35S	++	−	−	−	+
Ras12V,37G	−	+++	+++	+++	++
Ras12V,40C	−	−	−	−	−
Ras12V,33V	+++	+++	+++	−	+++

3. Resuspend the pellet in 0.3 ml of yeast lysis buffer [2.5 *M* LiCl, 50 m*M* Tris-HCl (pH 8.0), 40% (v/v) Triton X-100, 62.5 m*M* EDTA]. Transfer to a microcentrifuge tube. Add 0.3 g of acid-washed glass beads (0.45 to 0.55 mm in diameter) and 0.3 ml of phenol–chloroform–isoamyl alcohol (25:24:1, v/v/v) and vortex vigorously for 3 min.
4. Spin at full speed in a microcentrifuge for 5 min at room temperature.
5. Remove the aqueous phase to a new tube.
6. Transform elecrocompetent *E. coli* with 1–2 μl of supernatant.

The recovered plasmids are amplified in *E. coli*, and target interactions are retested by standard yeast two-hybrid reporter analysis to verify the original phenotypes scored in the screens. The Ras cDNAs are then sequenced to identify the mutations responsible for altering target recognition. Table I lists Ras effector mutants we have identified that can selectively interact with subsets of Ras-binding proteins.[8,15,16]

Summary

Ras mutants displaying selective target interactions will often display partial loss of function phenotypes when expressed in cells.[15,17] Thus the isolation of mutations in Ras and Ras family members has proved to be a productive approach for testing the requirement of specific target interactions to mediate downstream responses.[17–20] The procedures outlined here

[15] M. A. White, C. Nicolette, A. Minden, A. Polverino, L. Van Aelst, M. Karin, and M. H. Wigler, *Cell (Cambridge, Mass.)* **80,** 533 (1995).
[16] L. Shivakumar and M. A. White, unpublished observations (1999).
[17] T. Joneson, M. A. White, M. H. Wigler, and D. Bar-Sagi, *Science* **271,** 810 (1996).
[18] P. Rodriguez-Vicinia, P. H. Warne, A. Khwaja, B. M. Marte, D. Pappin, P. Das, M. D. Waterfield, A. Ridley, and J. Downward, *Cell (Cambridge, Mass.)* **87,** 519 (1997).
[19] N. Lamarche, N. Tapon, L. Stowers, P. D. Burbelo, P. Aspenstrom, T. Bridges, J. Chant, and A. Hall, *Cell (Cambridge, Mass.)* **87,** 519 (1996).
[20] A. S. Alberts, O. Geneste, and R. Treisman, *Cell (Cambridge, Mass.)* **92,** 475 (1998).

greatly simplify the isolation of such mutants, and it is hoped will contribute to a better understanding of Ras effector function.

Acknowledgments

This work was supported by grants from the National Institutes for Health and the Welch Foundation.

[22] Two-Hybrid Dual Bait System to Discriminate Specificity of Protein Interactions in Small GTPases

By ILYA G. SEREBRIISKII, OLGA V. MITINA, JONATHAN CHERNOFF, and ERICA A. GOLEMIS

Introduction

Two-hybrid systems[1] are useful tools with which to identify novel protein interactors for "bait" proteins of interest,[2–5] and to dissect the sequence requirements underlying previously defined protein–protein interactions. Because of the large number of proteins in the Ras superfamily, and because of the sometimes extensive sequence homology between different family members, a particular challenge for the two-hybrid system or any protein interactive technique is to devise a means of readily distinguishing interactions specific for individual members of the Ras group from interactions that do not discriminate between family members. To address such issues of specificity, a second-generation form of the two-hybrid system, which simultaneously compares the interaction of two distinct baits with one interactive partner, has been developed.[6–9] In the dual bait system[8] (shown in schematic form in Fig. 1), one protein of interest is expressed as a fusion to the DNA-binding protein cI (bait 1), while a second is expressed as a

[1] S. Fields and O. Song, *Nature* (*London*) **340,** 245 (1989).

[2] C. T. Chien, P. L. Bartel, R. Sternglanz, and S. Fields, *Proc. Natl. Acad. Sci. U.S.A.* **88,** 9578 (1991).

[3] T. Durfee, K. Becherer, P. L. Chen, S. H. Yeh, Y. Yang, A. E. Kilburn, W. H. Lee, and S. J. Elledge, *Genes Dev.* **7,** 555 (1993).

[4] J. Gyuris, E. A. Golemis, H. Chertkov, and R. Brent, *Cell* **75,** 791 (1993).

[5] A. B. Vojtek, S. M. Hollenberg, and J. A. Cooper, *Cell* **74,** 205 (1993).

[6] C. Inouye, N. Dhillon, T. Durfee, P. C. Zambryski, and J. Thorner, *Genetics* **147,** 479 (1997).

[7] R. Jiang and M. Carlson, *Genes Dev.* **10,** 3105 (1996).

[8] I. Serebriiskii, V. Khazak, and E. A. Golemis, *J. Biol. Chem.* **274,** 17080 (1999).

[9] C. W. Xu, A. R. Mendelsohn, and R. Brent, *Proc. Natl. Acad. Sci. U.S.A.* **94,** 12473 (1997).

0076-6879/00 $35.00

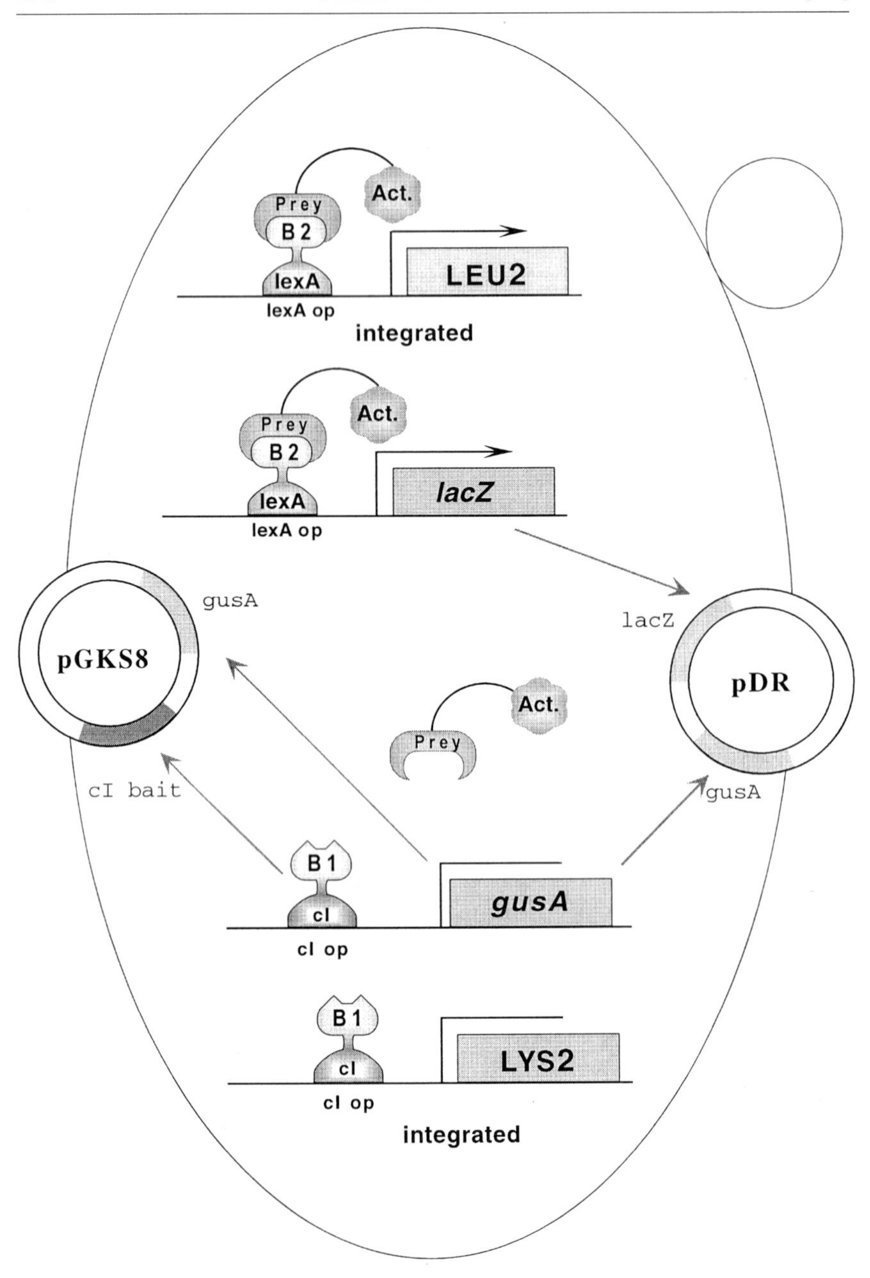
Prey
B2
Act.
lexA
LEU2
lexA op
integrated
Prey
B2
Act.
lexA
lacZ
lexA op
gusA
pGKS8
lacZ
pDR
Act.
Prey
gusA
cI bait
B1
cI
gusA
cI op
B1
cI
LYS2
cI op
integrated

fusion to the DNA-binding protein LexA (bait 2). *Saccharomyces cerevisiae* screening strains express these two baits, in conjunction with four separate reporter genes: *GusA* and *LYS2* transcriptionally responsive to an operator for cI (*cIop-GusA* and *cIop-LYS2*), and *lacZ* and *LEU2* transcriptionally responsive to an operator for LexA (*lexAop-lacZ* and *lexAop-LEU2*). A last plasmid expresses an activation domain (AD)-fused component, which is either a defined protein interactor or a cDNA library. Selective interaction of the AD fusion with one or the other of the two baits is scored by observing a transcriptional activation profile such that scored activity of LacZ = LEU2 $\ll$ LYS2 = GusA (preferential interaction with bait 1) or LacZ = LEU2 $\gg$ LYS2 = GusA (preferential interaction with bait 2).

In the initial validation of the approach, we analyzed the selective interactions of AD-fused Raf, RalGDS, and Krit1[10] (also known as the *CCM1* gene product[11,12]) with DNA binding domain fusions to related Ras family GTPases, LexA–Ras and cI–Krev-1.[8] Raf interacts preferentially with Ras; Krit1 with Krev-1; and RalGDS with both.[10,13–15] The dual bait system successfully discriminated affinity of interactions among these proteins, and was able to select specific high-affinity interacting pairs from

[10] I. Serebriiskii, J. Estojak, G. Sonoda, J. R. Testa, and E. A. Golemis, *Oncogene* **15,** 1043 (1997).

[11] S. Laberge-le Couteulx, H. H. Jung, P. Labauge, J. P. Houtteville, C. Lescoat, M. Cecillon, E. Marechal, A. Joutel, J. F. Bach, and E. Tournier-Lasserve, *Nat. Genet.* **23,** 189 (1999).

[12] T. Sahoo, E. W. Johnson, J. W. Thomas, P. M. Kuehl, T. L. Jones, C. G. Dokken, J. W. Touchman, C. J. Gallione, S. Q. Lee-Lin, B. Kosofsky, J. H. Kurth, D. N. Louis, G. Mettler, L. Morrison, A. Gil-Nagel, S. S. Rich, J. M. Zabramski, M. S. Boguski, E. D. Green, and D. A. Marchuk, *Hum. Mol. Genet.* **8,** 2325 (1999).

[13] M. Frech, J. John, V. Pizon, P. Chardin, A. Tavitian, R. Clark, F. McCormick, and A. Wittinghofer, *Science* **249,** 169 (1990).

[14] C. Herrmann, G. Horn, M. Spaargaren, and A. Wittinghofer, *J. Biol. Chem.* **271,** 6794 (1996).

[15] X. F. Zhang, J. Settleman, J. M. Kyriakis, E. Takeuchi-Suzuki, S. J. Elledge, M. S. Marshall, J. T. Bruder, U. R. Rapp, and J. Avruch, *Nature* (*London*) **364,** 308 (1993).

FIG. 1. Schematic of dual bait system. In yeast cells engineered to express dual bait components, interaction between an activation domain-fused prey and a LexA-fused bait drives transcription of *lexA* operator (*lexAop*)-responsive *LEU2* and *lacZ* reporters, whereas interaction between the activation domain-fused prey and a cI-fused bait would be required to drive *cI* operator (*cIop*)-responsive *LYS2* and *gusA* reporters. As shown here, the prey does not interact with a cI-fused bait and thus does not turn on transcription of *cIop*-responsive *LYS2* and *gusA* reporters, and hence is specific for the LexA-fused bait. However, baits can also be chosen so that prey interacts with both baits. *Note:* The system can be configured such that the *gusA* reporter is either on the same plasmid as the cI fusion-expressing moiety (pGKS8) or in combination with the LexA-responsive *lacZ* reporter (pLacGUS).

seeded pools of low-affinity interacting proteins,[8] suggesting that these reagents would be useful for more demanding library screening and mutational applications.

As an example of the utility of the dual bait approach, we have analyzed effectors for the Ras-related GTPase Cdc42. Both Cdc42 and the related protein Rac are known to be required for transformation by Ras.[16,17] In their activated forms, Rac and Cdc42 can themselves induce DNA synthesis, cell cycle progression, and, ultimately, unrestrained growth.[18–20] These effects require the "insert region,"[21] a short sequence motif that is present in all Rho family GTPases but absent from Ras and other small GTPases. Deletion of the insert region from Rac1 or Cdc42 leads to loss of their ability to transform cells, but not to induce actin reorganization or to activate stress-activated protein kinases. These results imply that the insert region is required for the binding of one or more effectors that mediate the effects of Rac and Cdc42 on transformation. To find effectors specific for interaction with the insert region, we adapted the dual bait interaction trap system as follows.

1. We created two baits: LexA-Cdc42 L28 (encoding an activated, transforming allele of Cdc42 as fusion domain),[22] and cI-Cdc42 L28-Δ8 (activated Cdc42, insert region replaced with a sequence from Ras).[20]
2. An AD-HeLa library was introduced into yeast containing both these baits and cognate reporter constructs.
3. Yeast containing baits, library, and reporters were plated to Leu− medium, to select for proteins that interact with activated Cdc42 (i.e., Cdc42 L28). Interactions were confirmed with the second LexA-responsive reporter, β-galactosidase.
4. These confirmed interactors were replica plated to Lys− medium, to identify those proteins that require the Cdc42 insert region for binding. Interactors that stimulate growth on Leu− but not Lys− medium potentially encode such proteins.

From an initial pool of 41 interactors with activated Cdc42, we found 2 clones that appeared to require the Cdc42 insert region for binding, as confirmed by subsequent and biochemical tests. Thus, by using the dual

[16] R. G. Qiu, J. Chen, D. Kirn, F. McCormick, and M. Symons, *Nature* (*London*) **374,** 457 (1995).

[17] R. G. Qiu, A. Abo, F. McCormick, and M. Symons, *Mol. Cell. Biol.* **17,** 3449 (1997).

[18] T. Joneson and D. Bar-Sagi, *J. Biol. Chem.* **273,** 17991 (1998).

[19] M. F. Olson, A. Ashworth, and A. Hall, *Science* **269,** 1270 (1995).

[20] W. J. Wu, R. Lin, R. A. Cerione, and D. Manor, *J. Biol. Chem.* **273,** 16655 (1998).

[21] J. L. Freeman, A. Abo, and J. D. Lambeth, *J. Biol. Chem.* **271,** 19794 (1996).

[22] R. Lin, S. Bagrodia, R. Cerione, and D. Manor, *Curr. Biol.* **7,** 794 (1997).

bait system, we were able to isolate differential binding partners for Cdc2 in a straightforward and facile manner.

Another potential application currently under exploration is to use the dual bait system to carry out structure–function analyses of Cdc42- and Rac-interacting proteins. In this screen, we have randomly mutagenized the amino-terminal Rac/Cdc42-binding segment of the kinase Pak1,[23] creating an activation domain fusion library of mutant variants. We then introduced this mutant library into a yeast strain bearing as baits activated Cdc42 and Rac1. Preliminary data suggest it will be possible to isolate mutants of Pak1 that selectively associate with Cdc42 but not Rac, and vice versa (M. Reeder and J. Chernoff, unpublished results, 1999).

The following represents a protocol written in general form to allow similar experiments with other GTPases, or unrelated proteins. A flow chart illustrating the chain of events described in the text is presented in Fig. 2. Finally, as noted above, the following protocol is a second-generation development of an earlier two-hybrid system, the interaction trap.[4] Space limitations here do not allow detailed presentation of basic auxiliary protocols related to execution of the technique (e.g., preparation of yeast medium and Western blotting): for information on these techniques, it is recommended that the investigator consult a basic protocol manual describing the interaction trap,[24,25] in parallel with the specific instructions here describing the dual bait system.

Cloning into Bait Vectors

To begin, clone DNAs (include a carboxy-terminal translational stop) encoding proteins of interest into the polylinker of the two bait vectors. It is important to perform this cloning in frame with the LexA and cI parts of the fusion protein. Table I describes currently available plasmids; restriction maps for the basic bait vectors are shown in Fig. 3. The recommended standard set of plasmids is the combination of pMW103 (for a LexA-fused protein) and pGKS6 (for a cI-fused protein), to be used in conjunction with the reporter plasmid pLacGUS (containing a *lexAop-lacZ* and *cIop-GusA* reporter cassette). However, when choosing which fusion bait plasmids to use, consider the following.

[23] S. Ottilie, P. J. Miller, D. I. Johnson, C. L. Creasy, M. A. Sells, S. Bagrodia, S. L. Forsburg, and J. Chernoff, *EMBO J.* **15,** 5908 (1995).

[24] E. Golemis and I. Serebriiskii, *in* "Cells: A Laboratory Manual" (D. L. Spector, R. Goldman, and L. Leinward, eds.). Cold Spring Harbor Laboratory Press, Cold Spring Harbor, New York, 1997.

[25] E. A. Golemis, I. Serebriiskii, J. Gyuris, and R. Brent, *in* "Current Protocols in Molecular Biology" (F. M. Ausubel, ed.), pp. 20.1.1–20.1.35. John Wiley & Sons, New York, 1997.

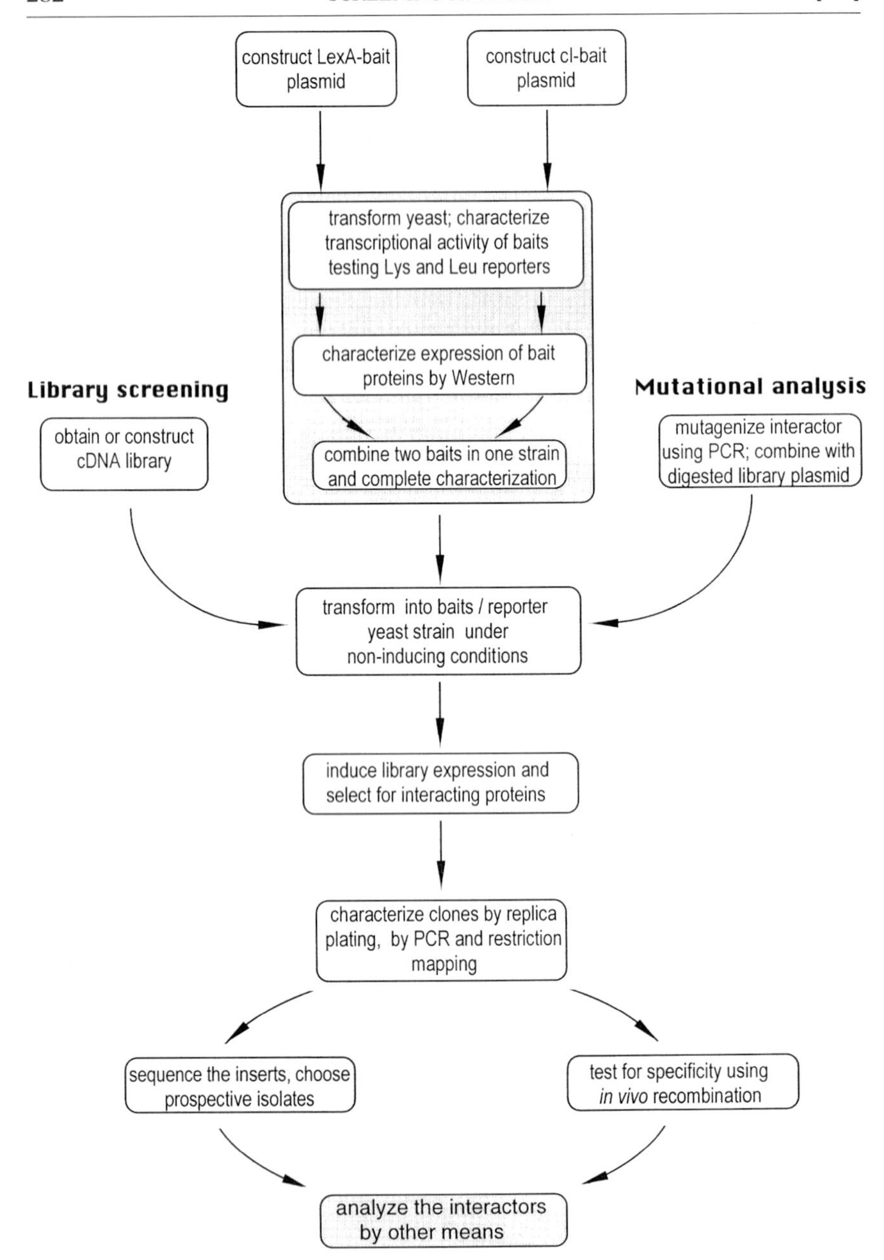

FIG. 2. Flow chart of library screening using the dual bait system. See text for details.

The choice between bait plasmids with the GAL1 or ADH1 promoter should be based on the known or predicted characteristics of bait proteins in yeast; ADH1 is most often used if no problems are expected, and allows more ready separation of bona fide positives from false positives: GAL1 is recommended if the bait protein is somewhat toxic to the yeast. For LexA-fused baits, there are more alternative bait expression vectors described in Refs. 24 and 26.

If it is desirable for the cI-fused bait to be used in two-hybrid systems other than the interaction trap, this is feasible; however, in this case pGKS8, containing both cI fusion and *cIop-GusA* reporter, should be chosen for it is the only bait plasmid with compatible selection markers for other two-hybrid systems. *Note:* the *GusA* reporter on pGKS8-1 produces somewhat higher background levels of β-glucuronidase than that on pLacGUS, and thus requires more careful initial adjustment of assay conditions, to exclude transcriptional activation domains on baits.

The suggested combination of fusion expression vectors and pLacGUS reporter does not allow for adjustment of sensitivity of the LacZ versus GusA reporter via selection of plasmids with different numbers of operators for LexA or cI, an option that is sometimes useful if one bait (e.g., the LexA fusion) has weak background transcriptional activation, while the other does not. If this is likely to be a concern, it is suggested that the combination of pMW103, pGKS8, and a suitable member of the LacZ reporter series pMW112, pMW109, or pMW111 be used.

Note: It may be useful to generate a small number of nonspecific negative control constructs in the vector selected for DBD expression (see step 4 under Second Confirmation of Positive Interactions).

For plasmids with the zeocin resistance marker, the same antibiotic is used for selection in both bacteria and yeast. In cloning, select for bacteria containing the Zeo^R gene on low-salt LB plates (concentration of NaCl 5 g/liter, not 10 g/liter) with zeocin added to a final concentration of 50 μg/ml. When cloning into LexA bait plasmid pMW103, select for bacteria resistant to kanamycin at 50 μg/ml.

Confirm constructed plasmids by sequence analysis with the following primers: forward for cI-insert cDNA junction (ATG ATC CCA TGC AAT GAG AG) and forward for LexA-insert cDNA junction (CGT CAG CAG AGC TTC ACC ATT G). The reverse primer to sequence insert is the same for LexA and cI fusion plasmids (TTC GCC CGG AAT TAG CTT GG).

[26] E. A. Golemis and R. Brent, *in* "The Yeast Two Hybrid System: A Practical Approach" (P. Bartel and S. Fields, eds). Oxford University Press, New York, 1997.

TABLE I
DUAL BAIT TWO-HYBRID SYSTEM PLASMIDS

Plasmid name and source	Selection in yeast/	in *E. coli*	Comment/description
LexA Fusion Plasmids			
pMW103[37]	*HIS3*	Km^R	*ADH* promoter expresses LexA followed by polylinker: basic plasmids used for cloning bait 1
pEG202[25]		Ap^R	
pGilda[38]			*GAL1* promoter expresses LexA followed by same polylinker: for use with baits whose continuous presence is toxic to yeast
pRFHM1[25]			*ADH1* promoter expresses LexA-Bicoid fusion protein. (Use as negative control for activation assay and to check specificity of interaction)
pEG202-Ras[8]			*ADH1* promoter expresses LexA-Ras(ΔCys) fusion protein. (Use as negative control for activation assay and to check specificity of interaction)
pSH17-4[25]			*ADH1* promoter expresses LexA fused to GAL4 activation domain. Use as a strong positive control for activation assay

cI Fusion Plasmids			
pGKS-8[8]	Zeo^R		Dual purpose vector: ADH promoter expresses cI followed by polylinker, while cI-responsive gusA reporter cassette is integrated into the same backbone
pGKS6[8]			*ADH1* promoter expresses cI followed by polylinker
pGKS6-Krev[8]			Expresses cI-Krev fusion protein (Δ*CaaX*). Use as negative control for activation assay
pGKS6-Krit[8]			Expresses cI-Krit fusion protein. Use as positive control for activation assay
Activation Domain Fusion Plasmid			
pJG4-[54]	*TRP1*	Ap^R	*GAL1* promoter expresses nuclear localization domain, transcriptional activation domain, HA epitope tag, cloning sites: used to express cDNA libraries

TABLE I. (*continued*)

Reporter Plasmids				
			Number of operators	
pMW112[37]	*URA3*	Km^R	8 lexA	LexA operators direct transcription of the *lacZ* gene: the most sensitive indicator plasmid for transcriptional activation has 8 operators, intermediate reporter 2, and the most stringent reporter 1 operator
pMW109[37]			2 lexA	
pMW111[37]			1 lexA	
pLacGUS (InVitrogen)			8 lexA, 3 cI	LexA-responsive *lacZ* reporter, same cassette as in pMW112, while cI-responsive gusA reporter has lower background level that in pGKS8
LEU2/LYS2 **Selection Strains**				
	Genotype		Number of operators	
SKY48[8]	*MATα trp1 his3 ura3 lexAop-LEU2 cIop-LYS2*		6 lexA, 3 cI	Provides a more stringent selection for interaction partners of cI-fused baits, and more sensitive *lexA*-responsive LEU2 reporter than the one in SKY191
SKY191[8]			2 lexA, 3 cI	Provides a more stringent lexA-responsive LEU2 reporter, and more sensitive cI-responsive LYS2 reporter than the one in SKY48

Transformation into Yeast: Characterization of Bait Transcriptional Activation Profile and Expression

After the two bait plasmids have been constructed, it is important to test whether both bait proteins are properly expressed in yeast, and make certain that this expression does not result in the activation of the reporter genes in the absence of the interaction with activation domain-fused protein. The two bait vector series for LexA and cI fusions have been previously characterized to ensure that LexA- and cI-fused baits are expressed at comparable levels to each other.[8]

We recommend initial transformation of yeast with bait plasmids incorporating the gene for resistance to zeocin (e.g., pGKS6 or pGKS8 derivatives), with selection on YPD supplemented with zeocin to 200 μg/ml. At this stage a first quick characterization of the cI bait is performed; if it activates transcription of the LYS2 reporter, conditions can be adjusted to reduce sensitivity or modify the bait. If the initial characterization indicates no problems, yeast with the cI fusion are subsequently cotransformed with plasmids expressing the LexA fusion and reporter plasmids, which are selected on Ura− His− minimal plates, while maintaining selective pressure for zeocin (now at 350 μg/ml; generally, zeocin is more effective as a selective agent on YPD than minimal plates). This strategy tends to minimize the number of zeocin-resistant background colonies, which occur with greater frequency on minimal plates than YPD plates (for unknown reasons). Complete characterization of the expression and transcriptional activation profile of the two baits is performed at this stage. The specific steps are as follows.

1. Transform *S. cerevisiae* strain SKY191 or SKY48 (see Table I) with the following bait plasmids (to do so, grow yeast in YPD and follow the library transformation protocol below. Scale the protocol down by harvesting only 50 ml of exponentially growing yeast, and resuspend the pellet from centrifugation in 0.5 ml of TE–lithium acetate: this will be enough for 10 transformations, ~50 μl each):

   Test for activation: pBait1 (a cI fusion)
   Positive control for activation: pGKS6-Krit
   Negative control for activation: pGKS6-Krev

   Note: Allowing yeast transformed with cI fusion plasmids to grow in nonselective medium for 6 hr prior to plating on YPD/Zeo (200 μg/ml) greatly increases the efficiency of transformation. Incubate the plates for 2–3 days at 30°.

2. Make a master plate from which colonies will be taken for activation and expression assays.

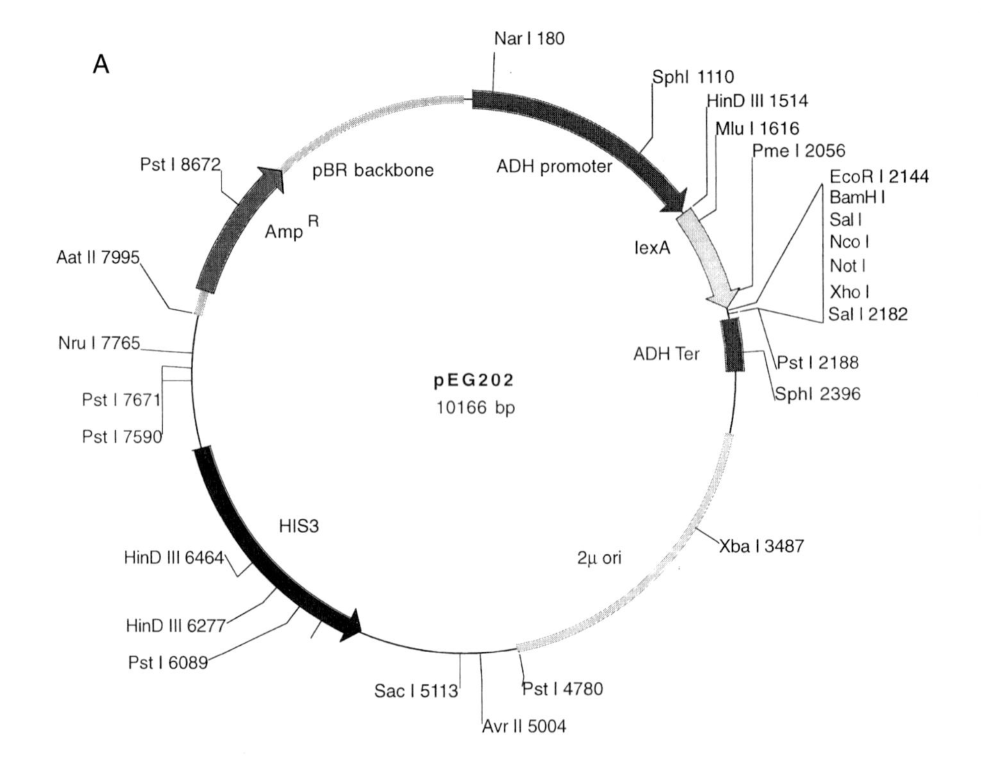
A
Nar I 180
SphI 1110
HinD III 1514
Mlu I 1616
Pme I 2056
EcoR I 2144
BamH I
Sal I
Nco I
Not I
Xho I
Sal I 2182
Pst I 2188
SphI 2396
Xba I 3487
Pst I 4780
Avr II 5004
Sac I 5113
Pst I 6089
HinD III 6277
HinD III 6464
Pst I 7590
Pst I 7671
Nru I 7765
Aat II 7995
Pst I 8672
pBR backbone
ADH promoter
lexA
ADH Ter
Amp R
pEG202
10166 bp
HIS3
2μ ori

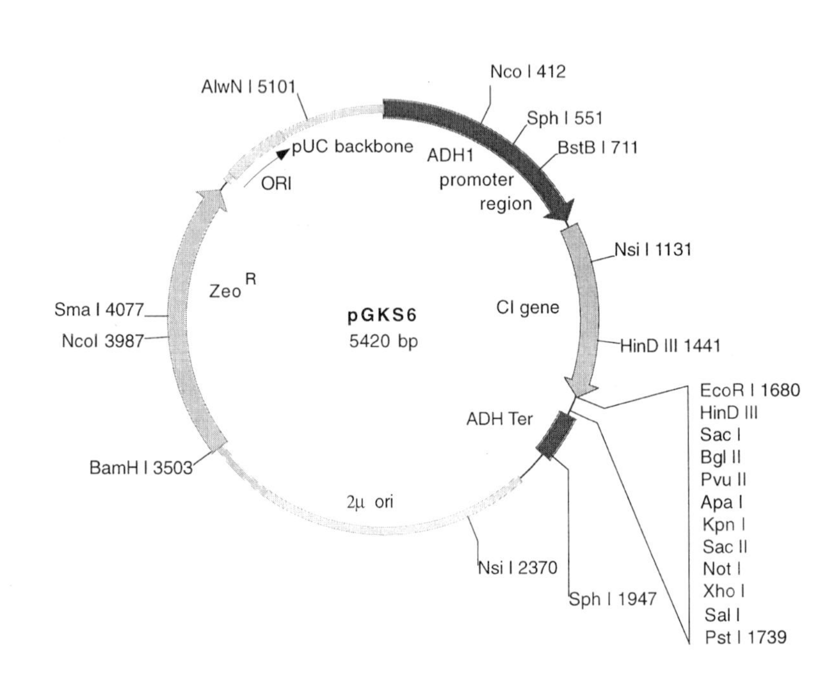
B
AlwN I 5101
Nco I 412
Sph I 551
BstB I 711
pUC backbone
ORI
ADH1
promoter
region
Nsi I 1131
Zeo R
pGKS6
5420 bp
CI gene
Sma I 4077
NcoI 3987
HinD III 1441
EcoR I 1680
HinD III
Sac I
Bgl II
Pvu II
Apa I
Kpn I
Sac II
Not I
Xho I
Sal I
Pst I 1739
ADH Ter
BamH I 3503
2μ ori
Nsi I 2370
Sph I 1947

3. *Initial Activation Assay: Bait 1.* To test whether the cI bait activates, replica plate[24] or use toothpicks to transfer colonies to the following set of plates. Generally, test at least four to six colonies for each construction, to ensure homogeneous behavior in the assay:

Lys− Glu: Test for activation (activation reflected as growth)
Leu− Glu: Negative control for growth
CM Glu Zeo+: Positive control for growth

Incubate the plates at 30° and monitor growth for up to 4 days. The positive control should grow on medium lacking lysine within 1–2 days, while the negative control should show no discernible growth on this medium in 4 days. Test baits that show no growth, or weak growth, at 4 days should be acceptable for use.

If pGKS8 is being used as bait vector, it is possible to test immediately transcriptional activation of the GusA reporter at this step (see below, steps 8 and 9).

4. *Expression Assay: Bait 1.* Confirm the expression of the bait by Western blot analysis (this is particularly important if the bait does not activate transcription of the *LYS2/GusA* reporter genes), using antibody to cI. Yeast transformed with cI-Krit or cI-Krev-1 can be used as positive controls (molecular mass of Krev-1 and Krit fusions to cI are ~50 and

FIG. 3. Two-hybrid basic vectors. (A) LexA fusion vector (GenBank U89960 for pEG202; in pMW103 Ap^R was replaced with Km^R). The strong *ADH1* promoter is used to express bait proteins as fusions to the DNA-binding protein LexA. Sequencing primer (forward) CGT CAG CAG AGC TTC ACC ATT G can be used to confirm correct reading frame for LexA fusions. Alternative and specialized LexA fusion vectors are available; see Ref. 36 for details. (B) cI fusion vector (for maps and sequences, see Ref. 36). To allow introduction of all system components into a single yeast strain, a combined cI-bait expression/*cIop-GusA* reporter with a single selectible marker (Zeo^R) was constructed. To create this plasmid, pGKS8, a *3cIop-gusA* reporter cassette was inserted into pGKS6 downstream of the Zeo resistance cassette. The map of pGKS8 is available on the web.[36] (C) *lacZ/gusA* reporter. The pLacGUS reporter plasmid is available from Invitrogen. Eight operators for LexA binding in the context of a minimal *GAL1* promoter direct transcription of the *lacZ* reporter, while three operators for cI in a second copy of a minimal *GAL1* promoter direct transcription of the *gusA* gene. (D) Activation domain fusion vectors: pJG4-5 (GenBank U89961). The library plasmid expresses inserted cDNAs as a translational fusion to a cassette consisting of the SV40 nuclear localization sequence, the acid blob B42, and the hemagglutinin (HA) epitope tag.[4] Expression of sequences is under the control of the *GAL1* galactose-inducible promoter. Sequencing primer (forward): CTG AGT GGA GAT GCC TCC. Alternative AD fusion vectors are available.[36] (E) Polylinker maps of dual bait plasmids. Sites no longer suitable for the insertion of coding sequences on either pMW103 or pGKS8 are indicated by asterisks (*) on the maps. Alternative reading frames are available for pJG4-5, and the Invitrogen library plasmid pYesTrp2 has an extended polylinker (*www.invitrogen.com*).

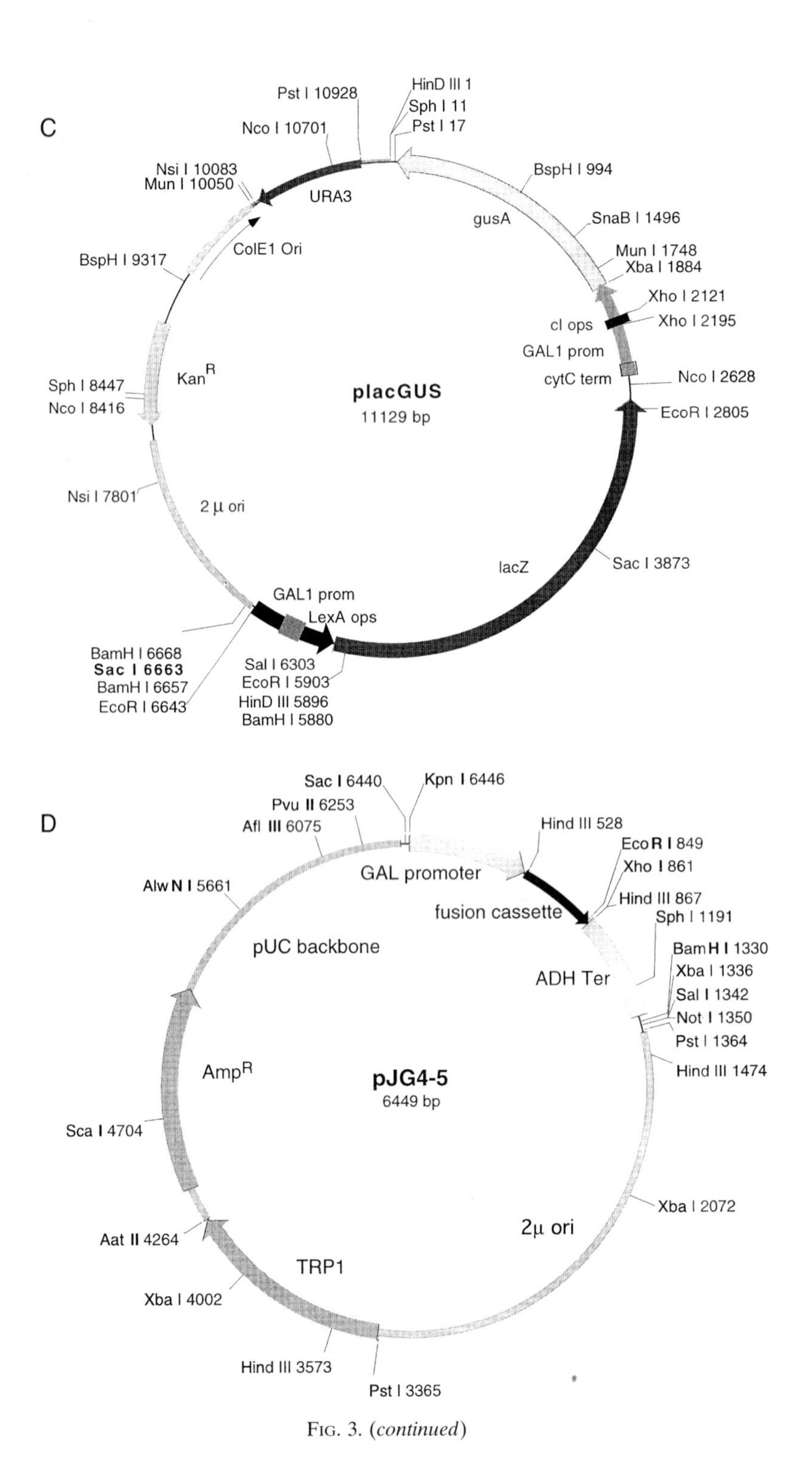

FIG. 3. (*continued*)

E

DBD-fusion (bait) vectors' polylinkers:

pEG202; pGilda; pMW103 (*)

```
                         SalI        NotI          SalI
 EcoRI      BamHI          NcoI*       XhoI
GAA TTC CCG GGG ATC CGT CGA CCA TGG CGG CCG CTC GAG TCG AC
```

pGKS6/pGKS8(*)

```
EcoRI          BglII          ApaI*           NotI        SalI*
        SacI          PvuII          KpnI          XhoI*        PstI*
GAA TTC GAG CTC AGA TCT CAG CTG GGC CCG GTA CCG CGG CCG CTC GAG TCG ACC TGC AG
E   F   E   L   R   S   Q   L   G   P   V   P   R   P   L   E   S   T   C
```

AD-fusion (library) vector's polylinker:

pJG4-5

```
                                            EcoRI           XhoI
ATG GGT GCT CCT CCA AAA AAG AAG ...  CCC GAA TTC GGC CGA CTC GAG AAG CTT ...
M   G   A   P   P   K   K   K   ...  P   E   F   G   R   L   E   K   L   ...
```

FIG. 3. (*continued*)

~75 kDa, respectively). To test expression of bait proteins, perform the following steps.

a. Starting from at least two independent colonies of yeast containing the test and control cI fusions, grow ~2-ml cultures overnight in YPD/Zeo (200 μg/ml).
b. In the morning dilute overnight cultures to an OD_{600} of ~0.15–0.25 in YPD with the same antibiotic and grow for 4–6 hr to an OD_{600} of ~0.45–0.7.
c. Spin down the cells in 1.5-ml Eppendorf tubes at full speed for 3 min and pour off/aspirate the supernatant.
d. Add 50 μl of 2× Laemmli buffer to each yeast pellet, vortex, and immediately place on dry ice until ready for use. Samples can be stored at −70°.
e. Remove the samples from dry ice or freezer directly to a 100° water bath and boil for 5 min.
f. Microcentrifuge the samples for 15 sec at full speed to pellet large cell debris. Load 5–20 μl on a sodium dodecyl sulfate (SDS)–polyacrylamide gel.

g. Perform Western analysis, using standard protocols[27] with antibodies to cI (Invitrogen, Carlsbad, CA).

5. Prepare yeast frozen stocks, using at least two colonies for which expression of cI fusion protein has been confirmed.

6. *Introducing Bait 2 and Colorimetric Reporters.* Transform yeast containing pBait1 (for which expression of cI fusion protein has been confirmed) with the following combinations of plasmids:

Test for activation: pBait2 and double reporter plasmid pLacGUS*

Positive control: pSH17-4 (LexA-Gal4p) and double reporter plasmid pLacGUS

Negative control: pEG202-Ras and double reporter plasmid

Plate transformation mixtures on His− Ura− Zeo+ (350 μg/ml) Glu plates and incubate for 2–4 days at 30°.

7. Make a master plate from which colonies will be taken for activation and expression assays.

8. *Testing for Activation of All Reporters.* To test if the baits activate transcription of reporter genes, replica plate or streak colonies on the following plates.

His− Ura− Zeo+, Glu or Gal** (3 plates: one for a new master plate; two others to test for cI bait activation of GusA and for LexA bait activation of LacZ)

His− Ura− Leu− Zeo+, Glu or Gal** (test for LexA bait activation of Leu2)

His− Ura− Lys− Zeo+, Glu or Gal** (test for cI bait activation of Lys2)

To assess activation of auxotrophic reporters, monitor growth for up to 4 days on plates lacking leucine or lysine. Results for cI should parallel those seen at step 3; for the LexA bait, it is also desirable that little or no growth should be observed after 4 days.

To assess activation of colorimetric reporters, plates used for characterization of activation of LacZ and GusA are initially incubated overnight at 30°.

9. Test activation of LacZ and GusA reporter genes by performing a chloroform overlay assay.[28]

[27] F. M. Ausubel, R. Brent, R. Kingston, D. Moore, J. Seidman, J. A. Smith, and K. Struhl, "Current Protocols in Molecular Biology." John Wiley & Sons, New York, 1999.

* Alternatively, if pGKS8 has been chosen as bait plasmid, a reporter plasmid containing only a LexA-responsive *lacZ* reporter gene should be used (e.g., pMW109, pMW112).

** Choice of glucose or galactose is determined by whether the LexA bait protein being tested is expressed under the control of the ADH1 or GAL1 (pGilda) promoter.

[28] H. M. Duttweiler, *Trends Genet.* **12,** 340 (1996).

a. In a chemical fume hood, overlay two plates with chloroform to completely cover colonies.
b. Incubate the plates for 5 min at room temperature. If necessary, add more chloroform to keep the colonies completely covered.
c. Pour off the chloroform and let the plates dry upside down for 5 min in the hood and then for 5 min at 37°.
d. Prepare two aliquots (7–10 ml each) of 1% (w/v) low melting point agarose in 0.1 *M* $KHPO_4$, pH 7.0, containing either 5-bromo-4-chloro-3-indolyl-β-D-galactopyranoside (X-Gal) or X-Gluc (Biosynth International, Naperville, IL). To do so, add substrates from concentrated stock solutions [20–40 mg/ml in dimethylformamide (DMF)] to agarose cooled below 45°, to the final concentration of 0.25 mg/ml.
e. Overlay one plate with X-Gal agarose (to assess LacZ activity), and the other plate with X-Gluc agarose (to assess GusA activity), making sure that all colonies are covered completely.
f. Incubate the plates at 30° and monitor for color changes. Blue color usually develops within 30 min (for the LexA-Gal4p and cI-Krit1-positive controls) to several hours. Suitable baits will develop little or no color relative to the negative controls.

Note, there are alternative approaches to assessing activity of colorimetric reporters, including growth on plates with incorporated X-Gal or X-Gluc, summarized in Refs. 24 and 25.

10. Confirm expression of lexA-fused bait by Western blot analysis, as described in step 4. Use anti-LexA antibodies (e.g., from Invitrogen, Carlsbad, CA) instead of anti-cI antibodies. Reconfirmation of the expression of the cI-fused bait at this time is prudent.

11. On the basis of the results of activation and expression tests, choose at least two colonies demonstrating appropriate behavior, and prepare frozen stocks.

If baits are not well expressed, activate transcription too strongly, or cause toxicity in yeast, they may be inappropriate for use in two-hybrid library screening. A detailed discussion of these issues, and suggested corrective actions, are presented in Refs. 24, 25, and 29.

Transforming and Selecting Interactors from a Library

Libraries compatible with the dual Bait system are the same as those compatible with the two-hybrid interaction trap.[4] Currently, the largest

[29] E. A. Golemis, I. Serebriiskii, and S. F. Law, *in* "Gene Cloning and Analysis" (B. Schaefer, ed.), pp. 11–28. Horizon Press, Wymondham, England, 1997.

number of libraries are available commercially from Invitrogen, Clontech (Palo Alto, CA), OriGene (Rockville, MD), and Display Systems Biotech (Vista, CA), and described by these companies at their web sites. Libraries are usually cloned in the pJG4-5 (Fig. 3D) or pYesTrp (Invitrogen) vectors, which drive expression of the library under the control of the GAL1 promoter.

Two approaches can be utilized to introduce the library into yeast containing baits and reporters. In the first approach, library DNA is transformed directly into yeast containing both bait plasmids and the reporter plasmid. In the second approach, library DNA is transformed into a yeast strain whose mating type is the opposite of the yeast strain containing the baits. Then the bait strain is mated with the yeast pretransformed with the library.[30,31] Both approaches are quite efficient; the former requires fewer steps to perform, while the latter is preferred if the library is to be used to screen a large number of distinct baits. A protocol for the first approach is described below.

Transforming Library in Bait Strain

1. Select a colony of yeast demonstrated to appropriately express baits 1 and 2, and grow an ~20-ml culture in His− Ura− Zeo+ (350 g/ml) liquid dropout medium overnight at 30° on a shaker.
2. Dilute the overnight culture into ~300 ml of the same medium so that the diluted culture has an OD_{600} of ~0.15. Incubate at 30° on an orbital shaker until the culture has reached an OD_{600} of ~0.5–0.7.
3. Collect the cells by spinning for 5 min at 1000–1500*g* at room temperature, and pouring off the supernatant. Wash the yeast by resuspending the pellet in sterile water (equal to the starting volume of medium) and spinning again as before. Repeat the water wash. Pour off the water and resuspend the yeast in 1.5 ml of TE buffer–0.1 *M* lithium acetate.
4. Mix 30 μg of library DNA and 1.5 mg of freshly denatured carrier DNA (see Refs. 24, 25, 32, and 33 for preparation of carrier) in an Eppendorf tube, and add the DNA mix to the yeast. Mix gently and aliquot the DNA–yeast suspension into 30 microcentrifuge tubes (~60 μl each).
5. To each tube, add 300 μl of sterile 40% (w/v) polyethylene glycol (PEG) 4000–0.1 *M* lithium acetate–TE buffer, pH 7.5. Mix by gently inverting the tubes a number of times (do not vortex). Place the tubes at 30° for 30–60 min.

[30] P. L. Bartel, J. A. Roecklein, D. SenGupta, and S. Fields, *Nat. Genet.* **12,** 72 (1996).
[31] R. Finley and R. Brent, *Proc. Natl. Acad. Sci. U.S.A.* **91,** 12980 (1994).
[32] R. H. Schiestl and R. D. Gietz, *Curr. Genet.* **16,** 339 (1989).
[33] D. Gietz, A. St. Jean, R. A. Woods, and R. H. Schiestl, *Nucleic Acids Res.* **20,** 1425 (1992).

6. To each tube, add ~40 μl of dimethyl sulfoxide (DMSO), and again mix by inversion. Then place the tubes in a heat block set to 42° for 10 min.

7. For each tube, pipette the contents onto a 24 × 24 cm Trp− Ura− His− Glu plate (without zeocin), and spread the cells evenly. One efficient means to accomplish spreading is to add one or two dozen sterile glass beads, 3–4 mm (e.g., Fisher, Pittsburgh, PA), and rotate/agitate the plate vigorously for several minutes (inspect the plates visually to ensure the spread is even; add sterile water if necessary to ensure adequate cell dispersion). Incubate at 30° until colonies appear. Count colonies to estimate the efficiency of library transformation; it should be ~10^5 transformants per microgram of library DNA. (The total number of transformants required to screen a cDNA library derived from a mammalian genome should be about 1–2 × 10^6; scale down for less complex genomes.)

Note: While addition of zeocin on the large plates will keep selective pressure for the presence of cI-fused bait, it also tends to slow down growth of yeast colonies and reduce plating efficiency (it is, in addition, extremely expensive, given the large volume of medium in transformation plates). Omitting zeocin from the library transformation plating medium will result in an ~2.5-fold increase in plating efficiency accompanied by only a ~12% rate of bait loss (I. G. Serebriiskii, unpublished results, 1999), so the net result is positive.

8. Harvest and pool primary transformants by pouring 10 ml of sterile water on each of five 24 × 24 cm plates containing transformants, shaking with glass beads and collecting the resulting yeast slurry into 50-ml conical tubes. This will result in up to six full conical tubes.

9. Spin and wash the yeast cells with water twice as described in step 3 above. Resuspend the packed cell pellet in 1 volume of glycerol solution. Combine the contents of the different tubes and mix thoroughly. Disperse as 0.2- to 1.0-ml aliquots in a series of sterile Eppendorf tubes; freeze at −70° (for yeast frozen for less than 1 year, viability will be greater than 90%).

If desired, leave one aliquot unfrozen and proceed directly to the screen for interacting proteins, below.

Screening for Interacting Proteins

In this procedure, expression of library cDNA transformants is induced by initial growth in medium containing galactose. Induced yeasts are plated on selective plates lacking leucine or lysine to select LexA fusion- or cI fusion-dependent interactions. Positive colonies containing library clones are then further tested for activation of LEU2, LYS2, LacZ, and GusA reporter genes.

1. Thaw an aliquot of the yeast transformed with the library DNA (or use unfrozen cells directly). Dilute 100 μl into 10 ml of His− Ura− Trp− Zeo+ galactose–raffinose (Gal/Raff) liquid medium. Incubate with shaking for 5–6 hr at 30° to induce the expression of library cDNA (and in some cases bait proteins) under the control of the GAL1 promoter.

2. Determine the OD_{600} of the culture, and dilute in sterile His− Ura− Trp− Zeo+ Gal/Raff until the OD_{600} is 1.0 (reflecting approximately 2×10^7 yeast/ml). Plate 50 μl of transformants (10^6 cells) on as many His− Ura− Trp− Leu− Zeo+ Gal/Raff and His− Ura− Trp− Lys− Zeo+ Gal/Raff plates as are necessary to fully represent the library on each selection medium. Generally, 3–10 cells should be plated for each original transformant obtained. Incubate at 30° for 2–5 days.

Plate no more than 10^6 cells per 100-mm plate. Plating of larger amounts of cells will result in cross-feeding of colonies and increased numbers of false positive. As a general strategy, it is recommended to mark colonies with different markers as they appear on selection plates (e.g., colonies appeared on the first day—red, on the second day—blue), and pick them in order of their appearance.

3. Pick Leu- and Lys-positive colonies and replica plate them on the following plates:

Plate	Composition
a. Master plate	His− Ura− Trp− Zeo+ Glu
b. Test plate for activation of LEU2	Leu− Gal/Raff
	Leu− Glu
c. Test plate for activation of LYS2	Lys− Gal/Raff
	Lys− Glu
d. Test plate for activation of *LacZ*	His− Ura− Trp− Glu
	His− Ura− Trp− Gal/Raff
e. Test plate for activation of *GusA*	His− Ura− Trp− Zeo+ Glu
	His− Ura− Trp− Zeo+ Gal/Raff

Note that we omit zeocin for the plates used to characterize binding to LexA-fused baits, or when an alternative selection is available to maintain pressure on positive interactors for cI fusions (as on the Lys− medium, which provides selection for growth) to speed up growth of yeast patches (and reduce the cost of the experiment). While plates for auxotrophy are described as Leu− or Lys−, these can be substituted at the investigator's discretion for His− Ura− Trp− Leu− or His− Ura− Trp− Lys−.

4. Incubate the plates for tests a, b, and c for 2–5 days at 30° and monitor yeast growth.

5. Incubate the plates for tests d and e overnight at 30°. Perform X-Gal and X-Gluc overlay assays as described in step 9 of Transformation into Yeast: Characterization of Bait Transcriptional Activation and Expression (above).

6. Wrap the master plates (test a) in Parafilm and keep at 4°. These will be used for further confirmation of interactions.

At this point, colonies are tentatively considered positive if colorimetric analysis shows blue color on Gal/Raff plates, but white, or only faintly blue, after culture on Glu plates; and if they grow on auxotrophic Gal/Raff but not Glu plates. If a prey associates uniquely with a LexA-fused primary bait but not with a cI-fused alternative bait, yeast would turn blue on medium containing X-Gal but not X-Gluc, and grow on medium lacking leucine, but fail to grow on medium lacking lysine. The opposite should be true for a prey that interacts with a cI-fused bait only. Promiscuously interacting clones, likely to be false or familial positives, would be revealed by their growth on medium lacking both leucine and lysine, and blue color with both X-Gal and X-Gluc. A more thorough discussion of false positives and other variables is found in Refs. 25 and 29.

Second Confirmation of Positive Interactions

After identification of colonies demonstrating behavior suggestive of a library-encoded clone that interacts specifically with bait 1 and bait 2 (or with both baits), it is of interest to determine further the specificity of the interaction and characterize the DNA sequence of the clone. One standard method of analyzing positive interactions involves isolation of library plasmids from yeast, retransformation into *Escherichia coli*, characterization of the inserts by PCR and/or restriction analysis, retransformation into yeast, and testing for specificity against multiple baits. This approach has been described.[25] Below we describe an alternative approach in which PCR is used to amplify library cDNA fragments directly from yeast. These fragments are then analyzed by restriction analysis. Finally, PCR products are transformed into yeast containing baits and reporters, along with linearized library plasmid: highly efficient recombination reconstitutes the library plasmid in a single step.[34] Yeast with recombinant plasmids are tested for activation of the four reporter genes. In this approach, identical clones are identified and excluded prior to plasmid isolation from yeast and retransformation into bacteria, which is frequently the most time-consuming step.

[34] R. Petermann, B. M. Mossier, D. N. Aryee, and H. Kovar, *Nucleic Acids Res.* **26,** 2252 (1998).

To Amplify Library cDNA from Yeast by Polymerase Chain Reaction

1. Add 25 μl of water into each PCR tube. Use a toothpick to collect a small aliquot of yeast from a colony (just enough so that cells are visible), and place the cells in the PCR tubes. Allow yeast to disperse for about 1 min.
 As controls, use the following combinations of templates:
 Empty library plasmid or library plasmid with an insert of known size (positive control)
 Library plasmid with the insert of the known size plus yeast from one of the analyzed colonies (control for inhibition of PCR by lysed yeast)
 Untransformed yeast of the same strain as that used to pretransform the library, or no template (negative control)
 a. Prepare master mix for a standard *Taq* polymerase PCR (such that the final volume of PCR mix is 50 μl per sample) and distribute it to PCR tubes. If the insert to be amplified is in pJG4-5, use
 Forward primer, FP1: 5′-CTG AGT GGA GAT GCC TCC
 Reverse primer, FP2: 5′-CTG GCA AGG TAG ACA AGC CG
 For inserts in pYesTRP, see the Invitrogen web page (*http://www.invitrogen.com*).
 b. Carry out PCR amplification as follows:

Temperature	Time
94°	10 min
94°	45 sec
56°	45 sec
72°	45 sec
72°	5 min

2. Run 30 μl of PCR product on a 1% (w/v) agarose gel and identify inserts of apparently the same size. Perform restriction analysis of the remaining 20 μl of the PCR product with *Hae*III in a total volume of 25 μl. Load the digestion products on the gel so that digests of PCR fragments of the same size run side by side.

3. Choose independent clones. Purify PCR products from the gel by standard molecular biological techniques.[27] At this step one option is to sequence PCR products to select clones of interest, although it is desirable first to complete the following steps to confirm the specificity of interaction.

4. Transform the yeast bait/reporter strain that was used to transform the library [stored as frozen stock at the end of Transformation into Yeast (above), freshly grown prior to transformation], with 200 ng of PCR product

and 50 ng of *Eco*RI–*Xho*I-linearized library vector, using the lithium acetate procedure described above for bait transformation. Plate transformants on His− Ura− Trp− Zeo+ Glu plates. Homology between plasmid sequences in the PCR fragment and linear vector mediates the recombination process in yeast.[34] As a result, circular recombinant plasmid is generated. Pick about 10 colonies for each characterized clone and use them for replica plating on selective plates to test the activation of the reporter genes (see above, Screening for Interacting Proteins). As a control, it is recommended that a number of nonspecific, unrelated baits be transformed in parallel, using the same PCR fragment–vector mixtures, to provide additional confidence in the specificity of the interaction. Some candidate molecules that can be used for this purpose [pRFHMI (LexA-Bicoid), pEG202-Ras, pGKS6-Krev-1] are described in Table I.

5. When a specific pattern is observed, sequence the PCR fragments. (*Note:* Only the forward primer FP1 works well in cycle sequencing of PCR fragments.)

Toward Physiological Validation

As with other interaction trap screens[29,35] or any method of identifying novel protein associations, it is important to validate the biological significance of positives, and exclude false positives. While some frequently recorded false positive clones have been publically posted,[36] and others are being noted through the efforts of the two-hybrid based Genome Projects [in the laboratories of Fields (*http://depts.washington.edu/sfields/projects/YPLM/*), Brent (*http://www.molsci.org*), Vidal (*http://www.mgh.harvard.edu/depts/cancercenter/vidal/index.html*), Finley (*http://cmmg.biosci.wayne.edu/finlab/finlab-home.html*), and others], there is no substitute for empirical determination of interaction validity by means other than the two-hybrid system. Biochemical means include such techniques as glutathione *S*-transferase (GST) pulldowns or overlays, or coimmunoprecipitation with antibodies to interacting partners, performed on endogenously expressed proteins; optimal biological validation would be demonstration of alteration of the physiological activity of one interactive partner on the basis of variance in the expression of the second. For the Cdc42- and Rac-interacting

[35] I. Serebriiskii, J. Estojak, M. Berman, and E. A. Golemis, *BioTechniques* **28,** 328 (2000).

[36] I. Serebriiskii and E. A. Golemis, *http://www.fccc.edu/research/labs/golemis/InteractionTrapInWork.html.*

[37] M. A. Watson, R. Buckholz, and M. P. Weiner, *BioTechniques* **21,** 255 (1996).

[38] D. A. Shaywitz, P. J. Espenshade, R. E. Gimeno, and C. A. Kaiser, *J. Biol. Chem.* **272,** 25413 (1997).

proteins noted above, one obvious test for those in which the interaction is dependent on the presence of the insert region would be to examine whether these proteins are required for Cdc42- or Rac-dependent cellular transformation.

In spite of these caveats, "traditional" two-hybrid systems have long since proved their value in speeding the development of understanding of genetically and physically interacting protein networks. Two-bait two-hybrid systems promise to provide a rapid and convenient means of developing specific mutations that allow the dissection of unique contributions of individual effectors to the biological output of genes of interest, and the establishment of hierarchies of preferentially interacting proteins within protein superfamilies.

Acknowledgments

E.A.G. and I.G.S. were supported in work on the dual bait two-hybrid system by a grant from the Merck Genome Research Institute, and by core funds CA-06927 (to Fox Chase Cancer Center). J.C. and O.V.M. were supported by RO1 GM1127 from the NIH and by American Cancer Society grant 3018.

[23] Functional Proteomics Analysis of GTPase Signaling Networks

By GORDON ALTON, ADRIENNE D. COX, L. GERARD TOUSSAINT III, and JOHN K. WESTWICK

Proteomics Analysis of Ras Signaling Networks

Signal transduction is increasingly being viewed as a complex web of interacting proteins and protein complexes in which small GTPases play a crucial mechanistic and physiological role. Methods for analysis of gene expression regulated by small GTPases are described elsewhere (e.g., see [1] and [19] in this volume[1]). However, while crucial for understanding biology, it is notable that analysis of regulated gene expression rarely yields drug targets amenable to screening for therapeutic agents. Also, changes in gene expression seldom correlate with changes in protein expres-

[1] J. J. Fiordalisi, R. L. Johnson III, A. S. Ülkü, C. J. Der, and A. D. Cox, *Methods Enzymol.* **332,** [1], 2001 (this volume); G. G. Habets, M. Knepper, J. Sumortin, Y-J. Choi, T. Sasazuki, S. Shirasawa, and G. Bollag, *Methods Enzymol.* **332,** [19], 2001 (this volume).

0076-6879/00 $35.00

sion.[1a,2] For novel pharmacologic target discovery, it is desirable to identify the proteins involved, both upstream and downstream, in Ras signaling. Posttranslational modifications of GTPases have been found to be an essential component in GTPase action (reviewed in Refs. 3–6). Protein kinases are critical upstream and downstream regulators of GTPase signals.[7] Thus, sensitive methods to assess targets of lipid modification, kinase levels, activation, and phosphorylation sites are required. Most importantly, multiprotein complexes, not individual proteins, are the most physiologically relevant mediators of signal transduction.[8,9] Therefore, methods are needed to isolate native complexes from biologically appropriate cells and to identify the individual components.[10] Advances in methodology and detection allow assessment of changes in protein expression, modification, and protein–protein interactions both in cultured cells and *in vivo*.

Several key questions can be probed with the technologies described below. For example, farnesyltransferase inhibitors (FTIs) are rationally designed drugs intended to inhibit Ras protein function, and are now in clinical trials.[11–13] However, despite the original design of FTIs as peptidomimetics of the Ras farnesylation motif, the most physiologically relevant target of inhibiting the modification of proteins by inhibition of farnesyltransferase is now thought to be not Ras but another farnesylated protein(s) ("target X").[11–13] Use of FTIs concurrently with metabolic labeling and proteomic analysis (as described below) can potentially be used to identify such a "target X" protein. Other potential applications of proteomics technology in Ras family biology include (1) determining which proteins are activated in a given cell type after Ras transformation, including both immediate downstream effectors and elements further downstream (i.e.,

[1a] M. Mann, *Nat. Biotechnol.* **17,** 954 (1999).

[2] S. P. Gygi, Y. Rochon, B. R. Franza, and R. Aebersold, *Mol. Cell. Biol.* **19,** 1720 (1999).

[3] A. D. Cox and C. J. Der, *Crit. Rev. Oncog.* **3,** 365 (1992).

[4] P. J. Casey, *Science* **268,** 221 (1995).

[5] M. H. Gelb, *Science* **275,** 1750 (1997).

[6] G. W. Reuther and C. J. Der, *Curr. Opin. Cell Biol.* **12,** 157 (2000).

[7] A. B. Vojtek and C. J. Der, *J. Biol. Chem.* **273,** 19925 (1998).

[8] M. Wartmann and R. J. Davis, *J. Biol. Chem.* **269,** 6695 (1994).

[9] A. J. Whitmarsh, J. Cavanagh, C. Tournier, J. Yasuda, and R. J. Davis, *Science* **281,** 1671 (1998).

[10] G. Rigaut, A. Shevchenko, B. Rutz, M. Wilm, M. Mann, and B. Seraphin, *Nat. Biotechnol.* **17,** 1030 (1999).

[11] A. D. Cox and C. J. Der, *Biochim. Biophys. Acta* **1333,** F51 (1997).

[12] A. D. Cox L. G. Toussaint III, J. J. Fiordalisi, K. Rogers-Graham, and C. J. Der, *in* "Farnesyltransferase and Geranylgeranyltransferase I: Targets for Cancer and Cardiovascular Therapy" (S. M. Sebti and A. D. Hamilton, eds.). Humana Press, Totowa, New Jersey, 1999.

[13] A. Oliff, *Biochim. Biophys. Acta* **1423,** C19 (1999).

not detectable by yeast two-hybrid, coimmunoprecipitation, or other analyses), (2) ascertaining which particular set of GEF (GTP/GDP exchange factor)/GTPase pairs or GTPase/effector pairs interacts in a given cell, (3) identifying individual components in a multiprotein signaling complex, including both signaling and scaffolding proteins, and (4) determining the consequences of inhibiting the function of Ras family or other signaling molecules to identify their mechanism of action.

Protein Labeling and Acquisition

Metabolic Labeling with Isoprenoids

Posttranslational modification by isoprenoids is critical to Ras family biology and is currently the most tractable route to pharmacologic intervention.[11–13] The C_{15}-farnesyl isoprenoid, a product of mevalonic acid metabolism, is an obligate intermediate in cholesterol biosynthesis and is also covalently attached to members of the Ras family and to some Rho family proteins. Geranylgeranyl isoprenoids, which are C_{20} moieties formed by condensation of two C_{10} geraniol groups, are attached to members of the R-Ras, Rap, and Ral families of Ras-related proteins, as well as to most Rho and Rab family proteins. In addition, many known signaling molecules are modified by prenylation,[11] and the existence of other unidentified proteins so modified is demonstrated both by expression cloning (see Refs. 1, 14, and 15) and by the demonstration of additional prenylated spots on two-dimensional gel analyses of total cellular protein in mevalonate-labeled cells.[16] Metabolic labeling of cells with tritiated isoprenoids and their precursors therefore allows detection of both known and unknown posttranslationally modified proteins.

Pharmacological inhibitors of the enzymes farnesyltransferase (FTase) and geranylgeranyltransferase I (GGTase I) are now commercially available from several sources, especially Biomol (Plymouth Meeting, PA) and Calbiochem (San Diego, CA). Treatment of cells with such pharmacologic agents concurrently with metabolic labeling allows for elucidation of cellular proteins whose isoprenoid modification has been altered. Farnesylated and geranylgeranylated proteins can be labeled with tritiated forms of the prenyl alcohols farnesol (FOH) or geranylgeraniol (GGOH),

[14] B. J. Biermann, T. A. Morehead, S. E. Tate, J. R. Price, S. K. Randall, and D. N. Crowell, *J. Biol. Chem.* **269,** 25251 (1994).

[15] D. A. Andres, H. Shao, D. C. Crick, and B. S. Finlin, *Arch. Biochem. Biophys.* **346,** 113 (1997).

[16] G. L. James, J. L. Goldstein, R. K. Pathak, R. G. Anderson, and M. S. Brown, *J. Biol. Chem.* **269,** 14182 (1994).

respectively, or with their prenyl pyrophosphate analogs farnesyl pyrophosphate (FPP) or geranylgeranyl pyrophosphate (GGPP). (Although less expensive, the prenyl pyrophosphates are not taken up from the culture medium as efficiently as the prenyl alcohols.[17]) In addition, labeling with the tritiated mevalonic acid (MVA) precursor can be used to label proteins containing either modification; however, because the labeling method itself can influence prenylation,[18] it is a good idea to confirm results using more than one radiolabel. We have used a combination of these types of radiolabels to study cellular proteins whose farnesylation is abrogated or is switched to geranylgeranylation in the presence of FTI.

A labeling protocol modified for use in MCF-7 breast carcinoma cells is described below. (For additional protocols, see Refs. 19–23.) Proteins labeled in this manner can be separated by two-dimensional polyacrylamide gel electrophoresis (2D-PAGE), as described in a later section and illustrated in Fig. 1.

MCF-7 Radiolabeling Protocol. Labeling medium for isoprenoid labeling of mammalian cells consists of the following components: Iscove's modified Eagle's medium (or other culture medium as appropriate for the specific cell type being labeled); dialyzed fetal bovine serum, 10% (v/v); bovine insulin, 0.01 mg/ml (cell type dependent—required for MCF-7 cells; penicillin, 100 U/ml; streptomycin, 100 μg/ml; sodium pyruvate, 5 mM; nonessential amino acids, 400 μM; L-glutamine, 2 mM. All reagents are from GIBCO-BRL (Gaithersburg, MD).

At the time of use, add 50 μM compactin (Calbiochem) to block endogenous MVA metabolism and increase utilization of exogenously supplied tritiated MVA or isoprenoids. Also at the time of use, add any desired study compounds such as inhibitors to the labeling medium at appropriate concentrations. It is important to use tritiated isoprenyl alcohols and their MVA precursor that have high specific activity; these are obtained from American Radiolabeled Chemicals (St. Louis, MO): *trans,trans*-[1-^{3}H] far-

[17] D. C. Crick, D. A. Andres, and C. J. Waechter, *Biochem. Biophys. Res. Commun.* **211,** 590 (1995).

[18] H. C. Rilling, E. Bruenger, L. M. Leining, J. E. Buss, and W. W. Epstein, *Arch. Biochem. Biophys.* **301,** 210 (1993).

[19] A. D. Cox, *Methods Enzymol.* **250,** 105 (1995).

[20] J. F. Hancock, *Methods Enzymol.* **255,** 237 (1995).

[21] R. Danesi, C. A. McLellan, and C. E. Myers, *Biochem. Biophys. Res. Commun.* **206,** 637 (1995).

[22] A. Corsini, C. C. Farnsworth, P. McGeady, M. H. Gelb, and J. A. Glomset, *Methods Mol. Biol.* **116,** 125 (1999).

[23] D. A. Andres, D. C. Crick, B. S. Finlin, and C. J. Waechter, *Methods Mol. Biol.* **116,** 107 (1999).

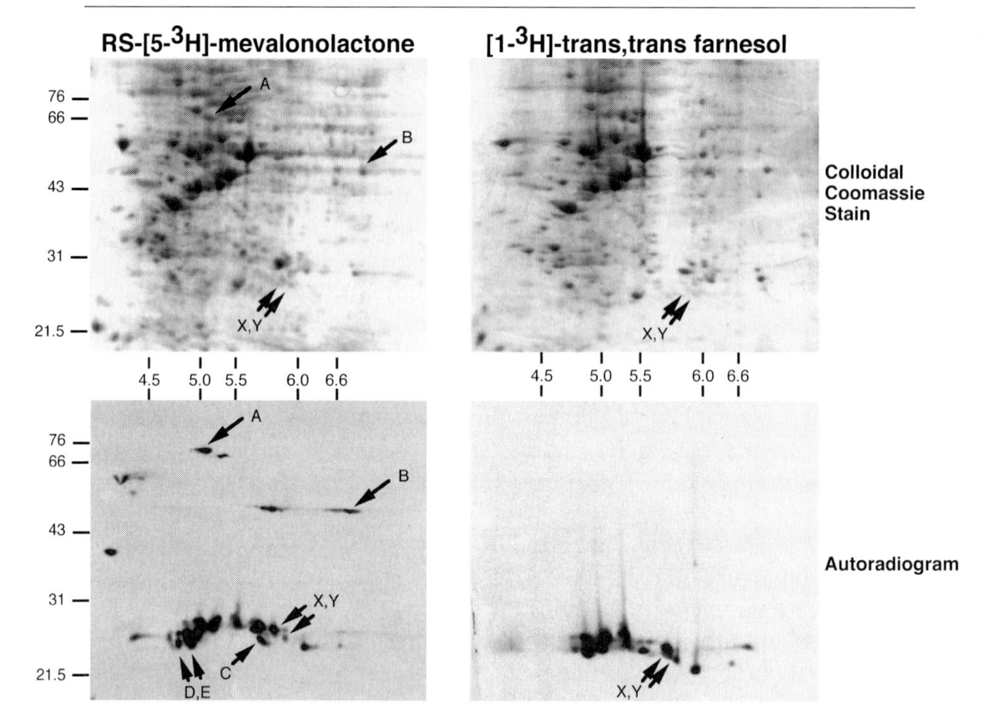

FIG. 1. Visualization of posttranslationally modified proteins by tritium labeling of prenyl groups followed by 2D-PAGE. MCF-7 cells were metabolically labeled with tritiated mevalonic acid or farnesol and total cellular proteins were subjected to two-dimensional gel electrophoresis as described in text. Gels were stained with colloidal Coomassie blue followed by treatment with an autoradiography enhancer (Enlightning; NEN, Boston, MA). Dried gels were exposed to film (BioMax; Eastman Kodak, Rochester, NY) for 2 days at −80°. *Top:* Stained gels; *bottom:* autoradiography of these gels. Identical IPG strips and gels were run with 2D standards (Bio-Rad) to derive the p*I* and molecular mass (kDa) values shown. The majority of labeled proteins are found in the low molecular weight range, and include small GTPases. Illustrated are several higher molecular weight protein spots (e.g., A and B) visible in both stained gels, but labeled only in [^{3}H]MVA-treated cells; thus, these high-abundance proteins are likely modified by geranylgeranylation (GG), not farnesylation. Spots C, D, and E, among others (*left*), are likewise specifically labeled by [^{3}H]MVA but not [^{3}H]FOH; these proteins are also GG modified but are present in relatively low abundance. The majority of spots visible by autoradiography in the molecular mass range between 21.5 and 31 kDa are not visualized by colloidal Coomassie blue staining, indicating the relatively low abundance and high level of posttranslational modification of these proteins. A small subset of sites (e.g., X and Y) is clearly visible by colloidal staining on both maps, and this subset is also visible on both autoradiograms, indicating that they are both relatively abundant and farnesylated. Use of this technology for determining a potential "target X" of farnesyltransferase inhibitors (FTIs) should show a protein(s) that was labeled with [^{3}H]MVA and [^{3}H]FOH in the absence but not the presence of FTI. The ability to detect it on colloidal staining would facilitate the obtaining of the protein sequence.

nesol (50–60 Ci/mmol), all-trans-[1-^{3}H]geranylgeraniol (50–60 Ci/mmol), (*RS*)-[5-^{3}H]mevalonolactone (15–30 Ci/mmol). (Mevalonolactone is converted intracellularly to mevalonic acid.)

Tritiated labeling medium is prepared as follows: for each type of radiolabel, place aliquots of tritiated FOH, GGOH, or MVA sufficient for the assay in separate, open 60-mm tissue culture dishes, to allow the ethanol solvent to evaporate in the sterile atmosphere of the tissue culture hood (this takes 20 min or less). When the solvent has evaporated, add labeling medium (1 ml per 60-mm dish in the assay; 3 ml per 100-mm dish) to the dishes and swirl or pipette gently up and down to dissolve and distribute the radiolabel evenly. Use immediately by pipetting onto prepared dishes as described below. Approximately 250 μCi of [^{3}H]farnesol or [^{3}H]geranylgeraniol per dish is sufficient, but 600 μCi is required for [^{3}H]MVA labeling.

Plate cells in 60-mm dishes at the density required to achieve 85–90% confluence 72 hr later. This assures an adequate number of cells to be labeled and allows for continued cell growth during overnight labeling. For MCF-7 cells, this requires 8.8×10^5 cells per dish in 3 ml of medium. One day after plating, treat the cells with an appropriate range of concentrations of study compounds such as FTIs. Sixty-four to 68 hr after plating, or 4–8 hr prior to labeling, pretreat the cells with 50 μM compactin or lovastatin to reduce pools of MVA and its products, and to increase uptake of radiolabel. At the time of labeling, aspirate the culture medium from each dish and add the labeling medium containing radiolabel (see preparation above) to the cells. Incubate overnight at normal temperature and CO_2 conditions.

Although these tritiated compounds are not as volatile as [^{35}S]methionine,[24] labeling chambers (CBS Scientific, San Diego, CA, and others) can be used to prevent their dissemination in the tissue culture incubator. If a homemade airtight box is used, then care must be taken to provide appropriate levels of humidity and CO_2. This can be accomplished by including a small piece of wet tissue or filter paper in the bottom of the labeling chamber, along with a small (~0.8-g) piece of dry ice, kept well away from direct contact with the dishes.

To harvest labeled cells, on the following day (84 hr after plating) remove the tritiated labeling medium and store at −80° for another use (the uptake of tritiated FOH, GGOH, and MVA is inefficient, such that the medium can be reused on the same type of cells once or twice more without significant loss of activity). Rinse the cells containing the radiolabeled proteins twice with cold phosphate-buffered saline (PBS); discard these washes safely to tritiated waste. Store cell pellets or lysates at −80° or use immediately for the first-dimension separation of 2D-PAGE (see

[24] J. Meisenhelder and T. Hunter, *Nature* (*London*) **335,** 120 (1988).

below). If the labeled cells are to be lysed at this stage, see the section on 2D-PAGE for details; in brief, add lysis buffer (400 μl) and mechanically disrupt the cells with either the bent tip or the flat bottom of a disposable pipette tip prior to storage at $-80°$ or to immediate use as described.

Affinity Methods: Affinity Purification of Ras-Associated Proteins, Kinases, and Phosphorylated Substrates

Affinity chromatography is a useful method for purifying proteins or peptides. This method has been used to purify Ras-modifying enzymes such as FTase, by virtue of its ability to bind Ras as its substrate.[25] Ras-associated proteins such as immediate or downstream effector kinases, exchange factor/GTPase pairs, and even components of multiprotein signaling complexes can be purified through the use of antibodies to surface-specific epitopes, small molecule ligands that interact with the ATP-binding site of kinases, or with protein–protein interaction motifs such as PH, SH2, or SH3 domains.

Currently, the state-of-the-art in affinity approaches is the use of receptor activity-directed affinity tagging, or ReTagging.[26] A ReTag is a tripartite molecule containing a photoactive labeling group, a bioaffinity group that interacts with the molecule to be tagged, and a retrieval tag, such as biotin. The ReTag and the sample are incubated to form a complex and then irradiated with ultraviolet light to initiate covalent photo-cross-linking. The tagged sample is then purified by chromatography, for example, avidin–agarose. A protocol for use of this method to isolate Ras-associated proteins is described below.

ReTag Protocol for Isolation of Ras-Associated Proteins. Sulfo-SBED (Pierce, Rockford, IL), which will covalently couple to any amine-containing molecule, has a biotin tag and a photo-cross-linking group. One additional feature of sulfo-SBED is a cleavable linker that permits separation of the bioaffinity tag (in this case, Ras itself) from the biotinylated, cross-linked target protein(s). Dissolve approximately 200 μg of recombinant Ras[27] in PBS (100 μl) and then add 20 μg of sulfo-SBED in 20 μl of dimethyl sulfoxide (DMSO). Incubate for 30 min at room temperature. Because the reaction is light sensitive, dark or foil-covered tubes should be used. Also, do not include a reducing agent [such as dithiothreitol (DTT) or mercaptoethanol], which will cleave the SBED linker. Use a desalting

[25] Y. Reiss, M. C. Seabra, M. S. Brown, and J. L. Goldstein, *Biochem. Soc. Trans.* **20,** 487 (1992).

[26] D. Ilver, A. Arnqvist, J. Ogren, I. M. Frick, D. Kersulyte, E. T. Incecik, D. E. Berg, A. Covacci, L. Engstrand, and T. Boren, *Science* **279,** 373 (1998).

[27] S. L. Campbell-Burk and J. W. Carpenter, *Methods Enzymol.* **255,** 3 (1995).

column, such as Presto (Pierce), and collect the void fraction, which will contain the SBED-modified Ras protein affinity tag. Incubate the SBED-modified Ras with cellular extracts or other samples in PBS (200 μl) for 2 min at room temperature, followed by 10 min on ice. Irradiate with a long-wavelength UV lamp (365 nm) at a distance of 5 cm for 15 min to cross-link the SBED-modified Ras with proteins in the cell extracts. Apply the cross-linked sample to avidin–agarose (Sigma, St. Louis, MO) and wash as necessary (salt, urea, etc.) to remove undesirable background proteins. Then either elute from the avidin–agarose with low pH or boil in sample dissociation buffer to run one-dimensional gels. Most sample dissociation buffers contain mercaptoethanol or DTT, which will cleave the linker of the SBED molecule, hence separating Ras from the biotinylated, cross-linked proteins. These proteins can be identified by Western blotting with avidin-coupled detection reagents. Critical factors are the purity of the bioaffinity tag (e.g., Ras, GEF, or downstream kinase) and the stringency of washing of the avidin-bound proteins [e.g., Ras-associated proteins, GTPase, or phosphorylated substrate(s)]. Always check all fractions by gel analysis and Western blotting.

Immobilized Metal Affinity Chromatography for Phosphopeptide Recovery. Ras activation leads to activation of several downstream protein kinases. These kinases phosphorylate their targets on serine, threonine, or tyrosine residues. One of the approaches to phosphorylation site determination is to isolate and sequence proteolytically generated phosphopeptides derived from the target protein. This is a difficult task because of substoichiometric phosphorylation of the substrate peptide, even for *in vitro* phosphorylated proteins.

Immobilized metal affinity chromatography (IMAC) has been used successfully to purify phosphopeptides.[28] A technical advance is the use of Ga^{3+} as the metal for the IMAC procedure, as described in detail.[29] This allows recovery of subpicomole amounts of the phosphopeptides from complex digests. The peptides are subsequently sequenced or identified by mass spectrometry. Briefly, a chelating resin such as Poros MC (PE Biosystems, Foster City, CA), which contains imidodiacetate groups, is prepared with a $GaCl_3$ solution and washed with acetic acid, and the acidified phosphopeptide digest is loaded onto the resin. This is washed with acetic acid and the phosphopeptides are eluted with pH 10 ammonium hydroxide. Prior to mass spectrometry a desalting step may be necessary. In our laboratory we can routinely detect 150 fmol of phosphopeptides recovered from picomole quantity peptide mass fingerprinting digestions.

[28] L. Andersson and J. Porath, *Anal. Biochem.* **154,** 250 (1986).

[29] M. C. Posewitz and P. Tempst, *Anal. Chem.* **71,** 2883 (1999).

Protein Fractionation and Visualization

Two-Dimensional Polyacrylamide Gel Electrophoresis

Proteomics has come to be synonymous with two-dimensional electrophoretic separation of proteins. The first-dimension separation exploits differences in the isoelectric point of proteins and the second dimension is a traditional Laemmli protocol that resolves proteins on the basis of their molecular weight.[30] Advances in equipment and methodology have made this once laborious and artful technique reproducible.[31] The most notable advance has been the development of immobilized pH gradient (IPG) strips for the first-dimension isoelectric focusing step. These replace the cumbersome tube gels originally used for this purpose. Several wide and narrow pH gradients are available from commercial sources (Pharmacia, Piscataway, NJ; Bio-Rad, Hercules, CA).

Protocols for 2D-PAGE are described in detail elsewhere.[31a] For GTPase and signaling analyses, we have developed the protocol described below, which utilizes the Pharmacia IPGphor first-dimension apparatus and precast 18-cm IPG strips. This apparatus substantially simplifies the process of strip rehydration, sample loading, and isoelectric focusing. We also use the Bio-Rad PROTEAN II vertical gel electrophoresis apparatus for the second dimension. We can simultaneously run up to 12 first-dimension strips and 8 second-dimension gels in a 2-day total cycle time. Several companies offer imaging systems designed specifically for 2D-PAGE analysis; for comparison, we suggest consulting one of the proteomics web sites listed at the end of this chapter. The most important point is that the gels must be at least 20 cm wide to easily run the standard large-format 18-cm strips (useful for maximum resolution). Minigels may be used with the smallest (7-cm) strips, but resolution is significantly decreased.

A crucial concept for 2D-PAGE separations is that extracts must be free of salt for the isoelectric focusing step. DNA is not usually a problem, but if high protein loads are attempted it can adversely affect the first-dimension resolution. DNA and lipids can be removed with Polymin 10.[32]

Cell pellets can be solubilized directly in rehydration buffer after a PBS wash, but it is important to remove as much residual PBS as possible. If cell extracts are in a salt-containing buffer, start with a minimum of 100

[30] U. K. Laemmli, *Nature* (*London*) **227,** 680 (1970).

[31] A. Gorg, W. Postel, and S. Gunther, *Electrophoresis* **9,** 531 (1988).

[31a] M. R. Wilkins, K. L. Williams, and R. D. Appel (eds.), "Proteome Research: New Frontiers in Functional Genomics." Springer-Verlag, New York, 1997.

[32] R. R. Burgess and J. J. Jendrisak, *Biochemistry* **14,** 4634 (1975).

μg of protein (up to a total of 3–5 mg of protein, depending on the length of strip). Proteins can be precipitated with 10% (w/v) trichloroacetic acid (TCA)–0.12% (w/v) DTT in acetone. Precipitate at −20° for more than 2 hr. Pellet the precipitated protein at maximum speed in a chilled microcentrifuge (2 min at 12,000*g*) and wash the pellets twice with ice-cold acetone. Resuspend the pellets in 400 μl of rehydration buffer (below). Let the pellets sit for up to 1 hr at room temperature if necessary to ensure dissolution of precipitates and then spin for 5 min at maximum speed in a microcentrifuge to pellet insolubles. Do not heat the samples (this avoids carbamylation of the proteins). Alternatively, extracts or column fractions can be dialyzed (Slide-a-lyzer; Pierce) against rehydration buffer lacking ampholytes and 3-[(3-cholamidopropyl)-dimethyl-ammonio]-1-propanesulfonate (CHAPS). These will be added before strip rehydration.

Stock IPG rehydration buffer [CHAPS, Triton X-100, Orange G (10 μl of 1% solution in H_2O per 5 ml of buffer)] may be prepared in advance and stored for several weeks at 4°. Urea, thiourea, Pharmalyte (Pharmacia), and DTT should be added immediately before use. Complete rehydration buffer contains 2 *M* thiourea, 5–7 *M* urea (5 *M* for general use, 7 *M* for difficult-to-solubilize samples), 4% (w/v) CHAPS, 0.5% (v/v) Triton X-100, 0.4% (w/v) ampholine/Pharmalyte pH 3–10 (Pharmacia), 20–50 m*M* DTT (20 for pH 4–7, 50 for pH 3–10), 0.002% (v/v) Orange G dye, [0.1% (w/v) taurodeoxycholate; optional, in which case reduce Triton X-100 to 0.2% (v/v)].

Add 350 μl of the rehydrated protein solution to the center of the IPG chamber and allow to diffuse (for shorter IPGs, use a 200-μl loading volume for 11-cm strips; 170 μl for 7-cm strips). Peel the plastic backing off the IPG strips and lay the acrylamide side down in the proper orientation (the low-pH end of the strip is the pointed end and is oriented toward the positive anode). Bow the strip up and down to spread sample throughout the chamber. Ensure that there are no bubbles under the strip and that no fluid is on top of the strip. In addition, ensure that the acrylamide on the bottom of the strip is touching both electrodes (i.e., shove the strip forward in the chamber).

For silver staining, use no more than 300 μg of total protein with long strips or 150 μg with short strips. Maximal loading is 5 mg of total protein for long strips or 3 mg for short strips. Overlay the strips with 2.5 ml of mineral oil, apply the plastic covers, and place the rehydration chambers in the IPGphor apparatus. Rehydration and isoelectric focusing steps are conducted at a controlled temperature of 20° (higher temperatures may induce carbamylation of the protein, whereas lower temperatures may result in urea crystallization).

TABLE I
Voltage Ramp Program for First-Dimension Immobilized pH Gradient Separation

Voltage	Time (hours)
150	0.5
300	2.5
700	0.5
1000	0.5
2000	0.5
3000	0.5
4000	0.5
5000	18 (up to 30 for high protein load)

First Dimension Separation. The voltage ramp program for the first-dimension IPG separation is shown in Table I. (*Note:* Current should be at 50 μA/strip.) The typical program terminates at 100 kV · hr.

The IPGphor can provide up to 8000 V but we find that the strips can burn; therefore 5000 V is safer. If the voltage does not ramp up during the run, it may indicate that the salt concentration in the sample is too high. In this case, prepare extracts by a different protocol. Handle thiourea with care and dispose of buffer and used strips safely.

After fractionation of proteins in the first dimension, we find that strips can be carefully frozen in large borosilicate screw-cap tubes at −80° without any subsequent loss of reproducibility. On thawing, subject strips to reduction and alkylation (see the next section) prior to loading onto the top of the second-dimension gel.

Reduction and Alkylation. Stock buffer is composed of 50 m*M* Tris–acetate (pH 8.0), 30% (v/v) glycerol, 6 *M* urea, and 2% (w/v) sodium dodecyl sulfate (SDS) (can be stored for several weeks at room temperature). To make reduction solution, add DTT just prior to use to give a final concentration of 0.8% (w/v). Place the IPG strips with the gel side up in capped 200-mm borosilicate screw-cap tubes, add 3.5 ml of the reduction solution per tube, and incubate at room temperature for 15 min on a rocking platform. Pour off the reduction solution and add iodoacetamide [final concentration, 4% (w/v)] to additional stock buffer to make alkylation solution. Add 3.5 ml of the alkylation solution per tube and incubate for 15 min at room temperature. Remove the solution by aspiration.

Second Dimension Electrophoresis. We currently use the "extended" Bio-Rad PROTEAN II apparatus, which has wider seals and narrower spacers to increase the width of the gel. After strip alkylation, dip the strips briefly in running buffer.

Second-dimension gel casting: Pour a standard 10% (w/v) acrylamide SDS–PAGE resolving gel, using ProSieve 50 acrylamide gel solution (FMC BioProducts, Rockland, ME), in a Tris-Tricine buffer (Sigma). Tris–glycine buffers are less expensive and may be used if resolution of low molecular mass proteins (<20 kDa) is not important. For Ras family GTPases (~18–25 kDa), we recommend and routinely use Tris–Tricine buffer. A stacking gel is not necessary if care is taken in pouring the resolving gel. Pour the gel solution to the top of the notch on the casting stand, leaving 0.5 cm to the top of the plate. Alternatively, precast large-format gels are available from Bio-Rad. Add 1 ml of butanol to the top and allow the gel to polymerize. After polymerization, rinse the well with water. Gels may be stored at 4° for up to 3 days by covering with paper towels soaked in running buffer and then wrapping in Saran Wrap.

The gels must be at room temperature prior to loading the strip. Overlay the gels (one at a time) with hot 1% (w/v) agarose in running buffer, containing 0.1% (w/v) bromphenol blue as a tracking dye, and ensure that it has filled to the top of the plates. If necessary, cut off the basic (not-pointed) end of the strip so that it fits into the space just above the second-dimension gel. Be sure to note the orientation of the strip on the gel. Quickly load the first-dimension strip with the low-pH (pointed) end oriented to the left-hand side of the gel. We use a small, thin ruler to push down the strip so it is flush with the top of the gel. The agarose overlay hardens rapidly, at which point the complete gels are inserted into the running apparatus. Cooling at 12° is appropriate for a constant current setting of 30 mA/gel and 6-hr run time. Gels can also be run at 15 mA/gel overnight.

Prior to staining, cut off the strip and agarose, and notch the top left-hand corner of the gel. This is the acidic (pH 3) edge. For optimal sensitivity, utilize the Rabilloud staining method.[33] For subsequent extraction from the gel for protein digestion, analysis, and identification, we suggest the modified silver stain from Protana (Odense, Denmark; *http://www.protana.com/services/protocols*), or a conventional colloidal Coomassie blue staining (e.g., Novex, San Diego, CA).

Protein Stains, Visualization Techniques, and Alternatives

Visualization of proteins separated by 2D-PAGE can be accomplished by any of three methods, singly or in combination. These are chemical staining (colorimetric or fluorometric), covalent fluorophore tagging, and radioisotope labeling.[34] A large number of colorimetric protein stains have been developed for visualization of proteins in polyacrylamide gels. Sensi-

[33] T. Rabilloud, *Methods Mol. Biol.* **112,** 297 (1999).

[34] P. J. Wirth and A. Romano, *J. Chromatogr. A* **698,** 123 (1995).

TABLE II
SENSITIVITY AND PROCEDURAL DIFFICULTY OF PROTEIN STAINING TECHNIQUES

Stain or dye	Sensitivity	Difficulty of protocol[a]
Coomassie blue	100 ng	Low
Silver	1 ng	Moderate
Fluorometric dye	10 ng	Low
Fluorescent labeling	100 pg	Difficult
Radioactive labeling	10 pg	Difficult

[a] This parameter involves the length of time the protocol takes, the number of steps, and the attention to detail required to obtain good reproducibility.

tivity and dynamic range are the most important considerations when choosing between various staining protocols (Table II). In our laboratory we find that Coomassie blue and a modified silver-staining protocol work well. The protocol for Coomassie staining is the least sensitive but is simple, fast, and reliable, has a large dynamic range, and is compatible with mass spectrometry. The silver-staining protocol provides extremely high sensitivity but requires attention to detail. Also, this particular protocol is not compatible with mass spectrometry. The group of M. Mann has described a mass spectrometry-compatible silver-staining protocol.[35] However, this is not much more sensitive than colloidal Coomassie blue (at least in our hands) and has a higher background compared with other silver stains.

Although the sensitivity of fluorometric stains (such as Sypro Orange; Molecular Probes, Eugene, OR) is reported by their manufacturers to be excellent, significant drawbacks include a high background level and the requirement for specialized imaging equipment such as a phosphoimager. Also, the gel itself does not have visible protein spots, so it is difficult to precisely locate a given spot for removal for downstream processing.

An alternative is to use a fluorescent tag. Covalent tagging or labeling of the protein is performed prior to gel electrophoresis.[36,37] Typically, these tags are reactive to either amino or thiol groups of the proteins. Because virtually all 2D-PAGE protocols involve reduction and alkylation of thiols it is convenient to perform the alkylation step with a fluorescent tag.

Individual fluorescent tags may be intrinsically fluorescent or become fluorescent after covalent attachment to protein.[38] In either case the most

[35] A. Shevchenko, M. Wilm, O. Vorm, and M. Mann, *Anal. Chem.* **68,** 850 (1996).

[36] K. L. Hsi, S. A. O'Neill, D. R. Dupont, and P. M. Yuan, *Anal. Biochem.* **258,** 38 (1998).

[37] J. J. Gorman, *Anal. Biochem.* **160,** 376 (1987).

[38] T. Toyooka and K. Imai, *Anal. Chem.* **56,** 2461 (1984).

significant drawback to this approach is that some proteins will contain a larger number of reactive functional groups, compared with other proteins, and so will be more heavily labeled, thus overestimating their relative abundance. Also, some proteins will not be observed because they lack a functional group to which the tag can react.

Despite this, fluorometric tagging can be useful for comparison of two different sets of proteins, such as those derived from normal versus malignant tissue or stimulated versus unstimulated cells. This is known as protein differential display or difference gel electrophoresis.[39] One set of proteins is covalently tagged with one dye and the other set with a different dye. The two samples are then mixed and run on the same first- and second-dimension gel. Postrun imaging of each fluorophore is performed and the two different colors are merged to provide the final image. Each pair of fluorescent tags used should have nearly identical chemical structure, especially with regard to ionizable groups. Thus, these tag pairs should not impart a differential mobility to the two sets of proteins. In practice this is difficult to achieve, as the chemical structure of the tags must be at least minimally different to confer the required spectral shifts to attain the two-color separation. However, the use of appropriate imaging software can correct for global mobility shifts. In addition, there will be a mobility shift, in either dimension, for the tagged proteins but not untagged protein. This may prevent meaningful comparison with gels containing the same set of proteins that were not treated with the fluorescent tags. A publication using the CY3 and CY5 pair of dyes illustrates the potential of this technique.[40]

The most sensitive protein visualization methods involve radioactive labeling of the protein. This approach works well for proteins derived from cells or synthesized by *in vitro* translation. For protein labeling, cells are treated with [^{35}S]methionine prior to protein isolation.[41,42] This incorporates the radioactive methionine into newly synthesized proteins. Gel images can be prepared by film autoradiography or by use of phosphoimager technology.

As described in the example above for prenylation, the study of post-translational modifications is greatly facilitated by this labeling approach. This is especially true for phosphorylation and lipidation events. For phosphorylation, cells are labeled with ortho[^{32}P]phosphate (ICN, Irvine, CA) in phosphate-free medium prior to protein isolation. The resulting phosphoprotein display is evaluated as a function of time to delineate signaling

[39] M. Unlu, M. E. Morgan, and J. S. Minden, *Electrophoresis* **18,** 2071 (1997).

[40] V. Glaser, *Genet. Eng. News* **20,** 1 (2000).

[41] A. D. Cox, P. A. Solski, J. D. Jordan, and C. J. Der, *Methods Enzymol.* **255,** 195 (1995).

[42] J. F. Hancock, *Methods Enzymol.* **255,** 60 (1995).

networks. The combination of traditional colorimetric staining with ^{32}P labeling is powerful, as protein expression levels can be correlated with the amount of radioactive phosphate labeling. In addition, many other posttranslational modifications such as glycosylation[43] and lipidation[20,41,44] are easily monitored by tritiated labeling techniques. It should be noted that silver-staining techniques may decrease the level of radioactive labels in the gels, and should be avoided if autoradiography is the goal.

An alternative to 2D-PAGE and protein differential display is a mass spectrometric technique using isotope-coded affinity tags, or ICATs.[45] This is a quantitative technique that provides a meaningful comparison of the protein expression levels between two samples. The two different samples are covalently derivatized through the protein side-chain thiols with molecules that contain a retrieval tag, such as biotin. One of the tags is unlabeled and the other is labeled with deuterium. After derivatization the samples are mixed, proteolyzed, and purified by avidin chromatography. Biotinylated peptides are observed in the mass spectrum. The ratio of the unlabeled to deuterium-labeled peptides reveals the relative expression levels of the parent proteins.

Analysis of Proteins from Two-Dimensional Polyacrylamide Gel Electrophoresis

Protein Identification–Peptide Mass Fingerprinting

Two approaches to identification of a protein directly from a gel spot include microsequencing and mass spectrometry of in-gel proteolytically generated peptides. For microsequencing, it is necessary to purify the peptides by chromatography to obtain a single peptide per sequencing run. In contrast, the mixture of peptides can be analyzed directly by mass spectrometry. A sensitive and rapid technique is to use matrix-assisted laser desorption/ionization time-of-flight (MALDI-TOF) mass spectrometry. Using this technique all peptides in the digest are analyzed simultaneously, thereby providing a peptide mass fingerprint of the protein that was proteolyzed. This "fingerprint" is a list of peptide masses and depends on the actual protein sequence and the protease used to digest the protein. This peptide mass fingerprint (list of masses) can be used to search a database that contains theoretical mass fingerprints of virtually all known proteins.

[43] R. S. St Jules, S. B. Smith, and P. J. O'Brien, *Exp. Eye Res.* **51,** 427 (1990).

[44] A. Wolven, W. van't Hof, and M. D. Resh, *Methods Mol. Biol.* **84,** 261 (1998).

[45] S. P. Gygi, B. Rist, S. A. Gerber, F. Turecek, M. H. Gelb, and R. Aebersold, *Nat. Biotechnol.* **17,** 994 (1999).

Several parameters must be optimized in order to obtain reliable protein identification.[46]

One of the most critical parameters is mass accuracy of the spectrometer. Ideally, each sample is calibrated with internal mass standards to obtain accuracy within 0.01%. This is a stringent requirement and cannot always be obtained, especially for samples that give weak signals. It is optimal to increase the number of peptide masses used in the search and this depends on the amount of protein digested. Spots containing 1 μg of protein will typically yield 10–20 intense tryptic peptides. Spots containing 10 ng may only yield two or three peptides. With a mass accuracy of 0.01% as few as three or four peptides can provide a confident result. As the accuracy decreases the number of peptide masses required increases. Note that it is always necessary to proteolyze a blank gel spot and to remove these masses from the sample mass list. Additional criteria that should be used in the search are the molecular weight of the protein and the source organism. These can dramatically improve the identification of the protein. In many cases a list of possible proteins is returned from the database. These results are ranked according to the number of peptides matched between the query mass list and the theoretical protein digest. Also, the deviation of experimental masses from the theoretical masses is considered.

If the sample is a novel protein, that is, it is not in the database, this approach may yield an ambiguous result. If there is too little protein in a given spot to yield enough peptides for a given mass, accuracy will suffer. An alternative approach is *de novo* sequencing by nanoelectrospray liquid chromatography mass spectrometry. The peptide mixture is separated by microbore or capillary high-performance liquid chromatography (HPLC) and the eluant is directly infused into a triple quadrupole mass spectrometer. This method is also quite sensitive and only one or two peptides need to be sequenced to identify a protein in the database. However, if the sample is a novel protein, a more powerful method is to use the peptide sequences to generate oligonucleotide probes for screening cDNA libraries. It is much easier to sequence an isolated (and amplifiable) cDNA clone than to obtain the complete amino acid sequence of a protein from a gel spot.

Peptide Mass Fingerprinting Protocol. Several excellent protocols for peptide mass fingerprinting can be found on the internet. We closely follow the protocol found at the University of California (San Francisco, CA; USCF) mass spectrometry facility (*http://donatello.ucsf.edu/ingel.html*). Briefly, protein is excised from the gel and the gel slice is washed to remove residual SDS and to bring the pH to a level optimum for the subsequent proteolytic digestion. A protease is added and after a certain period of

[46] P. Berndt, U. Hobohm, and H. Langen, *Electrophoresis* **20,** 3521 (1999).

incubation the peptides are recovered and subjected to mass spectrometry. An innovation is the use of C_{18} ZipTips (Millipore, Bedford, MA). These small reversed-phase pipettor tips permit desalting and near quantitative recovery of peptides from digests. The enhancement of the signal-to-noise ratio is dramatic.

Databases

Once the peptide mass fingerprint has been obtained, the mass list is used to query databases containing theoretical digests of proteins. Many different databases are available on the internet. An excellent review of these is found in Kellner *et al.*[47] Currently, one of the most comprehensive databases can be found at ProteinProspector at UCSF (*http://prospector.ucsf.edu/*).

Other databases of importance for protein analysis include the ExPASy (Expert Protein Analysis System) proteomics server of the Swiss Institute of Bioinformatics (*http://www.expasy.ch/* or *http://expasy.cbr.nrc.ca/*) and the Protein Data Bank (*http://nist.rcsb.org/pdb/*), which contains three-dimensional structures of proteins. Each of the sites mentioned in this chapter is comprehensive and extensively hyperlinked to other useful sources.

Acknowledgments

A.D.C. acknowledges support from NIH grants CA76092 and CA67771. L.G.T. received support from an NIH short-term research fellowship and the Distinguished Medical Scholar Program of UNC-CH.

[47] R. Kellner, F. Lottspeich, and H. E. Meyer (eds.), "Microcharacterization of Proteins," 2nd Ed. Wiley-VCH, New York, 1999.

Section III

Analyses of Mitogen-Activated Protein Kinase Cascades

[24] Analyzing JNK and p38 Mitogen-Activated Protein Kinase Activity

By ALAN J. WHITMARSH and ROGER J. DAVIS

Introduction

The JNK and p38 mitogen-activated protein (MAP) kinase groups are important for many physiological processes including cell growth, oncogenic transformation, cell differentiation, apoptosis, and the immune response.[1-3] In particular, these protein kinases mediate the cellular response to stress signals. JNK and p38 MAP kinases are activated by signaling modules that consist of a MAP kinase kinase (MKK) and a MAP kinase kinase kinase (MKKK).[1-3] The Rho family small GTPases have been demonstrated to trigger activation of the JNK and p38 signaling modules and appear to coordinate the activation of these protein kinases with changes in the cytoskeleton.[4-8]

To understand the role of JNK and p38 MAP kinases in the cell, several protocols for measuring their activation state and protein kinase activity have been used. These include immune-complex protein kinase assays, solid-phase protein kinase assays, *in situ* detection of protein kinase activity after sodium dodecyl sulfate–polyacrylamide gel electrophoresis (SDS–PAGE), and immunoblotting or immunocytochemistry using phosphorylation-specific antibodies. In this chapter, we describe these different techniques and the use of reporter gene assays to monitor JNK and p38 activity. In addition, a number of tools have been developed to specifically activate or inhibit the JNK and p38 MAP kinase signaling pathways *in vivo* and we discuss their relevance for analyzing particular aspects of these signaling pathways.

[1] Y. T. Ip and R. J. Davis, *Curr. Opin. Cell Biol.* **10,** 205 (1998).
[2] L. A. Tibbles and J. R. Woodgett, *Cell. Mol. Life Sci.* **55,** 1230 (1999).
[3] T. S. Lewis, P. S. Shapiro, and N. G. Ahn, *Adv. Cancer Res.* **74,** 49 (1998).
[4] O. A. Coso, M. Chiariello, J.-C. Yu, H. Teramoto, P. Crespo, N. Xu, T. Miki, and J. S. Gutkind, *Cell* **81,** 1137 (1995).
[5] A. Minden, A. Lin, F.-X. Claret, A. Abo, and M. Karin, *Cell* **81,** 1147 (1995).
[6] M. F. Olson, A. Ashworth, and A. Hall, *Science* **269,** 1270 (1995).
[7] S. Bagrodia, B. Dérijard, R. J. Davis, and R. Cerione, *J. Biol. Chem.* **270,** 27995 (1995).
[8] S. Zhang, J. Han, M.-A. Sells, J. Chernoff, U. G. Knaus, R. J. Ulevitch, and G. M. Bokoch, *J. Biol. Chem.* **270,** 12665 (1995).

METHODS IN ENZYMOLOGY, VOL. 332
0076-6879/00 $35.00

Measuring JNK and p38 Protein Kinase Activity

The standard assay that is used to measure the protein kinase activity of JNK and p38 in cells is the immune-complex protein kinase assay.[9,10] This involves immunoprecipitating JNK or p38 from cells, using appropriate antibodies that are bound to protein G or A–Sepharose (the choice of protein G or A–Sepharose depends on the species and subclass of the antibody[11]). The immunoprecipitates are then incubated with substrate protein and [γ-^{32}P]ATP. The incorporation of [^{32}P]phosphate into the substrate protein is quantitated after SDS–PAGE. This is the method of choice for measuring the protein kinase activity of exogenously expressed JNK and p38 in tissue culture cell lines. The expression plasmids for JNK or p38 are constructed to allow their fusion to an amino-terminal epitope tag [e.g., FLAG-Tag or hemagglutinin (HA)-Tag]. The protein kinases are then immunoprecipitated with antibodies raised against the epitope tag [M2 monoclonal antibody recognizes FLAG and 12CA5 monoclonal antibody recognizes HA; available from Sigma (St. Louis, MO) and Boehringer Mannheim (Indianapolis, IN) respectively]. To measure the protein kinase activity of endogenous JNK we use either a rabbit polyclonal antibody raised against the full-length JNK1α1 isoform[10] or a goat polyclonal antibody (C17; available from Santa Cruz Biotechnologies, Santa Cruz, CA) that is raised against the carboxyl terminus of the JNK1α1 isoform. To measure endogenous p38 protein kinase activity, we use a rabbit polyclonal antibody raised against the full-length p38α isoform.[10] The specificity of the antibody used will determine which particular isoforms of JNK or p38 are immunoprecipitated. We are not aware of any commercial antibodies that immunoprecipitate all the JNK or p38 isoforms, therefore immune-complex protein kinase assays measure the protein kinase activity of particular isoforms as opposed to the activity of all the JNK or p38 isoforms in the cell. While the antibodies we use predominantly immunoprecipitate JNK1α1 and p38α, it should be noted that many commercial antibodies that are raised against particular isoforms are not completely specific for those isoforms and show some cross-reactivity with additional isoforms.

The combined protein kinase activities of p38α and p38β2 isoforms can be measured indirectly by the immune-complex protein kinase assay. The mitogen-activated protein kinase-activated protein kinase 2 (MAP-

[9] B. Dérijard, M. Hibi, I.-H. Wu, T. Barrett, B. Su, T. Deng, M. Karin, and R. J. Davis, *Cell* **76,** 1025 (1994).

[10] J. Raingeaud, S. Gupta, J. Rogers, M. Dickens, J. Han, R. J. Ulevitch, and R. J. Davis, *J. Biol. Chem.* **270,** 7420 (1995).

[11] E. Harlow and D. Lane, "Using Antibodies: A Laboratory Manual." Cold Spring Harbor Laboratory Press, Cold Spring Harbor, New York, 1999.

KAPK-2) is a specific substrate for p38α and p38β2 *in vivo*[12] and can be immunoprecipitated from cells with a polyclonal antibody (available from Upstate Biotechnology, Lake Placid, NY) and its activity measured *in vitro* using either a synthetic peptide substrate or recombinant heat shock protein 27 (Hsp27). The protein kinase activity of MAPKAPK-2 is a measure of the p38α and p38β2 protein kinase activity in the cell. This assay is useful if the experiment involves inhibitors of p38 protein kinase activity (see below), as the direct immunoprecipitation of p38 may result in the inhibitor being washed away so its effect on p38 protein kinase activity in the cell cannot be assessed.

An alternative to the immune-complex protein kinase assay is the solid-phase protein kinase assay.[13] This assay relies on the ability of the protein kinase to bind to its substrate prior to phosphorylation. It also requires that the substrate used binds to a unique kinase. JNK appears to be the only protein kinase that forms a stable association with the amino-terminal *trans*-activation domain of the transcription factor c-Jun. Indeed, this property of JNK led to the initial characterization of its activity.[13] On the other hand, the known substrates of p38 are either also substrates for other protein kinases *in vitro*, or do not form stable complexes with p38. For these reasons, the assay can be used to measure JNK protein kinase activity but is not suitable for measuring p38 protein kinase activity. An important distinction between this assay compared with the immune-complex protein kinase assay is that, as all the JNK isoforms bind to c-Jun,[14] it is a measure of their combined activities. In this assay the fusion protein glutathione *S*-transferase (GST)–cJun (amino acids 1–79) is immobilized on glutathione–Sepharose and then incubated with cell lysates to allow endogenous JNK to bind to the immobilized GST–cJun. The protein kinase activity of the bound JNK is measured by adding [γ-^{32}P]ATP and quantitating the [^{32}P]phosphate incorporation into GST–cJun.

A third type of protein kinase assay, which unlike those previously described does not require the isolation of the protein kinase from the cell lysates, is the *in situ* detection of protein kinase activity after SDS–PAGE (often referred to as the "in gel" protein kinase assay).[13,15,16] In this assay the substrate protein is polymerized into an SDS–polyacrylamide gel. Cell lysates are separated on the gel and treated to a denaturation/renaturation

[12] P. Cohen, *Trends Cell Biol.* **7,** 353 (1997).

[13] M. Hibi, A. Lin, T. Smeal, A. Minden, and M. Karin, *Genes Dev.* **7,** 2135 (1993).

[14] S. Gupta, T. Barrett, A. J. Whitmarsh, J. Cavanagh, H. K. Sluss, B. Dérijard, and R. J. Davis, *EMBO J.* **15,** 2760 (1996).

[15] I. Kameshita and H. Fujisawa, *Anal. Biochem.* **183,** 139 (1989).

[16] Y. Gotoh, E. Nishida, T. Yamashita, M. Hoshi, M. Kawakami, and H. Sakai, *Eur. J. Biochem.* **193,** 661 (1990).

cycle prior to incubation with [γ-^{32}P]ATP. Protein kinases phosphorylate the polymerized substrate in the area of the gel where they migrate. The substrate does not have to be unique to a particular protein kinase if the protein kinases that phosphorylate it migrate at different positions in the SDS–polyacrylamide gel. It should be noted that signals may be observed independently of the polymerized substrate, because of the autophosphorylation of protein kinases from the lysate. Therefore an important control experiment is to separate the lysates on a gel that does not include the polymerized substrate. This will allow the signals arising from autophosphorylating protein kinases and those arising from phosphorylation of the polymerized substrate to be distinguished.

This assay can be used to analyze JNK protein kinase activity and measures the activity of all the isoforms in the cell. The JNK isoforms comigrate in two groups at 46 and 55 kDa, respectively, and in most cell types the 46-kDa activity is mainly contributed by JNK1 isoforms and the 55-kDa activity by JNK2 isoforms (JNK3 isoforms are generally present only in brain, testis, heart, and neuronal-like cell lines). Therefore, the relative activities of these isoforms can be assessed by this assay. In our hands, the assay is not useful for measuring p38 protein kinase activity, presumably because p38 is poorly refolded during the renaturation step.

The protein kinase assays that have been discussed are described in detail below for measuring JNK protein kinase activity from COS-7 cell lysates, but the methods can equally well be applied to measuring JNK (and p38, using the immune-complex protein kinase assay) activity from other immortalized cell lines, primary cells, or animal tissues.

Materials and Reagents

Buffers

Triton lysis buffer (TLB): 20 m*M* Tris–HCl (pH 7.4), 137 m*M* NaCl, 25 m*M* sodium β-glycerophosphate, 2 m*M* sodium pyrophosphate, 2 m*M* EDTA, 1 m*M* sodium vanadate, 10% (v/v) glycerol, 1% (v/v) Triton X-100, 1 m*M* phenylmethylsulfonyl fluoride (PMSF), leupeptin (5 μg/ml), aprotinin (5 μg/ml)

Kinase assay buffer (KB): 25 m*M* HEPES (pH 7.4), 25 m*M* sodium β-glycerophosphate, 25 m*M* $MgCl_2$, 0.1 m*M* sodium vanadate, 0.5 m*M* dithiothreitol (DTT)

Buffer A: 50 m*M* Tris–HC1 (pH 8.0), 5 m*M* 2-mercaptoethanol

Buffer B: 40 m*M* HEPES (pH 7.4), 5 m*M* $MgCl_2$, 0.1 m*M* EGTA, 2 m*M* DTT

Phosphate-buffered saline (PBS): 20 m*M* sodium phosphate (pH 7.4), 150 m*M* NaCl

2-Propanol (20%, v/v)–50 m*M* Tris–HCl (pH 8.0)
Trichloroacetic acid (TCA, 5%, w/v)–sodium pyrophosphate (1%, w/v)
SDS–PAGE sample buffer
Coomassie stain and destain

Reagents

Protein G–Sepharose (Pharmacia-LKB, Piscataway, NJ) resuspended in 10 m*M* HEPES (pH 8.0)–0.025% (v/v) Triton X-100 to form a 50% slurry
Glutathione–Sepharose 4B (Pharmacia-LKB)
GST and GST–cJun (amino acids 1–79) proteins; expressed in *Escherichia coli* and purified by affinity chromatography using glutathione–Sepharose 4B[17]
Goat anti-JNK1 antibody (C17; Santa Cruz Biotechnologies)
[γ-^{32}P]ATP (10 Ci/mmol; New England Nuclear, Boston, MA)
ATP (Sigma)
Polyacrylamide (10%, w/v) gels
Guanidine hydrochloride (Sigma)
Tween 40 (United States Biochemical, Cleveland, OH)
COS-7 cells (American Type Culture Collection, Manassas, VA)

Equipment

Microcentrifuge and tubes (1.5 ml)
Rotating platform
Shaking platform
Heating blocks, 30 and 100°
Gel dryer
PhosphorImager (Molecular Dynamics, Sunnyvale, CA)

Immune-Complex Protein Kinase Assay

The protocol described below measures endogenous JNK1 protein kinase activity in COS-7 cells that have been grown in six-well dishes (35-mm diameter). The same protocol, with the indicated modifications, can be used to measure p38α protein kinase activity.

1. Wash the COS-7 cells in ice-cold PBS (3 ml per well) and then scrape into TLB (200 μl per well). Transfer the lysate to 1.5-ml microcentrifuge tubes. Allow lysis to proceed on ice for 10 min. After lysis, remove the insoluble material by centrifugation at 15,000 rpm for 15 min at 4°. (To

[17] S. B. Smith and K. S. Johnson, *Gene* **67,** 31 (1988).

lyse tissue samples, homogenize the tissue in TLB, using a Dounce homogenizer, and allow lysis to proceed on ice for 30 min prior to centrifugation).

2. Prebinding of antibody to protein G–Sepharose: Incubate 1 μg of anti-JNK1 antibody in a microcentrifuge tube with 400 μl of TLB and 20 μl of a 50% slurry of protein G–Sepharose (use 1 μg of p38α antibody with protein A–Sepharose to precipitate p38α, and 1 μg of M2 or 12CA5 antibody with protein G–Sepharose to precipitate FLAG or HA epitope-tagged protein kinases, respectively). Perform binding for 30 min at 4° on a rotating platform. Wash the antibody-bound protein G–Sepharose twice with 1 ml of TLB. Between the washes, collect the Sepharose beads at the bottom of the microcentrifuge tube by pulse centrifugation and aspirate the buffer.

3. Add the clarified cell lysate obtained in step 1 to the antibody-bound protein G–Sepharose and make up the total volume to 400 μl with TLB. Perform the immunoprecipitation for 3 hr at 4° on a rotating platform (overnight incubation is also acceptable).

4. Wash the protein G–Sepharose beads three times with 1 ml of TLB and twice with 1 ml of KB. Make sure all the excess buffer is removed from the beads after the final wash.

5. Protein kinase assay: Add 2 μg of GST–cJun (amino acids 1–79) and 50 μ*M* [γ-^{32}P]ATP (10 Ci/mmol) to the 10 μl of compacted protein G–Sepharose beads from step 4, and adjust the total volume to 25 μl with KB (to measure p38 protein kinase activity use GST–ATF2 (amino acids 1–109) as a substrate). Incubate the reaction mix at 30° for 30 min. Gently tap the tubes every 10 min to mix the contents. Terminate the protein kinase reaction by adding 10 μl of SDS–PAGE sample buffer.

6. Heat the reaction mix at 100° for 5 min and then examine by 10% (w/v) SDS–PAGE. After electrophoresis, stain the gel with Coomassie stain to visualize the substrate, destain, and dry the gel. Quantitate the incorporation of [^{32}P]phosphate into GST–cJun (amino acids 1–79) by analysis on a PhosphorImager.

Solid-Phase Protein Kinase Assay

The protocol we describe below measures the overall JNK protein kinase activity in COS-7 cells grown in six-well plates (35-mm diameter). This protocol is adapted from Hibi *et al.*[13] It is not suitable for measuring p38 protein kinase activity.

1. Binding of substrate to glutathione–Sepharose: Add 10 μg of GST–cJun (amino acids 1–79) to 400 μl of TLB in a 1.5-ml microcentrifuge tube with 25 μl of a 50% slurry of glutathione–Sepharose (that has been washed

with several changes of TLB). Incubate for 60 min at 4° on a rotating platform and then wash the GST–cJun-bound glutathione–Sepharose twice with 1 ml of TLB.

2. Prepare COS-7 cell lysates as described in step 1 of the immune-complex protein kinase assay protocol. Incubate the clarified lysates with the GST–cJun-bound glutathione–Sepharose in a total volume of 400 μl of TLB overnight at 4° on rotating platform. The next day wash the glutathione–Sepharose beads three times with 1 ml of TLB and twice with 1 ml of KB.

3. Protein kinase assay: Add 10 μl of KB supplemented with 50 μM [γ-^{32}P]ATP (10 Ci/mmol) to the GST–cJun-bound glutathione–Sepharose. Incubate the reaction mix at 30° for 30 min. Gently tap the tubes every 10 min to mix the contents. Terminate the reaction by adding 10 μl of SDS–PAGE sample buffer.

4. Process the assay samples as described in step 6 of the immune-complex protein kinase assay protocol and quantitate the incorporation of [^{32}P]phosphate into GST–cJun (amino acids 1–79) by PhosphorImager analysis.

In Situ Detection of Protein Kinase Activity after Sodium Dodecyl Sulfate–Polyacrylamide Gel Electrophoresis

The protocol described below is for measuring JNK protein kinase activity in COS-7 cells grown in six-well dishes (35-mm diameter). This protocol is adapted from Gotoh *et al.*[16] It is not suitable for the analysis of p38 protein kinase activity.

1. Prepare COS-7 cell lysates as described in step 1 of the immune-complex protein kinase assay protocol.

2. Prepare 10% (w/v) polyacrylamide gels that include either GST–cJun (amino acids 1–79) (0.25 mg/ml) or GST (0.25 mg/ml) in the gel mix (the gel with polymerized GST acts as a control to distinguish which signals arise from the phosphorylation of the polymerized GST–cJun). Heat the cell lysate samples containing between 20 and 50 μg of protein (from step 1) in SDS–PAGE sample buffer at 100° for 5 min, prior to loading on the gels.

3. Remove the SDS from the gels after electrophoresis by washing twice in 100 ml of 20% (v/v) 2-propanol–50 mM Tris–HCl (pH 8.0). Then wash the gels twice in 100 ml of buffer A. Perform all the washes for 30 min each at room temperature with gentle shaking.

4. Denaturation/renaturation: Incubate the gels for two 30-min washes in 100 ml of 6 M guanidine hydrochloride in buffer A (it is important to

filter this solution prior to incubation with the gel) at room temperature with gentle shaking. Renature the gels by gentle shaking overnight at 4° with several changes of 200 ml of 0.04% (v/v) Tween 40 in buffer A.

5. The next day wash the gels in 25 ml of buffer B for 30 min at room temperature with gentle shaking.

6. Protein kinase assay: Incubate the gels in 25 ml of buffer B plus 0.25 μM [γ-^{32}P]ATP (10 Ci/mmol) for 60 min at room temperature with gentle shaking.

7. Wash the gels by gentle shaking at room temperature with several changes of 200 ml of 5% (w/v) TCA–1% (w/v) sodium pyrophosphate until the radioactivity from around the edges of the gel reaches background levels (detected with a Geiger counter). This may take from 12 to 24 hr. Dry the gels and measure the incorporation of [^{32}P]phosphate, using a PhosphorImager.

Measuring JNK and p38 Activation Using Phosphorylation-Specific Antibodies

In response to stress signals, JNK and p38 are phosphorylated by MKKs on threonine and tyrosine residues within a Thr-X-Tyr motif located in protein kinase subdomain VIII.[1-3] The phosphorylation of these residues is required for the activation of JNK and p38 protein kinase activity.[1-3] Antibodies that recognize only the phosphorylated (i.e., activated) forms of JNK and p38 are commercially available and can be used to assay the activation state of JNK and p38 in cells. The phosphorylation sites and the sequences surrounding them are conserved within the JNK and p38 families, respectively. Therefore the antibodies should recognize all the isoforms present in the cell. The advantage of this technique (in addition to not requiring radioactivity) is that it can be used to measure JNK and p38 activation in whole cell lysates by immunoblotting, and to examine the localization of activated JNK and p38 in cells or tissues by immunocytochemistry. It should be noted, however, that this technique is not a measure of JNK and p38 protein kinase activity. In many cases, the level of phosphorylation of JNK and p38 will correlate with the level of protein kinase activity, but other factors in the cell may alter JNK and p38 protein kinase activity without affecting their phosphorylation. Because of the unique specificity of JNK for its substrate cJun, the protein kinase activity of JNK can be measured by using antibodies that specifically recognize c-Jun phosphorylated at the JNK phosphorylation sites.

Potential caveats associated with the use of phosphorylation-specific antibodies are (1) it is a less quantitative technique than the protein kinase assays described previously, and (2) the antibodies may not be totally specific. The second point should be borne in mind particularly for immunocytochemistry studies using these antibodies. Control competition experiments using blocking peptides should be performed to verify the specificity of the antibody.

Below we describe immunoblotting and immunocytochemistry protocols for detecting JNK phosphorylation using phospho-SAPK/JNK (Thr-183/Tyr-185) antibody (New England BioLabs, Beverly, MA) and p38 phosphorylation using Anti-ACTIVE p38 (pTGpY) antibody (Promega, Madison, WI).

Materials and Reagents

Buffers

Western transfer buffer (WTB): 25 m*M* Tris, 200 m*M* glycine, 20% (v/v) methanol, 0.1% (w/v) SDS
Western blotting buffer (WBB): 15 m*M* Tris–HCl (pH 7.4), 150 m*M* NaCl
Phosphate-buffered saline (PBS): 20 m*M* sodium phosphate (pH 7.4), 150 m*M* NaCl
Paraformaldehyde (3%, w/v)–PBS (per 10 ml): Dissolve 0.3 g of paraformaldehyde in 8.5 ml of water and 20 μl of 10 *M* NaOH. Once dissolved, add 1 ml of 10× PBS, 50 μl of $MgCl_2$ and 210 μl of 1 *M* HCl. Check that the pH is between pH 7.2 and 7.4
Tris-buffered saline (TBS): 50 m*M* Tris–HCl (pH 7.4), 150 m*M* NaCl
Tris-buffered saline plus Triton X-100 (TBST): 50 m*M* Tris–HCl (pH 7.4), 150 m*M* NaCl, 0.1% (v/v) Triton X-100
Blocking buffer: 3% (w/v) bovine serum albumin–TBS

Reagents

Methanol
Horse serum (gamma globulin free; Life Technologies, Rockville, MD)
Tween 20 (United States Biochemical)
SDS–10% (w/v) polyacrylamide gels
Enhanced chemiluminescence (ECL) reagents (New England Nuclear)
Paraformaldehyde (Sigma)
Bovine serum albumin (Boehringer Mannheim)
Vectashield (Vector Laboratories, Burlingame, CA)
Rabbit polyclonal phospho-SAPK/JNK (Thr-183/Tyr-185) (New England BioLabs) and Anti-ACTIVE p38 (pTGpY) (Promega) antibodies
Texas Red-conjugated anti-rabbit Ig secondary antibody (Jackson Immunoresearch, West Grove, PA)

Anti-rabbit Ig secondary antibody (Amersham, Arlington Heights, IL)
4,6-Diamidino-2-phenylindole (DAPI; Molecular Probes, Eugene, OR)

Equipment

Semidry Western transfer apparatus (Hoefer Scientific, San Francisco, CA)
Parafilm (American National Can, Chicago, IL)
Coverslips (Corning, Acton, MA)
Glass slides (VWR, San Francisco, CA)
Immobilon-P transfer membrane (Millipore, Bedford, MA)
Gel blot paper (Schleicher & Schuell, Keene, NH)
Zeiss (Thornwood, NY) Axiophot microscope
X-ray film

Immunoblotting Protocol

The method we describe is an alternative to the antibody manufacturer protocol for detecting activated JNK using phospho-SAPK/JNK (Thr-183/Tyr-185) antibody. We recommend adhering to the manufacturer protocol for detecting activated p38 using the Anti-ACTIVE p38 (pTGpY) antibody.

1. Perform SDS–PAGE [10% (w/v) gel] using 50–100 μg of cell or tissue lysate per sample.
2. Soak the gel in WTB for 5 min. Immerse the Immobilon-P transfer membrane in methanol for a few seconds to wet it and then soak it in WTB for 5 min. Set up the Western transfer. Sandwich the gel and membrane with two sheets of gel blot paper on either side, and perform the transfer (2–3 hr at 15 V, using the semidry apparatus).
3. Blocking: Incubate the membrane with 200 ml of 20% (v/v) horse serum–WBB for 2 hr at room temperature with gentle shaking.
4. Primary antibody incubation: Incubate the membrane overnight at 4° with gentle shaking in a 1:1000 dilution of phospho-SAPK/JNK (Thr-183/Tyr-185) antibody in 20% (v/v) horse serum–WBB.
5. The following day, wash the membrane twice for 10 min in 100 ml of 0.5% (v/v) Tween 20–WBB at room temperature with shaking.
6. Secondary antibody incubation: Incubate the membrane with anti-rabbit Ig secondary antibody diluted 1:10,000 in 20% (v/v) horse serum–WBB for 30 min at room temperature with gentle shaking.
7. Wash the membrane four times for 15 min with 0.5% (v/v) Tween 20–WBB at room temperature with shaking, and then for 5 min with WBB.

8. Develop the immunoblot using ECL reagents and detect the signal by exposing to X-ray film.

Immunocytochemistry Protocol

The method described below can be used to detect activated JNK, using phospho-SAPK/JNK (Thr-183/Tyr-185) antibody, and activated p38, using Anti-ACTIVE p38 (pTGpY) antibody, in cells grown on glass coverslips in six-well dishes (35-mm diameter). This protocol is adapted from the manufacturer protocol for immunocytochemistry using the phospho-SAPK/JNK (Thr-183/Tyr-185) antibody.

1. Aspirate the medium and wash the cells with PBS (3 ml per well).
2. Fix the cells with 1 ml of 3% (w/v) paraformaldehyde–PBS for 10 min at 4°. Wash the cells with TBST (2 ml per well) twice for 5 min at room temperature with shaking, and then rinse the cells with TBS.
3. Permeabilize the cells by incubating with 1 ml of methanol (prechilled at −20°) for 10 min at −20°, and then wash twice for 10 min in 2 ml of TBST at room temperature with shaking.
4. Blocking: Incubate the cells with 2 ml of blocking buffer for 60 min at room temperature with gentle shaking.
5. Primary antibody incubation: Spot 35 μl of either diluted phospho-SAPK/JNK (Thr-183/Tyr-185) antibody or Anti-ACTIVE p38 (pTGpY) antibody (diluted 1:100 with blocking buffer) onto Parafilm and carefully lay the coverslips on top. Place the Parafilm with the coverslips in a damp environment (to prevent drying out, a 150-mm-diameter tissue culture dish covered in aluminum foil is useful) overnight at 4°.
6. The next day place the coverslips back into the six-well tissue culture dish and wash three times for 5 min with 2 ml of TBST with shaking, and once for 5 min with TBS.
7. Secondary antibody incubation (in the dark): Spot 35 μl of diluted Texas red-conjugated anti-rabbit Ig secondary antibody (diluted 1:100 in blocking buffer) onto Parafilm and carefully lay the coverslips on top. Incubate in the dark at room temperature for 45 min. All the subsequent steps must be performed in the dark because of the light sensitivity of the secondary antibody.
8. DNA staining: Place the coverslips back in the six-well dish and rinse with TBS. Incubate the coverslips with DAPI DNA stain (diluted 1:10,000 in TBS) for 2 min at room temperature.
9. Wash the coverslips three times for 10 min in 2 ml of TBST at room temperature with shaking, and then for 5 min in 2 ml of TBS. Wrap the six-well dishes in aluminum foil during the washes.

10. Lay the coverslips on top of a drop of Vectashield that is spotted onto glass slides and visualize the staining pattern of phosphorylated JNK by fluorescence microscopy.

Reporter Gene Assays for Assessing JNK and p38 Activity in Cells

The activation of the JNK and p38 MAP kinase signaling pathways in cells can be measured by reporter gene assays. The principle of the assay is that the protein kinase phosphorylates the transcriptional activation domain of an exogenously expressed transcription factor substrate that is fused to the DNA-binding domain of the yeast transcription factor GAL4. These fusion proteins bind to a cotransfected reporter plasmid that features five GAL4-binding sites upstream of a minimal promoter and the firefly luciferase gene. The phosphorylation of the fusion protein by the protein kinase leads to increased transcription of the luciferase gene and the resulting luciferase protein can be assayed enzymatically with a luminescent substrate. JNK signaling activity can be measured by transfecting cells with a construct expressing a GAL4–cJun (amino acids 1–79) fusion protein and the reporter plasmid.[13] Using this assay to measure p38 activity is more problematic as most p38 transcription factor substrates are targeted by other protein kinases. GAL4–ATF2 (amino acids 1–109) works well for measuring either JNK or p38 activity if specific activators of these signaling pathways are used[18,19] (see below). The combined activities of p38α and p38β2 isoforms can be measured with GAL4–MEF2A (amino acids 266–413).[20] While other protein kinases may target MEF2A (e.g., ERK5), control experiments using a specific inhibitor of p38α and p38β2 (see below) can be performed to verify the assay. A reporter system that measures the *trans*-activation of the p38 substrate CHOP[21] fused to GAL4 can also be used to assess the activity of the p38 signaling pathway (available from Stratagene, La Jolla, CA).

In the protocol described below, Chinese hamster ovary (CHO) cells expressing GAL4–cJun (amino acids 1–79) are used to measure the activity of the JNK signaling pathway. To control for differences in expression levels and transfection efficiency between samples, we also transfect a plamid featuring a β-galactosidase reporter gene and measure β-galactosidase activity in the cell lysates. A similar protocol can be performed with

[18] S. Gupta, D. Campbell, B. Dérijard, and R. J. Davis, *Science* **267,** 389 (1995).

[19] J. Raingeaud, A. J. Whitmarsh, T. Barrett, B. Dérijard, and R. J. Davis, *Mol. Cell. Biol.* **16,** 1247 (1996).

[20] S.-H. Yang, A. Galanis, and A. D. Sharrocks, *Mol. Cell. Biol.* **19,** 4028 (1999).

[21] X. Wang and D. Ron, *Science* **272,** 1347 (1996).

other GAL4 fusion proteins to monitor JNK and p38 activity, and with other cell lines and primary cells that are capable of being transfected.

Materials and Reagents

Buffers

Phosphate-buffered saline (PBS): 20 m*M* sodium phosphate (pH 7.4), 150 m*M* NaCl
Potassium phosphate buffer, pH 7.8
Luciferase assay buffer: 15 m*M* potassium phosphate (pH 7.8), 25 m*M* glycylglycine (pH 7.8), 15 m*M* Mg_2SO_4, 4 m*M* EGTA, 2 m*M* ATP (made as 100 m*M* stock and adjusted to pH 7.8), 1 m*M* DTT
Buffer Z: 100 m*M* sodium phosphate (pH 7.0), 10 m*M* KCl, 1 m*M* Mg_2SO_4
Na_2CO_3, 1 *M*

Reagents

D-Luciferin (PharMingen, San Diego, CA): Make a 1 m*M* stock solution by dissolving 10 mg of D-luciferin in 31.5 ml of water and adjusting to pH 6–6.3 by adding 100 m*M* NaOH solution dropwise. Store frozen
o-Nitrophenyl-β-D-galactopyranoside (ONPG; United States Biochemical): 4 mg/ml in 100 m*M* potassium phosphate, pH 7.0; filter sterilize and store frozen
pSG424–cJun: Encodes a fusion of cJun amino acids 1–79 with the GAL4 DNA-binding domain[13]
pG5E1bLuc: Firefly luciferase gene reporter plasmid featuring five GAL4-binding sites and the minimal E1b promoter[22]
pCH110: β-Galactosidase reporter plasmid (Pharmacia-LKB)
Chinese hamster ovary (CHO) cells: American Type Culture Collection
LipofectAMINE (Life Technologies)

Equipment

Microcentrifuge tubes (1.5 ml)
luminometer cuvettes
luminometer
Heating block, 37°
spectrophotometer

[22] S. Gupta, A. Seth, and R. J. Davis, *Proc. Natl. Acad. Sci. U.S.A.* **90,** 3216 (1993).

Luciferase and β-Galactosidase Assays

1. Transfect CHO cells that are grown to 70% confluence in six-well dishes (35-mm diameter) with 0.25 μg of pSG424–cJun, 0.25 μg of pG5E1bLuc, and 0.25 μg of pCH110, using LipofectAMINE (as described by the manufacturer). Forty-eight hours after transfection, wash the cells three times with 3 ml of ice-cold PBS. Scrape the cells into 200 μl of ice-cold 15 m*M* potassium phosphate buffer, pH 7.8, and transfer to a 1.5-ml microcentrifuge tube.

2. Lysis by freeze–thawing: Incubate the cells on dry ice for 5 min and then thaw at 37° for 5 min. Repeat this three times. Remove the insoluble material by centrifugation at 15,000 rpm for 15 min at 4°.

3. Aliquot the luciferase assay buffer into luminometer cuvettes (300 μl per cuvette) and adjust the luminometer to inject 100 μl of the D-luciferin stock solution per sample.

4. Add the clarified cell lysate from step 2 (usually between 2 and 20 μl, depending on the expected signal) to the luciferase assay buffer in the cuvettes and read the luminescence values 10 sec after injection of the D-luciferin stock solution.

5. β-Galactosidase assay: Add 15 μl of the cell lysate from step 2 to 200 μl of buffer Z and 100 μl of ONPG. Incubate the reaction mix at 37° until a yellow color appears. Terminate the reaction by adding 500 μl of 1 *M* Na_2CO_3. Clarify the samples by centrifugation at 15,000 rpm for 5 min at room temperature and measure the optical density of the supernatants by spectrophotometry (420 nm).

Examination of Role of JNK and p38 Activity in Cells, Using Specific Activators and Inhibitors

The ability to specifically activate or inhibit the JNK and p38 signaling pathways in cells is critical for our understanding of their physiological roles. The overexpression of MAPKKKs such as the catalytic domain of MEKK1, the mixed-lineage protein kinases (MLKs), and TGF-β-activated kinase 1 (TAK1) results in constitutive JNK and p38 activation.[23] As these MKKKs, and other signaling molecules upstream of them, are capable of activating both JNK and p38 to some extent,[23] they are not that useful for uncovering the individual functions of these protein kinases. Similarly, overexpression of members of the dual-specificity MAP kinase phosphatase

[23] G. R. Fanger, P. Gerwins, C. Widmann, M. B. Jarpe, and G. L. Johnson, *Curr. Opin. Genet. Dev.* **7,** 67 (1997).

family[3] results in promiscuous dephosphorylation and inactivation of MAP kinases, rather than the specific inactivation of a particular group.

A number of tools have now been developed to probe the specific functions of JNK and p38 MAP kinases in cells. These include activated mutant alleles of components of these signaling pathways, dominant interfering mutant alleles of these protein kinases, and inhibitors (small molecules and proteins). These tools have been used to specifically activate or inhibit p38 or JNK in cells (or in animals using transgenes) and studies using them have provided important information on the function of these MAP kinases.

Activated Mutant Alleles

To specifically activate p38 in cells, activated alleles of the protein kinases that phosphorylate it (i.e., MKK3 and MKK6)[19] can be used. This is achieved by mutating the amino acids in these MKKs that are phosphorylated by upstream kinases (i.e., MKKKs) from serine and threonine to acidic residues such as glutamate. These amino acid substitutions mimic the negative charge that is associated with phosphorylation. The expression of the mutated forms of MKK3 and MKK6 in cells potently activates both exogenously expressed and endogenous p38.[19,24,25] In addition, mutant MKK6 transgenic animals have been used to address the function of p38 *in vivo*.[26] In our experience, similar mutations in MKK7 (a specific activator of JNK) and MKK4 (an activator of both JNK and p38) do not activate JNK or p38 any better than their wild-type counterparts. Transient overexpression of MKK7 leads to the activation of coexpressed JNK,[27] while transient overexpression of MKK4 leads to the activation of coexpressed JNK and p38.[28] In most expression systems, MKK4 and MKK7 are poor activators of endogenous MAP kinases. However, it has been reported that MKK4 can activate endogenous JNK when stably transfected into

[24] S. Huang, Y. Jiang, Z. Li, E. Nishida, P. Mathias, S. Lin, R. J. Ulevitch, G. R. Nemerow, and J. Han, *Immunity* **6,** 739 (1997).

[25] Y. Wang, S. Huang, V. P. Sah, J. Ross, Jr., J. H. Brown, J. Han, and K. R. Chien, *J. Biol. Chem.* **273,** 2161 (1998).

[26] M. Rincón, H. Enslen, J. Raingeaud, M. Recht, T. Zapton, M. S.-S. Su, L. A. Penix, R. J. Davis, and R. A. Flavell, *EMBO J.* **17,** 2817 (1998).

[27] C. Tournier, A. J. Whitmarsh, J. Cavanagh, T. Barrett, and R. J. Davis, *Proc. Natl. Acad. Sci. U.S.A.* **94,** 7337 (1997).

[28] B. Dérijard, J. Raingeaud, T. Barrett, I.-H. Wu, J. Han, R. J. Ulevitch, and R. J. Davis, *Science* **267,** 682 (1995).

fibroblasts,[29] and MKK7 can activate endogenous JNK when cardiac myocytes are infected with adenoviral expression vectors.[30]

The mutation of the phosphorylation sites on JNK and p38 to acidic residues does not result in activated alleles. However, a constitutively activated mutant of JNK has been described.[31] This was achieved by fusing MKK7 (also called JNKK2) to JNK1. This strategy was based on previous studies demonstrating that the fusion of the MKK MEK1 to the MAP kinase ERK2 resulted in a constitutively activated mutant ERK2[32]; therefore it is possible that a similar approach may produce a constitutively activated mutant p38.

Dominant-Interfering Mutant Alleles

Mutant MKK proteins can also be used to specifically block MAP kinase signaling pathways. Either the activating phosphorylation sites on the MKKs can be substituted with alanine residues, or catalytically inactive mutants can be engineered by mutating the catalytic site lysine to arginine or leucine. The expression in cells of mutant forms of the p38 activators MKK3 and MKK6 can selectively block the activation of p38 by extracellular stimuli or upstream signaling molecules.[19,33] Similar mutations in the JNK activator MKK7 can block JNK activation.[34] Mutated MKK4 (an activator of JNK and p38) predominantly blocks signaling to JNK,[35] although it has been reported to block p38 activation in some cell lines.[29] The mutation of the phosphorylation sites on JNK and p38 from threonine and tyrosine to alanine and phenylalanine, respectively, also results in mutant alleles that can block signaling through these pathways when expressed in cells.[9,25,26,36]

These dominant-negative mutants exert their effect by binding and sequestering upstream signaling components or substrates. Their efficiency in blocking JNK and p38 signaling can vary widely depending on the type

[29] Z. Guan, S. Y. Buckman, A. P. Pentland, D. J. Templeton, and A. R. Morrison, *J. Biol. Chem.* **273,** 12901 (1998).

[30] Y. Wang, B. Su, V. P. Sah, J. H. Brown, J. Han, and K. R. Chien, *J. Biol. Chem.* **273,** 5423 (1998).

[31] C. Zheng, J. Xiang, T. Hunter, and A. Lin, *J. Biol. Chem.* **274,** 28966 (1999).

[32] M. J. Robinson, S. A. Stippec, E. Goldsmith, M. A. White, and M. H. Cobb, *Curr. Biol.* **8,** 1141 (1998).

[33] F. Cong and S. P. Goff, *Proc. Natl. Acad. Sci. U.S.A.* **96,** 13819 (1999).

[34] T. Moriguchi, F. Toyoshima, N. Masuyama, H. Hanafusa, Y. Gotoh, and E. Nishida, *EMBO J.* **16,** 7045 (1997).

[35] A. J. Whitmarsh, S.-H. Yang, M. S.-S. Su, A. D. Sharrocks, and R. J. Davis, *Mol. Cell. Biol.* **17,** 2360 (1997).

[36] M. Rincón, A. J. Whitmarsh, D. D. Yang, L. Weiss, B. Dérijard, P. Jayaraj, R. J. Davis, and R. A. Flavell, *J. Exp. Med.* **188,** 1817 (1998).

of experiment, the stimulus, and the cell line that is used. Therefore, more than one mutant may need to be tested in a particular experiment and suitable controls performed to verify the specificity of the block that is observed.

Small Molecule and Protein Inhibitors

Two commonly used methods for inhibiting JNK and p38 signal transduction in cells involve overexpressing JNK-interacting protein 1 (JIP-1)[37] and treating cells with a family of pyridinylimidazole-based drugs[38] (the most widely used of which is SB203580, available from Calbiochem, La Jolla, CA), respectively. JIP-1 is a scaffold protein that binds components of the JNK MAP kinase pathway and enhances signaling through the pathway.[39] However, the overexpression of either the domain within JIP-1 that binds to JNK, or the full-length protein, potently inhibits JNK signaling in cells.[37,39] This is because JIP-1 blocks nuclear translocation of JNK and also sequesters JNK pathway components into different JIP-1 complexes. To date, scaffold proteins that regulate p38 signaling have not been characterized. Therefore, similar methods for inhibiting p38 signaling are not in use.

The pyridinylimidazole-based drug SB203580 inhibits the protein kinase activity of p38 isoforms α and $\beta 2$ by competing with ATP for binding to the ATP-binding pocket.[38] This drug has proved useful in understanding the role of these p38 isoforms in a number of physiological processes, as it can either be added to cells in culture or administered intravenously to animals.[12,38] It should be noted that the drug inhibits the protein kinase activity of p38α and p38$\beta 2$ but not their phosphorylation and activation by upstream signaling molecules.[40] Therefore, the *in vitro* measurement of the protein kinase activity of p38 isolated from cells does not correlate with the effect of the drug on p38 protein kinase activity *in vivo*, as the drug will be washed away during the immunoprecipitation of p38. Instead, effects downstream of p38 activity can be measured, such as MAPKAPK-2 protein kinase activity or reporter gene expression (see previous sections). Some care must be taken when interpreting data obtained with SB203580, as studies indicate that it is not completely specific for p38 enzymes. The drug

[37] M. Dickens, J. S. Rogers, J. Cavanagh, A. Raitano, Z. Xia, J. R. Halpern, M. E. Greenberg, C. L. Sawyers, and R. J. Davis, *Science* **277,** 693 (1997).

[38] J. C. Lee, S. Kassis, S. Kumar, A. Badger, and J. L. Adams, *Pharmacol. Ther.* **82,** 389 (1999).

[39] A. J. Whitmarsh, J. Cavanagh, C. Tournier, J. Yasuda, and R. J. Davis, *Science* **281,** 1671 (1998).

[40] S. Kumar, M. S. Jiang, J. L. Adams, and J. C. Lee, *Biochem. Biophys. Res. Commun.* **263,** 825 (1999).

can also cause some inhibition of other protein kinases[41,42] (including some JNK isoforms when used at high concentrations[35]) as well as other enzymes such as cyclooxygenases[43] and hepatic cytochrome *P*-450 isozymes.[44] However, the *in vivo* specificity of SB203580 in a number of cellular events has been validated with cell lines that inducibly express a drug-resistant mutant p38α.[45] In future, small molecule inhibitors of JNK protein kinase activity should prove equally useful for analyzing the role of JNK in the cell.

Concluding Remarks

In this chapter, we have discussed many of the techniques that are currently in use for analyzing the activities of JNK and p38 MAP kinases *in vitro* and *in vivo*. The choice of a particular technique will depend on a number of factors, including whether (1) the overall activity or the activity of particular isoforms is to be measured; (2) the effect of inhibitors on JNK and p38 activation or protein kinase activity is being measured; or (3) the signaling events downstream or upstream of these MAP kinases are being analyzed. In future the development of improved antibodies to individual JNK and p38 isoforms as well as novel phosphorylation-specific antibodies that recognize the JNK and p38 target phosphorylation sites in substrates will greatly aid the dissection of the roles of JNK and p38 family members in the cell. In addition, the promise of highly specific drugs that target particular JNK and p38 enzymes will allow a more complete understanding of these complex signaling pathways.

Acknowledgments

We thank Cathy Tournier, Hervé Enslen, and Jennifer Lamb for helpful discussions. R.J.D. is an investigator of the Howard Hughes Medical Institute.

[41] P. A. Eyers, M. Craxton, N. Morrice, P. Cohen, and M. Goedert, *Chem. Biol.* **5,** 321 (1998).
[42] C. A. Hall Jackson, M. Goedert, P. Hedge, and P. Cohen, *Oncogene* **18,** 2047 (1999).
[43] A. G. Börsch-Haubold, S. Pasquet, and S. P. Watson, *J. Biol. Chem.* **273,** 28766 (1998).
[44] J. L. Adams, J. C. Boehm, S. Kassis, P. D. Gorycki, E. F. Webb, R. Hall, M. Sorenson, J. C. Lee, A. Ayrton, D. E. Griswold, and T. F. Gallagher, *Bioorg. Med. Chem. Lett.* **8,** 3111 (1998).
[45] P. A. Eyers, P. van den Ijssel, R. A. Quinlan, M. Goedert, and P. Cohen, *FEBS Lett.* **451,** 191 (1999).

[25] Phospho-Specific Mitogen-Activated Protein Kinase Antibodies for ERK, JNK, and p38 Activation

By SAID A. GOUELI and BRUCE W. JARVIS

Introduction

Protein phosphorylation plays a critical role in the mechanism of action of a variety of hormones, neurotransmitters, growth factors, mitogens, and cytokines, and in response to stress and irradiation, etc. The phosphorylation of cellular proteins or enzymes may result in alteration of their conformation and/or their enzyme activity. This is best illustrated by a group of enzymes collectively known as mitogen-activated protein (MAP) kinases, whose dual phosphorylation by upstream protein kinases results in their activation.[1] These kinase cascades are found in all eukaryotic organisms and consist of a three-kinase module. A canonical MAPK module consists of three protein kinases: an MAPK kinase kinase (MEKK) that phosphorylates and activates an MAPK kinase (MEK) that, in turn, phosphorylates and activates an MAPK/ERK enzyme.[2,3] There are at least three main modules in mammalian cells that fall under this category, and these are conserved across all eukaryotes, indicating that they are involved in mediating a diverse array of extracellular signals. The three modules contain the extracellular signal-regulated kinase (ERK), c-Jun N-terminal kinase (JNK), and p38 MAPK (p38) pathways (Fig. 1). The first enzyme in the module, MEKK, represents a group of enzymes that are serine/threonine protein kinases and phosphorylate MEK enzymes on two serine or threonine residues within an Ser-X-X-X-Ser/Thr motif. The latter enzymes in turn phosphorylate the MAPK on Thr and Tyr residues within the Thr-X-Tyr consensus sequence. An important feature of the MAPK superfamily of enzymes is that they are activated on dual phosphorylation of the TXY motif present in the activation loop of the catalytic domain. The middle amino acid varies among the three MAPKs: glutamic acid in ERKs, proline in JNKs, and glycine for p38. A single phosphorylation on Thr or Tyr is not sufficient to activate MAPKs. Dual phosphorylation is required for their activation, and the phosphorylation on tyrosine usually but not invariably precedes that of theonine in a two-step reaction.[4] Assessment of MAPK

[1] T. Hunter, *Cell* **80,** 225 (1995).
[2] M. H. Cobb, *Promega Notes* **59,** 37 (1996).
[3] T. S. Lewis, P. S. Shapiro, and N. G. Ahn, *Adv. Cancer Res.* **74,** 49 (1998).
[4] J. E. Ferrell and R. R. Bhatt, *J. Biol. Chem.* **272,** 19008 (1980).

0076-6879/00 $35.00

activation was carried out by either of two methods: immunoprecipitation of the enzyme by an MAPK-specific antibody and determination of the activity of the active enzyme in the precipitate, using myelin basic protein and radioactive ATP as substrates, or by band shift in gels of MAPKs after their activation. The former is a slow procedure and requires multiple steps, none of which can be precisely controlled for accurate and reproducible measurement of enzyme activity; this method also requires the use of radioactivity. The latter does not usually result in accurate assessment of the active enzyme and it is difficult to interpret the results when other protein bands with similar molecular weights appear in the vicinity of the MAPKs. Thus the need for a simple method that can accurately determine the activation of the enzyme was apparent. Toward this goal, we developed a new class of antibodies that can selectively detect only the presence of the dual phosphorylated form of each MAPK (ERK or JNK or p38) with minimal interference from other proteins. This is of paramount importance because the three MAPK signaling modules act in concert and with other cell signaling systems. This novel class of antibodies provides the selective tools that are required to study the role of each of the MAP kinase signaling pathways in response to extracellular stimuli. In addition to detection of the dual phosphorylated form (active) of MAPK, *in situ* localization of MAPKs with these novel antibodies enables scientists to determine the subcellular localization of each MAPK before and after activation.[5]

Antibody Production

Antigen Selection

Polyclonal antibodies to dually phosphorylated ERK, JNK, and p38 kinases are typically raised in rabbits that have been immunized with peptide antigens containing the pT-X-pY motifs. A potential peptide epitope of 10 to 20 amino acids can be analyzed with software such as Protean from DNAStar to select a sequence with maximal antigenicity. This sequence is then compared, using a BLAST search algorithm, with a database of protein sequences, such as SWISSPROT, to ensure minimal similarity to other nonhomologous proteins. The BLAST search can be done over the Internet, using the website of the National Center for Biotechnology Information at the National Library of Medicine at *http://www.ncbi.nlm.nih.gov/*. When the peptide sequence of the immunogen has been optimized in this manner, a cysteine residue is added to one terminus to facilitate later coupling to

[5] A. Brunet, D. Roux, P. Lenormand, S. Dowd, S. Keyse, and J. Pouyssegur, *EMBO J.* **18,** 664 (1999).

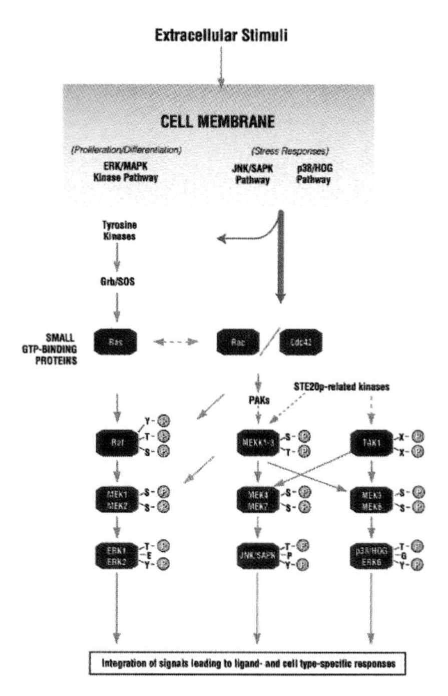

FIG. 1. Activation of different ERK/MAPK signaling cascades by different extracellular stimuli.

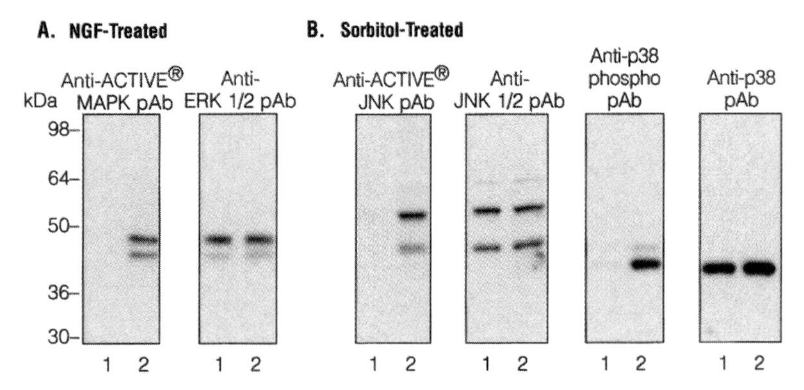

FIG. 2. Western blots of PC12 extracts probed with phospho-specific and anti-pan antibodies. Western blots were made from SDS–10% (w/v) polyacrylamide minigels. (A) Lanes contain 2 μg of unstimulated (lanes 1) or NGF-stimulated (lanes 2) PC12 cell extract. Blots were probed with Anti-ACTIVE MAPK antibody (Promega) and Anti-pan ERK1/2 antibody, as indicated. (B) Lanes contain 10 μg of unstimulated (lanes 1) or sorbitol-stimulated (lanes 2) PC12 cell extract. Blots were probed with Anti-ACTIVE JNK antibody (Promega), anti-pan JNK-1 antibody, antiphospho-p38 antibody (New England BioLabs), and an anti-pan p38 antibody. The secondary antibody was donkey anti-rabbit, Anti-ACTIVE Qualified antibody conjugated with AP (Promega). Chemiluminescence detection was by means of a Tropix Western Star kit.

affinity resin. The finalized peptide can be synthesized by a commercial vendor. Both the phospho and nonphospho forms of the peptide are made for use on affinity columns, and the phospho form is also used to prepare the immunogen. It is recommended to request the synthesis of a sufficient amount (25 mg) of each peptide for use as immunogen, affinity purification, and for blocking experiments in *in situ* localization of the activated MAPK.

Peptide Conjugation and Immunization

Peptide immunogens must be coupled to a carrier protein in order to elicit an immune response in rabbits. The carrier protein of choice is keyhole limpet hemocyanin (KLH), which is cross-linked to the peptide by maleimide chemistry. A reliable source for KLH and a coupling method is the Imject maleimide-activated mcKLH (Pierce, Rockford, IL).

Immunization of the rabbit is generally as follows. A small (~2 ml) blood sample is taken to serve as an indicator of the preimmunization status of the rabbit serum. The phosphopeptide–KLH immunogen (200 μg) is then emulsified in Freund's complete adjuvant and injected subcutaneously on the rabbit's dorsum.[6] Two weeks later (week 2) another 200-μg amount of immunogen is emulsified in Freund's incomplete adjuvant (FIA) and similarly injected into the rabbit. In weeks 4 and 6, another "boost" injection is made with immunogen in FIA. During week 7 the rabbit is bled to obtain ~20 ml of serum. In week 8 the rabbit receives a boost as described above and is again bled in week 9. This boost-and-bleed cycle subsequently continues on a monthly basis: boost in week 1, bleed in week 2, rest in week 3, and bleed in week 4. To monitor the titer and specificity of serum reactivity to the immunogen, dot blots can be done using the phospho- and nonphosphopeptides. A 10-ng amount of peptide in a 1-μl "dot" on nitrocellulose membrane is then probed with a 1:1000 dilution of the rabbit serum. (The probing method is the same as discussed below for Western blot probing.) If under these "dot-blot" conditions, the serum cannot detect the phosphopeptide dot with greater intensity than the nonphosphopeptide dot, then further purification will not improve the antibody specificity, and the yield will be too low to warrant affinity purification. Usually by week 9 with its second 20-ml bleed, one can obtain such predictive data on serum titer and specificity by use of the dot blot.

Antibody Purification

Once a favorable determination of the titer and specificity of the serum has been made, the affinity columns can be prepared. Couple each phospho-

[6] E. Harlow and D. Lane, "Antibodies: A Laboratory Manual." Cold Spring Harbor Laboratory, Cold Spring Harbor, New York, 1988.

and nonphosphopeptide via the terminal cysteine residue to a column matrix using the SulfoLink resin and instructions (Pierce). Approximately 8 mg of SulfoLink with 8 mg of peptide are needed to purify the antibody from 40 to 50 ml of serum. The efficiency of coupling can be monitored with Ellman's reagent, which reacts with free sulfhydryl groups. Before adding the serum to the columns, it is useful to remove some of the extraneous serum proteins by ammonium sulfate precipitation. To 40 to 50 ml of rabbit serum add dropwise, while stirring, an equal volume of saturated ammonium sulfate, pH 7. Continue to stir overnight in the cold. Centrifuge at 1400*g* in a swing rotor for 30 min at 10°. Resuspend the pellet in one-third the original volume with TBS [50 m*M* Tris-HCl (pH 7.5), 75 m*M* NaCl]. Dialyze the resuspended pellet in 50,000 molecular weight cutoff tubing versus three changes of 1 liter of TBS at 4°, twice for 4 hr each, and then overnight. Add to the resulting dialysate protease inhibitors at the following final concentrations: 1 m*M* benzamidine, 1 m*M* EDTA, 100 μM phenylmethylsulfonyl fluoride (PMSF) (made from a fresh stock in ethanol).

Preequilibrate the nonphosphopeptide column with TBS. Rock the dialysate with nonphosphopeptide-coupled SulfoLink for 1 hr at room temperature. Pour into the column and save the flowthrough. Rinse the column with 8 ml of TBS and add to the nonphosphopeptide flowthrough. Preequilibrate the phosphopeptide column with TBS. Rock the nonphosphopeptide column flowthrough with phosphopeptide-coupled SulfoLink for 1 hr at room temperature. Pour into the column and rinse the phosphopeptide column with TBS until the A_{280} decreases to about 0.01. Rinse the phosphopeptide column with 1 *M* NaCl in TBS until the A_{280} decreases to 0.01. Elute the phosphopeptide column with 0.1 *M* glycine, pH 3, collect fractions, and continue rinsing until the A_{280} decreases to about 0.01. Each collection tube contains 0.1 volume of 2 *M* Tris, pH 7.3, to neutralize the pH. Dialyze the eluted antibody versus phosphate-buffered saline (PBS: 137 m*M* NaCl, 27 m*M* KCl, 15 m*M* KPO_4, pH 7.4), three times with 1 liter each for 4 hr each time at 4°. Another suggested eluant that may be needed, if glycine does not elute the antibody, is 3 *M* $MgCl_2$ in 20 m*M* Tris-HCl, pH 7.3.

Western Blot Probing

One stringent test of the quality of an affinity-purified, anti-phosphopeptide antibody is the probing of a Western blot containing cell extracts. A highly purified, phospho-specific antibody will not detect nonspecific bands on such a Western. Of course, the phospho-specific antibody must also detect the phospho form of the purified kinase and not the nonphospho form. In the case of a dual phospho-specific antibody, judicious use of phosphatases can be made to strip off either the threonyl or tyrosyl phosphate from the kinase target and monitor decreased antibody binding on a Western blot.[7] A reliable method for probing Western miniblots follows. Block a nitrocellulose membrane with 1% (w/v) bovine serum albumin (BSA) in 2× TBS [100 m*M* Tris-HCl (pH 7.5), 150 m*M* NaCl] for 1 hr at 37° or overnight at 4°. Incubate the membrane with the phospho-specific antibody in TBST [TBS with 0.05% (v/v) Tween 20] for 2 hr at room temperature with agitation. Wash 3 times with 75 ml of TBST for 15 min each with agitation. Incubate the membrane with the secondary antibody conjugate in TBST for 1 hr at room temperature with agitation. (The secondary antibody conjugate alone should be tested in a control incubation to ensure that no extraneous bands arise from it.) Wash three times with 75 ml of TBST for 15 min each with agitation. Detect with chemiluminescent reagents such as ECL (Amersham-Pharmacia, Piscataway, NJ) or with a Western Star kit (Tropix, Bedford, MA). If, on the other hand, a polyvinylidene difluoride (PVDF) membrane is available, block it in PVDF buffer [TBS with 0.2% (w/v) I-Block and 0.1% (v/v) Tween 20; I-Block is purified casein, and is available from Tropix] for 1 hr at 37° or overnight at 4°. Incubate the membrane with the phospho-specific antibody in PVDF buffer for 2 hr at room temperature with agitation. Wash three times with 75 ml of PVDF buffer for 15 min each with agitation. Incubate the membrane with the secondary antibody conjugate in PVDF buffer for 1 hr at room temperature with agitation. Wash three times with 75 ml of PVDF buffer for 15 min each with agitation. Detect as described previously. Figure 2 contains an example of Western blots of PC12 cell extracts probed with affinity-purified, phospho-specific ERK, JNK, and p38 antibodies.

Preparation of Active and Inactive MAPK-Containing PC12 Cell Extracts

The following method for preparing PC12 cell extracts stimulated by either nerve growth factor (NGF) or sorbitol is adapted from previous publications.[8–10] PC12 cells are grown in tissue culture flasks that were precoated with rat tail collagen, 6 μg/cm^2 for 1 hr. The medium is RPMI 1640 with 25 m*M* HEPES and L-glutamine, 10% (v/v) horse serum, 5%

[7] A. Khokhlatchev, S. Xu, J. English, P. Wu, E. Schaefer, and M. H. Cobb, *J. Biol. Chem.* **272,** 11057 (1997).

[8] R. Meier, J. Rouse, A. Cuenda, A. R. Nebreda, and P. Cohen, *Eur. J. Biochem.* **236,** 796 (1996).

[9] X. Zhengui, M. Dickens, J. Raingeaud, R. J. Davis, and M. E. Greenberg, *Science* **270,** 1326 (1995).

[10] T. G. Boulton, J. S. Gregory, and M. H. Cobb, *Biochemistry* **30,** 278 (1991).

(v/v) fetal bovine serum, 0.5 m*M* EGTA. During growth at 37° in 5% CO_2 the medium needs to be changed every other day, until the cells reach ~80% confluence. For NGF treatment to activate ERK on the day before harvest, add fresh medium with serum. On the day of harvest add NGF at 50 ng/ml in RPMI for a 5-min incubation with cells at 37°, and then harvest immediately. For sorbitol treatment to activate JNK and p38 kinase, on the day before harvest add fresh medium without serum. On the day of harvest add sorbitol in the medium to a final concentration of 0.5 *M* for 5 min at 37°, and then harvest immediately. To harvest the cells, wash cells twice with Dulbecco's phosphate-buffered saline (DPBS) without Ca^{2+} or Mg^{2+}, and add cold harvest buffer [20 m*M* Tris-HCl (pH 7.3), 20 m*M* *p*-nitrophenyl phosphate, 1 m*M* EGTA, 50 m*M* NaF, 50 μ*M* *o*-vanadate, and 5 m*M* benzamidine]. In the first flask, scrape cells into buffer with a cell scraper, transfer cells and buffer to other flasks, and scrape. Break cell membranes with a Dounce type A glass homogenizer on ice. Spin the lysate in a microcentrifuge for 10 min at 10,000 rpm at 4° to pellet the nuclei and membranes. Aliquot and store at −70°.

Characterization of Phosphorylation State-Specific Antibodies

To check for the dual phosphorylation and activation of MAPK it is recommended that an active MAPK be used as a standard with inactive enzyme as control. But the most rigorous test for antibody specificity is cell extracts from stimulated and native cells. For our studies we used PC12 cells, as mentioned previously, that were stimulated with NGF for detection of activated ERKs and cells that were grown in the presence of sorbitol for detection of activated JNKs and p38 protein kinase. Extracts prepared from cells that were left overnight in medium with serum served as a control for the inactive MAPKs. As shown in Fig. 2A, NGF activated both ERK1 and ERK2, whereas the basal level of these enzymes in the extract of control cells showed negligible response to the antibody. Both ERK1 and ERK2 can be detected equally in extracts of both NGF-treated and untreated cells when an antibody that recognizes the enzyme regardless of its phosphorylation status (pan antibody) is used as a probe. The sorbitol treatment of PC12 cells results in remarkable activation of JNKs and p38 as detected by the anti-dual phospho-specific anti-JNK antibodies and anti-dual phospho-specific anti-p38 antibodies, respectively (Fig. 2B). The use of pan antibodies that recognize equally the inactive as well as the active forms of JNK or pan antibodies that detect equally the inactive and active forms of p38 shows equal amounts of these enzymes in control and sorbitol-treated cells. Thus the activation of all MAPKs results in the appearance of only the phospho forms of the enzymes without affecting their total cellular level (Fig. 2B), indicating postsynthetic phosphorylation of the enzymes and not an increase in their transcription/translation rate. It is noteworthy that the anti-dual phospho-specific MAPK antibodies were successfully used in *in situ* localization of activated MAPKs in several model systems,[5,11,12] indicating its usefulness in studying the subcellular localization as well as kinetics of activation of the various MAPKs *in vivo*.[13,14] These reagents have also been useful in the development of selective inhibitors of early activation of MAPKs.[13,15]

Conclusion

The availability of antibodies that recognize the active form of the various members of MAPKs (ERKs, JNKs, and p38) has made it possible to study the activation of these enzymes in response to a variety of stimuli in an accurate and time-dependent fashion. Similar to the anti-phosphotyrosine antibodies, the dual phospho-specific antibodies have proved to be valuable reagents in the pursuit of selective inhibitors of these enzymes. Finally, studies of these enzymes with these novel antibodies resulted in the discovery of novel MEK inhibitors such as U0126.

[11] K. M. Walton, R. DiRocco, B. A. Bartlett, E. Koury, V. R. Marcy, B. Jarvis, E. M. Schaefer, and R. V. Bhat, *J. Neurochem.* **70,** 1764 (1998).

[12] P. Lenormand, J.-M. Brondello, A. Brunet, and J. Pouyssegur, *J. Cell Biol.* **142,** 625 (1998).

[13] M. F. Favata, K. Y. Horiuchi, E. J. Manos, A. J. Daulerio, D. A. Stradley, W. S. Feeser, D. E. Van Dyk, W. J. Pitts, R. A. Earl, F. Hobbs, R. A. Copeland, R. L. Magolda, P. A. Scherle, and J. M. Trzaskos, *J. Biol. Chem.* **273,** 18623 (1998).

[14] N. Tolwinski, P. S. Shapiro, S. Goueli, and N. G. Ahn, *J. Biol. Chem.* **274,** 6168 (1999).

[15] J. S. Sebolt-Leopold, D. T. Dudley, R. Herrera, K. V. Becelaere, A. Wiland, R. C. Gowan, H. Tecle, S. D. Barrett, A. Bridges, S. Przybranowski, W. R. Leopold, and A. R. Saltiel, *Nat. Med.* **5,** 810 (1999).

[26] Immunostaining for Activated Extracellular Signal-Regulated Kinases in Cells and Tissues

By DANIEL GIOELI,* MAJA ZECEVIC,* and MICHAEL J. WEBER

Introduction

The mitogen-activated protein (MAP) kinases, or extracellular signal-regulated kinases (ERKs), are a family of conserved serine/threonine kinases that are implicated in the control of cell growth, differentiation,

* These authors contributed equally to this work.

0076-6879/00 $35.00

migration, and secretion.[1–4] Traditionally, ERK activity has been assessed by electophoretic mobility shift and immune-complex kinase assays. Although these are powerful techniques, they are limited to analysis of cell lysates that of necessity are prepared from cell populations. Thus, information on heterogeneity of activation within cells and between cells is lost when these methods are used.

On activation, MAP kinases become dually phosphorylated on tyrosine and threonine.[5–7] It has been possible to generate antibodies that react specifically with the phosphorylated, active form of the enzyme.[8,9] Because dual phosphorylation of MAP kinase on tyrosine and threonine residues is both necessary and sufficient for activation of the enzyme, reactivity with these antibodies can be used as a surrogate for enzyme assays and electophoretic mobility shift analysis. Use of these activation state-specific antibodies allows determination of the intracellular localization of the active MAP kinase pool, without confounding signals from the large pool of inactive enzyme.[9] It also allows determination of the active MAP kinase distribution between cells of different types of physiologies within tissues and organs.[10,11] In addition, immunoblotting with these antibodies provides a procedure for assaying the activation of the enzyme that is quicker and simpler than immune-complex kinase assays, although not as quantitative. This chapter details the techniques necessary for analysis of MAP kinase activation in cells and tissues with phosphorylation-state specific antibodies. The techniques described should be applicable to use of other "activation state"-specific antibodies.

[1] L. B. Ray and T. W. Sturgill, *Proc. Natl. Acad. Sci. U.S.A.* **84,** 1502 (1987).

[2] A. J. Rossomando, D. M. Payne, M. J. Weber, and T. W. Sturgill, *Proc. Natl. Acad. Sci. U.S.A.* **86,** 6940 (1989).

[3] T. G. Boulton, S. H. Nye, D. J. Robbins, N. Y. Ip, E. Radziejewska, S. D. Morgenbesser, R. A. DePinho, N. Panayotatos, M. H. Cobb, and G. D. Yancopoulos, *Cell* **65,** 663 (1991).

[4] G. L'Allemain, *Prog. Growth Factor Res.* **5,** 291 (1994).

[5] N. G. Anderson, J. L. Maller, N. K. Tonks, and T. W. Sturgill, *Nature* (*London*) **343,** 651 (1990).

[6] D. M. Payne, A. J. Rossomando, P. Martino, A. K. Erickson, J. H. Her, J. Shabanowitz, D. F. Hunt, M. J. Weber, and T. W. Sturgill, *EMBO J.* **10,** 885 (1991).

[7] J. H. Her, S. Lakhani, K. Zu, J. Vila, P. Dent, T. W. Sturgill, and M. J. Weber, *Biochem. J.* **296,** 25 (1993).

[8] M. Zecevic, A. D. Catling, S. T. Eblen, L. Renzi, J. C. Hittle, T. J. Yen, G. J. Gorbsky, and M. J. Weber, *J. Cell Biol.* **142,** 1547 (1998).

[9] L. Gabay, R. Seger, and B. Z. Shilo, *Science* **277,** 5329 (1997).

[10] J. W. Mandell, I. M. Hussaini, M. Zecevic, M. J. Weber, and S. R. VandenBerg, *Am. J. Pathol.* **153,** 1411 (1998).

[11] D. Gioeli, J. W. Mandell, G. R. Petroni, H. F. Frierson, Jr., and M. J. Weber, *Cancer Res.* **59,** 279 (1999).

Materials and Methods

Phospho-MAP Kinase-Specific Antibody

For our experiments, antibody specific for the dually phosphorylated mitogen-activated protein kinases, ERK1 and ERK2, has been developed as follows.[9] After collection of preimmune rabbit serum, polyclonal antiserum is raised against the phosphopeptide CHTGFLpTEpYVATR conjugated to keyhole limpet hemocyanin (KLH) (Quality Controlled Biochemicals, Hopkintown, MA). This peptide is identical for both ERK1 and ERK2 and conserved in vertebrates. Affinity purification of the antibody is performed by negative selection over a column of nonphosphorylated peptide conjugated to KLH and subsequent positive selection over a column of dually phosphorylated peptide conjugated to bovine serum albumin. Antibody is used at a concentration of 1.7 μg/ml for immunoblotting, 8.5 μg/ml for immunocytochemistry, and 0.5 μg/ml for immunohistochemistry. Presently, phospho-specific antibodies that recognize a variety of signal transduction proteins are available from commercial vendors.

Characterization of Antibody

For immunostaining of cells and tissues it is critical that the specificity of the antibody be rigorously determined so that reactivity with the antibody can be used as a surrogate for enzyme activity measurements. To determine whether our antibody reacts with proteins other than ERKs, we use whole cell lysates of PtK1 rat kangaroo epithelial cells (American Type Culture Collection, Rockville, MD) separated by sodium dodecyl sulfate–polyacrylamide gel electrophoresis (SDS–PAGE) and analyzed by Western blotting. These cells are also used for the immunofluorescence studies described below. Cells are grown as monolayers at 37° in 5% CO_2 in minimal essential medium (MEM) supplemented with 10% (v/v) fetal calf serum (FCS). At 80% confluency, cells are stimulated for 15 min with epidermal growth factor (EGF, 0.2 μg/ml); control cells are untreated. The cells are then lysed in Laemmli sample buffer, sonicated, clarified by centrifugation, and 100 μg of protein is electrophoresed on a 10% (w/v) SDS–polyacrylamide gel. The phospho-MAP kinase antibody recognizes the phosphorylated forms of the MAP kinases p44 ERK1 and p42 ERK2 on Western blot (Fig. 1). Furthermore, the phospho-MAP kinase antibody demonstrates an elevated intensity with lysates from EGF-treated cells. This is consistent with elevated MAP kinase activity in mitogen-stimulated cells. As expected, the activation of MAP kinase is inhibited when the MEK inhibitor PD 98059 (Biomol, Plymouth Meeting, PA) is added at a concentration of 50 μM for 30 min prior to EGF stimulation in PtK1 cells (Fig. 1A).

Reprobing the same blot with an antibody against total ERK2 (monoclonal ERK2 antibody 05-157, 0.1 μg/ml; UBI, Lake Placid, NY), which recognizes phosphorylated and nonphosphorylated p42 ERK2, demonstrates that the band detected with the phospho-MAP kinase antibody corresponds to the phosphorylated, activated, and mobility-shifted ERK2. The total ERK2 blot also shows that substantial amounts of the protein remain nonphosphorylated in these cells; the phospho-MAP kinase antibody does not recognize this protein.

The specificity of the phospho-MAP kinase antibody for the individual MAP kinase phospho residues is determined with a panel of hamster CCL39 fibroblasts expressing various hemagglutinin (HA)-tagged phosphorylation site mutants of ERK2: wild type (wt), T183A (TA), Y185F (YF), and the double mutant T183A/Y185F (TAYF). The cells are grown as monolayers at 37° in 7.5% CO_2, using DMEM supplemented with 10% (v/v) FCS. When cells reach 50% confluency, the growth medium is removed and cells are washed twice with phosphate-buffered saline (PBS) and then serum deprived for 25 hr in DMEM containing 0.2% (v/v) FCS. Cells are then serum stimulated for 5 min with DMEM containing 20% (v/v) FCS; control cells are untreated. Cells are washed twice in ice-cold PBS and then lysed in cold lysis buffer [50 m*M* 4-(2-hydroxyethyl)-1-piperazineethanesulfonic acid (HEPES, pH 7.5), 100 m*M* NaCl, 2 m*M* EDTA, 1% (v/v) nonidet p-40 (NP-40), 1μ*M* pepstatin, leupeptin (1 μg/ml), 1 m*M* phenylmethylsulfonyl fluoride (PMSF), 0.2 m*M* sodium vanadate, aprotinin (2 μg/ml), 40 m*M* 4-nitrophenyl phosphate disodium salt (PNPP), 50 m*M* β-glycerophosphate, 2 m*M* dithothreitol (DTT)]. Lysates are cleared by centrifugation at 14,000 rpm for 15 min in a microcentrifuge prior to immunoprecipitation. The tagged ERKs are immunoprecipitated with an HA-specific monoclonal antibody 12CA5 (BAbCo, Berkeley, CA) used at a concentration of 18 μg per immunoprecipitation. The antibody is preabsorbed for 1 hr at 4° to protein A–agarose (Boehringer Mannheim, Indianapolis, IN) before incubation with 500 mg of CCL39 cell extracts for 4 hr at 4°. Immunoprecipitates are washed four times with lysis buffer before Laemmli buffer is added. The samples are resolved by 10% SDS–PAGE and Western blotted with phospho-MAP kinase antibody (Fig. 1C). The antibody reacts strongly only with the wild-type enzyme from serum-stimulated cells. Weaker reactivity to the YF mutant is seen, but only when the amount of MAP kinase loaded is in great excess relative to the other mutants. As shown previously, phosphorylation at either T183 or Y185 is sufficient to cause MAP kinase to display an electrophoretic mobility shift, but insufficient to activate the enzyme.[7] The double TAYF mutant does not shift at all. These data demonstrate that our phospho-MAP kinase antibody does not react detectably with nonphosphorylated or tyrosine-phosphorylated MAP kinase,

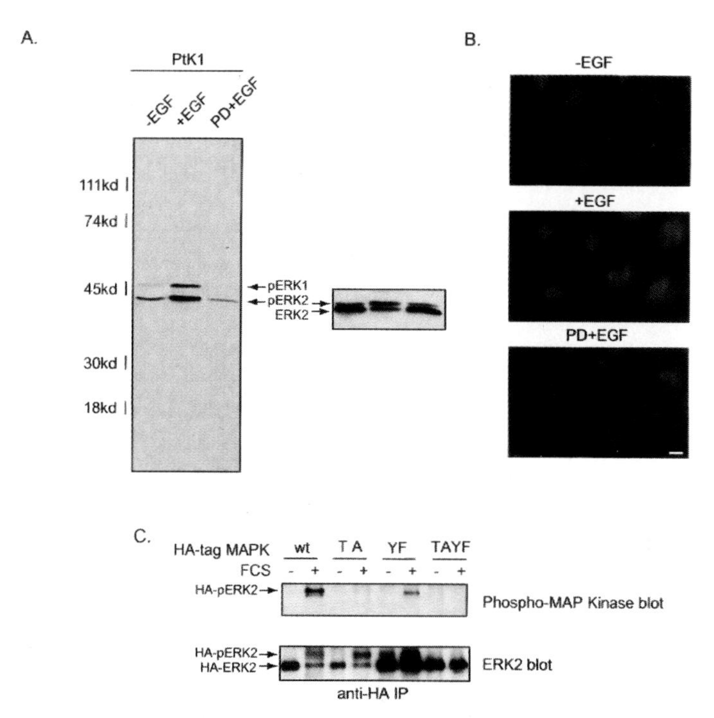

FIG. 1. Specificity of phospho-MAP kinase antibody. (A) Phospho-MAP kinase immunoblot of whole cell lysate from PtK1 epithelial cells maintained in medium with 10% (v/v) FCS (−EGF), EGF stimulated (0.2 μg/ml, 15 min) (+EGF), or treated with the MEK inhibitor PD 98059 (50 μ*M*) for 0.5 hr before EGF stimulation (PD + EGF). Next to the immunoblot is the corresponding monoclonal ERK2 antibody blot. (B) Indirect immunofluorescence of acrolein/formaldehyde-fixed and phospho-MAP kinase antibody-labeled PtK1 cells that were maintained in 10% (v/v) FCS medium (−EGF), stimulated with EGF (0.2 μg/ml, 15 min) (+EGF), or treated with PD 98059 (50 μ*M*) for 0.5 hr before EGF addition (PD + EGF). Bar: 10 μm. (C) Phospho-MAP kinase antibody and the corresponding ERK2 monoclonal immunoblot of hemagglutinin (HA)-immunoprecipitated wild-type HA-ERK2 (wt) and HA-ERK2 phosphorylation site mutants [T183A (TA), Y185F (YF), and T183A/Y185F (TAYF)] from stable expressing CCL39 fibroblasts. The cells were grown in 10% (v/v) FCS-containing medium and then were either serum deprived [0.2% (v/v) FCS] for 25 hr (−FCS) or serum deprived and then stimulated [20% (v/v) FCS] for 5 min (+FCS). Proteins were visualized by enhanced chemiluminescence.

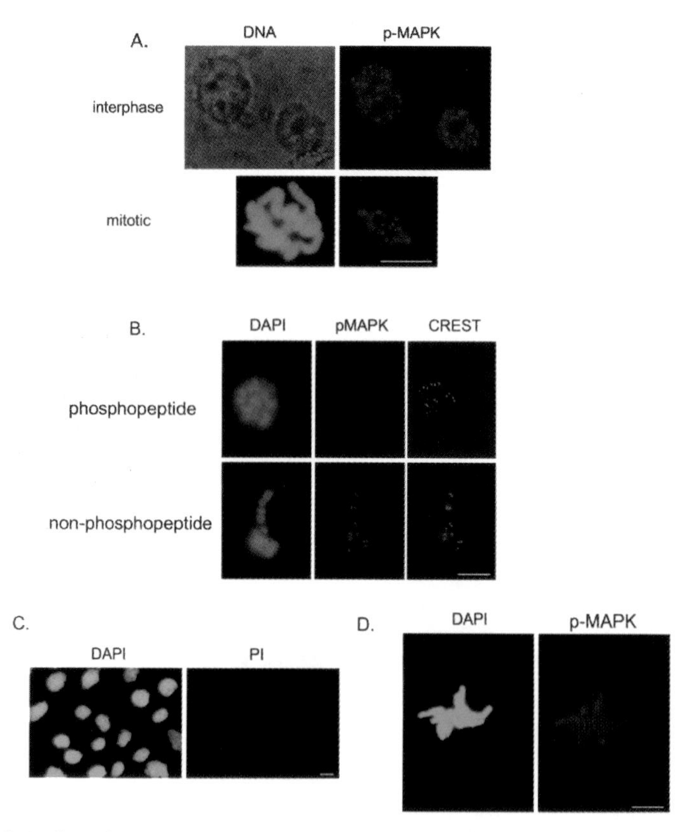

FIG. 2. Indirect immunofluorescence of somatic cells, using the phospho-MAP kinase antibody. (A) Phospho-MAP kinase antibody labeling (red) of interphase and mitotic PtK1 epithelial cells, extracted with 1% (w/v) CHAPS and fixed with 1% (v/v) formaldehyde. The cells are counterstained with the DNA dye, DAPI (blue). For the interphase cell, a phase-contrast picture is shown instead of DAPI. (B) Inhibition of phospho-MAP kinase antibody labeling by blocking with the phosphopeptide CHTGFLpTEpYVATR and not the nonphosphopeptide CHTGFLTEYVATR. PtK1 cells arrested in mitosis are extracted and fixed, after which they are stained with phospho-MAP kinase antibody (red), CREST autoimmune serum (green), and DAPI (blue). Both peptides are used at 10-fold excess (mol/mol) relative to the antibodies. (C) Staining of preimmune phospho-MAP kinase serum (PI) and DAPI in an asynchronous culture of PtK1 cells extracted and fixed as described. (D) Phospho-MAPK antibody (red) and DAPI (blue) staining of an extracted and fixed mitotic PtK1 cell without phosphatase inhibitors (vanadate and β-glycerophosphate). Bars: 10 μm.

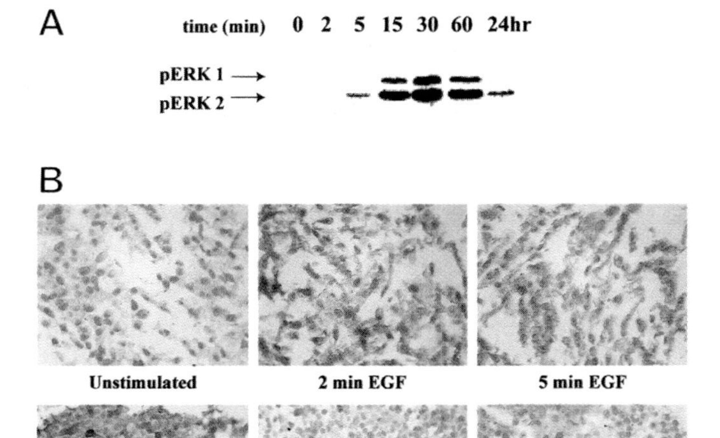

FIG. 3. Optimization of phospho-MAP kinase staining in LNCaP cells. (A) Immunoblot with phospho-MAP kinase antibody of LNCaP cells treated with EGF (100 ng/ml) for various times. (B) Phospho-MAP kinase staining of LNCaP cells treated with EGF (100 ng/ml) for various times. In the last two panels, after stimulation with EGF (100 ng/ml) for 30 min, LNCaP cells were incubated in PBS for various times. Original magnification: ×400.

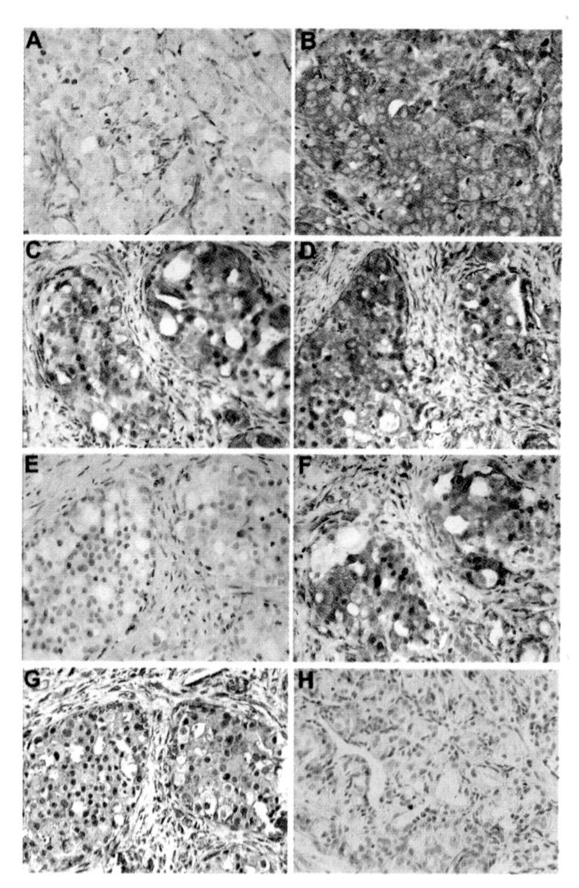

FIG. 4. Specificity of staining for activated MAP kinase in a prostate tumor. In (A–G), serial sections of the same neoplasm were stained. (A and C) Phospho-MAP kinase staining; (B and D) total MAP kinase staining. (A and B) Area of tumor with no activated MAP kinase in neoplastic cells. (C–F) Area of tumor with elevated activated MAP kinase. (E) Anti-phospho-MAP kinase antibody preincubated with dually phosphorylated immunizing peptide; (F) anti-phospho-MAP kinase antibody preincubated with nonphosphorylated peptide. (G) Total MAP kinase antibody preincubated with dually phosphorylated immunizing peptide; (H) preimmune staining of a tumor with high levels of active MAP kinase. Original magnification: ×400.

reacts weakly with the threonine-phosphorylated enzyme, and reacts strongly with the dually phosphorylated and activated MAP kinase. No reactivity with proteins other than activated ERK is detected in whole cell lysates. Thus, reactivity with this antibody is indicative of activated MAP kinase.

Immunostaining in Cells

Activation of Mitogen-Activated Protein Kinase in Cells. The phospho-MAP kinase antibody can be used to detect activated endogenous MAP kinase by indirect immunofluorescence under various physiologic conditions. Cells are grown on coverslips in six-well plates until 80% confluent. In the procedure described below, all wash steps are done with the coverslips in the six-well plate. The blocking and antibody incubation steps are done by inverting the coverslips onto Parafilm, cell side down, onto 25–50 μl of antibody solution in a humidified chamber. Humidified incubation chambers can easily be made from 150-mm dishes with a piece of moist Whatman (Clifton, NJ) filter paper cut to fit the bottom of the dish followed by a piece of Parafilm. A small-gauge needle can be used to lift and transfer coverslips between the six-well plate and humidified chamber.

To visualize active MAP kinase by immunofluorescence, cells are rinsed once with PBS and fixed with acrolein/formaldehyde [0.2% (v/v) acrolein (Polyscience, Warrington, PA), 4% (w/v) paraformaldehyde, 0.2% (v/v) Triton X-100, and 2 m*M* sodium vanadate in PBS] for 20 min at room temperature; all incubations are carried out at room temperature. Cells are then immediately quenched with 0.025% (w/v) $NaBH_4$, 2% (v/v) glycine in PBS twice for 5 min and then once for 10 min. Cells are then washed twice with PBS–0.05% (v/v) Tween 20 and blocked with 20% boiled normal goat serum (NGS; Sigma, St. Louis, MO) in MBST [10 m*M* 3-(*N*-morpholino) propanesulfonic acid (MOPS), 150 m*M* NaCl, 0.05% (v/v) Tween 20; pH 7.4] on a rocker. Cells are then quickly rinsed with PBS–0.05% (v/v) Tween 20, and incubated with the phospho-MAP kinase antibody (8.5 μg/ml) diluted in MBST containing 5% boiled NGS for 1 hr with rocking. After this, cells are washed by rocking for 15 min in PBS–0.05% (v/v) Tween 20 and incubated with Texas red-conjugated goat anti-rabbit secondary antibody (Jackson ImmunoResearch Laboratories, West Grove, PA) diluted in MBST containing 5% boiled NGS to a final IgG concentration of 7.5 μg/ml. The final wash is done for 15 min on a rocker with PBS–0.05% (v/v) Tween 20, after which each coverslip is mounted on a slide containing 10 μl Vectashield mounting medium (Vector Laboratories, Burlingame, CA) and 10 m*M* $MgSO_4$. Coverslips are then sealed twice with Cytoseal (Stephens Scientific, Riverdale, NJ).

In our studies, PtK1 cells were grown as monolayers on coverslips in six-well plates as described above. Cells are either left untreated or stimulated with EGF (0.2 μg/ml) for 15 min. Figure 1B shows immunofluorescence images of asynchronously growing PtK1 cells, EGF-stimulated cells, or cells that have been pretreated with PD 98059 prior to EGF stimulation. The intensity of the signal parallels the Western blot results described above: EGF induces overall (nuclear and cytoplasmic) MAP kinase activation that is inhibited by the PD compound. We have found that the acrolein fixation is the only fixation method used for immunofluorescence that clearly shows a dramatic increase in the level of MAP kinase after growth factor or serum stimulation. Neither fixation in methanol–acetone (1:1, v/v) for 20 min at $-20°$ nor a method in which cells are fixed first and then extracted with detergent (fix/extract) reveals the increase in MAP kinase activation in response to mitogen stimulation. The fix/extract method is done essentially as described below for the extract/fix method used in the subcellular localization of active MAP kinase; however, the order of fixation and extraction is reversed.

Subcellular Localization of Active Mitogen-Activated Protein Kinase. Specific subcellular localization of active MAP kinase can be detected by immunofluorescence. In our study of MAP kinase activity in mitosis, PtK1 cells have a different pattern of phospho-MAPK localization depending on the stage of the cell cycle the cells are in.[8] Figure 2A shows an interphase cell with punctate nuclear staining and a mitotic cell with staining at the kinetochores and asters. The procedure to detect active MAP kinase in mitotic cells is detailed below.

Cells are grown on coverslips until 80% confluent and rinsed with PHEM [60 mM piperazine-N,N-bis(2-ethanesulfonic acid) (PIPES), 25 mM HEPES (pH 6.9), 10 mM EGTA, 4 mM $MgSO_4$]. All steps are carried out at room temperature. The cells are then extracted for 5 min with PHEM plus 1% (w/v) 3-(3-cholamidopropyldimethylammonio)-1-propane sulfonate (CHAPS), 1 μM pepstatin, leupeptin (1 μg/ml), aprotinin (2 μg/ml), 50 mM β-glycerophosphate, and 0.2 mM sodium vanadate. After the extraction buffer is removed, cells are fixed immediately in 1% (w/v) formaldehyde in PHEM for 15 min. Cells are rinsed twice with MBST and incubated in blocking, primary, and secondary antibody solutions as described above. After the last wash, coverslips are dipped in a solution containing the DNA dye 4,6-diamidino-2-phenylindole (DAPI; Sigma) diluted in water to a final concentration of 0.5 μg/ml. Coverslips are then mounted as described above.

The specificity of the phospho-MAP kinase signal is confirmed by performing the following control experiments: immunofluorescence with primary or secondary antiserum alone, immunofluorescence with preimmune purified IgG, and peptide-blocking experiments. For the peptide-blocking experiments, either the phospho-peptide CHTGFLpTEpYVATR or the nonphosphopeptide CHTGFLTEYVATR is preincubated with phospho-MAP kinase antibody in an excess of 10-fold (mol/mol) for 1 hr at room temperature in PBS. After the incubation, immunofluorescence is done as described above using the primary antibody pre-incubated with peptide.

In our study of MAP kinase activity in mitosis, incubation with primary or secondary antibodies alone or with the preimmune purified IgG gives no immunoreactivity (Fig. 2C). Because we observed active MAP kinase immunoreactivity at the kinetochores, CREST autoimmune serum is used as a positive control for the localization and a negative control for the phosphopeptide blocking. CREST immunoreactivity is visualized by a fluorescein isothiocyanate (FITC)-conjugated goat anti-human secondary antibody used at a concentration of 7.5 μg/ml (Jackson ImmunoResearch Laboratories). Figure 2B shows that the phosphopeptide, while blocking reactivity of the cells with phospho-MAP kinase antibody, has no effect on reactivity with CREST antibody.

It is important to emphasize that the visualization of phospho-MAP kinase immunoreactivity in extracted and fixed cells is optimal when tyrosine and serine/threonine phosphatase inhibitors (vanadate and β-glycerophosphate) are included in the extraction solution. When the inhibitors are omitted, the labeling is undetectable (Fig. 2D). This suggests antibody reactivity depends on phosphoepitopes and is further evidence of the specificity of the phospho-MAP kinase antibody.

Immunostaining in Tissues

Tissue Preparation. In our study of MAP kinase activation in prostate cancer, a total of 82 formalin-fixed (zinc-buffered formalin with rare exceptions), paraffin-embedded human primary and metastatic prostate cancers and corresponding adjacent nonneoplastic tissues have been studied.[11] To facilitate statistical analysis, as large a sample group as practical should be used. In studies of human cancer, tumor staging and treatment information for each case studied can be obtained from tumor registry files and slides reviewed for tumor grade. In general, zinc-buffered formalin is superior to standard buffered formalin as a fixative; however, we have not observed any appreciable differences in immunostaining when comparing tissues prepared with these two fixatives. The ability to use formalin-fixed paraffin-embedded tissues allows investigators to use material already present in tissue registries. However, one should be mindful that phospho epitopes are labile (Fig. 3B). In our experience, postmortem tissue is not suitable for phospho-specific antiserum analysis but routinely collected surgical material is. Our experience is largely limited to human prostate tissue. However, the techniques described should be adaptable to formalin-fixed paraffin-embedded material from other sources.[10]

Immunohistochemistry. Discussed below are the optimal immunohistochemistry conditions used for our analysis of human prostate material. Different tissue types and sources may require some modification of the methodology described. Ideally, optimization should be done with material from as representative a source as possible. In our study of MAP kinase activity in prostate cancer, we have used the LNCaP prostate cancer cell line for optimizing conditions for immunohistochemistry. Parameters that are typically adjusted to optimize immunostaining are as follows: fixative, antigen retrieval, concentrations and incubation times of primary and secondary antibodies, and wash conditions after antibody incubations.

For our optimization experiments, serum-starved LNCaP cells are treated with EGF (100 ng/ml) or left untreated for various times and fixed in zinc-buffered formalin. To emulate tissue, LNCaP cells are centrifuged, embedded in agar, and processed routinely into paraffin. Under the conditions detailed below, untreated LNCaP cells show little to no phospho-MAP kinase staining while EGF-treated LNCaP cells show intense nuclear and cytoplasmic phospho-MAP kinase immunoreactivity (Fig. 3B). Immunostaining with preimmune purified IgG is used as a negative control in optimization experiments. The immunohistochemistry of LNCaP cells shows kinetics of MAP kinase phosphorylation similar to that seen in immunoblotting (Fig. 3). In addition, we have found that the phospho epitope is labile in cells, presumably because of cellular phosphatases. Incubating LNCaP cells in PBS after maximal MAP kinase activation leads to a progressive decrease in detection of the phospho epitope (Fig. 3B).

Unstained human tissue specimens are deparaffinized in three 5-min incubations in xylene and two 3-min incubations in 100% ethanol. All incubations are at room temperature unless otherwise stated. Endogenous peroxidase activity is quenched by a 30-min incubation in 0.5% (v/v) hydrogen peroxide–methanol. Sections are then hydrated by two 3-min incubations each in 95% (v/v) ethanol, 70% (v/v) ethanol, and distilled H_2O. After hydration, microwave epitope retrieval is performed in 10 m*M* citrate buffer, pH 6.0, for 10 min at 1.15 kW. Citrate buffer is prepared fresh prior to use from 9 ml of 0.1 *M* citric acid, 41 ml of 0.1 *M* sodium citrate, and 450 ml of distilled H_2O. Epitope retrieval has not been necessary in our experience to detect immunoreactivity; however, immunostaining is improved when this step is included. There may be circumstances under which it is essential. Sections are cooled by gradual replacement of citrate buffer with distilled H_2O. Sections are then air dried and circled with a hydrophobic pen. This allows use of small volumes (e.g., <100 μl) of antibody solutions per section; washes are still carried out in slide chambers with an excess of solution. Special care must be taken to ensure sufficient space between the tissue and the hydrophobic film in order to prevent any staining artifact.

Immunohistochemistry is performed by the avidin–biotin–peroxidase complex method, using commercially available reagents (Vectastain Elite kit; Vector Laboratories). Sections are blocked in 2% (v/v) normal goat serum–0.02% (v/v) Triton X-100–PBS for 30 min. Blocking agent is removed by aspiration and primary antibody solution is added [anti-phospho-MAP kinase antibody (0.5 μg/ml)–2% (v/v) normal goat serum–0.02% (v/v) Triton X-100–PBS]. Sections are incubated overnight at 4° in a humidified chamber. Humidified incubation chambers can easily be made from plastic slide boxes with moist paper towels placed at the bottom; slides are placed level across the top of the slide dividers. It is critical that the slides and chamber be level. The following day, sections are dipped in distilled H_2O and washed in PBS for 10 min. Sections are incubated in secondary antibody solution [biotinylated goat anti-rabbit diluted 1:200–2% (v/v) normal goat serum–0.02% (v/v) Triton X-100–PBS] for 1 hr. Once sections are incubating in secondary antibody, the avidin–biotin complex (ABC) solution should be prepared according to the manufacturer specifications so that it can incubate a minimum of 30 min prior to addition to the sections. Sections are incubated in the ABC solution for 30 min and then dipped in distilled H_2O, washed in PBS for 10 min, and rinsed briefly in distilled H_2O. Diaminobenzidine (DAB) chromogen solution is prepared according to the manufacturer instructions and sections are incubated with DAB for 5 min. The reaction is stopped in distilled H_2O. It is important to monitor the DAB reaction during optimization experiments: it should be stopped before saturation. Sections are counterstained with hematoxylin for 4 min and destained by sequential dips in distilled H_2O, 1% (v/v) HCl–ethanol, distilled H_2O, 1% (v/v) ammonia–distilled H_2O, and distilled H_2O. Sections are dehydrated by sequential 1-min incubations in 70%, 95%, and 100% ethanol, and cleared in xylene. Coverslips are mounted with Cytoseal (Stephens Scientific).

To confirm the specificity of the phospho-MAP kinase antibody, control experiments are performed on a subset of sections. These control experiments include immunohistochemistry with preimmune serum, total MAP kinase antibody (ZS61-7400; Zymed Laboratories, San Francisco, CA), and peptide competition experiments. The peptide competition experiments are performed by preincubating diluted phospho-specific MAP kinase antibody (0.5 μg/ml) or total MAP kinase antibody (2.5 μg/ml) for 1 hr at room temperature with a 100 μM concentration of either the immunizing dually phosphorylated peptide or the nonphosphorylated peptide. Immunohistochemistry is then carried out as described above.

In our study of active MAP kinase in prostate cancer, immunohistochemistry using preimmune purified IgG at 0.5 μg/ml is negative (Fig. 4H). Staining with antibody to total MAP kinase reveals cytoplasmic immunostaining in the majority of epithelial and stromal cells independent of phospho-MAP kinase activation (compare Fig. 4A and B with Fig. 4C and D). In addition, nuclear staining for total MAP kinase is occasionally found in areas of active MAP kinase (Fig. 4D). This subcellular distribution of total and active MAP kinase is consistent with the current understanding that MAP kinase is activated in the cytoplasm and much of it subsequently translocates to the nucleus.

Peptide competition experiments are critical controls for immunohistochemistry using phospho-specific antibodies. In our studies, preincubation of the phospho-MAP kinase antibody with the dually phosphorylated peptide virtually eliminates immunoreactivity (compare Fig. 4C and E); staining of activated MAP kinase is not blocked when the phospho-MAP kinase antibody is preincubated with the nonphosphorylated peptide (compare Fig. 4C and F). The ability of the dually phosphorylated peptide to block immunoreactivity is specific to the phospho-MAP kinase antibody, as the phosphorylated peptide is unable to block total MAP kinase staining (Fig. 4G). The results of these control experiments strongly indicate that the antibody specifically recognizes phosphorylated MAP kinase in tissues.

A scoring system will need to be developed to assess the role of active MAP kinase in biological processes when tissue samples are studied by immunohistochemistry, as this technique is difficult to quantify. In our study of MAP kinase activity in prostate cancer, the tumor specimens are grouped into three categories based on the percentage of tumor cell nuclei staining with the phospho-MAP kinase antibody: zero, low ($<$10% tumor cells positive), and high ($\geq$10% tumor cells positive). This scoring system is highly reproducible and virtually bimodal. Most samples with low phospho-MAP kinase staining display less than 5% positive tumor cell nuclei and most samples with high phospho-MAP kinase staining show greater than 20% positive tumor cell nuclei. For statistical analysis, we use a generalized rank test for trend with an assumed known order alternative to test, *a priori*, if (1) higher levels of activated MAP kinase in samples are associated with higher Gleason scores and (2) higher levels of activated MAP kinase in samples are associated with more advanced tumor stage. The generalized case III rank test for trend provides the most power to test the particular alternative of interest (i.e., higher levels of activated MAP kinase in tumors with higher Gleason scores).

Comments and Concluding Remarks

Activation state-specific antibodies have emerged as a means to test for activation of specific signal transduction proteins at the cellular level. This chapter describes immunostaining in cells and tissues using a phospho-MAP kinase specific antibody. The list of activation state-specific antisera available from commercial vendors is growing daily, and thus this approach should become accessible to a wide variety of enzymes. These activation state-specific reagents enable one to assess the intracellular localization of active MAP kinase and thus understand the ways signal transduction are regulated spatially and temporally. The activation state antibodies also can reveal cellular heterogeneity within tissues and organs and thus be informative about intercellular communication. Finally, the ability to assay the activation state of signaling molecules in paraffin-embedded human pathology specimens opens up the possibility of examining the dynamics of cellular regulation in human disease and development, providing a powerful adjunct to the traditional tools of the pathologist, which primarily examine protein expression. Use of these reagents should increase our knowledge of the cellular biochemistry of signal transduction proteins and the role these proteins play in normal physiology and disease processes.

Acknowledgments

This work was supported by grants GM47332, CA76500, and CA39076 from the USPHS and by a gift from CaP CURE (to M.J.W.). D.G. was supported by the American Foundation for Urologic Disease/Scott Fund.

[27] Dominant Negative Mutants of Mitogen-Activated Protein Kinase Pathway

By M. JANE ARBOLEDA, DEREK EBERWEIN, BARBARA HIBNER, and JOHN F. LYONS

Introduction

One of the gene families most commonly found mutated in human tumors is the *ras* gene family. Activated Ras protein mutations are found in 50% of colon carcinomas, 30% of lung carcinomas, 80% of pancreatic carcinomas, and approximately 20% of hematopoietic malignancies.[1,2] In

[1] J. L. Bos, *Cancer Res.* **49,** 4682 (1989).

[2] J. Lyons, J. W. G. Janssen, C. Bartram, M. Layton, and G. J. Mufti, *Blood* **71,** 1707 (1988).

0076-6879/00 $35.00

addition to mutations in the *ras* gene, activation of Ras may be mediated by several mechanisms. Either overexpression or amplification of growth factor receptors signaling through Ras, such as Neu/Erb2, epidermal growth factor (EGF), and platelet-derived growth factor (PDGF) receptors,[3] a reduction in expression or activity of the neurofibromatosis type 1 protein (NF-1), a GTPase-activating protein,[4] or an upregulation of Grb-2 or Grb-7 adaptor proteins[5] may lead to higher than normal levels of active Ras in the cell.

Ras has been shown to regulate several pathways that contribute to cellular transformation, including the Rac and Rho pathways[6] and the Raf–MEK–MAPK pathway (MEK, MAPK/ERK kinase; MAPK, mitogen-activated protein kinase; ERK, extracellular signal-regulated kinase).[7] Ras activates the Raf–MEK–MAPK pathway by first recruiting Raf-1 to the plasma membrane, where it is activated by an unknown mechanism.[8] Activated Raf-1 phosphorylates and activates MEK, which in turn phosphorylates and activates MAP kinase.[9] Activated MAP kinases then translocate from the cytoplasm into the nucleus, where they phosphorylate target proteins, including transcription factors. These events initiate a sequence of gene transcription, which ultimately governs cell growth and differentiation.

In rodent fibroblasts, constitutively active Ras, Raf-1, or MEK elevates MAP kinase activity and induces malignant transformation.[10] Conversely, expression of inhibitory mutants of Raf-1, MEK, or MAP kinases significantly reduces the transforming ability of mutant Ras.[6] Cumulatively, these studies suggest that the Raf–MEK–MAPK pathway is pivotal for mediating the transforming effects of Ras

To define the significance of the Raf–MEK–MAPK pathway in tumor growth, we sought to selectively block that pathway in human tumor cells. A MEK mutant in which Ser-218 and Ser-222 are substituted by alanine residues is unable to be activated by Raf-1 *in vitro* and *in vivo*. Thus, to block signaling through Raf-1, we attempted to stably express this dominant negative MEK mutant (MEK218A/222A) in 15 tumor cell lines harboring

[3] D. W. Stacey, M. Roudebush, R. Day, S. D. Mosser, J. B. Gibbs, and L. A. Feig, *Oncogene* **6,** 2297 (1991).

[4] T. N. Basu, D. H. Gutmann, J. A. Fletcher, T. W. Glover, F. S. Collins, and J. Downward, *Nature* (*London*) **356,** 713 (1992).

[5] R. J. Daly, M. D. Binder, and R. L. Sutherland, *Oncogene* **9,** 2723 (1994).

[6] R. G. Qiu, J. Chen, D. Kirn, F. McCormick, and M. Symons, *Nature* (*London*) **374,** 457 (1995).

[7] S. Cowley, H. Paterson, P. Kemp, and C. Marshall, *Cell* **77,** 841 (1994).

[8] D. Stokoe, S. G. Macdonald, K. Cadwallader, M. Symons, and J. F. Hancock, *Science* **264,** 1463 (1994).

[9] C. M. Crews, A. Alessandrini, and R. L. Erikson, *Science* **258,** 478 (1992).

[10] A. Brunet, G. Pages, and J. Pouyssegur, *Oncogene* **11,** 3379 (1994).

Ras mutations. We found that such genetic inhibition of the Raf–MEK–MAPK pathway reduced the endogenous activity of the pathway, inhibited the anchorage-independent growth of tumor cells, blocked tumor progression, and increased survival in mice.

Materials

Cell Culture

MIAPaCa-2, DLD-1, HT-1080, HCT116, SW480, SW620, PANC-1, NCl H460, BxPC-3, A549, NCl H522, C-33A, NCl H792, NCl H841, and SK-OV-3 cell lines are grown in Dulbecco's modified Eagle's medium with glucose at 4500 mg/liter (DMEM; Irvine Scientific, Irvine, CA) supplemented with 10% (v/v) fetal calf serum (FCS; Irvine Scientific) and penicillin G (50 U/ml)–streptomycin sulfate (50 μg/ml) (Irvine Scientific) at 37° in a humidified incubator containing 5% CO_2. Dominant negative MEK (MEK218/222A) transfectants are maintained in medium containing G418 (500 μg; GIBCO-BRL, Gaithersburg, MD).

Construction of Plasmids and Transfections

A dominant negative mutant of murine MEK1 (MEK218A/222A), encoding a Glu-Glu epitope tag (EYMPME) at the C terminus, is constructed by mutating Ser-218 and Ser-222 to alanine residues. The mutations are verified by dideoxynucleotide sequencing and subsequently subcloned into the pCDNA expression vector (Promega, Madison, WI).

Cells are plated at 50 to 80% confluency in 60-mm dishes and incubated overnight in growth medium. The following day, the medium is replaced with serum-free medium after washing the cells three times with phosphate-buffered saline (PBS; Irvine Scientific). Cells are incubated with a cocktail of 5 μl of Transfectam (Promega) and 5 μg of DNA (vector or dominant negative MEK construct) for 5–24 hr. After this, the medium is replaced with growth medium. The next day, cells are trypsinized, diluted, and incubated in medium containing a 300- to 800-μg/ml concentration of geneticin (G418; GIBCO, Gaithersburg, MD). DNA uptake by the cell lines is verified with a cytomegalovirus (CMV)-driven secretable alkaline phosphatase (SEAP) reporter construct (Clontech, San Diego, CA). SEAP activity is measured as specified by the manufacturer (Tropix, Bedford, MA).

Cell Proliferation

Proliferation rates are determined by the chemiluminometric CytoLite assay (Packard, Meriden, CT). Luminescence is detected in an ML2250 microtiter plate luminometer (Dynatech, McLean, VA). Approximately 1500 to 2000 cells are seeded into 96-well plates in a volume of 100 μl/well. Six wells per clone and 12 wells for the vector control are prepared. The CytoLite assay is performed 24 hr after plating to determine baseline values and subsequently on days 3, 5, and 7. Growth rates are calculated by determining the mean values from days 3, 5, and 7 and then dividing by the corresponding baseline count.

Anchorage-Independent Growth

Twelve-well dishes are coated with 0.6% (w/v) Bacto-agar(Difco, Detroit, MI)-containing growth medium. Cells ($1–3 \times 10^3$) are resuspended in selective growth medium containing 0.3% (w/v) Bacto-agar and plated in the wells. The DLD-1 cell line resuspended in medium is applied through several cell strainers to ensure that the cells are in a single-cell suspension. Cells are fed weekly with 0.5 ml of 0.3% (w/v) Bacto-agar-containing selective growth medium. At the end of 2 weeks, colonies are stained with 0.3% (w/v) Bacto-agar-containing medium supplemented with the viability stain, neutral red. SW620, C-33A, and NCl H522 cell lines are assayed for 4 weeks. Colonies of greater than 15 cells are counted and growth inhibition is scored only when there are fewer colonies than in the control, irrespective of their size. The efficiency of colony formation is calculated by dividing the number of colonies per well by the number of cells plated. Measurements are performed in quadruplicate.

In Vivo Growth

Cells are suspended in PBS and injected subcutaneously into the flanks of athymic mice. For the HCT116, MIAPaCa-2, and PANC-1 cell lines 5×10^6 cells are injected and for the DLD-1 cell line 1×10^6 cells are injected. Five mice are injected with each cell line. Tumor size is measured weekly with Vernier calipers and tumor volume (V) is calculated by $V = LWD$, where L and W are the long and short diameters (mm), respectively, of the tumor mass and D is the height of the tumor mass. In some instances, tumor measurement is determined by the formula $L\ (WW)/2$ and reported in milligrams. Body weight is measured weekly and animals are observed daily for signs of morbidity. Measurements are taken for 5 to 7 weeks in animals injected with the DLD-1 or HCT116 cell lines and for 10 weeks in animals injected with the MIAPaCa-2 or PANC-1 cell line.

As a surrogate metastasis model, eight animals per group are injected intraperitoneally with 1×10^6 DLD-1 cells and are observed once a week for tumor growth, ascites accumulation, body weight, and morbidity. At the end of 5 to 8 weeks, the animals are killed and examined for tumor spread in the peritoneal cavity.

Survival studies are done by injecting mice with an MiaPaCa-2 clone expressing dominant negative MEK or the vector control. Intraperitoneal injections are performed as described above. Mice are examined weekly over 35 weeks for the presence of ascites, cachexia, and tumor growth. Animals are euthanized according to the guidelines established by the American Association for Laboratory Animal Care (AALAC).

Immunoblotting Analyses

Cells are washed with cold PBS three times and lysed in 400 μl of 20 m*M* Tris (pH 8.0), 137 m*M* NaCl, 1% (v/v) Triton X-100, 10% (v/v) glycerol, 5 m*M* EDTA, and 1 m*M* phenylmethylsulfonyl fluoride (PMSF). Lysates are cleared by centrifugation at 10,000*g* for 20 min at 4°. Total protein concentration is determined by the bicinchoninic acid (BCA) assay (Pierce Biochemicals, Rockford, IL). Volumes of lysate containing 50 μg of protein are run on an 8% (w/v) sodium dodecyl sulfate (SDS)–polyacrylamide gel and transferred to nitrocellulose. The MEK218/222A mutant is detected with the Glu-Glu monoclonal antibody.[11] Endogenous MEK-1 and MEK-2, Raf-1, A-Raf, and B-Raf are detected with specific rabbit polyclonal antibodies (Santa Cruz Biotechnology, Santa Cruz, CA). All proteins are visualized by chemiluminescence using the enhanced chemiluminescence (ECL) kit (Amersham, Arlington Heights, IL).

Kinase Assays

Cells are grown to 80% confluency in 10% serum. Lysates are prepared as described above excepting the addition of 1 m*M* sodium orthovanadate and 5 m*M* sodium fluoride in the lysis buffer. Raf-1 is immunoprecipitated from 300 μg of lysate by incubating with 3 μl of C-20 Raf-1 antibody (Santa Cruz Biotechnology) and 40 μl of protein A–Sepharose (40% slurry) for 2 hr at 4°. Immune complexes are washed three times in lysis buffer and once with kinase buffer [30 m*M* Tris (pH 8), 20 m*M* $MgCl_2$, and 2 m*M* $MnCl_2$]. Kinase reactions are performed in a total volume of 30 μl of kinase buffer containing 2 μg of catalytically inactive MEK-1 (MEKB) as substrate. MEK immunoprecipitations are performed either by first preclearing the Glu-Glu-tagged MEK218/222A mutant through successive immunoprecipitations with Glu-Glu-conjugated beads followed by immunoprecipitation of endogenous MEK-1 with an MEK-1 antibody (Santa Cruz Antibodies,

[11] T. Grussenmyer, K. H. Scheidtmann, M. A. Hutchinson, and G. Walter, *Proc. Natl. Acad. Sci. U.S.A.* **82,** 7952 (1985).

Santa Cruz, CA) or by immunoprecipitation of total MEK-1 with 10 μl of antibody for 6 hr at 4°. Quantitative immunoprecipitation of endogenous MEK-1 and the MEK mutants is verified by Western blot analysis of the supernatants with an MEK-1 antibody. Kinase reactions are then performed in 20 mM HEPES (pH 7.5), 20 mM β-glycerophosphate, 10 mM $MgCl_2$, 10 mM $MnCl_2$, 1 mM dithiothreitol (DTT), 50 μM sodium orthovanadate, and 2 μg of catalytically inactive ERK-2 (GSTK63M ERK-2). All reactions are incubated at 30° for 30 min. Phosphorylated substrates are resolved on 10% (w/v) SDS polyacrylamide gels, stained, dried, and exposed to X-ray film. After autoradiography, radiolabeled protein bands are excised and incorporated radioactivity is quantitated by liquid scintillation counting.

Statistical Analyses

All statistical analyses are performed using the Mann–Whitney U test (StatView; Abacus Concepts, Berkeley, CA). Survival analysis and Kaplan–Meier plots are also generated using the StatView program.

Reduction in Endogenous MEK Activity by Expression of Dominant Negative MEK Mutant

To determine the relative contribution of the Raf–MEK–MAPK pathway to the tumorigenic phenotype of human tumor cells, an epitope-tagged dominant negative MEK mutant (MEK218A/222A) is stably transfected into human tumor cell lines of epithelial or mesenchymal origin (Table I). The most extensively characterized are the MIAPaCa-2 cell line, derived from a human pancreatic tumor; the DLD-1 cell line, derived from a human colon carcinoma; and the HT-1080 fibrosarcoma cell line. They harbor activating K-Ras mutations at codons 12 and 13 and an N-Ras mutation at codon 61, respectively. Expression of the dominant negative MEK in the cells is confirmed by immunoblot analysis (data not shown).

To show that the expression of dominant negative MEK interferes with endogenous MEK-1 activity in these cells, total MEK-1 protein is quantitatively immunoprecipitated and kinase activity is assayed with catalytically inactive ERK as a substrate. Cells are assayed under proliferating conditions. To exclude the possibility that cellular responses may differ because of differences in levels of endogenous MEK in each cell line, cell lysates are analyzed for MEK expression by immunoblotting. It has been found that MEK expression in each of the cell lines is approximately equivalent. Expression of dominant negative MEK in each of the three cell lines results in a reduction in endogenous MEK-1 activity when compared with vector controls (Table II).

TABLE I
EXPRESSION OF DOMINANT NEGATIVE MEK IN HUMAN TUMOR CELL LINES

Cell line	Tumor type	Ras[a] mutation	Expression of MEK218A/222A	
			Toxicity[b]	Frequency[c]
DLD-1	Colon	K-Ras13D	None	11/71 (15%)
MIAPaCa-2	Pancreas	K-Ras12C	None	14/54 (26%)
HT-1080	Fibrosarcoma	N-Ras61K	10/80 (88%)	6/10 (60%)
HCT116	Colon	K-Ras13D	None	6/58 (10%)
SW480	Colon	K-Ras12V	None	6/45 (13%)
SW620	Colon	K-Ras12V	None	5/46 (11%)
PANC-1	Pancreas	K-Ras12D	None	11/43 (26%)
BxPC-3		wt	20/43 (53%)	2/6 (17%)
NCI H460	Lung	K-Ras61H	None	6/71 (8.4%)
A549	Lung	K-Ras12S	None	5/30 (17%)
NCI H522	Lung	wt	None	10/11 (91%)
NCI H792	Lung	wt	Yes	0
NCI H841	Lung	wt	Yes	0
C-33A	Cervix	wt	None	10/13 (77%)
SK-OV-3	Ovary	wt	Yes	0

[a] Reported mutational status of the cell lines; wt, wild-type sequence.
[b] Toxicity was determined by comparing the number of G418-resistant colonies for the vector control and the dominant negative MEK-transfected cells.
[c] Expression frequency was determined by the number of clones expressing the epitope-tagged dominant negative MEK construct.

Because dominant negative MEK may function by titrating out endogenous activators of MEK in cells, such as Raf, MEK immunoprecipitates are immunoblotted to test for the presence of Raf-1. Raf-1 coimmunoprecipitates with dominant negative MEK. To determine whether this interaction results in a catalytically nonproductive complex, endogenous Raf-1 activity is also examined in the transfected cell lines. Total Raf-1 protein is quantitatively immunoprecipitated and kinase activity is assayed with catalytically inactive MEK as a substrate. In the three cell lines assayed, there is a reduction in endogenous Raf-1 activity. Cumulatively, the results suggest that dominant negative MEK inhibits endogenous MEK activity by sequestering endogenous Raf-1.

Effect on Proliferation and Inhibition of Anchorage-Independent Growth by Reduction in Endogenous MEK Activity

We have determined over a 7-day period the proliferation rates of the tumor cell clones expressing dominant negative MEK. In the DLD-1 clones

TABLE II
MEK AND Raf ACTIVITY ASSAYED IN CLONES EXPRESSING DOMINANT NEGATIVE MEK

Clone	MEK activity[a]	Raf activity[a]
DLD-1/pCAN vector control	100[b]	100[c]
DLD-1 C35	63.8[d] ± 11.5	28 ± 10
DLD-1 C37	37.2[d] ± 21	26 ± 4
DLD-1 C58	65.6 ± 28	66 ± 19
MIAPaCa-2/pCAN vector control	100[e]	100[c]
MIAPaCa-2 C34	78.5[d] ± 0.7	66 ± 9[f]
HT-1080/pCAN vector control	100[c]	100[c]
HT-1080 C1	40 ± 9	52 ± 21
HT-1080 C2	92 ± 6	63 ± 53
HT-1080 C4	66	56 ± 29

[a] Kinase activity correlates with Raf or MEK per unit protein in the extract and is expressed as a percentage of vector control. Data are expressed as a mean reduction in MEK activity for all experiments.
[b] The DLD-1 experiments on MEK activity represent a total of $n = 13$ from five different experiments.
[c] This number represents a single experiment done in duplicate.
[d] $p < 0.05$; denotes a significant reduction in activity.
[e] The MIAPaCa-2 experiments represent a total of $n = 6$ from two different experiments.
[f] This number represents a single experiment done in sextuplet.

expressing dominant negative MEK, a significant decrease ($p < 0.05$) in the rate of cell proliferation is found (Table III). In low serum (0.5%) the rate of proliferation of the MIAPaCa-2 clones expressing dominant negative MEK is 20% lower than that of the vector control cell line, although the results are not statistically significant. The differences in the effect of dominant negative MEK expression on proliferation in the cell lines cannot be attributed to differences in expression of endogenous Raf, because the relative expression levels of the MEK activators Raf-1, A-Raf, and B-Raf, in each cell line have been determined by immunoblot analysis to be approximately equivalent (data not shown).

To explore further the cellular effects of dominant negative MEK in the cell lines, their ability to support anchorage-independent growth is examined. The MIAPaCa-2, DLD-1, and the HT-1080 clones expressing

TABLE III
EFFECTS OF DOMINANT NEGATIVE MEK ON CELL PROLIFERATION IN HUMAN TUMOR CELL LINES

Cell line	Clones analyzed	Average rate of proliferation[a] Vector	Average rate of proliferation[a] MEK 218A/222A	Reduction in rate of proliferation (%)
DLD-1	7	6.5	3.1	52[b]
MIAPaCa-2	2	1.9	1.52	20[c]
HT-1080	5	14.6	11.9	18
HCT116	6	7.6	2.7	64[b]
SW480	2	2.1	1.95	10[c]
SW620	5	12.7	9.9	22[c]
PANC-1	5	1.8	2.0	0[d]
BxPC-3	1	2.3	1.8	22
NCI H460	6	11.3	7.9	30[b]
A549	3	11.2	9.4	16
NCI H522	4	4.0	2.5	37[b]
C-33A	6	19.7	12.5	36[b]

[a] Cells (2×10^3) were plated into 6 wells of a 96-well plate and allowed to adhere overnight. Cytolite analysis was performed on days 1, 3, 5, and 7. Average proliferation was determined from two independent experiments.
[b] This number represents a significant reduction in proliferation rate.
[c] Experiments were done in 0.5% (v/v) FCS.
[d] There was no reduction noted in the presence of 0.5% (v/v) FCS.

dominant negative MEK show statistically significant decreases in anchorage-independent growth compared with the vector control cell lines (Table IV). Thus, while a reduction in MEK activity does not translate to a reduction in the rate of cell proliferation under all conditions, the inability of the clones expressing dominant negative MEK to support anchorage-independent growth correlates well with the reduction in MEK activity.

Inhibition of Tumorigenicity *in Vivo* by Reduction in Endogenous MEK Activity

While unregulated cell proliferation and anchorage-independent growth are hallmarks of the transformed cell phenotype, inhibition of one or both of these biological parameters may not be an accurate indicator of the tumorigenic potential of these cells *in vivo*. To explore their tumorigenic potential *in vivo*, DLD-1, MIAPaCa-2, PANC-1, and HCT116 clones expressing dominant negative MEK are injected subcutaneously into athymic

TABLE IV
EFFECTS OF DOMINANT NEGATIVE MEK EXPRESSED IN HUMAN TUMOR CELL LINES ON ANCHORAGE-INDEPENDENT GROWTH[a]

Cell lines	Numbers of clones analyzed	Cloning efficiency		Inhibition in soft agar (% ± SD)
		Vector	Clones	
DLD-1	7	1.9×10^{-2}	0.35×10^{-2}	84.3 ± 10
MIAPaCa-2	8	5.0×10^{-2}	1.4×10^{-2}	72 ± 28
HT-1080	5	27.0×10^{-2}	5.8×10^{-2}	78.5 ± 39
HCT116	6	8.8×10^{-2}	3.7×10^{-2}	55 ± 39
SW480	6	11.2×10^{-2}	9.4×10^{-2}	24.8 ± 28
SW620	5	0.8×10^{-2}	0.14×10^{-2}	83.2 ± 38
PANC-1	5	7.1×10^{-2}	3.3×10^{-2}	55 ± 51
BxPC-3	1	10.0×10^{-2}	0.52×10^{-2}	95.2
NCI H460	6	10.0×10^{-2}	7.5×10^{-2}	26.8 ± 27
A549	3	30.0×10^{-2}	14.8×10^{-2}	45 ± 14
NCI H522	4	8.4×10^{-2}	0.72×10^{-2}	92 ± 15
C-33A	6	13.9×10^{-2}	4.4×10^{-2}	68.8 ± 39

[a] The panel of human tumor cell lines in which dominant negative MEK was expressed for their ability to grow in soft agar. Cloning efficiency in each well is defined as the number of growing colonies at the end of 2 weeks per number of cells plated. Results are expressed as the percent inhibition of growth of dominant negative MEK-expressing clones, relative to the vector control. Data represent two separate experiments performed in triplicate for each clone analyzed.

mice (Table V). Tumor measurements are taken weekly over 7–10 weeks. At the end of this period, the mean tumor volume in mice injected with clones expressing dominant negative MEK is 17–22% compared with mice injected with parental vector controls. Furthermore, the vector control group develops palpable tumors within 2 weeks, whereas the mice injected with MIAPaCa-2 clones expressing dominant negative MEK require 6 weeks to develop palpable tumors. Mice subcutaneously injected with DLD-1 clones expressing dominant negative MEK show a similar prolonged latency in the development of palpable tumors. Therefore, the reduced MEK activity observed in the clones expressing dominant negative MEK correlates with their decreased ability to support anchorage-independent growth and a significant decrease in their ability to form tumors in athymic animals.

The formation of metastases is another biolgical parameter reflecting advanced tumor progression. One of the locations to which colon carcinomas frequently metastasize is the peritoneum, and therefore the ability of DLD-1 clones expressing dominant negative MEK to proliferate in the

TABLE V
GROWTH OF CLONES EXPRESSING DOMINANT NEGATIVE MEK INJECTED SUBCUTANEOUSLY INTO ATHYMIC MICE

Cell line	Inhibition in soft agar (%)	Number of sites injected	Weeks in animals	Average tumor size ± SE (mm^3)	Inhibition *in vivo* (%)[a]
DLD-1 pCAN[b]	—	8[c]	4	535.9 ± 64.3[d,e]	—
DLD-1 clone 37	69	8	4	215.1 ± 42.6	60[f]
DLD-1 clone 58	97	8	4	119.8 ± 15.3	78
MIAPaCa-2 pCAN	—	10	10	497 ± 74.8	—
MIAPaCa-2 clone 34	99	10	10	84.4 ± 42.8	83[f]
PANC-1 pCAN	—	8	10	41.5 ± 9.5	—
PANC-1 clone 6	99	10	10	1.2 ± 1	97[f]
HCT116 pCAN	—	8	8	3394.2 ± 526.8	—
HCT116 clone 42	62	10	8	2135.2 ± 303.2	37

[a] Inhibition is expressed as a percentage of vector control. Nude mice were inoculated subcutaneously into each flank with 5×10^6 cells resuspended in PBS. Five mice were injected for each group. Tumor size was measured weekly with Vernier calipers in three dimensions. Statistical analysis was determined using the Mann–Whitney test (StatView).
[b] Cells (1×10^6) were injected subcutaneously.
[c] Eight animals were injected, one site per animal.
[d] Measurements were reported in milligrams, which was determined by *LWD*/2.
[e] Measurements were taken three times weekly.
[f] $p < 0.05$.

peritoneal cavity is examined as a surrogate model for metastatic potential. The assay measures the ability of the tumor cells to survive in a nonadherent environment prior to adhesion to abdominal structures and so may represent metastatic potential. The animals are killed after 5 weeks and examined for the accumulation of ascites, the presence of nonadherent peritoneal tumor nodules, and the presence of adherent tumors at the site of injection. The primary manifestation of disease in the group of mice injected with clones expressing dominant negative MEK is the formation of small, nonadherent tumor nodules that are found distributed randomly in the peritoneal cavity (Table VI).

The most relevant end point in the treatment of cancer is extension of survival time, and *ras* mutations have been associated with poor prognosis in lung and colorectal cancers.[12] Therefore, the survival of athymic mice

[12] R. J. Slebos, R. E. Kibbelaar, O. Dalesio, A. Kooistra, J. Stam, C. J. Meijer, S. S. Wagenaar, R. G. Vanderschueren, N. van Zandwijk, W. J. Mooi, *et al., N. Engl. J. Med.* **323,** 561 (1990).

TABLE VI
INHIBITION OF METASTASES IN DLD-1 CLONES EXPRESSING DOMINANT NEGATIVE MEK PROTEIN

	in vitro			*in vivo*		
	Expression level of ΔMEK	Soft agar	Cell growth	Ascites volume	Primary[a] tumor	Degree of metastases[c]
Vector control DLD-1/pCAN	None	0%	N/A	1.7ml	579.6mgs 6 [b]	++++ 3 +++ 2 ++ 1 + 2
Low expressing clone DLD-1/MEKC37	+	69%	50%	0.75ml	336.8mgs (42%) 6 [b]	++++ 0 +++ 0 ++ 4 + 4
High expressing clone DLD-1/MEKC58	+++	97%	50%	0.1ml	55.62mgs (90%) 5[b]	++++ 0 +++ 0 ++ 0 + 7 - 1

[a] The number describes the mean tumor mass at the site of intraperitoneal injection and is expressed as the percentage inhibition compared with vector controls.
[b] Describes the number of animals with primary tumor mass ($n = 8$).
[c] Represents the number of animals and the degree of metastasis found in the cavity: +, low number of metastases; ++++, high number of tumor metastases in the cavity.

injected intraperitoneally with a MIAPaCa-2 clone expressing dominant negative MEK, or a vector control, is examined. Pre- and postmortem analyses include measurement of ascites accumulation, cachexia, and tumor growth. Mice injected intraperitoneally with a clone expressing dominant negative MEK lived at least twice as long ($p < 0.001$) as the animals injected with vector control cells (Fig. 1). Six of the eight mice in the group injected with cells expressing dominant negative MEK lived without signs of disease until they were euthanized at the termination of the study at 35 weeks. On autopsy these mice revealed no tumor growth in the peritoneal cavity. In contrast, within 4 to 8 weeks postinjection, clinical disease was evident in

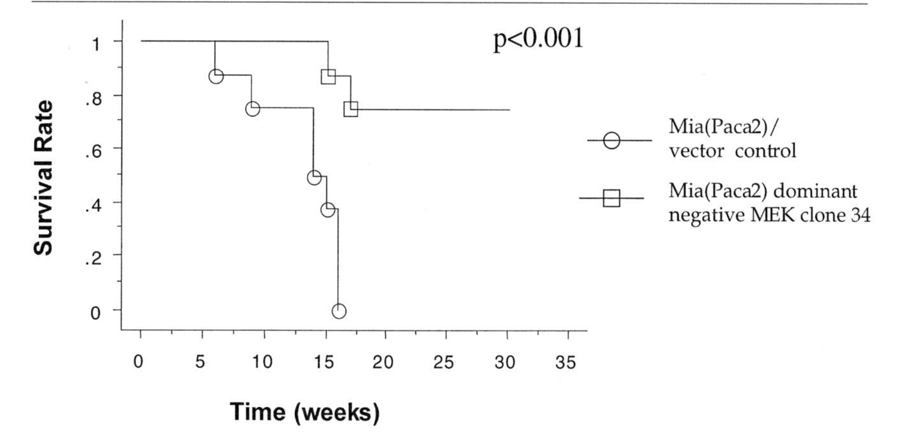

FIG. 1. Survival plot of animals bearing tumors expressing dominant negative MEK. Kaplan–Meier survival plot shows mice bearing the MIA cell line transfected with vector control (○) or the MIA cell line expressing the dominant negative MEK gene (□). Eight animals in each group were injected intraperitoneally with 5×10^5 cells and observed for 30 weeks. All six remaining animals injected with the dominant negative MEK clone remained disease free for more than 40 weeks.

all the animals injected with the cells containing the vector control. In this group, the first death occurred at 6 weeks and all remaining mice were euthanized at week 15 because of severe manifestation of disease. On autopsy, all these mice revealed the presence of tumors in the peritoneal cavity. Collectively, the data strongly suggest that a reduction in endogenous MEK function inhibits both tumor growth and metastases, and improves the survival of athymic mice bearing human tumors.

Discussion

We reasoned that the significance of the Raf–MEK–MAPK pathway in carcinogenesis could be further elucidated by specific inhibition of the pathway in tumor cells. To that end, we have shown that expression of dominant negative MEK in *ras*-mutated human tumor cell lines is sufficient to reduce endogenous Raf-1 and MEK activity, and inhibit the malignant phenotype *in vitro* and *in vivo*. We thus conclude that the Raf–MEK–MAPK pathway is necessary for Ras transformation in epithelial cells.

The immediate biochemical effect of expressing dominant negative MEK in the tumor cells was a reduction in endogenous Raf-1 and MEK-1 activity through the formation of nonproductive complexes between endogenous Raf-1 and dominant negative MEK. An unexpected finding was that there was no direct correlation between dominant negative

MEK expression levels and the percentage reduction of MEK activity in the cells, although all clones expressing dominant negative MEK showed reduced MEK activity under some conditions. Attempts by individual clones to compensate for the expression of dominant negative MEK or varying sensitivity to MEK inhibition due to clonal differences may account for these observations.

Anchorage-independent growth of the human tumor cell lines was inhibited by expression of dominant negative MEK. Certainly, the present study strongly suggests that reduction in MEK activity more accurately predicts inhibition of anchorage-independent growth of tumor cell lines of epithelial origin than a reduction in the rate of cell proliferation. Furthermore, examination of MEK activity and expression levels in clonal variants unequivocally showed that low-level expression of dominant negative MEK was sufficient to yield biochemical and biological effects in the tumor cell lines. These data are also consistent with the observation that inhibition of the Raf–MEK–MAPK pathway by antisense oligonucleotides to Raf-1[13] or by treatment with the small molecule MEK inhibitors PD098059 and U0126 also blocks anchorage-independent growth.[14,15] The fact that inhibition of only the Raf–MEK–MAPK pathway can mimic the effects of Ras inhibition of anchorage-independent growth implies that the Raf–MEK–MAPK pathway plays a pivotal role in governing this phenomenon.

Animal models of primary tumor growth, metastases, and survival revealed that mice injected with clones expressing dominant negative MEK consistently developed smaller tumors than those injected with parental vector controls. Mice injected with cells expressing dominant negative MEK formed fewer and smaller metastases and survived longer without evidence of morbidity than mice injected with the parental vector control cell lines. While part of these effects may be related to the ability of dominant negative MEK to inhibit anchorage-independent growth, the ability of the tumor cells to stimulate vascular development may also have been compromised. We noted that tumors formed from the parental vector control cells lines were highly vascularized, but those small tumors arising from cells expressing dominant negative MEK showed little, if any, vascularization. Supporting a role for MEK in the development of tumor vasculature is the finding that tumors formed in mice from fibroblasts expressing activated MEK exhibited a high degree of vascularization. Overall, these studies demon-

[13] B. P. Monia, J. F. Johnston, T. Geiger, M. Muller, and D. Fabbro, *Nat. Med.* **2,** 668 (1996).

[14] D. T. Dudley, L. Pang, S. J. Decker, A. J. Bridges, and A. R. Saltiel, *Proc. Natl. Acad. Sci. U.S.A.* **92,** 7686 (1995).

[15] M. F. Favata, K. Y. Horiuchi, E. J. Manos, A. J. Daulerio, D. A. Stradley, W. S. Feeser, D. E. Van Dyk, W. J. Pitts, R. A. Earl, F. Hobbs, R. A. Copeland, R. L. Magolda, P. A. Scherle, and J. M. Trzaskos, *J. Biol. Chem.* **273,** 18623 (1998).

strate a critical role for MEK in mediating tumor progression, perhaps by permitting anchorage-independent growth and stimulating the development of tumor vasculature.

Perhaps the most striking aspect of the *in vivo* studies is the finding that the inhibitory effect of dominant negative MEK on the measured end points was dose dependent. It was consistently observed that clones expressing the highest levels of dominant negative MEK formed the smallest tumors and the fewest metastases and exhibited significant increases in survival. However, it could be argued that high levels of dominant negative MEK might nonspecifically interfere with pathways other than the Raf–MEK–MAPK pathway and perhaps affect *in vivo* tumor growth. This possibility is negated by the observation that clones expressing low levels of dominant negative MEK were able to elicit all the biological responses measured, suggesting that altered MEK activity alone can have a profound effect on biological outcome.

Implications for Use of Dominant Negative Constructs in Cancer

This present study conclusively shows that the reduction of endogenous cellular MEK activity in various human tumor cells of epithelial origin impedes their ability to form large tumors and metastatic lesions, which leads to increased survival. Because activating Ras mutations are frequently found in human cancers, much effort has been directed toward inhibiting Ras function in the treatment of cancer. We have achieved stable expression of a dominant negative mutant of MEK in 12 human tumor cell lines. We have shown that manipulation of the Raf–MEK–MAPK pathway alone can affect both the tumorigenic and metastatic potential of human cancers. By genetically interfering with this signal transduction pathway, we have inhibited tumor growth and increased survival of animals and have further shown that these effects are specifically MEK dependent. Thus, our findings show that a block at the level of MEK activation is sufficient to impair tumor growth and progression and further imply that inhibition of MEK activity may be therapeutically useful in the treatment of cancer.

[28] Scaffold Protein Regulation of Mitogen-Activated Protein Kinase Cascade

By ANDREW D. CATLING, SCOTT T. EBLEN, HANS J. SCHAEFFER, and MICHAEL J. WEBER

Introduction

The mammalian mitogen-activated protein (MAP) kinases ERK1 and ERK2 (extracellular signal-regulated kinases 1 and 2) are activated in response to diverse extracellular signals that regulate many context-dependent responses, including cell growth and division, execution of differentiated cell function, cell motility, and programmed cell death. In concert with covalent changes that modify catalytic activity,[1] the ERKs are subject to regulation by translocation to the nucleus and other sites of action within stimulated cells.[2]

The ubiquitous activation of ERKs raises the important question of how specificity of signaling is established and maintained, and in particular how ERKs are targeted to upstream activators and downstream targets appropriate for the prevailing extracellular conditions. One mechanism by which signaling specificity can be imposed is through the formation of oligomeric protein complexes in which upstream activators and downstream targets are placed in physical proximity, enhancing the efficiency and specificity of signaling. There is compelling evidence for this mechanism in yeast, where pathway-specific scaffolding proteins such as STE5 direct signaling through discrete sets of kinases to effect specific biological end points.[3–7] In this way, stable physical interactions between the scaffold and the kinase module ensure a specificity and efficiency of response. These properties suggest a mechanism by which identical kinases can be used by the cell in functionally distinct pathways coupled to distinct upstream activators, because a scaffolding protein can select which upstream activator is used to activate the associated kinases, and determines where this activation

[1] D. M. Payne, A. J. Rossomando, P. Mastino, A. K. Erickson, J. H. Her, J. Shabanowitz, D. F. Hunt, M. J. Weber, and T. W. Sturgill, *EMBO J.* **10,** 885 (1991).

[2] P. Lenormand, C. Sardet, G. Pages, G. L'Allemain, A. Bruret, and J. Pouyssegur, *J. Cell Biol.* **122,** 1079 (1993).

[3] K. Y. Choi, B. Satterberg, D. M. Lyons, and E. A. Elion, *Cell* **78,** 499 (1994).

[4] S. Marcus, A. Polverino, M. Barr, and M. Wigler, *Proc. Natl. Acad. Sci. U.S.A.* **91,** 7762 (1994).

[5] E. A. Elion, *Trends Cell Biol.* **5,** 322 (1995).

[6] A. J. Whitmarsh, and R. J. Davis, *Trends Biochem. Sci.* **23,** 481 (1998).

[7] H. D. Madhani and G. R. Fink, *Trends Genet.* **14,** 151 (1998).

 0076-6879/00 $35.00

takes place. Different scaffolding proteins can therefore utilize common signaling components to respond to different stimuli, maximizing regulatory flexibility while minimizing the risk of unwanted cross-talk between pathways. Again, studies in yeast provided the first evidence for insulation of signaling in this manner.[8–10]

These properties of scaffolding proteins also suggest a mechanism for the temporal regulation of signaling. In particular, modification of a scaffold through feedback or signaling from parallel pathways could conceivably alter the stability or localization of the signaling complex, allowing temporal regulation of spatially distinct pools of the same enzyme. Feedback phosphorylation of STE5 and associated kinases has been reported.[3,11,12] Additional work in yeast reveals that molecular scaffolds are also important in determining which substrates are phosphorylated by FUS3/KSS1, the MAP kinases in the pheromone-response pathway, and hence dictate the biological response to activation of these MAP kinases.[13]

It is thus apparent that molecular scaffolds can dictate signaling specificity at four levels: by selecting which core kinases are used; by targeting these enzymes to specific activators or sites of activation; by regulating the time course of their activation; and by selecting downstream substrates for the MAP kinase. In addition to providing great specificity and regulatory flexibility, formation of such complexes likely increases the efficiency of signaling by placing each kinase in physical proximity to its substrate.

The role of scaffolding in mammalian MAP kinase signaling is less well understood, although a number of candidate scaffold proteins in the ERK[14–17] and JNK[18,19] pathways have been identified recently. MEK (MAPK/ERK kinase)-partner 1 (MP1) was identified in a yeast two-hybrid

[8] F. Posas and H. Saito, *Science* **276,** 1702 (1997).

[9] R. L. Roberts and G. R. Fink, *Genes Dev.* **8,** 2974 (1994).

[10] H. Liu, C. A. Styles, and G. R. Fink, *Science* **262,** 1741 (1993).

[11] B. Errede, A. Gartner, Z. Zhou, K. Nasmyth, and G. Ammerer, *Nature* (*London*) **362,** 261 (1993).

[12] J. E. Kranz, B. Satterberg, and E. A. Elion, *Genes Dev.* **8,** 313 (1994).

[13] D. M. Lyons, S. K. Mahanty, K. Y. Choi, M. Manandhar, and E. A. Elion, *Mol. Cell. Biol.* **16,** 4095 (1996).

[14] H. J. Schaeffer, A. D. Catling, S. T. Eblen, L. S. Collier, A. Krauss, and M. J. Weber, *Science* **281,** 1668 (1998).

[15] S. Stewart, M. Sundaram, Y. Zhang, J. Lee, M. Han, and K. L. Guan, *Mol. Cell. Biol.* **19,** 5523 (1999).

[16] M. Therrien, A. M. Wong, and G. M. Rubin, *Cell* **95,** 343 (1998).

[17] W. Yu, W. J. Fantl, G. Harrowe, and L. T. Williams, *Curr. Biol.* **8,** 56 (1998).

[18] A. J. Whitmarsh, J. Cavanagh, C. Tournier, J. Yasuda, and R. J. Davis, *Science* **281,** 1671 (1998).

[19] M. Dickens, J. S. Rogers, J. Cavanagh, A. Raitano, Z. Xia, J. R. Halpern, M. E. Greenberg, C. L. Sawyers, and R. J. Davis, *Science* **277,** 693 (1997).

screen[14] using MEK1 as bait. The structural and biochemical properties of MP1 are consistent with it functioning as a molecular scaffold in the ERK pathway. First, MP1 binds with striking specificity to MEK1 and ERK1, but not the closely related family members MEK2 and ERK2. Significantly, the MP1–MEK1 interaction is dependent on sequences within MEK1 previously found to be necessary for efficient *in vivo* coupling of MEK activity to ERK.[20,21] Second, in model cellular assay systems, MP1 functions as a coactivator for ERK signaling, consistent with its ability to act as a potent coactivator for MEK in *in vitro* reconstitution experiments.[14] Third, MP1 promotes the physical association of ERK1 and MEK1 in cells, consistent with our observations that MP1 function in reporter assays is dose dependent with respect to both MEK and ERK.[14] Taken together, these data strongly suggest that MP1 can influence signaling specificity and efficiency through the formation of oligomeric signaling complexes. However, the precise biological context within which MP1 functions remains elusive.

The purpose of this chapter is to detail the concepts and experimental approaches we have taken to assess the biochemical properties of MP1, with the expectation that such methods may be of general utility in the study of other scaffolding proteins functioning to modulate protein kinase activity.

Protein–Protein Interactions of MP1

A fundamental property of scaffolding proteins is that they interact with the kinases to be regulated. We have used two methods to investigate such protein–protein interactions: yeast two-hybrid assays and coimmunoprecipitation assays. Standard methodologies for the two-hybrid system are described in the literature[22,23] and are not discussed further here. Detailed materials and methods for the coimmunoprecipitation assays are described below.

Coimmunoprecipitation of MEK1 or ERK1 with MP1

Coimmunoprecipitation assays have been used to establish interactions between MP1 and its binding partners in mammalian cells. Recipient cells are transfected with appropriate expression constructs and transfection reagents. Hamster fibroblasts (ATCC CCL39) at approximately 60–70%

[20] A. D. Catling, H. J. Schaeffer, C. W. Reuter, G. R. Reddy, and M. J. Weber, *Mol. Cell. Biol.* **15,** 5214 (1995).

[21] A. Dang, J. A. Frost, and M. H. Cobb, *J. Biol. Chem.* **273,** 19909 (1998).

[22] C. T. Chien, P. L. Bartel, R. Sternglanz, and S. Fields, *Proc. Natl. Acad. Sci. U.S.A.* **88,** 9578 (1991).

[23] J. W. Harper, G. R. Adami, N. Wei, K. Keyomarsi, and S. J. Elledge, *Cell* **75,** 805 (1993).

confluency in 100-mm dishes are transfected with 2 μg of pLNC FLAG-MP1 (14), and 1 μg of either pCGN HA-ERK1[24] or pCMV HA-MEK constructs,[20] using 15 μl of LipofectAMINE (GIBCO-BRL, Gaithersburg, MD) for 5 hr in serum-free Dulbecco's modified Eagle's medium (DMEM; GIBCO-BRL). It is important that appropriate vector control experiments be performed to enable critical interpretation of each experiment. Thus, if FLAG-tagged MP1 is to be immunoprecipitated and analyzed for associated proteins, a control transfection in which empty vector substitutes for the FLAG–MP1 construct is required to demonstrate that precipitated ERK or MEK is recovered as a result of expression of the tagged MP1. Plates are washed with phosphate-buffered saline (PBS), and incubated in DMEM supplemented with 10% (v/v) fetal bovine serum (FBS) for 16–18 hr. The following procedures are carried out at 4°. Monolayers are washed with cold PBS and drained thoroughly. If necessary, monolayers can be quick frozen in liquid nitrogen and stored at −70° at this point. Cells are lysed (0.8 ml/100-mm plate) in FLAG buffer [50 m*M* Tris–HCl, 150 m*M* NaCl, 1% (v/v) Triton X-100, 10% (v/v) glycerol, 0.5 m*M* EDTA, 0.5 m*M* EGTA, 5 m*M* $Na_4H_2PO_7$, 50 m*M* NaF, pH 7.3 at 4°] supplemented with protease inhibitors [leupeptin (5 μg/ml), aprotinin (5 μg/ml), 3 m*M* benzamidine, 1 m*M* phenylmethylsulfonyl fluoride (PMSF)], and extracts are scraped into 1.5-ml tubes and clarified at 13,000 rpm in a microcentrifuge for 15 min at 4°. Extracts are then normalized with respect to protein concentration and immunoprecipitated with the appropriate anti-epitope antibody. Typically, we have used anti-FLAG monoclonal M2 covalently coupled to agarose (Sigma, St. Louis, MO). We use approximately 10 μg of conjugated M2 per immunoprecipitate and make the volume of beads manageable by supplementing with washed Sepharose (Sigma) to an approximate packed bead volume of 20–40 μl per tube. If the hemagglutinin (HA) epitope is to be immunoprecipitated, anti-HA monoclonal 12CA5 (BAbCo, Berkeley, CA; 5–10 μg per sample) is prebound to protein A agarose (Boehringer Mannheim, Indianapolis, IN), washed, and aliquoted to tubes for the immunoprecipitations. Immunoprecipitates are allowed to form for 2–4 hr on a rotating tube holder and then washed extensively. For analysis of MP1–MEK1 interactions, precipitates are washed four times with lysis buffer before boiling in sodium dodecyl sulfate–polyacrylamide gel electrophoresis (SDS–PAGE) sample buffer. For the MP1–ERK1 interaction, background binding of HA–ERK to the beads can be problematic, and hence these immunoprecipitates are washed twice with FLAG lysis buffer and twice with PBS containing 0.5 *M* NaCl. Immunoprecipitates are resolved by SDS–PAGE, using 10–20% gradient gels, and transferred to nitrocellu-

[24] Y. Chu, P. A. Solski, R. Khosravi-Far, C. J. Der, and K. Kelly, *J. Biol. Chem.* **271,** 6497 (1996).

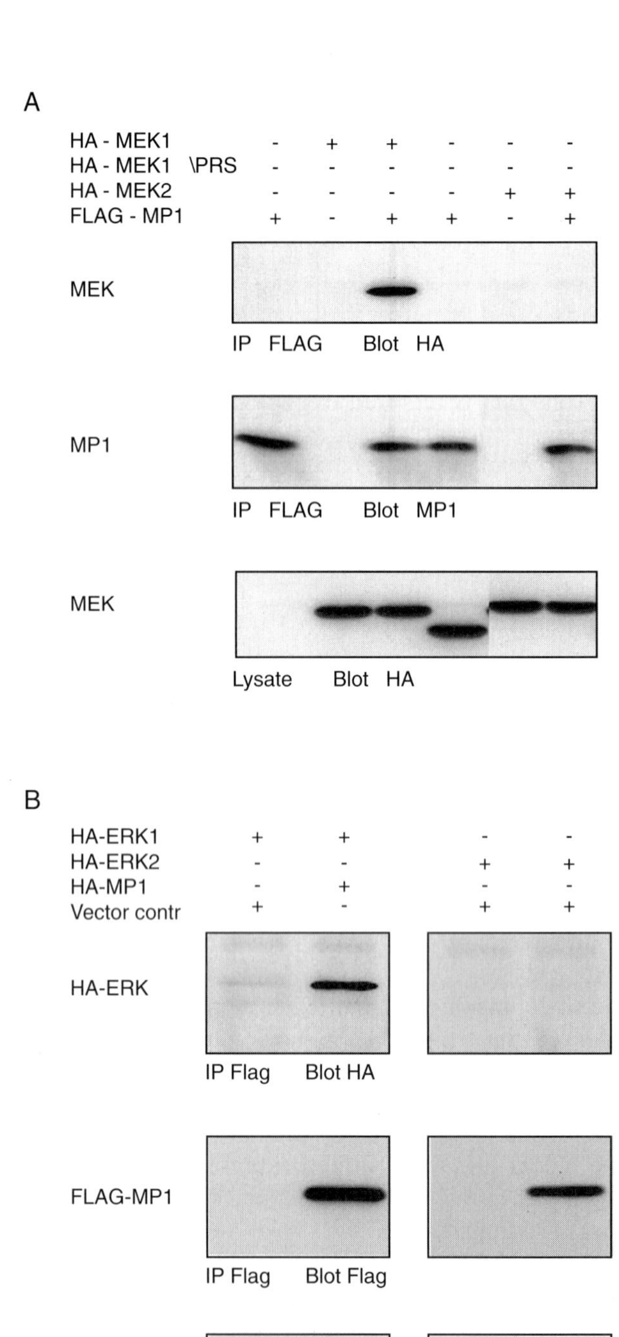

A
HA - MEK1
HA - MEK1 \PRS
HA - MEK2
FLAG - MP1
MEK
IP FLAG Blot HA
MP1
IP FLAG Blot MP1
MEK
Lysate Blot HA
B
HA-ERK1
HA-ERK2
HA-MP1
Vector contr
HA-ERK
IP Flag Blot HA
FLAG-MP1
IP Flag Blot Flag

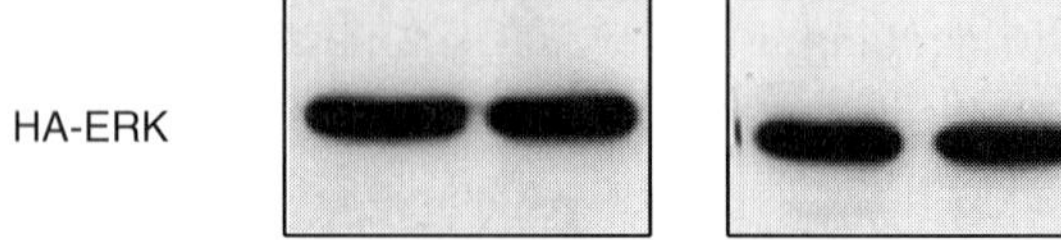

HA-ERK
Lysate Blot HA

lose (BA 83, 0.2-μm pore size; Schleicher & Schuell, Keene, NH), using a Trans-Blot cell (Bio-Rad, Hercules, CA). Immunoblots are performed by standard techniques. Typical results are shown in Fig. 1. These data indicate that MEK1 and ERK1 coimmunoprecipitate with MP1, whereas the closely related family members MEK2 and ERK2 are not detectably recovered. MEK1 ΔPRS deletes residues 270–307 of MEK1, a proline-rich sequence required both for functional coupling of MEK1 and ERK *in vivo*[20,21] and for interaction of MP1 and MEK1 in the two-hybrid system.[14]

Coimmunoprecipitation of MEK1 and ERK1 in Presence of MP1

The MEKs and ERKs are in inactive protein complexes in the cytoplasm of serum-starved cells, with the MEKs acting as cytoplasmic anchors for the ERKs.[25] To determine the effect of MP1 on the association of ERK1 and MEK1 in serum-starved cultured cells, untagged MP1 and epitope-tagged forms of MEK1 and ERK1 are transiently transfected into COS-1 cells (ATCC CRL1650), an African green monkey kidney epithelial cell line immortalized by simian virus 40 (SV40) large T antigen. These cells are chosen for their good transfection efficiency (20–30%) and for their high expression of genes controlled by a cytomegalovirus (CMV) promoter. The total amount of DNA in each transfection is kept constant (6 μg) by using the appropriate empty vector when needed. A constant amount of HA–MEK1 plasmid (3 μg) and FLAG–ERK1 plasmid (2 μg) is transfected as indicated in either the absence or presence of increasing amounts of MP1 plasmid. Transfected MP1 plasmid amounts are either 0, 0.1, 0.5, or 1.0 μg. In addition, transfections lacking either FLAG–ERK1 or HA–MEK1 are included to control for specific immunoprecipitation and for the background level of nonspecific coimmunoprecipitate protein binding. The cells are grown in DMEM containing 10% (v/v) fetal bovine serum at 37° and 5% CO_2. The cells are plated 24 hr prior to transfection at 8×10^5 cells per 10-cm dish. Transfection is carried out by the use of Lipofect-

[25] M. Fukuda, Y. Gotoh, and E. Nishida, *EMBO J.* **16,** 1901 (1997).

FIG. 1. (A) Specific binding of MP1 to MEK1 but not to MEK2. CCL39 cells were transiently cotransfected with FLAG-tagged MP1 and HA-tagged MEK constructs. FLAG–MP1 immunoprecipitates were immunoblotted for MEK with antibody to HA (*top*) or for MP1 with antibody to FLAG (*middle*). Identical amounts of cell lysates were blotted with antibody to HA to verify expression of comparable amounts of MEK (*bottom*). (B) Specific binding of MP1 to ERK1 but not ERK2. CCL39 cells were transiently cotransfected with FLAG-tagged MP1 and HA-tagged ERK constructs. Immunoprecipitates and blots were done as described in (A).

AMINE using a 1:4 ratio of DNA to LipofectAMINE. The plasmid DNA and LipofectAMINE for each transfection are brought up to 0.8 ml with serum-free DMEM in separate polystyrene snap-cap tubes (Fisher Scientific, Pittsburgh, PA). The LipofectAMINE solution is then added to the DNA solution, mixed gently, and incubated at room temperature for 15–45 min. During the incubation all cells are washed once with 8 ml of serum-free DMEM. After the incubation, the DNA–LipofectAMINE mixture is brought up to 8 ml total volume with serum-free DMEM, mixed in a pipette, and placed on the cells. Transfection is carried out for 5 hr at 37° and 5% CO_2.

After transfection the medium is removed, the cells are washed once with PBS, and placed in DMEM–10% (v/v) FBS overnight. The following day, at 20 hr after the start of the transfection, the growth medium is removed and the cells are washed twice with PBS to remove all the serum before placing them in serum-free medium to inactivate the ERK pathway. At this cell density, a 4-hr starvation is sufficient to inactivate the ERK pathway and cause the cytoplasmic localization of the ERKs. More prolonged incubation of these cells in serum-free medium results in a slow reactivation of the ERKs, presumably by the production and accumulation of autocrine growth factors (our unpublished observation, 1998). After the 4-hr starvation the cells are removed from the incubator and placed on ice. All subsequent manipulations are performed on ice or in a 4° cold room. The medium is aspirated and the cells are washed once with cold PBS. The cells are harvested and immunoprecipitations are performed by a slightly modified version of the technique used by Fukuda *et al.*[25] The cells are harvested by scraping into 1 ml of hypotonic buffer consisting of 20 m*M* 4-(2-hydroxyethyl)-1-piperazineethanesulfonic acid (HEPES, pH 7.4), 2 m*M* $MgCl_2$, 2 m*M* EGTA, 200 μM Na_3VO_4, aprotinin (5 μg/ml), 40 m*M* β-glycerophosphate, and 2 m*M* PMSF. The harvested cells are placed into a 1.5-ml microcentrifuge tube and left on ice for 5 min. The cells are then lysed by spinning in a microcentrifuge at 13,000 rpm for 20 min at 4°. The supernatant is transferred to a fresh tube and the pellet is discarded. A portion of the lysate (800 μl) is transferred to a separate tube for immunoprecipitation with anti-FLAG antibody (see above). The remainder of the lysate is saved (see below).

FLAG–ERK1 is immunoprecipitated for 2 hr on a rotating tube holder. Because of the lack of detergent in the buffer, a volume of at least 800 μl is required during the immunoprecipitation in order to break the surface tension of the solution and allow distribution throughout the tube as the tube rotates. Immunoprecipitates are pelleted by centrifugation at 7000 rpm for 1 min at 4°, and the supernatant is then aspirated down to the top of the beads. The remaining supernatant located in the bead volume is removed by vacuum aspiration through a 27-gauge needle attached to a 1-ml syringe. This allows removal of all the residual lysate without aspirating the beads. Each immunoprecipitate is then suspended in 100 μl of cold lysis buffer and placed into a Wizard minicolumn on a vacuum manifold (both from Promega, Madison, WI) at 4°. Another 100 μl of buffer is then used to wash the remaining beads from the tube and the pipette tip and is added to the column. A vacuum is then applied to the manifold to remove the buffer from the columns. The sample are then washed three times with 200 μl of buffer for each wash. After the last wash the columns are placed into microcentrifuge tubes and pulsed briefly (to about 5000 rpm) in a 4° microcentrifuge to remove residual buffer from the bottom of the columns. The columns are then placed into fresh microcentrifuge tubes and 100 μl of FLAG peptide (0.1 mg/ml; Sigma) in hypotonic buffer (without inhibitors) is added to the columns, covering the beads. The tops of the columns are covered with Parafilm and allowed to sit overnight at 4°. This peptide incubation allows for specific release of immunoprecipitated complexes from the antibody and is required because the nonstringent washing system results in considerable nonspecific protein binding to the beads. Simple addition of SDS–PAGE sample buffer is not favorable at this step because it removes all proteins associated with the beads, both specific and nonspecific, increasing the level of nonspecific HA–MEK1 protein detected in the immunoprecipitate.

After an overnight incubation with peptide at 4°, the resin is briefly pelleted in a microcentrifuge and 35 μl of 4× SDS–PAGE sample buffer is added to the recovered eluate. The samples are then boiled for 3 min and run on a 10–15% gradient gel. The gel is then transferred to a nitrocellulose membrane and immunoblotted with monoclonal 12CA5 anti-HA antibodies (BAbCo) and monoclonal M5 anti-FLAG antibody (Sigma). The M5 anti-FLAG blot of the immunoprecipitate will allow detection of the amount of FLAG–ERK1 protein that was immunoprecipitated, while the anti-HA blot will detect the amount of HA–MEK1 coimmunoprecipitated with the FLAG–ERK1. The antibodies are detected with anti-mouse secondary antibodies conjugated to horseradish peroxidase and visualized by enhanced chemiluminescence (Amersham, Piscataway, NJ).

It is necessary to determine the total amount of HA–MEK1 in each lysate to be able to interpret differences in the amounts of HA–MEK1 that coimmunoprecipitate with the FLAG–ERK1. To do this, a portion of each lysate remaining after the initial spin is subjected to SDS–PAGE, transferred, and immunoblotted with 12CA5 antibody. We typically use about 20 μl of lysate for these samples. By comparing the amount of HA–MEK1 in each lysate with the amount of coimmunoprecipitated HA–MEK1, the effect of MP1 on ERK1–MEK1 association can be determined.

A typical result is shown in Fig. 2. These data suggest that the expression of MP1 promotes the interaction between MEK1 and ERK1, consistent with a scaffolding function for this protein in the mammalian ERK pathway.

Advantages and Limitations

These assays serve to confirm the MP1–MEK1 interaction discovered using the yeast two-hybrid system, and enable an analysis of the structural and regulatory aspects of MEK–MP1–ERK complex formation in a mammalian cell background. However, these experiments do not indicate whether association is direct or indirect. Indeed, attempts to define sequences of ERK1 required for the MP1–ERK1 interaction through the use of an extensive panel of ERK1–ERK2 chimeric molecules (provided by P. Shaw, University of Nottingham Medical School, Nottingham, UK) have been uninformative, perhaps as a consequence of there being unidentified bridging molecules (our unpublished data, 1999).

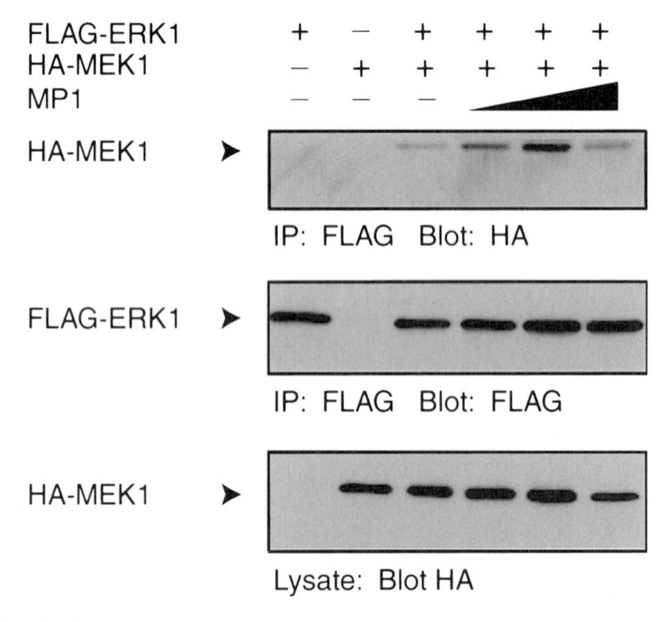

FIG. 2. MP1 increases the association between ERK1 and MEK1. COS-1 cells were transfected with FLAG–ERK1, HA–MEK1, and increasing amounts of untagged MP1 expression constructs. After 5 hr, the cells were washed and placed in DMEM supplemented with FCS (10%, v/v). The following day, the cells were washed and serum starved for 4 hr to inactivate the ERK pathway before harvesting in hypotonic lysis buffer. FLAG–ERK1 was immunoprecipitated from equal amounts of the soluble lysates and immunoprecipitated FLAG–ERK1 (*middle*) and associated HA–MEK1 (*top*) were analyzed by immunoblotting. Equal amounts of soluble lysate were immunoblotted for expression of HA–MEK1 (*bottom*).

A second limitation of the described methodologies is that they fail to establish an interaction between endogenous components, but rely on overexpression of the partners. Although endogenous complexes may be nonabundant and/or unstable and so fall below the detection level of this approach, it is conceivable that overexpression of binding partners might stimulate the formation of nonphysiological complexes.

Third, this approach works only if one is able to extract the complexes under study without promoting dissociation of the components. Hence, if complexes are found in compartments of the cell requiring stringent lysis conditions for extraction, native complexes may not be recovered. Similarly, if the off-rate of binding is rapid, it may be difficult to detect complexes. We suspect that this is why it was necessary to use a rapid and nonstringent wash procedure to detect the MEK–ERK complexes.

Biochemical Function of MP1

In vitro reconstitution assays using purified recombinant proteins can be used to demonstrate a function for MP1 in the enzymology of ERK activation. We have used these assays to show that MP1 can influence both the activation of MEK by Raf, and the activation of ERK by MEK.[14] Methods for preparing recombinant proteins and assaying MP1 function in reconstitution experiments are given below.

Purification of Recombinant Proteins

Recombinant MEK

Recombinant MEK1, MEK2, and mutants thereof are purified from *Escherichia coli* DH5α, using a strategy first employed by Marshall and Cohen and colleagues.[26] The MEK sequences are tagged at the amino terminus with glutathione *S*-transferase (GST) and at the carboxyl terminus with a polyhistidine sequence and expressed from the vector pGEX 2TK (Pharmacia, Piscataway, NJ). This enables full-length protein to be purified from a mixture of full-length and truncated products expressed in the bacteria by virtue of sequentially isolating protein with intact amino and carboxy termini. Minimal contamination with truncated products is probably the result of dimerization of the GST tag. A 5-ml stationary culture of recombinant *E. coli* is added to 500 ml of sterile Luria broth (LB) supplemented with ampicillin (100 μg/ml) and incubated at 37° and 250 rpm until

[26] D. R. Alessi, Y. Saito, D. G. Campbell, P. Cohen, G. Sithanandam, U. Rapp, A. Ashworth, C. J. Marshall, and S. Cowley, *EMBO J.* **13,** 1610 (1994).

A_{600} ~0.5 (1-cm path length). This typically takes 2–3 hr. The culture is removed from the incubator, isopropyl-β-D-thiogalactopyranoside (IPTG) is added to a final concentration of 30 μM, and incubation with shaking is continued at room temperature for 8–16 hr. Bacteria are harvested by centrifugation, and processed immediately or snap-frozen in liquid nitrogen and stored at −70°. The following steps are performed at 4°. Bacterial pellets are resuspended in 10 ml of TNT buffer [50 m*M* Tris–HCl, 50 m*M* NaCl, 0.1% (v/v) Triton X-100, pH 8.0 at 4°, supplemented with leupeptin (5 μg/ml), aprotinin (10 μg/ml), 0.3 *M* benzamidine, and 1 m*M* PMSF] and sonicated on ice to effect lysis. We use a Branson (Danbury, CT) sonifier 450, on setting 7, 80% output, and sonicate for three times for 2 min each. After centrifugation at 20,000*g* for 20 min at 4°, the supernatant is transferred to a 15-ml conical tube containing 0.5 ml of glutathione–Sepharose suspension (prewashed with TNT buffer) and rotated for ~4 hr. The matrix is then pelleted and washed three times with 15 ml of TNT buffer, twice with 15 ml of T/0.5 *M* buffer (50 m*M* Tris–HCl, 0.5 M NaCl, pH 8.0 at 4°, supplemented with 1 m*M* PMSF), and once with 15 ml of TN buffer (50 m*M* Tris–HCl, 50 m*M* NaCl, pH 8.0 at 4°), supplemented with 0.1 m*M* PMSF. Bound fusion protein is then eluted with 2 ml of 20 m*M* reduced glutathione in TN buffer for 15 min. The elution step is repeated a further two times and elutions are pooled. Before performing the elution be sure to adjust the elution buffer to pH 8: adding glutathione to TN causes a marked drop of pH. The pooled fusion protein (~68–70 kDa) can be analyzed by SDS–PAGE at this stage. Some preparations are contaminated with a bacterial protein of ~70 kDa, which can be dissociated from the GST–MEK by incubation of the eluate with ATP (see Pharmacia booklet 18-1123-20).

Full-length GST–MEK is purified from the pooled eluate by chromatography on Ni^{2+}-NTA resin (Qiagen, Valencia, CA). An 8-ml column is prepared in an HR10/10 column (Pharmacia, Piscataway, NJ) and equilibrated with 40 ml of TN buffer. All buffers used in Ni^{2+}-NTA chromatography are supplemented with 0.1 m*M* PMSF and 0.1% (v/v) 2-mercaptoethanol, and the flow rate is maintained at 0.5 ml/min. The pooled eluate described above is filtered and applied, and the column is washed sequentially with 40 ml each of TN and T/0.5 *M* buffer before elution with a 40-ml linear gradient of 0–200 m*M* imidazole in TN. Fractions of 1 ml are collected and analyzed by SDS–PAGE. The majority of truncated products are removed in the flow through and washes and full-length proteins elute early in the gradient. Appropriate fractions are pooled and dialyzed against storage buffer [50 m*M* Tris–HCl, 50 m*M* NaCl, 1 m*M* dithiothreitol (DTT), 0.1 m*M* each PMSF, EDTA, EGTA, 50% (v/v) glycerol, pH 8.0 at 4°] overnight. Purified MEK protein is stored at −20° and is stable for at least

1 year. For reasons that are not understood, we obtain 5- to 10-fold less MEK2 protein than MEK1 from these preparations. The column is washed with 0.2 *M* acetic acid–30% (v/v) glycerol between runs.

Recombinant Raf

B-Raf is purified as a GST fusion protein from baculovirus-infected Sf9 or Sf21 (*Spodoptera frugiperda* ovary) insect cells. The B-Raf baculovirus was a kind gift from W. Kolch, Beatson Laboratories, Glasgow, UK. Insect cells at $\sim 1 \times 10^6$/ml [100–400 ml in Grace's insect medium, supplemented (GIBCO-BRL), plus 10% (v/v) fetal calf serum] are infected with virus at a multiplicity of infection (MOI) of 5 and incubated for 72 hr. The following steps are performed at 4°. Cells are pelleted in a bench top centrifuge, washed with PBS, and either lysed immediately or frozen in liquid nitrogen and stored at −70°. Cell pellets are suspended in 10 ml of FLAG buffer supplemented with protease and phosphatase inhibitors,[14] and disrupted with 20 passes of a Dounce homogenizer. The extract is clarified at 20,000*g* for 30 min, and the supernatant is bound to 1 ml of prewashed glutathione-Sepharose for ~4 hr. The resin is washed three times with 13 ml of lysis buffer, and three times with 13 ml of TNE (50 m*M* Tris–HCl, 50 m*M* NaCl, 1 m*M* DTT, 0.1 m*M* each PMSF, EDTA, EGTA, pH 8.0 at 4°) before elution three times with 2 ml of 20 m*M* reduced glutathione. Typically, a great majority of the B-Raf is eluted in the first aliquot of elution buffer. This fraction is dialyzed against storage buffer overnight and stored at −20°. GST–B-Raf purified in this way contains a substantial amount of 14-3-3 protein.

ERK

The purification of native ERK2 has been described in detail previously.[27,28]

MP1

Histidine-tagged MP1 is purified from baculovirus-infected Sf21 cells. Cells at $\sim 1 \times 10^6$/ml are infected at an MOI of 5 for 48 hr, pelleted, and washed with PBS. Cell pellets are processed immediately or frozen in liquid nitrogen and stored at −70°. Cell pellets are suspended in 10 ml of TN buffer supplemented with 1% (v/v) Nonidet P-40 (NP-40), leupeptin (5 μg/ml), aprotinin (10 μg/ml), 3 m*M* benzamidine, 0.1% (v/v) 2-mercapto-

[27] J. Wu, A. J. Rossomando, J. H. Her, R. Del Vecchio, M. J. Weber, and T. W. Sturgill, *Proc. Natl. Acad. Sci. U.S.A.* **88,** 9508 (1991).

[28] C. W. Reuter, A. D. Catling, and M. J. Weber, *Methods Enzymol.* **255,** 245 (1995).

ethanol and 1 m*M* PMSF, and lysed with 20 passes of a Dounce homogenizer. The extract is clarified by centrifugation at 20,000*g* for 30 min and applied to an Ni^{2+}-NTA (8-ml) column (see above) equilibrated with 40 ml of TN buffer–0.1% (v/v) 2-mercaptoethanol. The column is then washed with 40 ml of TN buffer supplemented with 1% (v/v) NP-40 and 0.1% (v/v) 2-mercaptoethanol, followed by 40 ml of TN buffer–0.1% (v/v) 2-mercaptoethanol. His–MP1 is eluted with a linear gradient of imidazole (0–200 m*M*, 80 ml) in TN–0.1% (v/v) 2-mercaptoethanol. The flow rate is maintained at 0.5 ml/min, and 1-ml fractions are collected throughout the gradient. Because MP1 does not contain tryptophan residues, the broad peak of His–MP1 is not detectable by UV monitoring. Appropriate fractions are pooled, concentrated to 1–2 ml in a Centricon 10 (Amicon, Beverly, MA), and exchanged into 50 m*M* Tris–HCl, 50 m*M* NaCl, 0.1 m*M* EDTA, 0.1 m*M* EGTA (pH 8.0 at 4°) using a PD10 column (Pharmacia) for storage at 4°. His–MP1 prepared in this fashion is ~90% pure as judged by Coomassie blue staining, and is stable for at least 6 months. Preliminary data indicate that freezing and thawing may inactivate the protein (our unpublished observation, 1998).

Reconstitution of ERK Kinase Cascade

We reconstituted the three-kinase ERK cascade *in vitro* using the reagents described above. A number of assays were employed to measure the influence of MP1 on MEK phosphorylation, MEK activation, and ERK activation.

MEK Phosphorylation

After considerable experimentation, we determined that ~0.7 n*M* B-Raf and ~20 n*M* MEK gave a good signal in all assays, but did not saturate any assay. Under these conditions ~5 μ*M* MP1 was able to stimulate activity 5- to 10-fold. In preliminary experiments, we found that any effect of MP1 was diminished in the presence of excess Raf or MEK, as would be expected if MP1 serves as an enhancer of activation rather than being an obligate component of the cascade. Recombinant proteins were combined on ice in a total volume of 15 μl. Dilutions were made as necessary in 50 m*M* Tris–HCl, 50 m*M* NaCl, 0.1 m*M* EDTA, 0.1 m*M* EGTA (pH 8.0 at 4°), and the appropriate buffer control replaced protein solution when that enzyme was omitted. In some assays, we replaced MP1 with either lysozyme (Sigma) or boiled MP1 as controls. Duplicate reactions were initiated at 30-sec intervals by the addition of 15 μl of 2× reaction buffer containing 50 m*M* HEPES buffered with NaOH (pH 7.5), 20 m*M* $MgCl_2$, 2 m*M* DTT, and 200 μM [γ-^{32}P]ATP (~7500 cpm/pmol) and incubated at 30° for 8 min. Reactions were terminated by the addition of SDS–PAGE sample buffer, resolved by electrophoresis on 10% gels, and transferred to nitrocellulose. Incorporation of phosphate into MEK was quantitated by Cerenkov counting, or relative incorporation was estimated by phosphoimage analysis. A typical result is shown in Fig. 3A.

MEK Activation

The experimental protocol described above enabled us to demonstrate that MP1 could facilitate MEK phosphorylation in the presence of B-Raf. However, because MEK1 can be phosphorylated on multiple sites that do not cause activation of enzyme activity, it was important to ascertain whether enhanced phosphorylation of MEK resulted in enhanced catalytic activity. Reactions were initiated as described above, with the exception that radioactive ATP was omitted. After 8 min at 30°, 10 μl of 1× reaction buffer containing kinase-defective ERK2 ($p42^{KR}$)[27] and [γ-^{32}P]ATP was added to yield a reaction volume of 40 μl containing 0.6 μM $p42^{KR}$ and [γ-^{32}P]ATP at ~7500 cpm/pmol. Incubations were continued for a further 8 min, stopped by the addition of SDS–PAGE sample buffer, and processed as described above.

The use of kinase-defective ERK is recommended in these assays because it has little autophosphorylating activity and background incorporation of label into the substrate is low. Alternatively, recombinant wild-type p42 can be used in a modified assay that measures ERK activation. This is the preferred method, because autophosphorylation of wild-type ERK can mask phosphorylation by MEK. For these assays, 0.6 μM $p42^{WT}$ replaces $p42^{KR}$ and radioactive ATP is omitted at this step. Reactions are continued for 8 min, at which time 10 μl is withdrawn and added to a tube containing 30 μl of 1× reaction buffer supplemented with 20 μg of bovine myelin basic protein (MBP; Sigma or GIBCO-BRL) and [γ-^{32}P]ATP at 5000–8000 cpm/pmol. Reactions are continued for 8 min and terminated by the addition of SDS–PAGE sample buffer. Products are resolved by SDS–PAGE on 15% gels, and incorporation is measured by Cerenkov counting of excised nitrocellulose slices. A typical result is shown in Fig. 3B.

ERK Activation

Recombinant MP1 was found to enhance ERK activation by recombinant mutationally activated MEK1[14] (Fig. 3C). Reactions (40 μl) contained ~1 nM mutationally activated MEK (MEK1 S218/222D) and 0.6 μM $p42^{WT}$ in 1× reaction buffer, and were incubated for 8 min at 30°, at which time

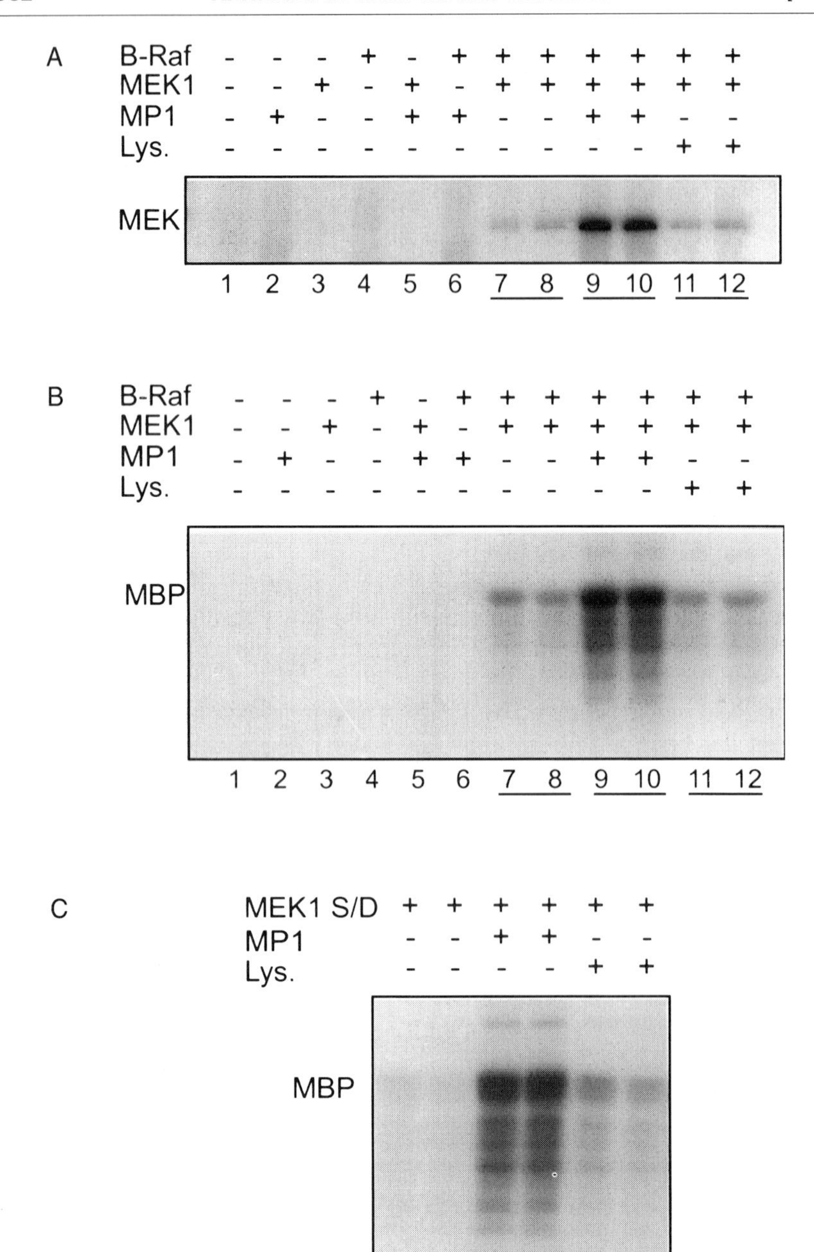
A
B-Raf - - - + - + + + + + + +
MEK1 - - + - + - + + + + + +
MP1 - + - - + + - - + + - -
Lys. - - - - - - - - - - + +
MEK
1 2 3 4 5 6 7 8 9 10 11 12
B
B-Raf - - - + - + + + + + + +
MEK1 - - + - + - + + + + + +
MP1 - + - - + + - - + + - -
Lys. - - - - - - - - - - + +
MBP
1 2 3 4 5 6 7 8 9 10 11 12
C
MEK1 S/D + + + + + +
MP1 - - + + - -
Lys. - - - - + +
MBP
1 2 3 4 5 6

10 μl is withdrawn and added to a tube containing 30 μl of 1× reaction buffer supplemented with 20 μg of bovine MBP and [γ-^{32}P]ATP at 5000–8000 cpm/pmol. After 8 min, reactions are terminated and processed as described above. Interestingly, MP1 was without effect when wild-type MEK1 stoichiometrically activated by B-Raf replaced MEK1 S218/222D (our unpublished data, 1998), suggesting that MP1 recognizes inactive or partially active conformations of MEK.

Advantages and Limitations

A potential role for MP1 in stimulation of the ERK pathway was discovered using these straightforward assays. The effects seen are dose dependent with respect to MP1 concentration, and further work will reveal the mechanistic basis for this activity. The magnitude of the stimulatory effect of MP1 seen *in vitro* is larger than we have observed in cellular assays in which MP1 is transiently overexpressed. The reasons for this difference are unclear at present, but it is likely that endogenous MP1 is present at an optimal level in most cell types and hence overexpression affects signaling only marginally. Additional possibilities include feedback or cross-pathway regulation that serves to inhibit MP1 activity. Thus, one advantage of the *in vitro* assays is that they facilitate a structure–function analysis of MP1 in the context of well-defined steps within the ERK pathway, absent the complicating effects of feedback/cross-pathway regulation.

There are, however, significant limitations to this approach. First, our preliminary data indicate that the binding specificity of MP1 toward MEK1 is not maintained in reconstitution experiments with purified recombinant proteins (our unpublished data, 1998). Indeed, MP1 is able to bind both MEK1 and MEK2, suggesting that additional proteins or posttranslational modifications regulate these interactions in both mammalian and yeast cells. As might be expected from these data, MP1 can stimulate the activation of both MEK1 and MEK2 in assays of the type described above. Second, the physiological relevance of the coactivator activity of MP1 is not addressed in these assays. Further work is required to demonstrate that mutants of MP1 affect signaling in intact cells and in reconstitution assays in a consistent manner. Third, of great practical significance is the finding that the compo-

---

FIG. 3. Effect of MP1 on MEK *in vitro*. Recombinant proteins were combined and assayed as stated in text. (A) Enhanced phosphorylation of MEK1 by B-Raf *in vitro* in the presence of MP1. (B) Enhanced activation of MEK1 by B-Raf in the presence of MP1. (C) Enhanced activation of ERK by mutationally activated MEK1 in the presence of MP1. Lys. denotes lysozyme, a size-matched control protein for MP1. Boiled MP1 was also without effect in these assays (not shown).

nents in these reactions must be carefully titrated in order to see any effect of MP1. We suspect that this will be the case when studying any putative molecular scaffold because the precise stoichiometry of the interacting partners is likely to dictate the biochemical outcome.

Reporter Assays

We have utilized reporter assays to measure the effects of overexpressing MP1 on ERK signaling in intact cells. We have used a bipartite reporter system in which the ERK-responsive *trans*-activation domain of ELK1 is fused to a heterologous DNA-binding domain from the yeast GAL4 protein. The second component is a luciferase reporter driven by a GAL4-responsive promoter. In this way, changes in ELK *trans*-activation activity are translated into changes in the cellular concentration of firefly luciferase, for which automated assay systems exist. We have used both activated Ras and activated MEK1 to stimulate ERK and ELK activity. COS-1 cells are plated at ~3.5×10^5 per 60-mm plate in DMEM–10% (v/v) FCS ~16 hr before transfection. Duplicate dishes are transfected for each combination of plasmids. Each dish is transfected with 50 ng of GAL4–ELK1 and 1 μg of 5× GAL4–E1b–luciferase reporter construct[29]; 2 μg of pLNC MP1 plasmid[14] or corresponding empty vector; 0, 25, 100, or 500 ng of pCMV HA–MEK1 S218/222D[20]; and 0, 50, or 100 ng of pCGN HA–ERK1 construct.[24] In all cases, the total amount of transfected DNA is kept constant by addition of the appropriate empty vector. DNA is diluted to 300 μl with serum-free DMEM (GIBCO-BRL) in polystyrene culture tubes (Fisher Scientific) and 300 μl of serum-free DMEM containing 7.5 μl of LipofectAMINE transfection reagent (GIBCO-BRL) is added and mixed gently. The DNA and transfection reagent are incubated together at room temperature for 15–20 min, during which time cells are washed with serum-free DMEM. After 15–20 min, 2.5 ml of serum-free DMEM is added to each transfection mix, mixed gently, and added to the washed, aspirated culture dishes. Cells are returned to a 37° incubator for 5 hr, and then washed with serum-free DMEM and fed 1% (v/v) FCS–DMEM to reduce the activity of endogenous ERK signaling. After an additional 19 hr, dishes are washed with PBS and assayed with an enhanced luciferase assay kit (PharMingen, San Diego, CA) according to the manufacturer instructions. Luciferase assays were performed with an Analytical Luminescence Laboratories Monolight 2010 luminometer (Ann Arbor, MI). Typically, dishes are lysed in 600 μl, and 5–20 μl of lysate is used per assay. To control for expression

[29] M. S. Roberson, A. Misra-Press, M. E. Laurance, P. J. Stork, and R. A. Maurer, *Mol. Cell. Biol.* **15,** 3531 (1995).

of MEK S218/222D, ERK1, and MP1 constructs, aliquots of this lysate are subjected to SDS–PAGE and processed for immunoblotting. These experiments verified that increasing the amount of MEK and ERK constructs transfected resulted in a corresponding increase in the expression of these proteins (our unpublished observation, 1998). Furthermore, expression of MP1 did not influence the expression of the other components. A typical result from one such experiment is shown in Fig. 4. Similar results were obtained when RasV12 replaced MEK1 S218/222D (not shown). From these experiments one can deduce that the effect of MP1 is seen only when the ratios of activator (MEK1 S218/222D) and ERK1 are within a certain range: MP1 apparently stimulates activation of ELK1 until ERK1 becomes limiting, and increasing the ERK1 concentration allows the MP1 effect

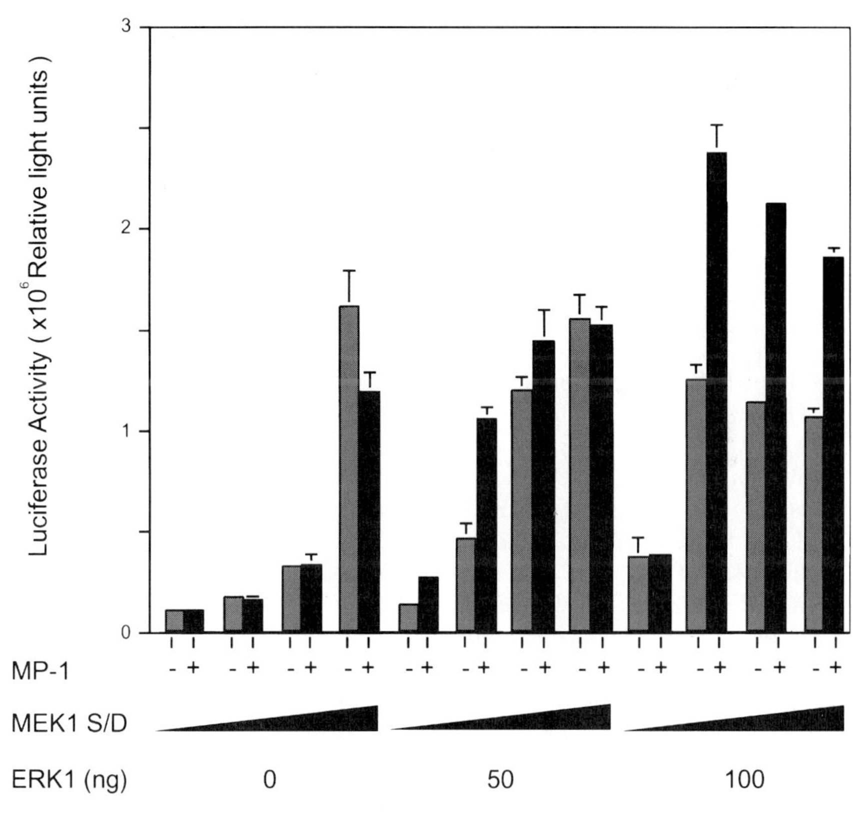

FIG. 4. MP1 enhances ERK1-stimulated gene transcription. COS cells were transfected with various amounts of activated MEK and ERK1 constructs, together with empty vector or MP1 as described in text. The activity of a cotransfected reporter construct was measured after 24 hr in 1% (v/v) FBS–DMEM. Duplicate dishes were used to generate each data point.

to occur at higher concentrations of MEK1. MP1 is without effect when exogenous ERK1 is omitted, suggesting that the ratio of endogenous MP1 and ERK1 is optimal. One problem we encountered with these assays was that the concentrations of MEK and ERK responsive to MP1 expression varied from experiment to experiment. Often, we found that the luciferase signal was maximal at the lowest concentrations used, precluding further stimulation by MP1. As a result of problems of this type, we routinely employed multiple doses of both MEK and ERK to maximize our chances of finding a window of concentration in which the effect of MP1 was detectable. It is unclear why this window shifted between experiments, although many factors including transfection efficiency, cell density, and the concentration of growth factors in any given batch of serum could contribute to this variability.

Advantages and Limitations

This approach to assaying ERK activation and function has the potential advantage that it measures a number of aspects of the signaling pathway not covered in conventional assays of MEK or ERK activity. Notably, if expression of MP1 or another component influences the transport of MEK or ERK to the cell nucleus, this assay may reveal such activity.

From a practical standpoint, this assay is simple to perform, and many combinations of components can be used in duplicate or triplicate in a given experiment. This enables one to titrate components and facilitates a statistical analysis of the data. Such comprehensive experiments are not feasible when immunecomplex kinase assays are to be performed.

However, there are limitations. First, ELK1 activity is regulated by both JNK and ERK,[30–32] and so controls are required to establish which pathway(s) is responsible for any effect seen. Second, this system might measure only the activity of ERK targeted to the nucleus, and hence if the scaffold of interest regulates a nonnuclear pool of ERK, it may be silent in this assay. Third, our assays have used constitutively active molecules to initiate signaling, and hence do not reflect physiological stimulation of ERK by extracellular agonists. In particular, this approach fails to assay the potential input of additional pathways downstream from growth factor receptors that modulate MP1 function, and furthermore, long-term expression of activated signaling molecules can institute feedback and autocrine/paracrine loops that complicate the interpretation of these data.

[30] M. Cavigelli, F. Dolfi, F. X. Claret, and M. Karin, *EMBO J.* **14,** 5957 (1995).

[31] E. Cano, C. A. Hazzalin, E. Kardalinou, R. S. Buckle, and L. C. Mahadevan, *J. Cell Sci.* **108,** 3599 (1995).

[32] A. J. Whitmarsh, P. Shore, A. D. Sharrocks, and R. J. Davis, *Science* **269,** 403 (1995).

Future Directions

The techniques described above have given some insight into the possible biochemical function of MP1 in scaffolding components of the mammalian ERK pathway. However, experiments of this type address the role of MP1 in generic ERK signaling and do not identify the circumstances and locations at which MP1 functions *in vivo*. An understanding of when and where MP1 functions will be necessary for us to appreciate the specific biological role of this protein in ERK signaling. Our experience with MP1 suggests that overexpression of this and perhaps other potential scaffolding molecules is not likely to yield these answers and, moreover, may indeed mislead researchers because the precise concentration of this protein with respect to its partners (identified and unidentified) may dictate the nature of any response. It is primarily for this reason that we believe that answers to these important questions will come from experiments in which our knowledge of the biochemistry of MP1 is combined with a genetic approach to identify the context of its involvement in the ERK pathway. Given the rapid advances made in understanding signaling in general and scaffold proteins in particular, using gene inactivation in yeast, it is perhaps fitting that we find ourselves once again looking to the yeast geneticists for direction.

[29] Bacterial Expression of Activated Mitogen-Activated Protein Kinases

By JULIE L. WILSBACHER and MELANIE H. COBB

Introduction

Mitogen-activated protein (MAP) kinases are serine/threonine protein kinases that are stimulated in response to activation of virtually all cell surface receptors. Included in the MAP kinase family are the extracellular signal-regulated protein kinases 1 and 2 (ERK1/2), ERK3, ERK4, ERK5, ERK7, four p38 MAP kinases, and three c-Jun N-terminal kinases/stress-activated protein kinases (JNK/SAPKs) (reviewed in Refs. 1–4). ERK1/2 are activated by growth factors and other mitogens. Activation of the

[1] T. S. Lewis, P. S. Shapiro, and N. G. Ahn, *Adv. Cancer Res.* **74,** 49 (1998).

[2] J. English, G. Pearson, J. Wilsbacher, J. Swantek, M. Karandikar, S. Xu, and M. H. Cobb, *Exp. Cell Res.* **253,** 255 (1999).

[3] J. M. Kyriakis and J. Avruch, *J. Biol. Chem.* **271,** 24313 (1996).

[4] M. Karin, *J. Biol. Chem.* **270,** 16483 (1995).

0076-6879/00 $35.00

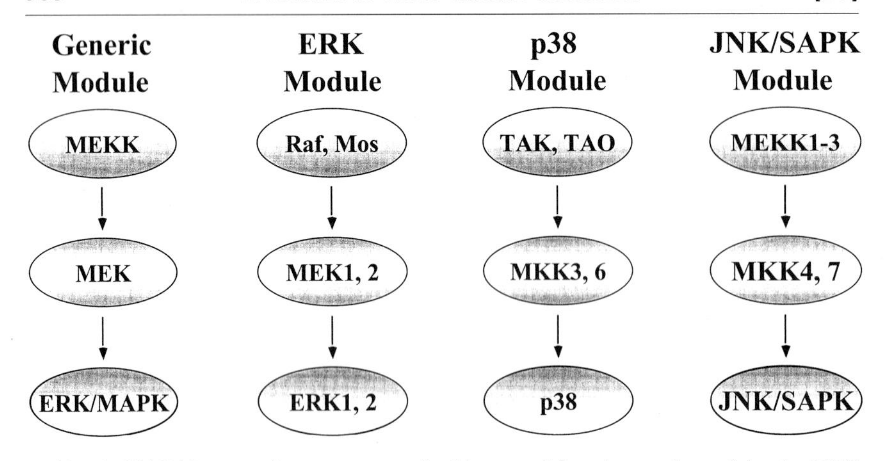

FIG. 1. MAP kinase pathways are organized into modules. A generic module, the ERK module, the p38 module, and the JNK module illustrate how the core enzymes of the MAP kinase cascade are grouped into sets of three kinases.

ERK1/2 cascade results in proliferative responses in many different cell types, and constitutive activation of the pathway has been implicated in cell transformation. ERK1/2 activation can also lead to differentiation in some cell types, and induction of homeostatic responses such as long-term potentiation in neurons.[5] The p38 MAP kinases include p38α, p38β, p38γ, and p38δ, and there are at least 10 splice variants of the JNK/SAPKs.[1,6] Both the p38 MAP kinases and the JNK/SAPKs are activated by cell stress and are collectively termed stress-activated protein kinases.

The enzymes in the different MAP kinase pathways are organized into modules consisting of three kinases: a MAP kinase, a MAP kinase/ERK kinase (MEK or MKK), and a MEK kinase (MEKK) (Fig. 1). Within these modules, the MEKK phosphorylates and activates the MEK, which in turn phosphorylates and activates the MAP kinase. For example, Raf is a MEKK that phosphorylates MEK1/2, and MEK1/2 activate ERK1/2. JNK/SAPKs are activated by MEK4 (also known as MKK4, SEK1, or JNKK1) and MKK7, while p38 MAP kinases are activated by MKK3 and MKK6. There are many MEKKs in the stress-activated pathways.

Unlike the MEKKs and MAP kinases, the MEKs are dual-specificity protein kinases. They activate their respective MAPK substrates by phosphorylating them on both the threonine and tyrosine residues of a specific TXY sequence within the MAP kinase phosphorylation lip (also known as

[5] C. M. Atkins, J. C. Selcher, J. J. Petraitis, J. M. Trzaskos, and J. D. Sweatt, *Nat. Neurosci.* **1,** 602 (1998).

[6] M. H. Cobb, *Prog. Biophys. Mol. Biol.* **71,** 479 (1999).

the activation loop), a regulatory loop that extends out of the active site. The intervening residue X is a glutamic acid in ERK1/2 and ERK5, a glycine in the p38 MAP kinases, and a proline in the mammalian JNK/SAPKs, suggesting that it has significance that is subgroup specific; however, this residue is not a major determinant of specificity for upstream MEKs. Phosphorylation of MAP kinases on both threonine and tyrosine residues in the phosphorylation lip is required for high activity. Phosphorylation of these residues is also required for formation of the dimeric form of ERK2 and for translocation of ERK2 into the nucleus.[7]

Overview of Expression Strategies

We have developed a method to produce large quantities of phosphorylated MAP kinases in *Escherichia coli*. The bacteria catalyze the dual phosphorylation events required to activate these kinases, thereby reducing the effort and reagents needed to activate the enzymes *in vitro*. Several permutations of the expression system have been developed and optimized to produce significant quantities of active forms of multiple MAP kinase and MEK family members.[8] The purified, active enzymes can then be used for a variety of applications, including crystallographic analysis, inhibitor screens, phosphorylation of substrates *in vitro*, or microinjection into cells to examine localization or to phosphorylate *in vivo* substrates.[7,9]

Multiple strategies for expression in *E. coli* were developed that have different features that impact their applicability. Depending on the anticipated use, the expression systems consist of one or two plasmids encoding two or three protein kinases that will be transformed into the same bacteria (Table I). To activate a MAP kinase, it is coexpressed with a constitutively active form of its upstream kinase, MEK, depending on the availability of a suitably active MEK mutant, or the MAP kinase and the wild-type MEK are coexpressed with a constitutively active form of an upstream MEKK. We have most often used the catalytic domain of MEKK1 (called MEKK-C), because this single kinase has activity toward many MEK family members, including MEKs 1–4, 6, and 7. To activate a MEK, it is coexpressed with the appropriate active form of MEKK. High expression of the MEKK is not necessary for efficient activation of the downstream MEK. Moreover, the fact that MEKs are generally less well expressed in bacteria than the MAP kinases has not limited the success of this method.

[7] A. Khokhlatchev, B. Canagarajah, J. Wilsbacher, M. Robinson, M. Atkinson, E. Goldsmith, and M. H. Cobb, *Cell* **93,** 605 (1998).

[8] A. Khokhlatchev, S. Xu, J. English, P. Wu, E. Schaefer, and M. H. Cobb, *J. Biol. Chem.* **272,** 11057 (1997).

[9] J. Wilsbacher, E. J. Goldsmith, and M. H. Cobb, *J. Biol. Chem.* **274,** 16988 (1999).

TABLE I
COMBINATIONS OF PLASMIDS USED TO EXPRESS PHOSPHORYLATED MAP KINASES AND MEKs[a]

Plasmid(s)	Proteins expressed	Method	Protein purified
MEK1R4F+ERK2 pETHis$_6$ (wild-type and mutants)	MEK1R4F, ERK2	1	Phosphorylated ERK2
MEK1R4F+ERK1pETHis$_6$	MEK1R4F, ERK1	1	Phosphorylated ERK1
MEK6DD+p38αpETHis$_6$	MEK6DD, p38α	1	Phosphorylated p38α
MEK1R4FpBB131 ERK1pT7-5	MEK1R4F, ERK1	2	Phosphorylated ERK1
MEK1R4FpBB131 ERK1pT7-5	MEK1R4F, ERK1	2	Phosphorylated ERK1
MEK4+SAPKαNpT7-5 MEKK-CpBB131	MEKK-C, MEK4, SAPKα	2	Phosphorylated SAPKα
MEK4+p38αpT7-5 MEKK-CpBB131	MEKK-C, MEK4, p38α	2	Phosphorylated p38α
MEK1pRSET1 MEKK-CpBB131	MEKK-C, MEK1	3	Phosphorylated MEK1
MEK2pRSET1 MEKK-CpBB131	MEKK-C, MEK2	3	Phosphorylated MEK2
MEK2pGEXKG MEKK-CpBB131	MEKK-C, GST-MEK2	3	Phosphorylated GST–MEK2
MEK3NpT7-5 MEKK-CpBB131	MEKK-C, MEK3	3	Phosphorylated MEK3

[a] The plasmids used, proteins expressed, and proteins purified are listed. Purification was as described in text.

In the first approach, bacteria are transformed with a single plasmid containing a gene conferring ampicillin resistance and encoding both a constitutively active mutant of an appropriate MEK family member and the tagged MAP kinase (Fig. 2A). This approach usually results in the best yield of active MAP kinase. In the second method, a similar plasmid that encodes the wild-type form of an appropriate MEK and the tagged MAP kinase is transformed into *E. coli* together with a second plasmid that encodes MEKK-C and a gene conferring resistance to a different antibiotic such as kanamycin (Fig. 2C and D). In a variation of the two-plasmid approach, the active MEK and the MAP kinase are used on separate plasmids (Fig. 2B and C). This variation is most useful if a series of MAP kinase or MEK mutants is being examined, because it does not require subcloning of the mutants into a single vector. Both hexahistidine (His$_6$) and glutathione *S*-transferase (GST) have been used as the purification tags, with the N terminus as the preferred location of the tag. We have also incorporated second tags to confer recognition by antiepitope antibod-

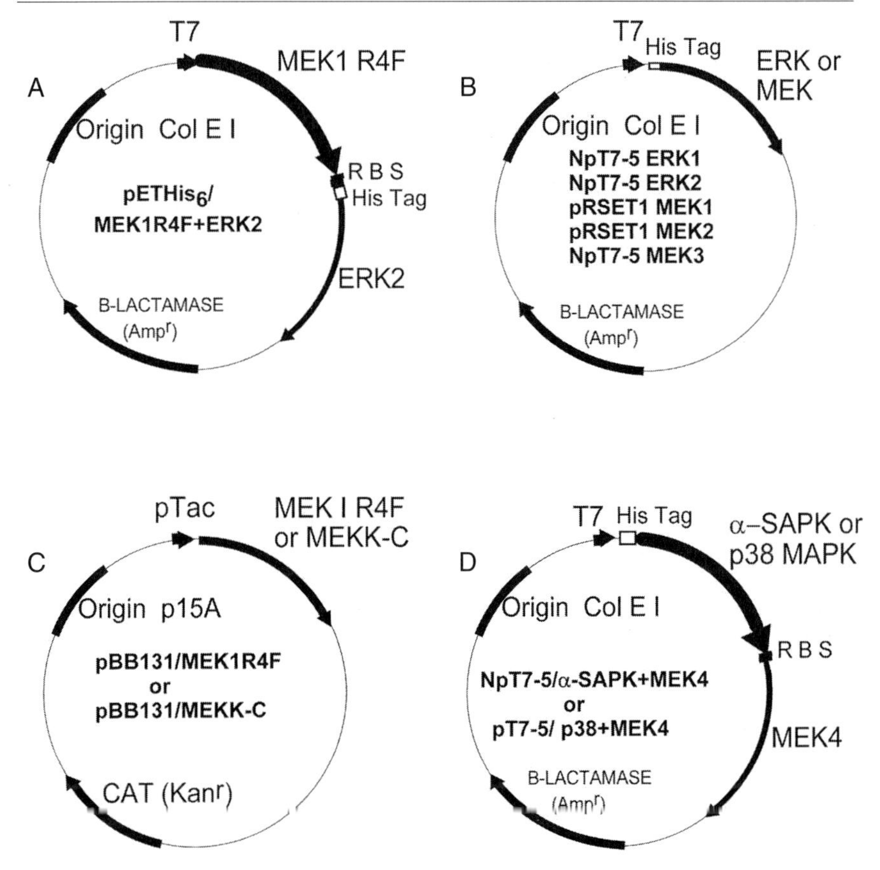

FIG. 2. Plasmids used for expression of phosphorylated MAP kinases and MEKs. (A) The plasmid used in method 1. The plasmids in (B) are used with the pBB131/MEK1 R4F, the plasmids in (C) are used to produce phosphorylated ERK1/2, and the plasmids in (D) are used with pBB131/MEKK-C in (C) to produce phosphorylated SAPK or p38, using method 2. The plasmids used for method 3 are the plasmid pBB131/MEKK-C in (C) and the MEK plasmids in (B). The plasmid encoding GST–MEK2, described in text for method 3, is not shown. [Plasmid maps adapted from Ref. 8.]

ies when desired. Purification of the phosphorylated proteins by epitope affinity chromatography and Mono Q chromatography is sufficient for most purposes.

Expression of Phosphorylated MAP Kinases from Single Plasmid

We have used the single-plasmid method to produce phosphorylated ERK2 and p38 for crystallographic studies. The plasmid used to make phosphorylated ERK2 encodes an untagged, constitutively active form of MEK1, called MEK1R4F, upstream of a His_6-tagged form of ERK2. The resulting plasmid, MEK1R4F + ERK2pETHis$_6$, contains a ribosomal binding site between each gene, in order to synthesize the proteins separately (Fig. 2A). The plasmid also contains an ampicillin resistance gene for selection by ampicillin or carbenicillin.

Method 1: Expression of Active, Phosphorylated His$_6$–ERK2

To express phosphorylated ERK2, transform MEK1R4F + ERK2-pETHis$_6$ into the *E. coli* strain BL21DE3. Plate the transformed bacteria onto dishes containing Luria broth (LB) and carbenicillin or ampicillin and incubate the plate at 37° overnight. To obtain a higher yield of phosphorylated ERK2, freshly transformed bacteria should be used for each expression experiment. After 1 week, expression levels of phosphorylated protein decrease substantially. On day 2, inoculate 100 ml of LB or Terrific broth (TB) containing ampicillin (100 μg/ml) with one or two colonies of the freshly transformed bacteria. Incubate the culture with shaking overnight at 37°.

The phosphorylated ERK2 is expressed on day 3. Add 25 ml of the overnight culture to four flasks containing 1 liter of TB plus ampicillin (100 μg/ml). Grow the cultures at 30° with shaking until the optical density at 600 nm (OD_{600}) is between 0.3 and 0.4. Be careful not to let the OD_{600} exceed 0.4, because the amount of phosphorylated protein decreases as the OD_{600} increases over 0.4. Once the cultures have reached the desired OD, add isopropyl-β-D-thiogalactopyranoside (IPTG) to a final concentration of 0.3 m*M*. Continue growing the cultures with shaking at 30° overnight. Make and chill 2 liters of dialysis buffer (Mono Q buffer A) and solutions for purification on nickel–nitriloacetic acid (NTA) agarose.

On day 4, harvest the bacteria by centrifugation in a Beckman J6B swinging bucket rotor or equivalent at 4000 rpm for 10 min at 4° in four 1-liter bottles. Resuspend each pellet in 40 ml of 4° sonication buffer, and transfer each resuspended pellet to a chilled, sterile 50-ml conical tube. Sediment the bacteria at 2500 rpm for 15 min at 4°. Discard the supernatants and snap freeze the pellets in liquid nitrogen. If desired, pellets can now be stored at −80°. Otherwise, proceed to the purification steps.

Expression of Phosphorylated MAP Kinases from Two Plasmids

The second method of coexpression can be adapted to express phosphorylated ERK1 and ERK2, p38 isoforms, isoforms of JNK/SAPK, or many other phosphoproteins. If a constitutively active form of the upstream MEK is available, only two proteins, the MEK and the MAP kinase, are expressed. We have used this method extensively to make phosphorylated ERK1 and ERK2. Otherwise, three proteins are expressed. These proteins are MEKK-C, a MEK, and a MAP kinase. Phosphorylated p38 and JNK/SAPK isoforms have been produced by this triple protein expression method.

The two plasmids used in this expression protocol are pBB131, which is a low-copy plasmid that contains a kanamycin resistance gene, and pT7-5, which encodes a His_6 tag and contains an ampicillin resistance gene. For the two-protein system, MEK1R4FpBB131 and ERK1pT75 or ERK2NpT7His_6 are used (Fig. 2B and C). For the triple expression system, MEKK-CpBB131 and p38 + MEK4pT7-5 or SAPK + MEK4pT7-5 are utilized (Fig. 2C and D). The p38 + MEK4pT7-5 encodes His_6-tagged p38 upstream of untagged MEK4. Again, there are ribosomal binding sites between each open reading frame. The SAPK + MEK4pT7-5 plasmid was constructed in the same manner (see Fig. 2).

One important step for expression of MAP kinases or MEKs using two plasmids is the optimization of activity produced in small cultures of bacteria before beginning the large-scale expression. Transformation of two types of plasmid at once may result in different amounts of each plasmid within a single bacterium. Therefore, several colonies of bacteria should be tested to determine which colony expresses the largest amount of phosphorylated protein.

Method 2: Expression of Phosphorylated ERK1 Using Two-Plasmid System

To optimize expression of phosphorylated ERK1, first transform MEK1R4FpBB131 and ERK1pT7-5 into the *E. coli* strain BL21DE3. Plate the transformed bacteria onto dishes containing LB plus carbenicillin and kanamycin and incubate the dishes at 37° overnight. Again, it is important to transform bacteria prior to each expression experiment. We have used DNA prepared on a small scale or by polyethylene glycol (PEG) purification for the transformation. Different ratios of the two plasmids should be tested to achieve the best transformation efficiency. Colonies from the plate with the most transformants are then used to inoculate 5 ml of TB containing 100 μg/ml each of ampicillin and kanamycin. Inoculate 10 to 15 cultures and allow the cultures to grow at 37° overnight with shaking. The next morning, inoculate 10-ml cultures of TB plus 100 μg/ml each of ampicillin and kanamycin with 1 ml of each overnight culture. Make frozen stocks of each overnight culture for later use by combining 600 μl of bacterial culture with 400 μl of sterile 50% (v/v) glycerol. Mix gently, snap freeze in liquid nitrogen, and store at −80°.

Grow the 10-ml cultures at 30° until the OD_{600} is 0.3 to 0.4. With this method, it is not as critical to stay within that range of cell density. Once this OD range is reached, add IPTG to achieve a final concentration of 300 μM, and allow the cultures to shake at 30° overnight. After the overnight incubation, pellet the cells at 2500 rpm in a J6B or equivalent swinging bucket rotor. Discard the medium and resuspend the pellets in 10 ml of cold sonication buffer. Collect the cell pellets by sedimentation a second time, and snap freeze the pellets in liquid nitrogen. The pellets can now be stored at −80°, or the next steps can be performed immediately.

Method for Analyzing Expression of Phosphorylated ERK1 Using Kinase Assay

To analyze the expression of phosphorylated ERK1, the pellets are lysed and tested for kinase activity. Resuspend each pellet in 500 μl of ERK purification buffer plus 1 μg/ml each of leupeptin and pepstatin, aprotinin (1.9 μg/ml), and 1 mM dithiothreitol (DTT) and transfer to microcentrifuge tubes. Sonicate each pellet for 40 sec to shear the DNA. Remove the cellular debris by sedimentation at 14,000 rpm in a 4° microcentrifuge for 15 min. Transfer the lysates to new tubes and dilute aliquots of each lysate 1:10 with ERK purification buffer. To test the kinase activity of each lysate, perform an *in vitro* kinase assay. For ERK1, we use myelin basic protein (MBP) as the substrate in the kinase assay. Make a master kinase mix containing the following components per reaction (number of reactions equals number of cultures in duplicate plus an enzyme blank):

Kinase buffer, 10×	5 μl
ATP, 1 mM	5 μl
[γ-^{32}P]ATP, 10 μCi/μl	1 μl
MBP, 5 mg/ml	5 μl
H_2O	29 μl
	45 μl

Start the reactions by adding 5 μl of each diluted lysate or buffer for the enzyme blank. Incubate the reactions at 30° for 15 min, and then add 15 μl of 5× sodium dodecyl sulfate (SDS) sample buffer to stop each reaction. Boil each sample for 2 min and load 20 to 30 μl of each reaction on a 12 to 15% (w/v) SDS–polyacrylamide gel. Stain the gel in Coomassie blue for 20 to 60 min and destain overnight. Expose the dried gel to film at room temperature for 1 hr. To determine the relative incorporation of phosphate catalyzed by each lysate, excise the MBP bands from the gel and measure the counts per minute of incorporated ^{32}P by liquid scintillation. Specific activities can be calculated, if desired. The lysate that incorporates the highest number of counts per minute into MBP should be used for the large-scale expression of phosphorylated ERK1.

For the large-scale expression, inoculate 100 ml of TB containing 100 μg/ml each of ampicillin and kanamycin with some of the frozen stock corresponding to the lysate that had the greatest kinase activity. Shake the culture at 37° overnight. The next morning, inoculate four 1-liter cultures of TB each with 25 ml of the overnight culture. Grow these large cultures at 30° until the OD_{600} is 0.3 to 0.4. Add IPTG to each culture to achieve a final concentration of 300 μM and continue shaking the cultures at 30° overnight. The bacteria should then be washed with sonication buffer and snap frozen as outlined for the one-plasmid method. If desired, the pellets can be stored at −80°, or the purification procedure can be performed right away.

Purification of Phosphorylated MAP Kinases

Purification: Day 1

Purification of phosphorylated MAP kinases by both the one-plasmid and two-plasmid methods are performed in the same manner. For many applications the initial affinity step is sufficient. Further purification may be achieved by ion-exchange and gel-filtration chromatography as described below.

Thaw the pellets on ice (this step may require up to 2 hr). Once the pellets are thawed, add to each pellet 30 ml of cold sonication buffer containing protease inhibitors: 1 μg/ml each of leupeptin and pepstatin, aprotinin (1.9 μg/ml), 100 μM phenylmethylsulfonyl fluoride (PMSF), and 1 mM benzamidine. Resuspend the pellets in the sonication buffer with a pipette, and then break up any clumps with a Polytron sonicator. Lyse the cells by adding 1 to 2 mg/ml of lysozyme (Amersham, Arlington Heights, IL) and incubate on ice for 20 min. Shear the DNA by sonication or add a pinch of DNase (Boehringer Mannheim, Indianapolis, IN) to each tube. Transfer each lysate to a chilled Ti35 tube on ice, and spin in a prechilled (4°) ultracentrifuge in a Beckman Ti35 or equivalent rotor at 27,000 rpm for 30 min at 4°.

While the lysates are spinning, load a 15-ml disposable column with 1 ml of packed Ni^{2+}–NTA agarose (2 ml suspended). Rinse the resin with 50 column volumes of sonication buffer. Once the lysates have been clarified, pool the supernatants in a chilled centrifuge bottle. Save some of the pellet and 100 μl of supernatant as the applied sample. Load the pooled supernatant on the column. Collect the solution that passes through the column as the unadsorbed material. Wash the column with 50 column volumes of 5 mM imidazole in cold sonication buffer, pH 8.0. (Higher imidazole concentrations cause ERK2 to be washed off of the column.) Elute the protein from the nickel column with a 40-ml gradient of 0 to 250 mM imidazole in water. Collect 1-ml fractions.

To determine which fractions contain recombinant ERK2, load 10 μl of the solubilized pellet, the sample of applied material, the unadsorbed material, and 20–30 μl of every third fraction on a 10% (w/v) polyacrylamide gel in SDS. Pool the fractions containing protein. ERK2 tends to precipitate out of solution if it is too concentrated. This problem is reduced if the pooled protein is diluted 1:1 with dialysis buffer. Dialyze the protein against 2 liters of cold Mono Q buffer A overnight at 4°. Chill a JA-20 or ultracentrifuge rotor overnight.

Purification: Day 2

Transfer the dialyzed protein to a chilled JA-20 or Ti35 tube. Set aside an aliquot (50 μl) of the dialyzed protein. Clarify the protein solution by sedimentation at 10,000g at 4° in a JA-20 or ultracentrifuge rotor for 10 min. Set aside a 50-μl aliquot of the clarified protein solution as an aliquot to assay later. Load the protein solution onto a Mono Q HR5/5 fast protein liquid chromatography (FPLC) column that has been equilibrated with Mono Q buffer A. The UV lamp should be set to a sensitivity of 0.5 AU. Run the Mono Q program listed below. This program has a hold after 10 min to inject the protein and then another hold after an additional 25 min to start the fraction collector.

Mono Q program:

0.00 conc %B 100	35.00 hold
0.00 ml/min 1.0	35.00 conc %B 0
0.00 cm/min 0.5	35.00 ml/min 1.0
10.00 conc %B 0.0	40.00 conc %B 25.0
10.00 ml/min 1.0	40.00 ml/min 1.0
10.00 hold	90.00 conc %B 75.0
10.00 conc %B 0.0	90.00 ml/min 75.0
10.00 ml/min 1.0	95.00 conc %B 100
35.00 conc %B 0.0	95.00 ml/min 0.0
35.00 ml/min 1.0	95.00 cm/min 0.0

After eluting the protein from the Mono Q column, measure the protein concentration of the peak fractions. Phosphorylated ERK2 usually elutes in fractions 10–16 as relatively pure protein and in fractions 20–25 along with a high molecular weight impurity. Phosphorylated ERK1 elutes with a similar profile; however, in some cases it is retained on the column slightly longer than ERK2. If any unphosphorylated protein is present, it elutes in the fractions just before phosphorylated ERK2. Some of the early phosphorylated ERK2 fractions may be contaminated with unphosphorylated protein. Dilute the peak fractions, along with samples of unphosphorylated ERK2 and phosphorylated ERK2 as the standards, to 50 to 60 ng/μl. Add 5× SDS sample buffer to each diluted fraction, and boil for 2 min. Load 1 μl of each diluted fraction and the ERK2 standards on a 10–15% gradient SDS Phast Gel (Pharmacia, Piscataway, NJ) with 8/1 combs. Silver stain the gel to visualize protein. Alternatively, analyze the samples on a 9% (w/v) polyacrylamide gel (acrylamide–bisacrylamide ratio of 29:1) in SDS. Phosphorylated ERK2 migrates more slowly, and therefore is recognized by a shifted electrophoretic mobility. Confirmation that the active fractions have been identified can be obtained by protein kinase activity assay or by transferring proteins to nitrocellulose and immunoblotting with antibodies that recognize the doubly phosphorylated epitope as discussed below. Combine the fractions containing pure, band-shifted protein as pool 1 and the fractions with less pure, band-shifted ERK2 as pool 2. Concentrate pool 1 to at least 0.4 mg/ml in a Microcon 10 microconcentrator (Amicon, Danvers, MA). Make at least 2 liters of gel-filtration buffer and chill overnight.

Purification: Day 3

Dialyze the concentrated protein against 1 liter of gel-filtration buffer for 8 hr. Change the dialysis buffer, and dialyze for another 8 or more hours. During dialysis, wash the FPLC pumps and equilibrate a 24-ml Superdex G-75 gel-filtration column with 2 column volumes (50 ml) of chilled gel-filtration buffer that has been filtered and degassed. Use a flow rate of 0.33 ml/min and set the pressure limit to 1.8 MPa. Column equilibration takes about 2.5 hr.

After dialysis, spin the sample at 10,000*g* for 10 min at 4° to remove any precipitated protein. Assay the protein concentration and dilute a sample of the protein to 0.4 mg/ml. To test dimerization under different buffer conditions, add concentrated potassium chloride or sodium chloride to attain the desired final concentration, or ethylenediaminetetraacetic acid (EDTA) to 1 m*M*. Spin the diluted sample at 10,000*g* for 10 min to remove any precipitated protein. Load 50 μl of the 0.4-mg/ml protein sample on the gel-filtration column, elute for 60 min, and measure the OD_{280} of the eluate. Collect 0.33-ml fractions. Combine 30 μl of each peak fraction with 7 μl of 5× SDS sample buffer and boil for 2 min. Analyze 20 μl of each peak fraction on a 10% (w/v) polyacrylamide gel in SDS and transfer to nitrocellulose. Perform immunoblotting with an anti-active ERK antibody

[Quality Controlled Biochemicals (Hopkinton, MA) or New England BioLabs (Beverly, MA)] to confirm that peak fractions contain phosphorylated ERK2. To check the activity of the phosphorylated ERK2 pools, perform a kinase assay. Dilute aliquots of each pool to 0.1, 1, and 10 μg/ml in 1× kinase buffer and perform the kinase assay, using the protocol outlined above. Specific enzyme activity is expressed as micromoles of phosphate transferred per minute per milligram of enzyme protein. For phosphorylated ERK2 the specific activity is in the range of 1–3 μmol/min/mg with MBP as substrate and for the unphosphorylated protein the activity is 1–2 nmol/min/mg. Snap freeze the protein solutions in liquid nitrogen and store at −80°.

Expression and Purification of Phosphorylated GST–MEK2

Phosphorylated GST–ERK2 is produced in bacteria from a combination of MEKK-CpBB131 and MEK2pGEX-KG. Transformation of the plasmids and optimization of expression from bacterial colonies should be performed by a procedure similar to that outlined for phosphorylated ERK1 produced from the double-plasmid system, except that the kinase-dead ERK2 mutant, ERK2 K52R, should be used in place of MBP as the substrate in the kinase assay.

Method 3: Expression of Phosphorylated GST–MEK2 Using Two-Plasmid System

For large-scale expression, inoculate 50 ml of TB containing 100 μg/ml each of ampicillin and kanamycin with some of the frozen stock of the optimized bacteria. Incubate the culture at 37° overnight with shaking. Use 25 ml of the overnight culture to inoculate two 1-liter cultures of TB plus 100 μg/ml each of ampicillin and kanamycin. Grow the large cultures at 30° until the OD_{600} is 0.6 to 0.8, and add IPTG to each culture to achieve a final concentration of 200 μM. Continue shaking the cultures at 30° overnight. Pellet the cells in a J6B or equivalent swinging bucket rotor in 1-liter centrifuge bottles at 4000 rpm for 10 min at 4°. Wash the cells as described above, using GST buffer A instead. Because the cells may not pellet in the GST buffer A at lower speeds, sediment at 5000 rpm for 15 min at 4°. Discard the wash buffer and snap freeze the pellets in liquid nitrogen. The pellets can be stored at −80° as desired.

Purification on glutathione agarose does not yield pure phosphorylated GST–MEK2. However, the eluted protein has high activity toward ERK1 and ERK2. Further purification can be achieved on Mono Q or Mono S as described.[10,11] Thaw the bacterial pellets on ice, and resuspend the cells from each pellet in 40 ml of GST buffer B. Add lysozyme to a final concentration of 2 to 3 mg/ml, and incubate on ice for 30 min. Then add 1.3 ml of 5 *M* NaCl and incubate on ice for 10 min. Finally, add 1.8 ml of 25% (v/v) Triton X-100 and incubate on ice for 10 min. Sonicate the lysates to shear the DNA and spin the lysates at 25,000 rpm in a Ti35 or equivalent ultracentrifuge rotor for 25 min at 4°. Divide the clarified supernatant between two 50-ml tubes per lysate. Add 0.75 ml of glutathione agarose beads (Sigma, St. Louis, MO) in GST buffer C to each tube and incubate on a nutator for 2 hr at 4°. Spin the tubes in a J6B swinging bucket rotor or equivalent at 2000 rpm for 2 min at 4° to pellet the beads. Save the supernatants as nonadsorbed material. Resuspend the beads in a total of 25 ml of GST buffer C and combine the two sets of beads from each liter of cells in one tube. Wash the beads four times in GST buffer C, and resuspend them in a total of 10 ml of glutathione in GST buffer C. For each sample, transfer the resuspended beads to a disposable 15-ml column (Bio-Rad, Hercules, CA) and collect 2-ml fractions. Perform a kinase assay as described above with ERK2 K52R as substrate and pool the fractions with activity. Phosphorylation of the MEK2 can also be detected as shifts in its electrophoretic mobility or using anti-phospho-MEK antibodies (Promega, Madison, WI).

Solutions Needed for Purification of Phosphorylated MAP Kinases and MEKs

TB (4 liters): Combine

Tryptone	48 g
Yeast extract	96 g
Glycerol	16 ml
Distilled H_2O	to 3600 ml

Autoclave for 40 min and store at room temperature until needed. Add 100 ml of potassium phosphate buffer before use

Potassium phosphate buffer for TB (500 ml): Combine 11.55 g of KH_2PO_4 and 62.70 g of K_2HPO_4. Bring to 500 ml with distilled H_2O. Autoclave for 40 min

Ampicillin: Dissolve ampicillin at 100 mg/ml in distilled H_2O; sterile filter. Aliquot into Eppendorf tubes and freeze at −20°

IPTG: 0.3 *M* IPTG in water; sterile filter. Store at −20°

[10] R. Seger, N. G. Ahn, J. Posada, E. S. Munar, A. M. Jensen, J. A. Cooper, M. H. Cobb, and E. G. Krebs, *J. Biol. Chem.* **267,** 14373 (1992).

[11] S. J. Mansour, J. M. Candia, J. E. Matsuura, M. C. Manning, and N. G. Ahn, *Biochemistry* **35,** 15529 (1996).

Sonication buffer: 50 m*M* $NaPO_4$ (pH 8.0), 0.3 *M* NaCl. Store at 4°
Imidazole buffers: Store both at 4°
5 m*M* imidazole in sonication buffer, pH 8.0; column wash buffer 250 m*M* imidazole in water, pH 7.0
ERK purification buffer: 20 m*M* Tris, 1 m*M* EGTA, 10 m*M* benzamidine, 10% (v/v) glycerol; adjust pH to 7.5. Add 100 μM PMSF just before use. Store buffer at 4°
Kinase buffer (10×): 200 m*M* HEPES (pH 8.0), 100 m*M* $MgCl_2$, 1 m*M* benzamidine, 1 m*M* DTT
Mono Q buffer A: 20 m*M* Tris–HCl (pH 8.0), 10% (v/v) glycerol, 50 m*M* NaCl, 1 m*M* EDTA, 0.2 m*M* EGTA, 1 m*M* DTT, 1 m*M* benzamidine. Make 500 ml with highly purified water
Mono Q buffer B: 20 m*M* Tris–HCl (pH 8.0), 10% (v/v) glycerol, 400 m*M* NaCl, 1 m*M* EDTA, 0.2 m*M* EGTA, 1 m*M* DTT, 1 m*M* benzamidine
Gel-filtration buffer: For dialysis—12.5 m*M* HEPES (pH 7.3), 100 m*M* KCl, 6.25% (v/v) glycerol, and 0.5 m*M* DTT. For column—12.5 m*M* HEPES (pH 7.3), 100 or 150 m*M* KCl, 6.25% (v/v) glycerol, and ±1 m*M* EDTA. Make 500 ml fresh with highly purified water and degas at 4° for at least 8 hr prior to use
SDS sample buffer (5×): 0.3 *M* Tris, 10% (w/v) SDS, 50% (v/v) glycerol, 0.3 *M* 2-mercaptoethanol; adjust to pH 6.8 and add bromphenol blue for color
GST buffer A: 50 m*M* Tris (pH 8.0), 25% (w/v) sucrose, 10 m*M* EDTA. Sterile filter and store at 4°
GST buffer B: 10 m*M* Tris (pH 7.4), 1 m*M* EDTA; sterile filter and store at 4°. Add just before use: 1 m*M* DTT, 1 m*M* PMSF, leupeptin (1 μg/ml), pepstatin A (1 μg/ml), 1 m*M* benzamidine
GST buffer C: 20 m*M* HEPES (pH 7.6), 100 m*M* KCl, 1 m*M* EDTA, 20% (v/v) glycerol; sterile filter and store at 4°. Add on the day of use: 1 m*M* DTT, 1 m*M* benzamidine, leupeptin (1 μg/ml), and pepstatin A (1 μg/ml)
Glutathione agarose beads (Sigma): Swell the beads overnight in GST buffer C at 4°. Swell in 0.25 g of beads in 50 ml of GST buffer C
Glutathione (5 m*M*) in GST buffer C: Combine 46 mg of reduced glutathione in 30 ml of GST buffer C (with inhibitors). Add 2 drops of 10 *N* NaOH to bring to pH 7.5. Adjust pH as necessary and store at 4°

[30] Steroid Receptor Fusion Proteins for Conditional Activation of Raf–MEK–ERK Signaling Pathway

By MARTIN MCMAHON

Introduction

Conditionally active oncogenes have had broad utility for exploring the mechanisms of neoplastic transformation of cells in culture. As genome sequencing efforts add to our knowledge of the multiplicity and complexity of intracellular signaling pathways it will become increasingly important to devise additional strategies for the conditional expression or activation of signaling proteins within cells. This chapter focuses on a strategy that exploits the properties of the hormone-binding domains (HBDs) of steroid receptors that, when fused directly to a signaling protein of interest, frequently generate a hormone-dependent form of the molecule that can be used to study the transmission of signals within the cell.[1]

Fusion of E1A or c-Myc to the HBDs of the human glucocorticoid (hbGR) or estrogen (hbER) receptors, respectively, yielded fusion proteins that were regulated by the presence of the cognate steroid hormone.[2,3] This conditional strategy has subsequently been applied to a wide variety of tyrosine, serine/threonine, and dual specificity protein kinases, rendering their activity hormone dependent when expressed in a wide variety of cell types.[4–6] Furthermore, such HBD fusion proteins also display conditional activation in yeast, arguing for their broad utility in studying protein kinase function in most eukaryotes.[1,7] The major advantages of this system are as follows: (1) activation of most protein kinase : HBD fusions is rapid because much of the regulation is posttranscriptional; (2) it is simple to titrate the level of activation of the protein kinase and its downstream effectors by varying the hormone concentration in the medium such that both qualitative and quantitative effects of signal pathway activation on cell physiology may be assessed; and (3) the manipulations required to construct and utilize

[1] T. Mattioni, J. F. Louvion, and D. Picard, *Methods Cell Biol.* **43,** 335 (1994).
[2] M. Eilers, D. Picard, K. R. Yamamoto, and J. M. Bishop, *Nature* (*London*) **340,** 66 (1989).
[3] D. Picard, S. J. Salser, and K. R. Yamamoto, *Cell* **54,** 1073 (1988).
[4] M. L. Samuels, M. J. Weber, J. M. Bishop, and M. McMahon, *Mol. Cell. Biol.* **13,** 6241 (1993).
[5] N. Aziz, H. Cherwinski, and M. McMahon, *Mol. Cell. Biol.* **19,** 1101 (1999).
[6] I. Treinies, H. F. Paterson, S. Hooper, R. Wilson, and C. J. Marshall, *Mol. Cell. Biol.* **19,** 321 (1999).
[7] P. Jackson, D. Baltimore, and D. Picard, *EMBO J.* **12,** 2809 (1993).

0076-6879/00 $35.00

such fusion proteins in cultured cells are relatively simple. In addition, the combinatorial use of different HBDs offers the possibility of studying the cooperative effects of several protein kinase signaling pathways on cell physiology. The use of such conditional alleles in conjunction with classic conditional expression systems, small molecule and antisense signal pathway inhibitors, and cell lines carrying defined genetic alterations will greatly facilitate our understanding of the biochemical specificity and selectivity of signaling pathways, and how such pathways influence the behavior of mammalian cells.

Ras-Activated Raf–MEK–ERK Pathway

The Ras-activated Raf–MEK–ERK pathway (MEK, MAPK/ERK kinase; MAPK, mitogen-activated protein kinase; ERK, extracellular signal-regulated kinase) is the archetypal mammalian MAP kinase module (Fig. 1). Recruitment of Raf-1 to the plasma membrane by activated Ras leads to activation of its protein kinase activity. Activated Raf-1 phosphorylates MEK1 on neighboring serine residues (S218 and S222 in mouse MEK1), leading to its catalytic activation. Activated MEK1 is a dual-specificity protein kinase that phosphorylates to activate ERK1 and ERK2 on neighboring threonine and tyrosine residues (T183 and Y185 in mouse ERK2). Activated ERKs are pleiotropic mediators of cell physiology that phosphorylate a large number of cellular substrates including many transcription factors that control patterns of cellular gene expression.

The mouse and human genomes contain three *Raf* genes encoding the proteins: Raf-1, A-Raf, and B-Raf. All the Raf proteins have three conserved regions (Fig. 2, CR1–CR3). The amino-terminal CR1 and CR2 regions serve to regulate the activity of the protein kinase domain located in the CR3 region of the protein. Previous mutational analysis indicated that deletion of the amino terminus of all three Rafs yielded constitutively activated forms that elicit oncogenic transformation of NIH 3T3 cells.[8]

Design and Construction of Conditionally Active Forms of Raf-1

Sequences encoding a constitutively active form of human Raf-1 (ΔRaf-1, amino acids 305–648) were amplified by polymerase chain reaction (PCR) to introduce an in-frame *Eco*RI site at the 3′ end of the coding sequence. ΔRaf-1:ER was created by fusing ΔRaf-1 in-frame at this *Eco*RI site to hbER (encompassing amino acids 282–595 of the human estrogen

[8] V. P. Stanton, Jr., D. W. Nichols, A. P. Laudano, and G. M. Cooper, *Mol. Cell. Biol.* **9,** 639 (1989).

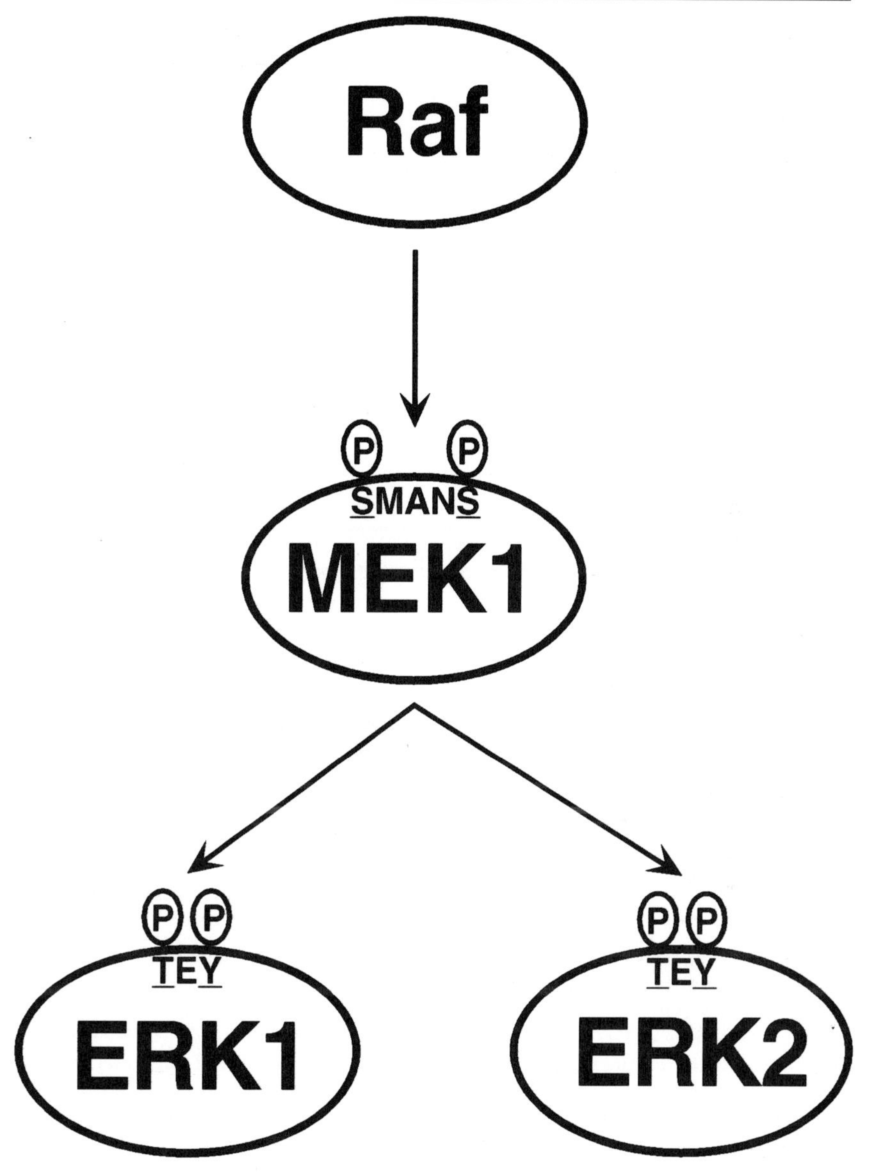

FIG. 1. The Raf–MEK–ERK pathway. Activated Raf phosphorylates MEK1 on serine residues 218 and 222 (underlined) in the sequence SMANS. Activated MEK1 in turn phosphorylates ERK1 and ERK2 on neighboring threonine and tyrosine residues (underlined) in the sequence TEY.

A

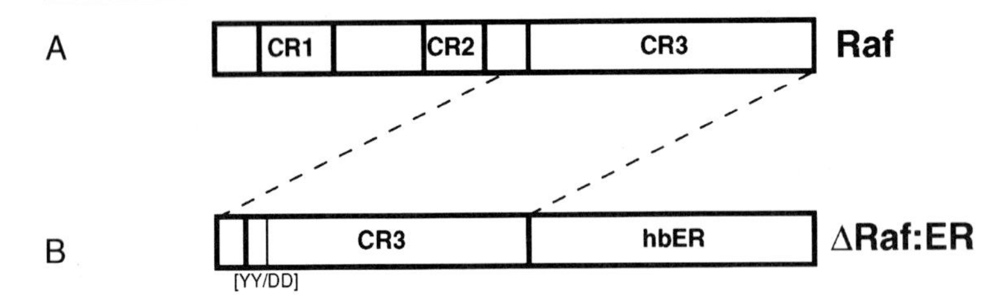

C

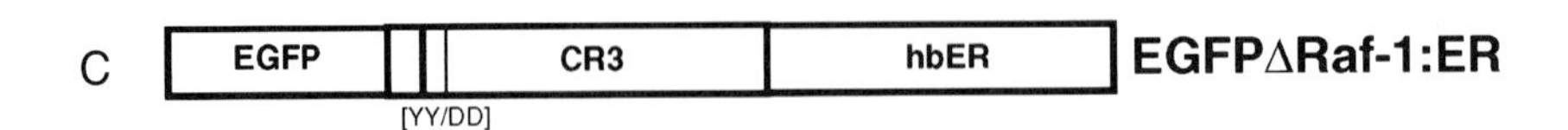

D

E

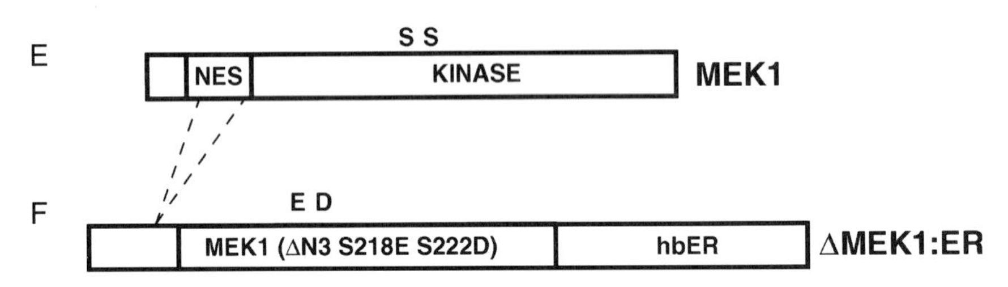

FIG. 2. Conditionally active forms of Raf and MEK1. (A) Schematic of mammalian Raf protein kinases. CR1 refers to the Ras-binding domain and the cysteine-rich region of the protein. CR2 refers to the serine/threonine-rich region of the protein that negatively regulates the catalytic activity of the protein kinase domain contained within CR3. (B) Schematic of ΔRaf:ER. The CR3 regions of human Raf-1 and mouse A-Raf and B-Raf were fused in frame to the hormone-binding domain of the human estrogen receptor (hbER). YY and DD refer to point mutations of two adjacent tyrosine residues to aspartic acid that potentiate the catalytic activity ΔRaf-1:ER and ΔA-Raf:ER. (C) Schematic of EGFPΔRaf-1:ER. The ΔRaf-1:ER fusion protein described in (B) was fused to the C terminus of an enhanced form of green fluorescent protein (EGFP), rendering the resulting chimeric protein constitutively fluorescent but retaining hormone-dependent regulation of Raf-1 catalytic activity. (D) Schematic of Raf-1:ER. Sequences encoding full-length Raf-1 were fused in frame with hbER. YY and DD refer to point mutations of two adjacent tyrosine residues to aspartic acid that potentiate the catalytic activity of full-length Raf-1, rendering it oncogenic in NIH 3T3 cells. (E) Schematic of MEK1. The nuclear export of MEK1 is promoted by a nuclear export signal

receptor-α; Fig. 2A and B). By chance, this form of hbER contains a point mutation (G400V) rendering it 10-fold less sensitive than the native estrogen receptor-α to hormone stimulation.[9,10] Maximal activation of ΔRaf-1:ER is obtained at 50–100 n*M* hormone. In addition, ΔRaf-1:ER is also activated by the estrogen analogs 4-hydroxytamoxifen (4-HT) and the ICI series of compounds [164,384 and 182,780; experimental compounds produced by Zeneca Pharmaceuticals (Cheshire, UK), code name ICI] that are antagonists of the native estrogen receptor. The use of both estrogen receptor agonists and antagonists to activate ΔRaf:ER proteins aids the design of appropriate controls to rule out a contribution of the native estrogen receptor to any biological effects observed.

ΔRaf-1:ER fusion proteins and their derivatives are expressed largely in the cytoplasm with little or no evidence of nuclear localization in the absence or presence of hormone. However, some hbER fusion proteins (Abl:ER and ER:E2F-1) show evidence of hormone-dependent nuclear localization, which likely plays a role in the conditionality observed with these proteins.[7,11]

The human hbER encompasses transcriptional activation function 2 (AF-2) of the estrogen receptor, which has complicated the interpretation of experiments using the Myc:ER protein.[12] To obviate concerns about the presence of the AF-2 transcription activation domain in the original ΔRaf-1:ER constructs, a second series of constructs was derived in which ΔRaf-1 was fused to the hormone-binding domain of a modified form of the mouse estrogen receptor (ER™ or hbER*) in which the AF-2 function is disabled by point mutation (G525R). ER™ is largely insensitive to 17β-estradiol at concentrations less than 100 n*M* but retains the capacity to be activated by 4-HT and ICI compounds.[13] Because ΔRaf-1:ER* does not contain the desensitizing G400V mutation found in ΔRaf-1:ER, it is 10

[9] L. Tora, A. Mullick, D. Metzger, M. Ponglikitmongkol, I. Park, and P. Chambon, *EMBO J.* **8,** 1981 (1989).

[10] V. Kumar, S. Green, A. Staub, and P. Chambon, *EMBO J.* **5,** 2231 (1986).

[11] E. Vigo, H. Muller, E. Prosperini, G. Hateboer, P. Cartwright, M. C. Moroni, and K. Helin, *Mol. Cell. Biol.* **19,** 6379 (1999).

[12] D. L. Solomon, A. Philipp, H. Land, and M. Eilers, *Oncogene* **11,** 1893 (1995).

[13] T. D. Littlewood, D. C. Hancock, P. S. Danielian, M. G. Parker, and G. I. Evan, *Nucleic Acids Res.* **23,** 1686 (1995).

(NES) upstream of the protein kinase domain of the molecule. Sites of Raf-mediated serine phosphorylation (S218 and S222) in subdomain VIII on the kinase domain are indicated. (F) Schematic of ΔMEK1:ER. Sequences encoding a constitutively active form of MEK1 with a deletion of the NES (ΔN3) and mutations of the Raf-mediated sites of serine phosphorylation to aspartate and glutamate (S218E and S222D), respectively, were fused in frame to hbER.

times more responsive to 4-HT and ICI and maximal activation of ΔRaf-1:ER* is obtained at 5–20 n*M* 4-HT. We have not detected any significant differences in the biochemical or biological effects of ΔRaf-1:ER compared with ΔRaf-1:ER*. Moreover, the complication of the AF-2 function in ΔRaf-1:ER is obviated by the fact that the protein is largely cytoplasmic and by the use of 4-HT and ICI compounds, which do activate the AF-2 function of the native estrogen receptor-α.

Conditional Transformation of Cells by ΔRaf-1:ER

For expression in cells, sequences encoding ΔRaf-1:ER were subcloned into a series of retrovirus vectors that also encode drug resistance to either G418, puromycin, or blasticidin, or in some cases encode marker genes such as green fluorescent protein (GFP). Replication-defective retroviruses were generated and tested for their ability to elicit transformed foci of NIH 3T3 cells. As expected from previous experiments a retrovirus expressing Neo alone or full-length Raf-1 did not transform NIH 3T3 cells (Table I) whereas a virus encoding ΔRaf-1 transformed ~100% of the drug resistant cells. Neither estrogen nor its analogs had any effect on transformation of cells by these vectors. Strikingly, cells expressing ΔRaf-1:ER formed few foci in the absence of estrogen but in the presence of estrogen almost all the drug resistant cells formed foci.

ΔRaf-1:ER-expressing cells expanded into mass culture displayed a normal flat morphology, contact inhibition, and had a well-ordered actin stress fiber network marshaled by abundant focal adhesions [Fig. 3, no estradiol ($-E_2$)]. After hormone addition the vast majority of cells displayed the well-defined characteristics of morphological oncogenic transformation: small, dense, refractile cell bodies; extensive cellular projections; loss of contact inhibition accompanied by the loss of focal adhesions and the actin stress

TABLE I
TRANSFORMATION OF CELLS BY Raf-1, ΔRaf-1, AND ΔRaf-1:ER

Virus stock	G418^R colonies/ml	Transformed foci (FFU)/ml (−E2)	Transformed foci (FFU/ml (+E2)
LNCX (Neo alone)	1×10^5	<10	<10
LNCX Raf-1 (full-length Raf-1)	1×10^5	<10	<10
LNCX ΔRaf-1 (Raf-1 aa 305–648)	1×10^5	1×10^5	1×10^5
LNCX ΔRaf-1:ER	1×10^5	<10	1×10^5

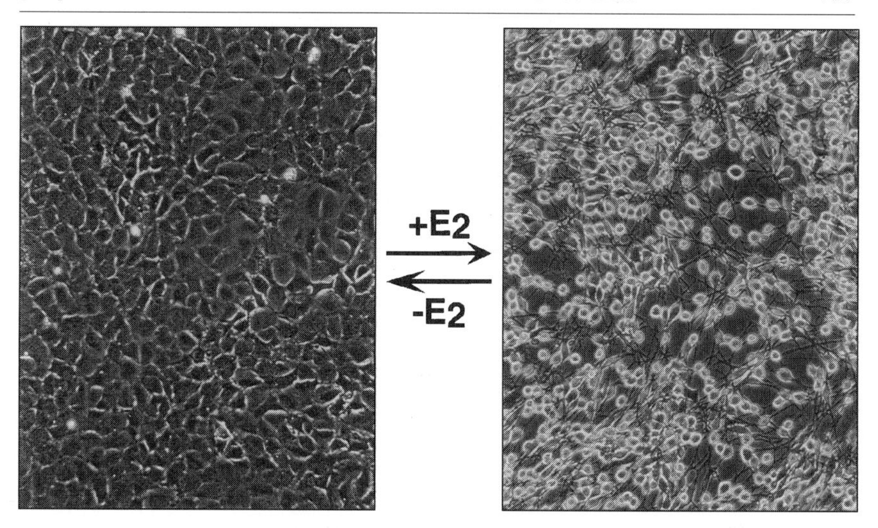

FIG. 3. Hormone-dependent morphological transformation of NIH 3T3 cells by ΔRaf-1:ER. Addition of 1 μM 17β-estradiol, 4-hydroxytamoxifen, or the ICI series of compounds to NIH 3T3 cells expressing ΔRaf-1:ER elicits striking alterations in cell morphology ($+E_2$) that are reversed on subsequent removal of the hormone ($-E_2$).

fiber network ($+E_2$). The remarkably tight conditionality of the ΔRaf:ER system allows the analysis of oncogenic transformation in individual cell clones and in pooled populations of cells derived from 10^5 separate virus infection events.[4]

Rapid, Protein Synthesis-Independent Activation of ERK–MAP Kinase Pathway and Gene Transcription by ΔRaf-1:ER

The biological effects of ΔRaf-1:ER require 16–24 hr to be manifest but the biochemical effects in cells are more rapidly detected. Addition of 25 nM 17β-estradiol to cells expressing ΔRaf-1:ER led to the activation of MEK and the ERKs within 10 min (Fig. 4A). ERK activation was followed within 30 min by phosphorylation of the nuclear transcription factor Ets-2 on Thr-72, a conserved ERK site (PLL$\underline{T}$P), detected with an anti-phospho-T72 Ets antiserum (Fig. 4B). Phosphorylated Ets-2 participates in the activation of gene expression through AP-1/Ets or tandem Ets–Ets cis-acting elements in the enhancer regions of a number of genes including heparin-binding epidermal growth factor (HB-EGF), the induction of which was detected within 30–60 min of ΔRaf-1:ER activation (Fig. 4C). A useful characteristic of this system is that activation of ΔRaf-1:ER is resistant to treatment of cells with cycloheximide. Consequently, cycloheximide did

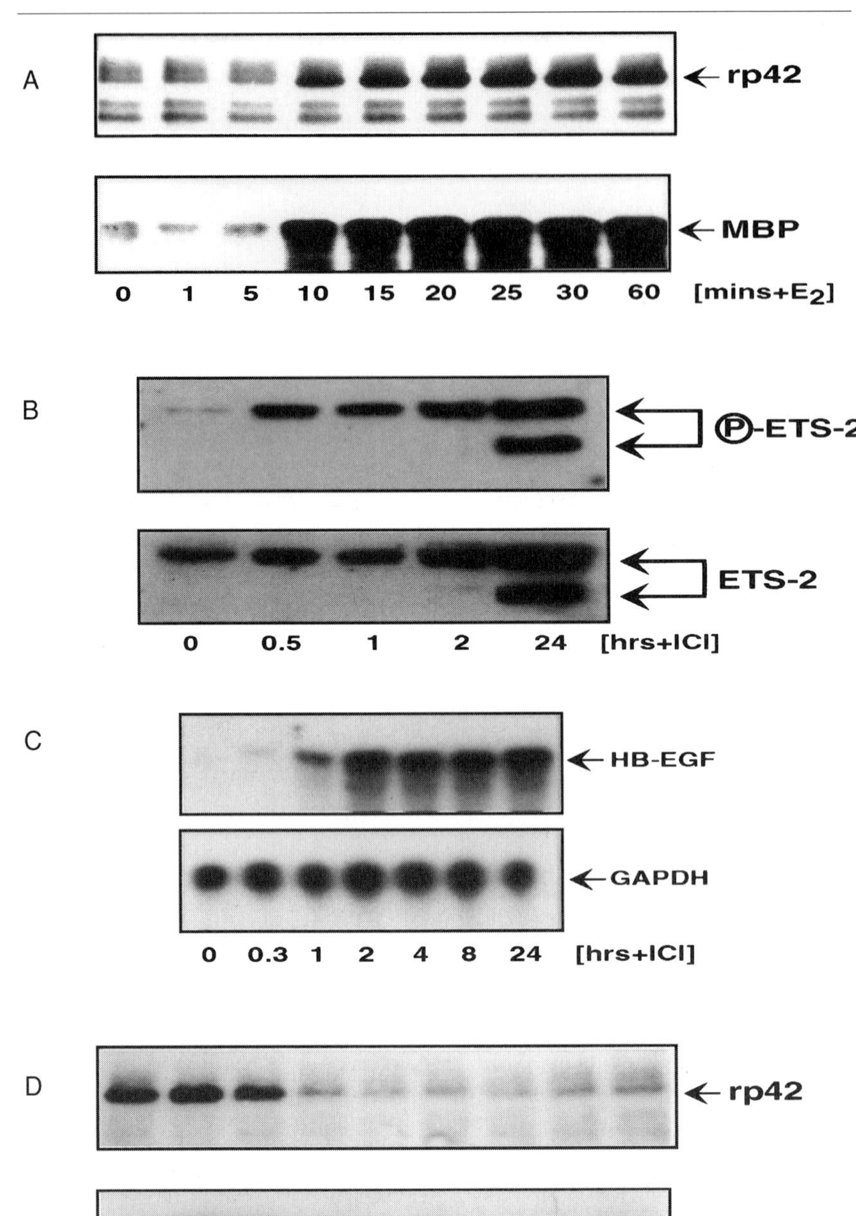
A
rp42
MBP
0 1 5 10 15 20 25 30 60 [mins+E2]
B
P-ETS-2
ETS-2
0 0.5 1 2 24 [hrs+ICI]
C
HB-EGF
GAPDH
0 0.3 1 2 4 8 24 [hrs+ICI]
D
rp42
MBP
0 0.25 0.5 1 2 3 4 5 U [hrs -E2]

not prevent HB-EGF mRNA induction after ΔRaf-1:ER activation. This property of ΔRaf-1:ER makes it possible to discriminate between immediate-early responses to Raf–MEK–ERK activation and those responses that occur with delayed kinetics.[4,14,15]

Rapid Inactivation of ΔRaf-1:ER after Hormone Withdrawal

The activity of ΔRaf-1:ER remains elevated as long as hormone is present in the culture medium but the protein is rapidly inactivated after hormone removal. Cells treated with 1 μM 17β-estradiol for 16 hr were washed extensively with warm phosphate-buffered saline (PBS) followed by culture in medium lacking hormone (Fig. 4D).Within 1 hr MEK and ERK activity had decreased to levels that were lower than that observed in untreated cells. After inactivation of ΔRaf-1:ER the cells revert to a normal morphology within 16–24 hr. The most efficient way to elicit morphological reversion is to trypsinize and replate the cells on a second plate in hormone-free medium. For reasons that are unclear the inactivation of ΔRaf-1:ER appears less efficient if 4-HT or ICI compounds are used in the

[14] S. A. McCarthy, D. Chen, B. S. Yang, J. J. Garcia Ramirez, H. Cherwinski, X. R. Chen, M. Klagsbrun, C. A. Hauser, M. C. Ostrowski, and M. McMahon, *Mol. Cell. Biol.* **17,** 2401 (1997).

[15] S. A. McCarthy, M. L. Samuels, C. A. Pritchard, J. A. Abraham, and M. McMahon, *Genes Dev.* **9,** 1953 (1995).

FIG. 4. Rapid activation and inactivation of ΔRaf-1:ER. (A) NIH 3T3 cells expressing ΔRaf-1:ER were treated with 25 nM 4-HT for various times as indicated, at which time the kinase activity of MEK and ERK was assessed with [γ-^{32}P]ATP and either inactive ERK2 (rp42) or myelin basic protein (MBP) as substrate, respectively. (B) Phosphorylation of Ets-2. NIH 3T3 cells expressing ΔRaf-1:ER were treated with 1 μM ICI 164,384 (ICI) for various times as indicated, at which time nuclear extracts were prepared. The phosphorylation of Thr-72 (P-ETS-2) in a conserved ERK phosphorylation motif (PLL$\underline{\text{T}}$P) was assessed with a phospho-T72-Ets-specific antiserum (*top*). Total expression of Ets-2 was assessed with an anti-Ets-2 antiserum (*bottom*). (C) Induced expression of HB-EGF mRNA. NIH 3T3 cells expressing ΔRaf-1:ER were treated with 1 μM ICI compound for the indicated times, at which time total cellular RNA was isolated and the expression of HB-EGF and GAPDH mRNAs was assessed by RNase protection, using the appropriate [α-^{32}P]CTP-labeled riboprobes. (D) Rapid inactivation of ΔRaf-1:ER after hormone withdrawal. Cells expressing ΔRaf-1:ER were treated with 1 μM estradiol for 16 hr, at which time the cells were washed and cultured in hormone-free medium. Cell extracts were prepared at various times after ΔRaf-1:ER inactivation, at which time the activity of MEK and ERK was assessed with inactive ERK2 (rp42) and myelin basic protein (MBP) as substrate, respectively. An extract from untreated cells (U) was processed in parallel as a control.

initial activation stage, hence the use of 17β-estradiol is recommended under these circumstances.

Mechanisms of ΔRaf-1:ER Activation

It is likely that multiple mechanisms account for the conditionality of the ΔRaf-1:ER protein in cells. The initial activation of the protein may involve the removal of the heat shock protein 90 (hsp90) chaperone from the hbER portion of the molecule, leading to dimerization of ΔRaf-1:ER. It is also possible that the binding of hsp90 sterically constrains access to substrates, prevents phosphorylation of sites essential for catalytic activity, or maintains the protein in a partially unfolded conformation that is catalytically incompetent.[16]

A second facet of the conditionality relates to the fact that 16–24 hr after hormone addition the level of expression of ΔRaf-1:ER is elevated 5- to 10-fold, at least in part, as a consequence of a 4- to 5-fold increase in ΔRaf-1:ER half-life. Consequently 16–24 hr after hormone addition the cells are expressing 5–10 times more ΔRaf-1:ER protein, the specific activity of which is 10–100 times greater.[17,18]

Conditional Transformation of Cells by A-Raf and B-Raf

Because all three mammalian *raf* genes have oncogenic potential, we constructed conditionally active forms of mouse A-Raf and B-Raf by fusing an analogous region of these protein kinases to hbER or hbER* as described above for Raf-1. *In vitro* B-Raf is the most potent MEK1 activator, Raf-1 is intermediate, and A-Raf is the least efficient. When expressed as ΔRaf:ER fusion proteins in mammalian cells this catalytic hierarchy was retained such that ΔB-Raf:ER activated MEK to a greater extent than ΔRaf-1:ER, which in turn was more efficient than ΔA-Raf:ER (Fig. 5). Such differences were also reflected in the level of HB-EGF mRNA induction.[15,19] Surprising, then, was the observation that activation of ΔA-Raf:ER led to a robust mitogenic response whereas activation of either ΔRaf-1:ER or ΔB-Raf:ER failed to do so. Rather, both ΔRaf-1:ER and ΔB-Raf:ER induced a form of G_1 arrest that could not be overcome by the exogenous addition of growth factors. These data indicated that the different Rafs could elicit different biological effects in NIH 3T3 cells and that this might be related

[16] M. A. Carson-Jurica, W. T. Schrader, and B. W. O'Malley, *Endocr. Rev.* **11,** 201 (1990).
[17] C. A. Pritchard, M. L. Samuels, E. Bosch, and M. McMahon, *Mol. Cell. Biol.* **15,** 6430 (1995).
[18] M. L. Samuels and M. McMahon, *Mol. Cell. Biol.* **14,** 7855 (1994).
[19] E. Bosch, H. Cherwinski, D. Peterson, and M. McMahon, *Oncogene* **15,** 1021 (1997).

ΔA-Raf:ER ΔRaf-1:ER

← rp42

0 0.2 2 24 48 72 0 2 [hrs+ICI]

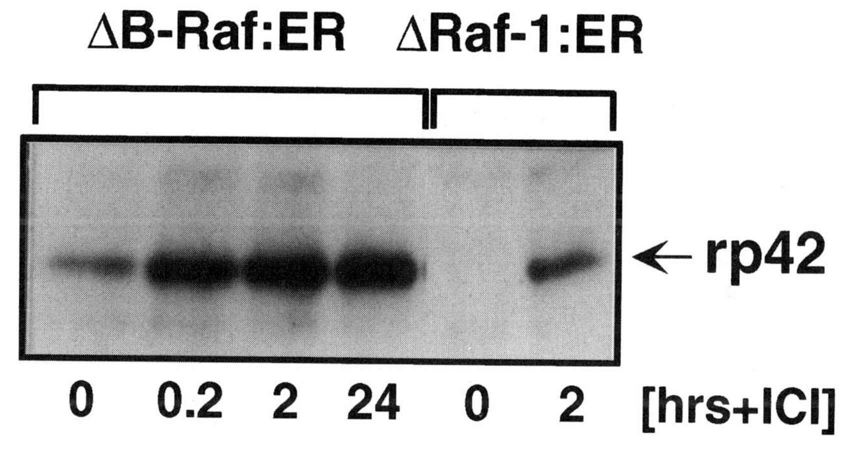

FIG. 5. Differential MEK activation by the different mammalian Raf protein kinases. NIH 3T3 cells expressing ΔA-Raf:ER, ΔB-Raf:ER, and ΔRaf-1:ER were treated with 1 μM ICI compound for various times as indicated, at which time the kinase activity of MEK in crude cell lysates was assessed with inactive ERK2 (rp42) and [γ-^{32}P]ATP as substrates. The levels of MEK activity in the 0- and 2-hr time points in extracts from ΔRaf-1:ER-expressing cells in the two panels are the same. The difference in band intensity is a consequence of the longer exposure time required to demonstrate the activation of MEK by ΔA-Raf:ER.

to the level of Raf–MEK–ERK activation. This hypothesis was borne out by subsequent hormone titration experiments using cells expressing ΔRaf-1:ER and ΔB-Raf:ER.[17,20]

Regulation of Full-Length Raf-1 by fusion to the Hormone-Binding Domain of Estrogen Receptor

The original ΔRaf-1:ER fusion protein was based on a truncated, constitutively active form of the protein. Subsequently we showed that fusion of hbER to full-length Raf-1 resulted in hormone-dependent regulation of Raf-1 kinase activity.[5] Two fusion constructs were created that consisted of either full-length Raf-1 or a more active variant in which tyrosine residues 340 and 341 (YY) were mutated to aspartic acid (DD) and then fused to hbER (Fig. 2C). Activation of these proteins in NIH 3T3 cells led to rapid activation of MEK1 and ERK1/2 and to induced HB-EGF expression as described above. However, only the DD form of full-length Raf-1:ER activated ERK1/2 to a level sufficient to elicit morphological oncogenic transformation in these cells.[5,21] These data are consistent with the work of others, who have shown that fusion of full-length Raf-1 to either FKBP12 or *Escherichia coli* gyrase B proteins yielded forms of Raf-1 that were regulated by the addition of an appropriate dimerizer.[22,23] Consistent also is the fact that the major hormone-dependent dimerization domain of the estrogen receptor is located in hbER, suggesting that dimerization may play a role in the conditionality of Raf:ER fusion proteins.

Properties of ΔRaf-1:ER Not Altered by Addition of Green Fluorescent Protein

Retrovirus vectors are efficient methods for generating stable cell lines expressing proteins of interest. However, in some cells the frequency of expression of the protein of interest is low even in a population selected for virus-infected cells by drug selection. Given the utility of the enhanced form of GFP (EGFP) for tracking protein expression in mammalian cells, we decided to test whether a chimeric protein consisting of EGFP fused to the N terminus of ΔRaf-1:ER would retain conditional activation of MEK and ERK but would be constitutively fluorescent when expressed

[20] D. Woods, D. Parry, H. Cherwinski, E. Bosch, E. Lees, and M. McMahon, *Mol. Cell. Biol.* **17,** 5598 (1997).

[21] J. R. Fabian, I. O. Daar, and D. K. Morrison, *Mol. Cell. Biol.* **13,** 7170 (1993).

[22] M. A. Farrar, I. Alberol, and R. M. Perlmutter, *Nature* (*London*) **383,** 178 (1996).

[23] Z. Luo, G. Tzivion, P. J. Belshaw, D. Vavvas, M. Marshall, and J. Avruch, *Nature* (*London*) **383,** 181 (1996).

in cells (Fig. 2D). In the absence of hormone, NIH 3T3 cells expressing EGFPΔRaf-1:ER were indeed fluorescent and were readily detected and isolated by fluorescence-activated cell sorting. After hormone addition for 24 hr the cells were 5- to 10-fold more fluorescent as a consequence of increased expression of the EGFPΔRaf-1:ER chimeric protein for reasons described above. Most importantly, the conditional biochemical and biological properties of ΔRaf-1:ER appeared to be unaffected by fusion of EGFP to the N terminus of the protein. We have found the properties of EGFP-ΔRaf-1:ER to be especially useful when working with primary mouse and human cells.[20,24] Furthermore the availability of EGFP variants with modified spectral properties makes it feasible to utilize multiple forms of EGFP in the same cell.

Construction of ΔMEK1:ER

To determine if the conditional strategy described above for Raf could be applied to its direct substrate MEK, sequences encoding a constitutively active form of MEK1 (ΔN3, S218D, S222E) were PCR amplified to introduce an *Nhe*I site and then fused in frame to hbER or hbER* as described above (Fig. 2E and F). This form of MEK1 has a small deletion (ΔN3) that removes the nuclear export sequence (NES) of the protein as well as mutations of the sites of Raf-induced phosphorylation (S218D, S222E) to negatively charged residues that partially mimic phosphorylation at these sites. Retrovirus-mediated expression of ΔMEK1:ER elicited hormone-dependent oncogenic transformation of NIH 3T3 and Rat-1 cells. Interestingly, however, the kinetics and fold activation of ERK1/2, as well as the rapidity of morphological transformation by ΔMEK1:ER, were slower compared with ΔRaf-1:ER. This may reflect the fact that, unlike ΔRaf-1:ER, the activation of ERK1/2 by ΔMEK1:ER is cycloheximide sensitive. This suggests that the preexisting protein is not activated by hormone binding and that *de novo* synthesis of ΔMEK1:ER is required for ERK activation. These properties of ΔMEK1:ER may make this conditional transformation system less useful than ΔRaf-1:ER-expressing cells but it is clear that cells expressing ΔMEK1:ER have already provided useful insights into cell signaling pathways.[5,6]

Alternative Steroid Receptor Hormone-Binding Domains

To data the HBDs of the receptors for thyroid hormone, glucocorticoids, estrogen, progesterone, androgens, and mineralocorticoids have been used

[24] J. Zhu, D. Woods, M. McMahon, and J. M. Bishop, *Genes Dev.* **12,** 2997 (1998).

to regulate the activity of heterologous proteins in the context of fusion proteins.[1,25,26] The choice of HBD may be dictated by a variety of concerns such as the expression of endogenous steroid receptors, the desire to use multiple HBDs to regulate more than one activity within the cell, or the tightness of regulation of any given protein kinase:HBD fusion. Such issues can be resolved only empirically.

Ideally it would be preferable to use an HBD that binds a ligand for which mammalian cells do not have an endogenous receptor. Despite the initial promise of using the HBD of the ecdysone receptor from *Drosophila*, it has not been possible to apply this HBD to regulate the activity of the Raf protein kinase.[27] This failure may reflect that rather than forming homodimers, the ecdysone receptor forms heterodimers with the Ultraspiracle protein. Efforts are currently being expended to engineer the specificity of mammalian steroid receptors such that they will accept synthetic hormones. The feasibility of this approach is evident from the generation of ER™, which is activated by 4-HT and ICI compounds but not by estradiol.

Where to Fuse: N Terminus or C Terminus?

Experience with Myc:ER fusion proteins and others has indicated that a HBD may be placed at either the N or C terminus of a protein and still confer steroid regulation on the resulting fusion protein.[2] There are suggestions that, at least for Raf, HBD fusion to either end of the molecule will generate useful conditional alleles. Fusion of the hormone-binding domain of the androgen receptor (hbAR) to the N terminus of ΔRaf-1 yields a protein that is regulated by testosterone and its analog R-1881.[28] A second construct in which hbAR was fused to the C terminus of ΔRaf-1 displays similar properties.[28a] Hence in the design of new protein kinase:HBD fusion proteins it is advisable to test both N- and C-terminal fusions unless there are other considerations such as membrane localization signals or other functionalities that must be retained at one end of the protein.

[25] S. M. Hollenberg, P. F. Cheng, and H. Weintraub, *Proc. Natl. Acad. Sci. U.S.A.* **90,** 8028 (1993).

[26] C. Kellendonk, F. Tronche, A. P. Monaghan, P. O. Angrand, F. Stewart, and G. Schutz, *Nucleic Acids Res.* **24,** 1404 (1996).

[27] K. S. Christopherson, M. R. Mark, V. Bajaj, and P. J. Godowski, *Proc. Natl. Acad. Sci. U.S.A.* **89,** 6314 (1992).

[28] A. Sewing, B. Wiseman, A. C. Lloyd, and H. Land, *Mol. Cell. Biol.* **17,** 5588 (1997).

[28a] A. M. Mirza, A. D. Kohn, R. A. Roth, and M. McMahon, *Cell Growth Diff.* **11,** 279 (2000).

Designing Correct Controls and Considerations for Cell Culture

A major concern in using steroid receptor fusion proteins is the possibility that the hormonal inducers may elicit additional biochemical and biological effects in the target cells that are independent of the fusion protein. Consequently the use of appropriate controls is essential to allow assessment of the validity of experimental results. An essential control is to construct a catalytically inactive form of the protein kinase:HBD fusion to confirm that all of the effects that are observed in cells are dependent on the kinase activity of the fusion protein.[20] In addition, cells expressing the relevant HBD in isolation may also be useful to determine if there are effects of the hormone that are not mediated by the kinase:HBD fusion. The appropriate solvent [ethanol or dimethyl sulfoxide (DMSO)] should also be used as controls in such experiments.

Multiple protein kinase:HBD fusions such as ΔRaf-1:ER and ΔRaf-1:AR can also be useful for confirming results with different steroid inducers. More recently the use of specific inhibitors of signaling pathways (e.g., the MEK inhibitors PD098059 and UO126) can be used to confirm the requirement for downstream signaling components for effects elicited by ΔRaf-1:ER. To date there are no reports of a protein kinase:hbER fusion protein having a dominant-negative effect in cells either in the absence or presence of hormone. However, consideration should be paid to such a possibility. Indeed, attempts to construct conditional dominant-negative forms of protein kinases by this strategy have been uniformly unsuccessful to date.

The use of HBD fusion proteins may be complicated by the presence of agonists in cell culture media that partially activate the HBD fusion protein in the absence of exogenously added hormone. For example, low levels of estradiol in fetal calf serum may, under some conditions, be sufficient to activate a protein kinase:hbER fusion protein to a low level. Stripping the serum with activated charcoal is a convenient method to remove steroids in fetal calf serum. Serum (500 ml) is incubated with 10 g of activated charcoal for 20 min at 4°, after which the charcoal is removed by centrifugation followed by filtration. However, charcoal stripping removes other nutrients and factors that support cell proliferation and cells often grow more slowly in this medium. Of more concern with the use of hbER fusion proteins is the presence of the weak estrogen agonist phenol red in culture media. This concern is obviated by the use of phenol red-free media. In our experience the use of charcoal-stripped serum is largely unnecessary if phenol red-free medium is used because the concentration of estradiol in fetal calf serum is usually in the picomolar range (K. Smith-McCune, personal communication, 1999). All the above concerns are largely obviated by the use of the ER™/hbER* domain, which is nonresponsive to both estrogens and phenol red.

Activation and Detection of hbER- and hbAR-Containing Fusion Proteins

Retrovirus vectors encoding Raf:ER, ΔMEK1:ER, hbER, hbER*, and hbAR are available from the author. 17β-Estradiol is from Sigma (St. Louis, Mo), (*Z*)-4-hydroxytamoxifen is from Research Biochemicals International/Sigma, and the ICI series of compounds (ICI 164,384 and ICI 182,780) were generously provided by K. Wakeling at Zeneca Pharmaceuticals. Estrogen and its analogs were made as 1 m*M* stock solutions in ethanol, stored in the dark at −20°, and diluted directly into cell culture media at the desired concentration. The optimal hormone concentration for activation of protein kinase:HBD fusion proteins and for any particular biological effect must be determined empirically.

Fusion proteins containing hbER or hbER* are readily detected by Western blotting or immunoprecipitation using a polyclonal antiserum that recognizes hbER (HC-20; available from Santa Cruz Biotechnology, Santa Cruz, CA). This antibody also supports protein kinase reactions. Fusion proteins containing hbAR are activated by testosterone (Sigma) or the synthetic testosterone antagonist R-1881 from New England Nuclear (Boston, MA). Because testosterone is a class III Drug Enforcement Agency (DEA)-controlled substance that requires a license from the DEA for its use, the use of R-1881 is recommended. Testosterone and its analogs are prepared, stored, and used as described above for estrogen. Fusion proteins consisting of hbAR are readily detected by Western blotting or immunoprecipitation using a polyclonal anti-peptide antiserum raised against the carboxy terminus of the human androgen receptor (C-19; Santa Cruz Biotechnology). This antibody will also support protein kinase reactions.

Future Prospects for Conditional Regulation of Protein Kinase Activity

To date there are conditionally active forms of at least five different enzymes that regulate MAP kinase modules: Raf-1, A-Raf, B-Raf, MEK1, and MEKK-3. It seems likely that this strategy will be broadly applicable to the study of many, if not all MAP kinase cascades. The description of conditionally active forms of phosphatidylinositol 3′-kinase and Akt/PKB suggests that this strategy may be more broadly applied to both protein and lipid kinases that participate in a variety of biological processes within the cell.[20,29,30]

Acknowledgments

I am particularly grateful to my colleagues T. Sreevalsan, J. M. Bishop, M. Weber, M. Ostrowski, E. Lees, M. Samuels, S. McCarthy, D. Woods, and N. Aziz, and many other colleagues and collaborators, for advice and assistance. I am grateful to D. Dankort and E. Hessel for critical review of this manuscript.

[29] A. Klippel, M. A. Escobedo, M. S. Wachowicz, G. Apell, T. W. Brown, M. A. Giedlin, W. M. Kavanaugh, and L. T. Williams, *Mol. Cell. Biol.* **18,** 5699 (1998).

[30] A. D. Kohn, A. Barthel, K. S. Kovacina, A. Boge, B. Wallach, S. A. Summers, M. J. Birnbaum, P. H. Scott, J. C. Lawrence, Jr., and R. A. Roth, *J. Biol. Chem.* **273,** 11937 (1998).

[31] Pharmacologic Inhibitors of MKK1 and MKK2

By NATALIE G. AHN, THERESA STINES NAHREINI, NICHOLAS S. TOLWINSKI, and KATHERYN A. RESING

Introduction

New approaches to identifying pharmacological agents involve screening of small molecule or peptide libraries, using assays for specific enzymes in critical signaling pathways. One such target is the MKK–ERK pathway, composed of mitogen-activated protein (MAP) kinases, extracellular signal-regulated kinases; 1 and 2 (ERKs 1 and 2), and MAP kinase kinases, (MKKs) (reviewed by Lewis *et al.*[1]). MKK1 and -2 are acutely stimulated by growth and differentiation factors in pathways mediated by receptor tyrosine kinases, heterotrimeric G protein-coupled receptors, or cytokine receptors, primarily through $p21^{ras}$-coupled mechanisms. These enzymes are ubiquitous and are generally expressed at micromolar levels in mammalian cells,[2] although variation in expression between different tissues has been noted.[3,4] Activation of MKK or ERK in response to cell stimulation involves phosphorylation at residues located within the activation lip of

[1] T. S. Lewis, P. S. Shapiro, and N. G. Ahn, *Adv. Cancer Res.* **74,** 49 (1998).

[2] C. Y. Huang and J. E. Ferrell, *Proc. Natl. Acad. Sci. U.S.A.* **93,** 10078 (1996).

[3] T. G. Boulton and M. H. Cobb, *Cell Regul.* **2,** 357 (1991).

[4] T. Moriguchi, H. Kawasaki, S. Matsuda, Y. Gotoh, and E. Nishida, *J. Biol. Chem.* **270,** 12969 (1995).

0076-6879/00 $35.00

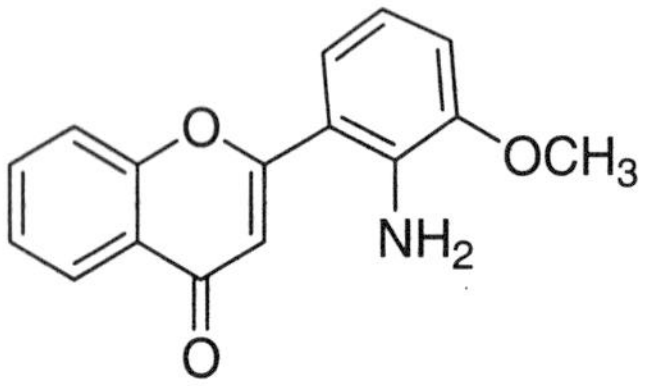

PD 98059

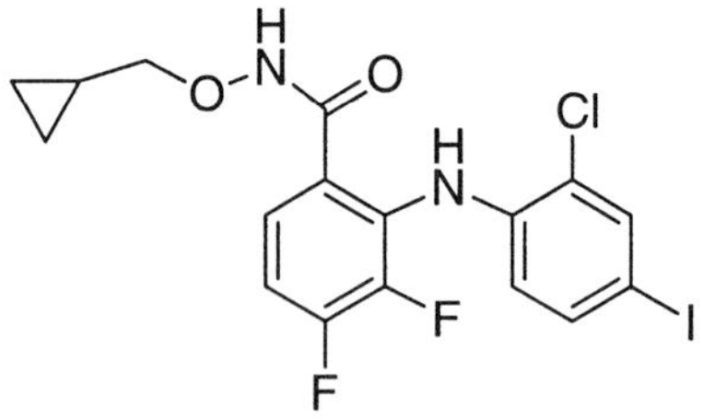

PD 184352

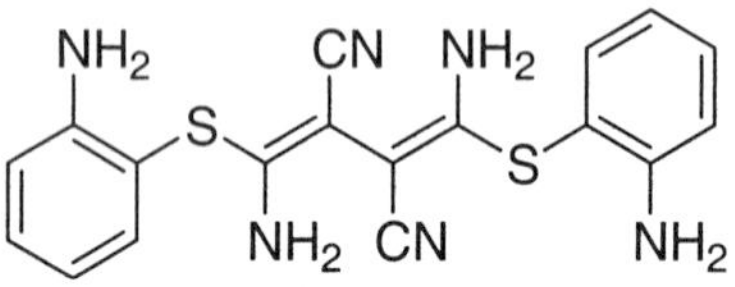

U0126

FIG. 1. Chemical structures of three MKK1/2 inhibitors.

each kinase. Raf-1, c-Mos, or MEKK1 phosphorylate two serine residues in MKK1/2 (residues 218 and 222 in hMKK1; residues 222 and 226 in hMKK2). MKK1/2 then activates ERK1/2 by phosphorylating Thr and Tyr residues within the sequence Thr-Glu-Tyr (residues 202–204 in hERK1; residues 185–187 in hERK2). Thus, MKK1 and -2 fall within a relatively rare class of protein kinases with dual specificity toward Ser/Thr and Tyr residues on exogeneous substrates. Furthermore, substrates for MKK1/2 other than ERK1/2 have not been identified, making MKK1/2 an excellent target for pharmacological intervention in hyperproliferative diseases. Small cell-permeable inhibitors provide an alternative strategy to the use of dominant negative and constitutively active mutants, in studying cellular responses controlled by the MKK–ERK pathway without encountering problems due to overexpression.

Three inhibitors of MAP kinase kinases have been reported, which inhibit MKK1 and MKK2 (Fig. 1). PD98059 [2-(2′-amino-3′-methoxyphenyl)oxanaphthalen-4-one] was identified by screening a small compound library, assaying for inhibition of ERK activation by constitutively active mutant MKK1 recombinant protein.[5] More recently, PD184352 [2-(chloro-4-iodophenylamino)-*N*-cyclopropylmethoxy-3,4-difluorobenzamide] was

[5] D. T. Dudley, L. Pang, S. J. Decker, A. J. Bridges, and A. R. Saltiel, *Proc. Natl. Acad. Sci. U.S.A.* **92,** 7686 (1995).

identified in a similar screen, and shown to inhibit ERK activation in colon carcinoma cells as well as colon tumor growth and invasiveness.[6] U0126 [1,4-diamino-2,3-dicyano-1,4-bis(2-aminophenylthio)butadiene] was identified in a screen for inhibitors of AP-1 *trans*-activation in a cell-based reporter assay, and blocks phorbol ester-dependent induction of transcription controlled by the TPA response element (TRE) by inhibiting MKK1/2.[7]

This chapter describes methods to measure drug inhibition of MKK activity and activation using purified recombinant proteins *in vitro*, and drug effects on ERK activation in intact mammalian cells. Additional characterization of mechanisms involved in inhibition is also included. At the time of this writing, two MKK1/2 inhibitors can be obtained from commercial sources. U0126 can be purchased from Calbiochem (San Diego, CA), LC Laboratories (Woburn, MA), Biomol (Plymouth Meeting, PA), and Promega (Madison, WI), and PD98509 can be purchased from Calbiochem, LC Laboratories, New England BioLabs (Beverly, MA), and Sigma (St. Louis, MO). In our experiments, U0126 was obtained from Promega and PD98059 from Calbiochem. The inhibitors were dissolved in dimethyl sulfoxide (DMSO, 50 and 100 m*M* for U0126 and PD98059, respectively) and stored protected from light at $-20°$.

Inhibitor Effects on Activation and Activity of MKK1 *in Vitro*

In vitro assays for MKK1/2 inhibitors should measure MKK1/2 phosphorylation and activation as well as direct inhibition of MKK1/2 activity. Activation assays require kinases that function upstream of MKK1/2, such as Raf-1, MEKK1, or Mos, prepared by recombinant expression in bacteria or insect cells or by immunoprecipitation from cell extracts.[7–11] Inhibitor is added to the reactions and the effect on the rate of MKK1/2 phosphorylation by upstream kinase is measured. Activity assays measure the rate of

[6] J. S. Sebolt-Leopold, D. T. Dudley, R. Herrera, K. Van Becelaere, A. Wiland, R. C. Gowan, H. Tecle, S. D. Barrett, A. Bridges, S. Przybranowski, W. R. Leopold, and A. R. Saltiel, *Nat. Med.* **5,** 810 (1999).

[7] M. F. Favata, K. Y. Horiuchi, E. J. Manos, A. J. Daulerio, D. A. Stradley, W. S. Feeser, D. E. Van Dyk, W. J. Pitts, R. A. Earl, F. Hobbs, R. A. Copeland, R. L. Magolda, P. A. Scherle, and J. M. Trzaskos, *J. Biol. Chem.* **273,** 18623 (1998).

[8] D. R. Alessi, A. Cuenda, P. Cohen, D. T. Dudley, and A. R. Saltiel, *J. Biol. Chem.* **270,** 27489 (1995).

[9] D. R. Alessi, P. Cohen, A. Ashworth, S. Cowley, S. J. Leevers, and C. J. Marshall, *Methods Enzymol.* **255,** 279 (1995).

[10] S. J. Mansour, J. M. Candia, J. E. Matsuura, M. C. Manning, and N. G. Ahn, *Biochemistry* **35,** 15529 (1996).

[11] S. Xu, D. Robbins, J. Frost, A. Dang, C. Lange-Carter, and M. H. Cobb, *Proc. Natl. Acad. Sci. U.S.A.* **92,** 6808 (1995).

ERK phosphorylation either by wild-type MKK1/2 phosphorylated with upstream kinase or by constitutively active mutant MKK1/2.

Recombinant Protein Preparation

Recombinant rat hexahistidine (His_6)-tagged ERK2 (His_6–ERK2), subcloned in wild-type or mutant forms into expression vector NpT7, is induced in *Escherichia coli* BL21-DE3 with isopropyl-β-D-thiogalactopyranoside (IPTG), purified by Ni^{2+}-nitrilotriacetic acid (NTA) agarose metal affinity chromatography, and dialyzed into 25 m*M* HEPES (pH 7.4), 50 m*M* NaCl, 2 m*M* dithiothreitol (DTT) to a final concentration of ~1 mg/ml.[12] We often use the mutant ERK2-T183A as substrate for MKK,[12] which eliminates the Thr-183 phosphorylation site (numbered as in the rat enzyme) and ensures linearity versus time in the phosphorylation reaction. Recombinant human His_6–MKK1, subcloned in wild-type or constitutively active mutant (ΔN3/S218E/S222D, R4F) forms into expression vector pRSET, is induced in *E. coli* with IPTG, and purified through Ni^{2+}-NTA agarose followed by DEAE-Sephacel, resulting in a final preparation of MKK1 at ~0.3 mg/ml in 25 m*M* Tris (pH 8), 100 m*M* NaCl, 2 m*M* DTT, 10% (v/v) glycerol.[10] Recombinant rat His_6–MEKK-C, containing the catalytic domain of MEKK1 subcloned into expression vector pET-His_6TEV, is induced in *E. coli* and purified on Ni^{2+}-NTA agarose to a final concentration of ~0.1 mg/ml in 25 m*M* Tris (pH 8), 100 m*M* NaCl, 2 m*M* DTT, 10% (v/v) glycerol.[11] Purified proteins are stored at −80°.

Activation of MKK1 by MEKK-C

The effect of inhibitor on MKK1 activation by MEKK-C can be assessed by measuring direct phosphotransfer from [γ-^{32}P]ATP to MKK1. Each 25-μl reaction contains MKK1-WT (30 μg/ml), MEKK-C (10 μg/ml), 20 m*M* HEPES (pH 7.4), 0.1 m*M* ATP, 2 μCi of [γ-^{32}P]ATP, 10 m*M* $MgCl_2$, 2 m*M* DTT. Reactions are run in the presence of inhibitor concentrations varying from 0.1 to 100 μ*M*, prepared by diluting the stock solution with 2% (v/v) DMSO in water. Inhibitor is added to MKK1 followed by MEKK-C, and reactions are initiated by addition of MgATP in the form of a 5× solution containing 100 m*M* HEPES (pH 7.4), 10 m*M* DTT, 50 m*M* $MgCl_2$, 0.5 m*M* ATP, and 2 μCi of [γ-^{32}P]ATP. Reactions are run at 30° for varying times, quenched with 7 μl of 5× Laemmli sample buffer [250 m*M* Tris (pH 6.8), 5% (w/v) sodium dodecyl sulfate, 5% (v/v) 2-mercaptoethanol, 50% (v/v) glycerol], and separated by sodium dodecyl sulfate–polyacrylamide

[12] D. J. Robbins, E. Zhen, H. Owaki, C. A. Vanderbilt, D. Ebert, T. D. Geppert, and M. H. Cobb, *J. Biol. Chem.* **268,** 5097 (1993).

gel electrophoresis (SDS–PAGE) [12% (w/v) acrylamide, 0.32% (w/v) bisacrylamide] followed by Coomassie staining and drying. Phosphate incorporation is quantified by PhosphorImager analysis or scintillation counting of excised proteins, using the specific activity of the tracer to convert radioactivity measurements to mole units.

Direct phosphorylation of MKK1 by MEKK-C is inhibited by PD98059 and U0126, with median inhibitory concentrations (IC_{50}) of 70 and 3 μM, respectively (Fig. 2A–C). In parallel reactions using myelin basic protein (2.5 μg/reaction) as a nonkinase substrate, we observe no inhibition of MEKK-C activity by either compound (Fig. 2A, B), suggesting that inhibition of phosphorylation is due to drug binding to MKK1 substrate, rather than direct inhibition of MEKK-C.

Activity of MKK1

The effect of inhibitor on MKK1 activity is measured by direct phosphotransfer from [γ-^{32}P]ATP to ERK2. MKK1 is preactivated in reactions containing MKK1-WT (30 μg/ml), MEKK-C (10 μg/ml), 20 mM HEPES (pH 7.4), 2 mM ATP, 15 mM $MgCl_2$, 2 mM DTT, and incubated at 30° for 2 hr. An aliquot of the preactivation reaction is diluted with buffer containing 10 mM HEPES (pH 7.4), 2 mM DTT, and 0.01% (v/v) Triton X-100, and then added to activity reactions containing ERK2 (42 μg/ml), 20 mM HEPES (pH 7.4), 0.1 mM ATP, 2 μCi of [γ-^{32}P]ATP, 10 mM $MgCl_2$, 2 mM DTT. The final concentration of activated MKK1 is 0.25 μg/ml, in order to maintain linearity of the reaction with time, and reactions are run with inhibitor concentrations ranging from 0.03 to 100 μM. Reactions are initiated with MgATP added as a 5× solution, and quenched with Laemmli sample buffer after 5 min at 30°. Phosphate incorporation is measured as described above. Reactions with constitutively active mutant MKK1-R4F are performed in an identical manner, except that MKK1-R4F is used at 0.1 μg/ml to maintain linearity of the reaction with time.

Similar to previous reports, the activity of phosphorylated MKK1-WT is unaffected up to 100 μM PD98059, which is the solubility limit of this drug in aqueous solution, whereas U0126 inhibits phosphorylated MKK1 with an IC_{50} of ~20 μM (Fig. 3A).[7,8] PD98059 and U0126 inhibit active mutant MKK1-R4F with IC_{50} values of ~70 and ~0.4 μM, respectively (Fig. 3B).

Inhibitor Effects in Intact Cells

The ability of U0126 and PD98059 to suppress ERK activation in mammalian cells illustrates a typical use of cell permeable inhibitors in block-

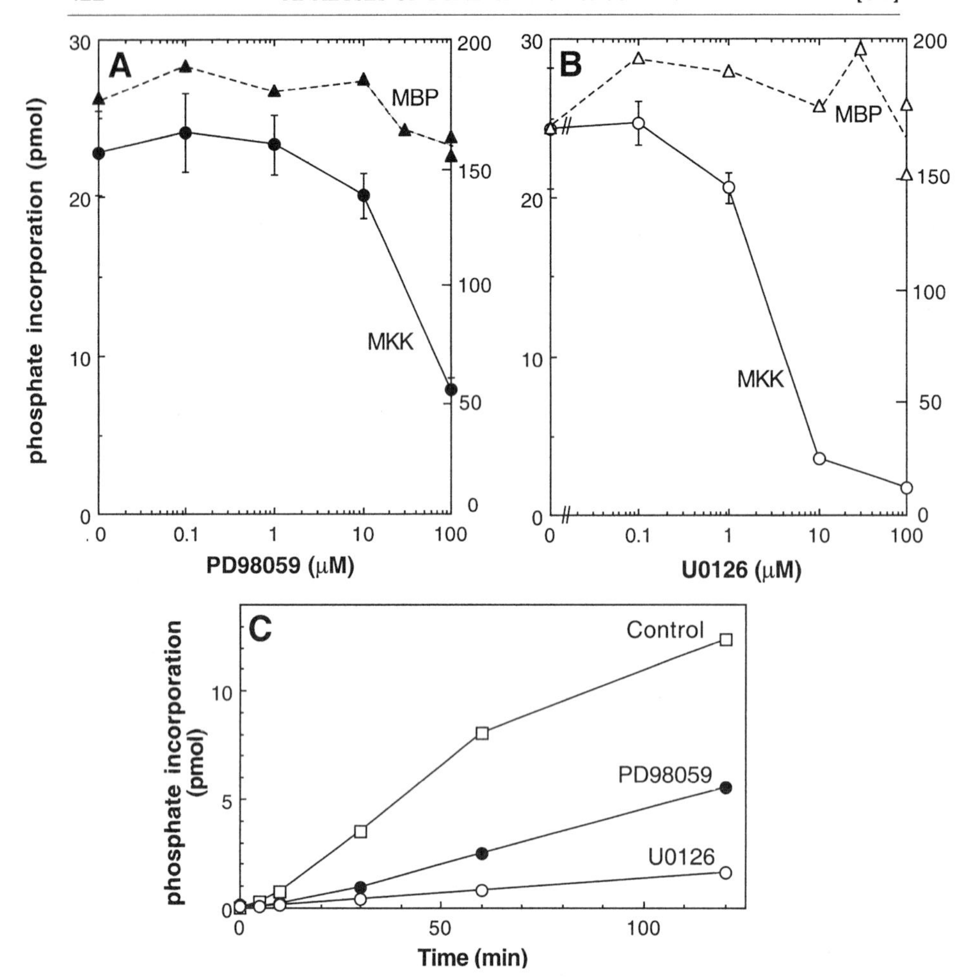

FIG. 2. Inhibition of MKK1 phosphorylation by MEKK-C. (A and B) Phosphotransfer from [γ-^{32}P]ATP to recombinant MKK1-WT (circles, scale on left axis) or myelin basic protein (triangles, scale on right axis), after a 1-hr reaction with recombinant MEKK-C, run in the presence of varying concentrations of PD98059 or U0126. (C) Time course of MKK1 phosphorylation by MEKK-C, in the presence of 0.5% (v/v) DMSO (squares), 100 μM PD98059 (closed circles), or 100 μM U0126 (open circles).

ing growth factor-mediated regulation of cellular targets. ERK1/2 and MKK1/2 activation in intact cells can be measured by immunoprecipitation from extracts and assay of their activity *in vitro*.[9] Alternatively, the phosphorylation state of ERK or MKK can be measured by reactivity on

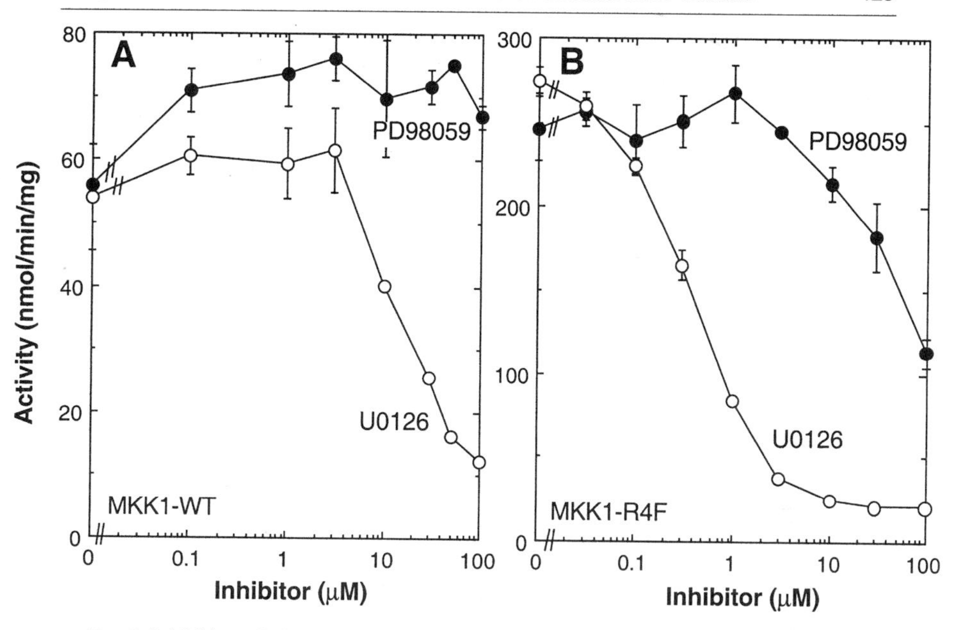

FIG. 3. Inhibition of MKK1 activity. Rate of phosphotransfer from [γ-^{32}P]ATP to recombinant ERK2-AEY, after 5-min reactions catalyzed by (A) MKK1-WT phosphorylated and activated by MEKK-C, or (B) constitutively active mutant MKK1-R4F, and run in the presence of varying concentrations of PD98059 (closed circles) or U0126 (open circles).

Western blots probed with phosphospecific antibodies (see [26] in this volume[13]).

Phospho-Specific Antibodies

Antibodies recognizing phosphorylated forms of ERK1/2 and MKK1/2 are available from several vendors. We currently use mouse monoclonal anti-ppERK antibody from Sigma or rabbit polyclonal anti-ppERK antibody from New England BioLabs, because they are unreactive with unphosphorylated ERK and have relatively low reactivity with the inactive monophosphorylated species. We use rabbit polyclonal anti-ppMKK1/2 antibodies from New England BioLabs, which react with recombinant diphosphorylated MKK1 about five times more effectively than monophosphorylated MKK1 (occupied at Ser-218), and show no detectable reactivity with unphosphorylated MKK1. Proteins are transferred to polyvinylidene difluoride (PVDF) membranes for Western blotting. Gel loading controls

[13] D. G. Giseli, M. Zeceure, and M. J. Weber, *Methods Enzymol.* **332,** Chap. 26, 2001 (this volume).

are performed by stripping and reprobing filters with anti-ERK1/2 (sc-154) and anti-MKK1 (sc-219) from Santa Cruz Biotechnology (Santa Cruz, CA), which recognize both phosphorylated and unphosphorylated ERK or MKK. Western blots are stripped with 2% (w/v) SDS, 62 m*M* Tris (pH 6.8), 100 m*M* 2-mercaptoethanol for 40 min at 62°, washed twice with phosphate-buffered saline (PBS), 0.05% (v/v) Tween 20 for 35 min at room temperature, and reblocked prior to addition of primary antibody.

Suppression of ERK Activation in Response to Serum Stimulation

NIH 3T3 cells are grown in Dulbecco's Modified Eagle's medium (DMEM) containing 10% (v/v) fetal bovine serum (FBS), penicillin (100 U/ml), and streptomycin (100 U/ml), and are passaged by a 1:10 split every 3–4 days. Cells are grown to confluence in six-well tissue culture dishes with 3 ml of 10% (v/v) FBS–DMEM per well, and then starved by switching medium to 2 ml of 0.1% (v/v) FBS–DMEM for 16 hr prior to treatment with inhibitor. Stock solutions of 50 m*M* U0126 and 100 m*M* PD98059 in DMSO are added to the medium to final concentrations of 3–100 μM with 0.5% (v/v) DMSO. It is useful to first remove part of the medium to a sterile tube into which each drug is mixed before transferring it back to the cells, because both compounds transiently precipitate and adhere to the cell layer on dilution into aqueous solution. Typically, studies report drug pretreatment times of 10–60 min, where 2–20 μM U0126 is satisfactory to achieve complete inhibition of MKK/ERK signaling, and PD98059 is effective between 10 μM and its solubility limit of 100 μM. Cytotoxicity is observed with 100 μM U0126 after 24 hr, and not observed with PD98059.

After 30 min of pretreatment with inhibitor, cells are stimulated by treatment with 5% (v/v) FBS plus 50 n*M* 12-phorbol 13-myristate acetate (PMA), incubated for 1 hr in a CO_2 incubator at 37°, and then washed twice with 2 ml of PBS and lysed. Cell lysis is achieved by addition of an extract buffer that enables assaying the response of interest. For example, buffers containing phosphatase and protease inhibitors are used when procedures require immunoprecipitation and enzymatic assay of kinase activity.[9] For Western blots shown in Fig. 4, cells were lysed by addition of 0.3 ml of 2× Laemmli sample buffer [100 m*M* Tris (pH 6.8), 2% (w/v) sodium dodecyl sulfate, 2% (v/v) 2-mercaptoethanol, 10% (v/v) glycerol], and lysates were collected and vortexed to shear DNA. Aliquots (20 μl, ~10 μg) were separated by SDS–PAGE [15% (w/v) acrylamide, 0.086% (w/v) bisacrylamide, for optimal separation of phosphorylated vs. unphosphorylated proteins], and analyzed by Western blotting using anti-ppERK1/2 primary antibody (New England BioLabs) at 1:2000 dilution. Blots were

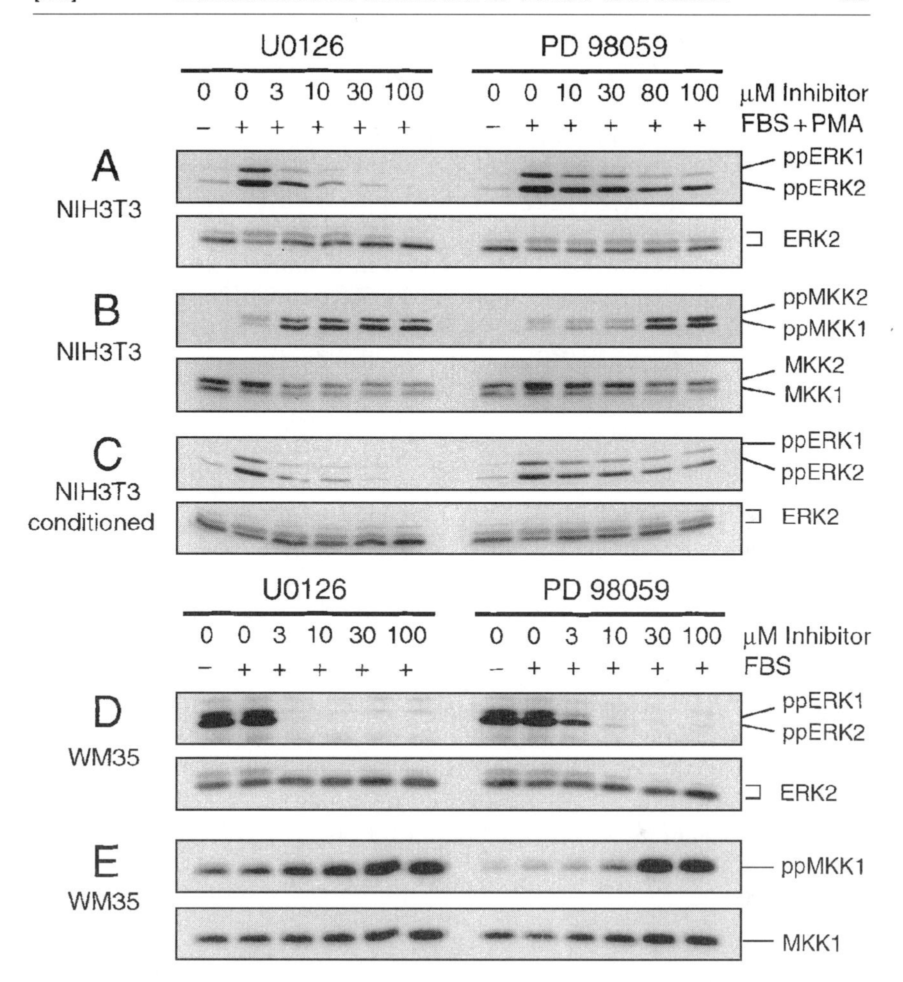

FIG. 4. Effects of MKK inhibitors on ERK and MKK phosphorylation in intact cells. (A) U0126 (*left*) or PD98059 (*right*) block ERK1/2 phosphorylation, measured in NIH 3T3 cells after a 30-min pretreatment with inhibitor and 1 hr of stimulation with 5% (v/v) FBS plus 50 n*M* PMA. Shown are reactivities of the same Western blot with anti-ppERK2 (*top*) and anti-ERK2 (*bottom*). (B) Neither inhibitor blocks MKK1/2 phosphorylation, measured by Western blotting of the same extracts as in (A), using anti-ppMKK1/2 (*top*), and anti-MKK1 (*bottom*). (C) Efficacy of U0126 and PD98059 after incubation with NIH 3T3 cells for 26 hr followed by transfer to naive cells for 30 min, 1 hr of stimulation with FBS plus PMA as in (A), and Western blotting with anti-ppERK1/2 and anti-ERK2. (D) Sensitivity of ERK1/2 phosphorylation to inhibitors in WM35 cells, after a 30-min pretreatment with inhibitor, 1 hr of stimulation with 2.5% (v/v) FBS, and Western blotting with anti-ppERK1/2 and anti-ERK2. (E) The same extracts as in (D), probed with anti-ppMKK1/2 and anti-MKK1.

reprobed with donkey anti-rabbit coupled horseradish peroxidase, and visualized by enhanced chemiluminescence detection (Amersham, Arlington Heights, IL).

The results show reduction of ERK phosphorylation to basal levels with 80–100 μM PD98059 (Fig. 4A). U0126 reduces ERK phosphorylation to below basal levels at 10 μM, and thus appears to be a more effective inhibitor than PD98059 in NIH 3T3 cells. Variability of the inhibitor response with a different cell type is shown in a similar experiment using WM35 melanoma cells, in which both PD98509 and U0126 suppress ERK phosphorylation to below basal levels at 3 μM (Fig. 4D). Blots were stripped and reprobed with anti-ERK2 antibody (1 : 2000 dilution) in order to monitor equal loading of protein in each lane.

Effects of Inhibitor on Phosphorylation of Endogenous MKK1

The activity state of MKK1/2 is examined by probing Western blots with anti-ppMKK1/2 primary antibody (1 : 2000) (Fig. 4B and E). In contrast to effects on ppERK1/2, neither inhibitor blocked MKK1/2 phosphorylation in response to serum; in fact, inhibiting ERK by drug treatment appears to enhance phosphorylation of MKK1/2. Similar effects were previously observed when inhibiting ERK by overexpressing MAP kinase phosphatase 1 (MKP-1).[14] Previous studies have suggested negative feedback of MKK1/2 and Raf activation by ERK, based on enhanced activation of Raf by MKP-1 transfection or PD98059 treatment.[8,14] Thus, although both inhibitors prevent activation of MKK1 *in vitro* by purified upstream kinases, neither compound effectively interferes with MKK1/2 phosphorylation by upstream kinases in intact cells.

Inhibitor Stability under Cell Culture Conditions

To determine drug stability under culture conditions, varying concentrations of inhibitors were added to NIH 3T3 cells that had been starved in 2 ml of 0.1% (v/v) FBS–DMEM, followed by further incubation at 37° for 26 hr. The resulting aliquots of conditioned medium containing inhibitors were removed to separate dishes of naive cells that had been starved overnight. Cells were then incubated at 37° for 30 min prior to treatment with 5% (v/v) FBS plus 50 nM PMA for 1 hr. Western blots of extracts probed with anti-ppERK antibody indicated that both compounds are still effective at inhibiting ERK phosphorylation, although PD98059 shows reduced efficacy after incubation with cells for 26 hr (Fig. 4C).

[14] P. S. Shapiro and N. G. Ahn, *J. Biol. Chem.* **273,** 1788 (1998).

Inhibition of Transfected Mutant MKK1

Consistent with its effects on constitutively active mutant MKK1 *in vitro*, U0126 inhibits activation of the MKK–ERK pathway on transient transfection with a plasmid enabling expression of hemagglutinin-tagged active mutant MKK1 (ΔN4/S218E/S222D, G1C) under the cytomegalovirus promoter.[10,15] Transient transfection is performed with the LipofectAMINE Plus reagent (GIBCO-BRL Gaithersburg, MD). NIH 3T3 cells are grown to 50% confluence in six-well tissue culture dishes, followed by replacement of medium with 0.8 ml of serum-free medium (Opti-MEM; GIBCO-BRL). During transfection, 1 μg of cDNA is mixed with 6 μl of Plus reagent and 100 μl of Opti-MEM and incubated at room temperature for 15 min, followed by addition of 4 μl of LipofectAMINE in 100 μl of Opti-MEM. The final mixture is allowed to stand for another 15 min, and then added to the cells, which are transferred to a CO_2 incubator. After 3 hr, the transfection solution is replaced with 3 ml of 10% (v/v) FBS–DMEM. MKK-G1C expression is apparent 12 hr after transfection, and maximal at 24 hr.

Twenty-four hours after transfection, cells are starved for 16 hr, stimulated with 5% (v/v) serum plus 50 n*M* PMA for 2 hr, and then processed for Western blotting probed with a mixture of anti-ppERK (1:2000) and anti-hemagglutinin (1 : 5000) antibodies. The basal phosphorylation state of ERK was comparable to that of cells challenged with serum plus PMA, because of active mutant MKK1 expression (Fig. 5, lanes 1 and 2). Pretreatment with U0126 for 30 min resulted in ERK dephosphorylation, whether or not cells were subsequently stimulated (Fig. 5, lanes 3 and 4), indicating that U0126 rapidly blocks signaling from active mutant MKK1.

Inhibition of Megakaryocyte Differentiation

K562 erythroleukemia cells provide an easily accessible model system with which to examine requirements for MKK–ERK signaling on cell differentiation. In response to PMA or transfection with active MKK1, several leukemia cell lines adopt characteristic megakaryocyte morphology and induce platelet-specific genes, including integrin $\alpha_{IIb}\beta_3$, a highly selective marker expressed only in megakaryocyte and platelets in humans.[15–17] The necessity for ERK activation during the response to mitogen- or cytokine-induced differentiation in these cells as well as in $CD34^+$ primary mouse

[15] A. M. Whalen, S. C. Galasinski, P. S. Shapiro, T. S. Nahreini, and N. G. Ahn, *Mol. Cell. Biol.* **17,** 1947 (1997).

[16] M. C. Rouyez, C. Boucheron, S. Gisselbrecht, I. Dusanter-Fourt, and F. Porteu, *Mol. Cell. Biol.* **17,** 4991 (1997).

[17] A. S. Melemed, J. W. Ryder, and T. A. Vik, *Blood* **90,** 3462 (1997).

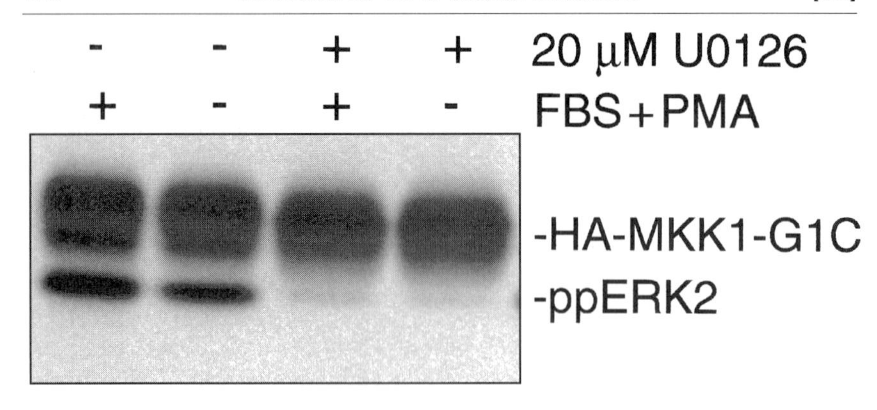

FIG. 5. U0126 inhibits constitutively active mutant MKK1 in intact cells. NIH 3T3 cells were transiently transfected with hemagglutinin-tagged MKK1-G1C for 24 hr, starved for 16 hr, and then preincubated with 0 or 20 μM U0126 for 30 min, and either treated with 5% (v/v) FBS plus 50 nM PMA for 2 hr (lanes 1 and 3) or left untreated (lanes 2 and 4). Western blots show expressed MKK1 and endogeneous ppERK, probed with anti-HA and anti-ppERK antibodies. [Reproduced from Promega Notes 71 by permission of Promega Corporation.]

myeloid cells can be demonstrated by suppressing megakaryocyte morphology with MKK inhibitors.[18]

K562 cells (CCL 243) obtained from the American Type Culture Collection (Rockville, MD) are grown in T flasks or spinner flasks at 37°, 5% CO_2 in medium containing RPMI, 10% (v/v) FBS, penicillin (100 U/ml), and streptomycin (100 U/ml). Cells should be maintained between 4×10^5 and 8×10^5 cells/ml, and are passaged by 1:1 dilution with medium every day. We have found that cells show a reduced capacity to differentiate on prolonged growth at densities much above or below this range.

K562 cells (20 ml) are pretreated for 60 min with 20 μM U0126 (or 50 μM PD98059), after which PMA is added to 50 nM. After PMA treatment, normal cell division arrest occurs within 7–12 hr; within 24–48 hr, cells show attachment to culture dishes and morphology characteristic of megakaryocyte differentiation (Fig. 6A,B). Induction of integrin $\alpha_{IIb}\beta_3$ can be observed on examination of mRNA by RT-PCR or of protein by Western blotting.[15] Both U0126 and PD98059 block characteristic changes in cell morphology induced by PMA (shown for U0126, Fig. 6B *vs* D), and suppress growth arrest and integrin $\alpha_{IIb}\beta_3$ induction.[15] In cells unstimulated with PMA, little effect of MKK inhibitor on morphology or growth is observed. However, examination of extracts by Western blotting shows that MKK inhibitors increase the abundance of hemoglobin subunits, presumably be-

[18] P. Rojnuckarin, J. G. Drachman, and K. Kaushansky, *Blood* **94,** 1273 (1999).

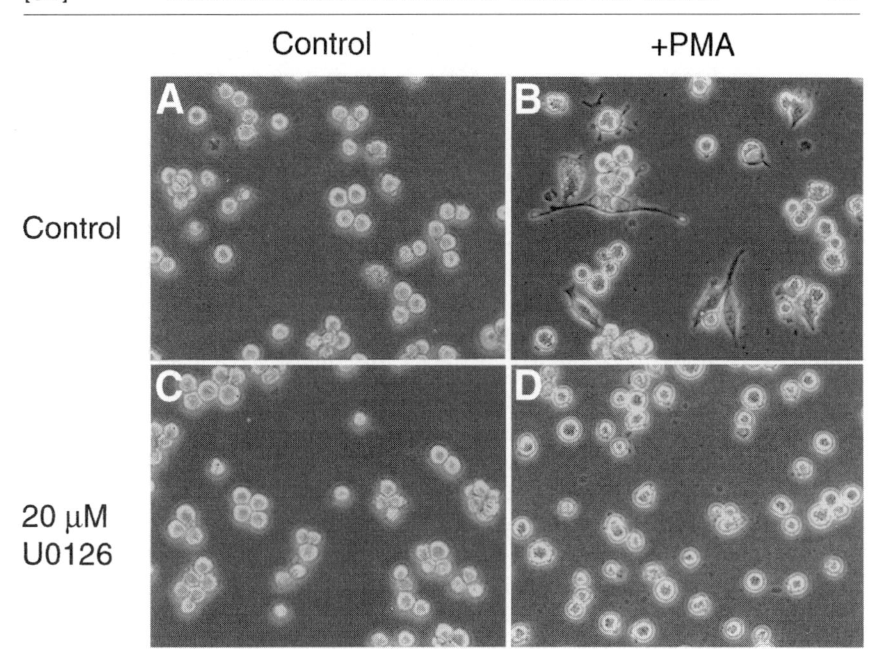

FIG. 6. U0126 inhibits PMA-induced megakaryocyte differentiation in K562 cells. K562 cells grown to 5×10^5 cells/ml in 10% (v/v) FBS–RPMI were preincubated with 0 μM (A and B) or 20 μM (C and D) U0126 for 30 min, and then (A and C) left untreated or (B and D) treated with 50 nM PMA for 48 hr. Cells were viewed and photographed by phase-contrast microscopy. Characteristic attachment and spreading of cells that accompanies megakaryocyte differentiation in response to PMA (B) is blocked by U0126 (D), because of inhibition of ERK1/2 activation (not shown). [Reproduced from Promega Notes 71 by permission of Promega Corporation.]

cause of derepression of basal ERK activity.[15] Thus, K562 cells provide a model system for assessing effects of inhibitors on basal as well as stimulated ERK activity.

Mechanism of Inhibitor Action

Previous studies have shown that although PD98059 inhibits constitutively active mutant MKK1 *in vitro*, it has little effect on activated MKK1-WT.[8] The ability of this compound to inhibit MKK1 phosphorylation by recombinant Raf-1 or MEKK-C *in vitro*, led to the hypothesis that its effect in intact cells occurs through inhibition of MKK1/2 phosphorylation by upstream kinases.[8] In contrast, U0126 has been reported to inhibit both active mutant and wild-type MKK1 but shows no effect on phosphorylation of MKK1 by Raf-1 immunoprecipitated from cell extracts, suggesting that it works by direct suppression of MKK1/2 activity.[7]

We find instead that U0126 and PD98059 act similarly, in that both prevent phosphorylation of MKK1 by upstream kinase in a manner that appears to be substrate directed. The difference in behavior of U0126 from the previous study might be explained by the use of purified recombinant MEKK-C versus immunoprecipitated Raf-1 as the upstream kinase.[7] Our results are consistent with the similarity between these inhibitors as noncompetitive inhibitors of MKK1, indicating that drug binding occurs outside the ATP- and substrate-binding sites on the kinase.[5,7] Equilibrium binding studies also show that PD98059 displaces binding of [^{3}H]U0126 to MKK1-R4F, indicating that the inhibitors bind MKK1 at the same or overlapping sites.[7]

Both compounds inhibit active mutant MKK1-R4F, although only U0126 is effective at inhibiting active phosphorylated MKK1-WT. However, the half-maximal inhibition of phosphorylated MKK-WT by U0126 is 50-fold weaker than MKK1-R4F, suggesting that inhibitor binding discriminates between wild-type and active mutant forms (Fig. 3 and Ref. 7). This difference suggests that inhibition of phosphorylated MKK1-WT by PD98059 would be observable at higher concentrations, were the experiments not limited by drug solubility.

In intact cells, PD98059 blocks MKK–ERK activation at concentrations ranging from 10 to 100 μM.[5,8] Interestingly, we find that neither PD98059 nor U0126 affects phosphorylation of MKK1, measured by reactivity with phosphospecific antibody, and in fact both drugs augment phosphorylation in parallel to ERK inhibition (Fig. 4B and E). This suggests that drug action occurs through inhibition of MKK activity in intact cells, which could conceivably involve other regulatory mechanisms, for example, sites of heterologous phosphorylation in the Raf-binding domain between subdomains 9 and 10, or scaffold proteins involved in MKK–ERK interactions, such as MP-1.[19,20]

Evaluating Inhibitor Specificity

Published reports on PD98059 or U0126 show little or no inhibition with protein kinases other than MKK1/2.[5,7,8] However, as with all inhibitors, verification of specificity requires further testing with other enzymes under

[19] J. A. Frost, H. Steen, P. Shapiro, T. Lewis, N. Ahn, P. E. Shaw, and M. H. Cobb, *EMBO J.* **16,** 6426 (1997).

[20] H. J. Schaeffer, A. D. Catling, S. T. Eblen, L. S. Collier, A. Krauss, and M. J. Weber, *Science* **281,** 1668 (1998).

a wide range of conditions. For example, PD98059 has been reported to act as a ligand for the aryl hydrocarbon receptor, displacing the binding of 2,3,7,8-tetrachlorodibenzo-*p*-dioxin (TCDD).[21] Both U0126 and PD98059 have been shown to inhibit MKK5 *in vitro* and *in vivo*, and U0126 has been stated to inhibit as many as 25 kinases.[22,23] Therefore, confirming experiments are needed to monitor potential nonspecific drug effects.

One approach is to compare the effects using more than one drug, and to compare the dose dependence of response inhibition with that of kinase inhibition, where discrepancies in IC_{50} may rule out regulatory relationships. In an example using LY294002 and wortmannin inhibitors of phosphatidylinositol 3-kinase (PI3K), evidence against ERK regulation by PI3K was shown by 10-fold differences between the drug concentrations needed to suppress ERK versus PI3K.[24]

A second approach is to test whether overexpression of constitutively active mutant MKK1 modulates cellular targets in a manner that opposes the effects of MKK1/2 inhibitors. Opposing behavior under conditions in which the pathway is turned on versus off provides confirmation of regulation by MKK–ERK. For example, in a functional proteomics study, we found that 24 of 41 proteins modulated by PMA in K562 cells were blocked by U0126.[25] Of the 24 proteins, all were modulated on overexpression of active MKK1/2 in the absence of PMA, and are good candidates for MKK–ERK targets.

A third approach, as yet unexplored with MKK1/2, is to identify mutations that render it impervious to inhibitors, and then examine the effects of overexpressing such mutants on responses that are normally blocked by inhibitor. Mutant overexpression should provide a dominant mechanism, bypassing inhibitor action of endogenous kinase. An example using a p38 MAPK mutant resistant to inhibition by the inhibitor SB203580 illustrates the success of this approach.[26]

[21] J. J. Reiners, J. Y. Lee, R. E. Clift, D. T. Dudley, and S. P. Myrand, *Mol. Pharmacol.* **53,** 438 (1998).

[22] S. Kamakura, T. Moriguchi, and E. Nishida, *J. Biol. Chem.* **274,** 26563 (1999).

[23] P. Cohen, *Curr. Opin. Chem. Biol.* **3,** 459 (1999).

[24] M. P. Scheid and V. Duronio, *J. Biol. Chem.* **271,** 18134 (1996).

[25] T. S. Lewis, J. B. Hunt, L. D. Aveline, K. R. Jonscher, D. F. Louie, J. M. Yeh, T. S. Nahreini, K. A. Resing, and N. G. Ahn, *Molecular Cell,* in press (2000).

[26] B. Frantz, T. Klatt, M. Pang, J. Parsons, A. Rolando, H. Williams, M. J. Tocci, S. J. O'Keefe, and E. A. O'Neill, *Biochemistry* **37,** 13846 (1998).

[32] Analysis of Pharmacologic Inhibitors of Jun N-Terminal Kinases

By BRION W. MURRAY, BRYDON L. BENNETT, and DENNIS T. SASAKI

Background

Jun N-terminal kinases (JNKs), also known as stress-activated protein kinases (SAPKs), are ubiquitously expressed serine/threonine protein kinases activated by proinflammatory cytokines, environmental stress, and UV irradiation as well as mitogens.[1–6] Activated JNK *trans*-activates the heterodimeric transcription factor AP-1 via phosphorylation of its cJun subunit on serine residues 63 and 73. The JNK pathway is downstream of the small GTPase Ras. In fact, both the discovery of the c-Jun phosphorylation sites for JNK[7] and the identification of JNK were enabled by analyzing Ha-*ras*-transformed cells.[1,8] The GTPases Rac and cdc42 are activated by inflammatory mediators, cell stresses, and by mitogens through Ras activity. One report has identified an EPS8, E3B1, Sos-1 G protein complex as the necessary signal transducers connecting Ras to the GTPase Rac.[9] Thus, the JNK pathway is directly downstream of Rac and cdc42, and by extension downstream of Ras.

Jun N-terminal kinases are members of the mitogen-activated protein kinase (MAPK) family of protein kinases. Signal transduction through an MAP kinase pathway proceeds through the sequential activation of kinases contained in a three-kinase cassette: an MAP kinase kinase kinase (MAPKKK) activates an MAP kinase kinase (MAPKK), which activates an MAP kinase. The JNK subgroup is composed of MAPKKKs (MEKK1,

[1] B. Derijard, M. Hibi, I. H. Wu, T. Barrett, B. Su, T. Deng, M. Karin, and R. J. Davis, *Cell* **76,** 1025 (1994).

[2] J. M. Kyriakis, P. Banerjee, E. Nikolakaki, T. Dai, E. A. Rubie, M. F. Ahmad, J. Avruch, and J. R. Woodgett, *Nature* (*London*) **369,** 156 (1994).

[3] A. Minden, A. Lin, T. Smeal, B. Derijard, M. Cobb, R. Davis, and M. Karin, *Mol. Cell. Biol.* **14,** 6683 (1994).

[4] V. Adler, A. Schaffer, J. Kim, L. Dolan, and Z. Ronai, *J. Biol. Chem.* **270,** 26071 (1995).

[5] E. Cano, C. A. Hazzalin, and L. C. Mahadevan, *Mol. Cell. Biol.* **14,** 7352 (1994).

[6] J. K. Westwick, C. Weitzel, A. Minden, M. Karin, and D. A. Brenner, *J. Biol. Chem.* **269,** 26396 (1994).

[7] T. Smeal, B. Binetruy, D. Mercola, A. Grover-Bardwick, G. Heidecker, U. R. Rapp, and M. Karin, *Mol. Cell. Biol.* **12,** 3507 (1992).

[8] M. Hibi, A. Lin, T. Smeal, A. Minden, and M. Karin, *Genes Dev.* **7,** 2135 (1993).

[9] G. Scita, J. Nordstrom, R. Carbone, P. Tenca, G. Giardina, S. Gutkind, M. Bjarnegard, C. Betsholtz, and P. P. Di Fiore, *Nature* (*London*) **401,** 290 (1999).

METHODS IN ENZYMOLOGY, VOL. 332
0076-6879/00 $35.00

-2, and -4; MLK-3; and Tpl-2),[10–13] two MAPKKs (JNKK1/MKK4[14–17] and JNKK2/MKK7[18–20]), and three MAPKs (JNK1, -2, and -3).[1,2,8,21–23] Each of these kinases exists in several isoforms whose distinct functions are not fully understood. For further reading, the authors recommend the following review articles on JNK/MAPKs[24–28] and protein kinase inhibition.[29–36]

Activation of the JNK pathway has been documented in a number of disease settings. Many of the genes encoding inflammatory molecules are

[10] C. A. Lange-Carter, C. M. Pleiman, A. M. Gardner, K. J. Blumer, and G. L. Johnson, *Science* **260,** 315 (1993).

[11] M. Yan, T. Dai, J. C. Deak, J. M. Kyriakis, L. I. Zon, J. R. Woodgett, and D. J. Templeton, *Nature (London)* **372,** 798 (1994).

[12] J. L. Blank, P. Gerwins, E. M. Elliott, S. Sather, and G. L. Johnson, *J. Biol. Chem.* **271,** 5361 (1996).

[13] A. Salmeron, T. B. Ahmad, G. W. Carlile, D. Pappin, R. P. Narsimhan, and S. C. Ley, *EMBO J.* **15,** 817 (1996).

[14] I. Sanchez, R. T. Hughes, B. J. Mayer, K. Yee, J. R. Woodgett, J. Avruch, J. M. Kyriakis, and L. I. Zon, *Nature (London)* **372,** 794 (1994).

[15] B. M. Yashar, C. Kelley, K. Yee, B. Errede, and L. I. Zon, *Mol. Cell. Biol.* **13,** 5738 (1993).

[16] B. Derijard, J. Raingeaud, T. Barrett, I. H. Wu, J. Han, R. J. Ulevitch, and R. J. Davis, *Science* **267,** 682 (1995).

[17] A. Lin, A. Minden, H. Martinetto, F. X. Claret, C. Lange-Carter, F. Mercurio, G. L. Johnson, and M. Karin, *Science* **268,** 286 (1995).

[18] Z. Wu, J. Wu, E. Jacinto, and M. Karin, *Mol. Cell. Biol.* **17,** 7407 (1997).

[19] C. Tournier, A. J. Whitmarsh, J. Cavanagh, T. Barrett, and R. J. Davis, *Proc. Natl. Acad. Sci. U.S.A.* **94,** 7337 (1997).

[20] T. Moriguchi, F. Toyoshima, N. Masuyama, H. Hanafusa, Y. Gotoh, and E. Nishida, *EMBO J.* **16,** 7045 (1997).

[21] T. Kallunki, B. Su, I. Tsigelny, H. K. Sluss, B. Derijard, G. Moore, R. Davis, and M. Karin, *Genes Dev.* **8,** 2996 (1994).

[22] H. K. Sluss, T. Barrett, B. Derijard, and R. J. Davis, *Mol. Cell. Biol.* **14,** 8376 (1994).

[23] S. Gupta, T. Barrett, A. J. Whitmarsh, J. Cavanagh, H. K. Sluss, B. Derijard, and R. J. Davis, *EMBO J.* **15,** 2760 (1996).

[24] R. Seger and E. G. Krebs, *FASEB J.* **9,** 726 (1995).

[25] Y. T. Ip and R. J. Davis, *Curr. Opin. Cell Biol.* **10,** 205 (1998).

[26] A. Minden, and M. Karin, *Biochim. Biophys. Acta* **1333,** F85 (1997).

[27] M. J. Robinson and M. H. Cobb, *Curr. Opin. Cell Biol.* **9,** 180 (1997).

[28] S. Lepp and D. Bohmann, *Oncogene* **18,** 6158 (1999).

[29] B. Stein and D. Anderson, *in* "The MAP Kinase Family: New "MAPs" for Signal Transduction Pathways and Novel Targets for Drug Discovery" (J. A. Bristol, ed.), p. 289. Academic Press, San Diego, California, 1996.

[30] P. R. Clarke, *Curr. Opin.* **4,** 647 (1994).

[31] J. C. Lee and J. L. Adams, *Curr. Opin. Biotechnol.* **6,** 657 (1995).

[32] S. S. Taylor and E. Radzio-Andzelm, *Curr. Opin. Chem. Biol.* **1,** 219 (1997).

[33] H. C. J. Hemmings, *Neuromethods* **30,** 112 (1997).

[34] G. McMahon, L. Sun, C. Liang, and C. Tang, *Curr. Opin. Drug Discov.* **1,** 131 (1998).

[35] S. Bhagwat, A. Manning, M. F. Hoekstra, and A. Lewis, *Drug Discov. Today* **4,** 472 (1999).

[36] B. W. Murray, Y. Satoh, and B. Stein, *in* "High Throughput Screening for Novel Anti-Inflammatories" (M. Kahn, ed.), p. 165. Birkhäuser Verlag, Basel, Switzerland, 2000.

regulated by the JNK pathway through activation of the transcription factors AP-1 and ATF-2, including tumor necrosis factor α (TNF-α), interleukin 2 (IL-2), and matrix metalloproteinases such as collagenase 1.[37] In heart disease, the JNK pathway is activated by atherogenic stimuli and regulates local cytokine and growth factor production in vascular cells.[38] JNK activation has been implicated in cancer through many studies including modulation of cell cycle progression through p53 phosphorylation[39] and by altered growth factor signaling in non-small cell lung cancer.[40] In addition, JNK may have a role in neurodegenerative disorders through JNK3, the brain-selective JNK isoform.[41] Numerous studies demonstrate that Jun kinases may be excellent therapeutic targets.

Because Jun kinases are only 3 of the estimated 2000 protein kinases in the human genome, the assembly of the proper combination of methods is critical to the discovery of selective kinase inhibitors. Phosphopeptide mapping of JNK-mediated phosphorylation, immune complex kinases assays, and transient transfection of cJun–GAL4 have been described in an earlier volume of this series.[42] This chapter focuses on both biochemical and cellular techniques essential to characterizing a kinase inhibitor as a JNK inhibitor. Methods described in this chapter include a more robust JNK *in vitro* assay, enzymatic analysis of inhibitors, evaluation of inhibitors on endogenous JNK, and functional analysis of JNK inhibitors (cytokine analysis and flow cytometry).

Enzymatic Analysis of JNK Inhibitors

In Vitro Kinase Assay

Reagents. Highly active JNK can be efficiently generated by coinfecting insect cells (Sf9 from *Spodoptera frugiperda*) with two baculoviral constructs, hexahistidine (His_6)–JNK and the untagged C terminus of MEKK1 (residues 1175–1493) in a 5:1 multiplicity of infection ratio.[43] A relevant

[37] A. M. Manning and F. Mercurio, *Exp. Opin. Invest. Drugs* **6,** 555 (1997).

[38] D. D. Yang, D. Conze, A. J. Whitmarsh, T. Barrett, R. J. Davis, M. Rincon, and R. A. Flavell, *Immunity* **9,** 575 (1998).

[39] T. K. Chen, L. M. Smith, D. K. Gebhardt, M. J. Birrer, and P. H. Brown, *Mol. Carcinog.* **15,** 215 (1996).

[40] T. Yin, G. Sandhu, C. D. Wolfgang, A. Burrier, R. L. Webb, D. F. Rigel, T. Hai, and J. Whelan, *J. Biol. Chem.* **272,** 19943 (1997).

[41] D. D. Yang, C. Y. Kuan, A. J. Whitmarsh, M. Rincon, T. S. Zheng, R. J. Davis, P. Rakic, and R. A. Flavell, *Nature* (*London*) **389,** 865 (1997).

[42] J. K. Westwick and D. A. Brenner, *Methods Enzymol.* **255,** 342 (1995).

[43] W. A. Gaarde, T. Hunter, H. Brady, B. W. Murray, and M. E. Goldman, *J. Biomol. Screening* **2,** 213 (1997).

substrate for JNK is a bacterially expressed truncated cJun–glutathione *S*-transferase (GST) fusion (pGEX; Amersham-Pharmacia, Piscataway, NJ) that contains both the phosphorylation sites (Ser-63, Ser-73) and the JNK docking site (residues 33–60).[43] The p38 inhibitor SB203580[44] (Calbiochem, La Jolla, CA) can be a positive control because it is known to be a weak JNK1 inhibitor yet a good inhibitor of JNK2 and JNK3.[45,46] In addition, the mechanism of inhibition by SB203580 is well characterized: reversible and ATP competitive for both p38[47,48] and JNKs (B. W. Murray, unpublished observation, 1999).

In Vitro Assay Protocol. Many JNK assay formats are possible utilizing different technologies to detect phosphate transfer to substrate: ^{32}P-based, time-resolved fluorescence, fluorescence, and colorimetric. A simple, cost-effective method for monitoring JNK activity relies on the solution-phase transfer of isotopically labeled phosphate from ATP to its protein target in 96-well assay plates.[43] Trichloroacetic acid (TCA)-precipitated proteins are collected on 96-well glass fiber plates and washed with phosphate-buffered saline to remove unreacted ATP. This method is described in detail below.

Enzymatic reactions (0.1 ml) are carried out in 96-well assay plates (Corning, Acton, MA) for 1 hr at room temperature. Assays contain the following: 20m*M* HEPES (pH 7.6), 0.5 μCi of [γ-^{32}P]ATP, 0.015 m*M* ATP, 0.1 m*M* EDTA, 15 m*M* $MgCl_2$, 0.005% (v/v) Triton X-100, 1.5 m*M* dithiothreitol (DTT), leupeptin (1 μg/ml), 10 m*M* β-glycerophosphate, 0.05 m*M* sodium vanadate, and 10m*M* *p*-nitrophenyl phosphate. Reactions are terminated with the addition of TCA [150 μl/well of 12.5% (w/v) TCA] and incubated for 30 min. The precipitate is then collected on 96-well glass fiber plates (Packard, Meriden, CT) and washed 10 times with approximately 0.3 ml per well of phosphate-buffered saline (PBS), pH 7.4 (Dulbecco's phosphate-buffered saline; Sigma, St. Louis, MO), using a cell harvester (Packard Filtermate 196). Scintillation fluid (0.05 ml, MicroScint20; Packard) is added to each well and the plate is analyzed with a plate format scintillation counter (Packard TopCount).

Essential for meaningful analysis of the generated data is incorporation

[44] A. Cuenda, J. Touse, Y. N. Doza, R. Meier, P. Cohen, T. F. Gallagher, P. R. Young, and J. C. Lee, *FEBS Lett.* **364,** 229 (1995).

[45] A. J. Whitmarsh, H.-H. Yang, M. S.-S. Su, A. D. Sharrocks, and R. J. Davis, *Mol. Cell. Biol.* **17,** 2360 (1997).

[46] A. Clerk and P. H. Sugden, *FEBS Lett.* **426,** 93 (1998).

[47] L. Tong, S. Pav, D. M. White, S. Rogers, K. M. Crane, C. L. Cywin, M. L. Brown, and C. A. Pargellis, *Nat. Struct. Biol.* **4,** 311 (1997).

[48] K. P. Wilson, P. G. McCaffrey, K. Hsiao, S. Pazhanisamy, V. Galullo, G. W. Bemis, M. J. Fitzgibbon, P. R. Caron, M. A. Murcko, and M. S. S. Su, *Chem. Biol.* **4,** 423 (1997).

of the proper controls into each experiment. Because the assay is based on purified proteins, the precipitated phosphoproteins result only from substrate phosphorylation and JNK autophosphorylation. The background could be determined by two methods: omission of c-Jun from the reaction and by totally inhibiting a complete reaction with a saturating level of inhibitor [20–50 times the concentration to achieve 50% inhibition (IC_{50})]. Velocity should be reported as picomoles per unit time. To convert from counts per minute (cpm) per hour to picomoles per hour, the ratio of picomoles to cpm is determined by scintillation counting a one-tenth amount of kinase assay buffer used in the reactions (the amount of nonradioactive ATP is known). Also, knowing the total amount of radioactivity added to each reaction also makes approximating the amount of substrate turnover simple: (cpm incorporated into c-Jun/total cpm) × 100%. This is critical because initial rate data should be taken from a reaction that has less than 5–10% substrate turnover.

Alternatively, kinase reactions can be terminated with a sodium dodecyl sulfate–polyacrylamide gel electrophoresis (SDS–PAGE) loading buffer and imaged via SDS–PAGE. The reaction volumes can be scaled down 3-fold for SDS–PAGE analysis. Quantitation can be achieved by either densitometry of an autoradiogram or phosphoimaging of the actual labeled gel.

Inhibitor Evaluation

The first step in the development of pharmacological JNK agents is enzymatic inhibitor analysis: evaluating the effect of an inhibitor on its target in a simple and well-defined system. Kinases can be inhibited at multiple sites: ATP-binding, phosphoacceptor-binding, and allosteric sites. In addition, kinase activity can be modulated by many mechanisms: reversible inhibition, tight inhibition, irreversible inactivation, and others. Enzymatic analysis of an inhibitor is critical for subsequent inhibitor characterization in a cellular context. In addition, the interactions of an inhibitor with the enzyme must be characterized for meaningful comparison of different inhibitors in both structure–activity relationship (SAR) analysis and structure-based drug discovery (SBDD).

Site of Action

Double-Reciprocal Analysis. The most accurate method to determine both the inhibition constant (K_i) and the type of inhibitor is double-reciprocal analysis: varying c-Jun, ATP, and inhibitor concentrations (Fig. 1). The substrate concentrations should surround the K_m level. The K_m values of JNK1α1, JNK2α2, and JNK3α1 are as follows: ATP (4.9, 2.2, and 7.0 μM)

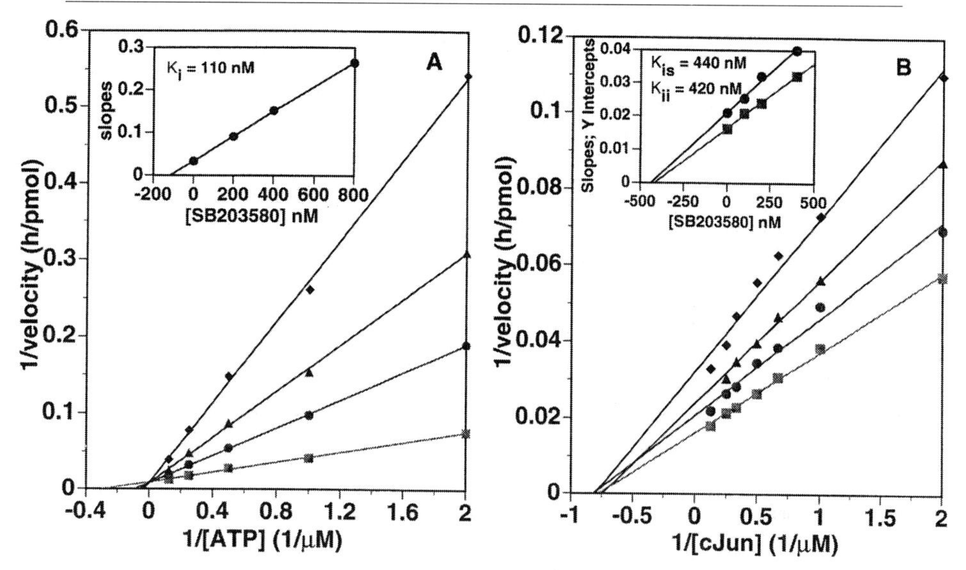

FIG. 1. Double-reciprocal analysis of JNK2 inhibition by SB203580. The inverse of velocity is plotted as a function of the inverse of either (A) the ATP concentrations at fixed concentrations of c-Jun or (B) the c-Jun concentrations at a fixed concentration of ATP. (A) The family of lines generated have a common *y* intercept, which indicates that the inhibitor is competitive with ATP. *Inset*: Replot of the double-reciprocal slopes as a function of SB203580 concentration. The inhibition constant is minus the *x* intercept (110 n*M*). (B) The family of lines generated does not have a common *y* intercept, which indicates that the inhibitor is not competitive with c-Jun. *Inset*: Replot of the double-reciprocal slopes (●) or *y* intercepts (■) as a function of SB203580 concentration yields K_{is} and K_{ii}, respectively.

and c-Jun (11.0, 8.6, and 8.3 μ*M*) (B. W. Murray, unpublished observation, 1998). Two experiments are performed. First, the inhibitor and c-Jun (1.25, 2.5, 5, 10, and 15 μ*M*) concentrations are varied at a fixed saturating concentration of ATP (25 μ*M*). Keeping the substrate concentration fixed at a saturating level allows for the pseudo-single substrate approximation. Second, inhibitor and ATP (1, 2, 5, 10, and 25 μ*M*) concentrations are varied at a fixed concentration of c-Jun (11 μ*M*, 0.40 mg/ml). The inhibitor concentrations should be less than the IC_{50} concentrations or multiples of the K_i level to minimize error. For example, SB203580 has a K_i of 110 n*M* and an IC_{50} value of 600 n*M* at 10 μ*M* ATP for JNK2α2 (B. W. Murray, unpublished observations, 1998). Suggested SB203580 concentrations could be 0, 200, 400, and 600 n*M*. The corrected data (see *In Vitro* Kinase Assay, above) are plotted as follows: 1/velocity versus 1/[substrate] for each inhibitor concentration. A family of lines is generated. If the lines are not linear, make sure that less than 5–10% of the substrate has been turned over.

Another potential cause could be that the inhibitor is not a simple, reversible inhibitor (see the next section). A common y intercept demonstrates that the inhibitor is competitive with the varied substrate. The example of the analysis of pyridinylimidazole SB203580 as a function of ATP and c-Jun concentrations is shown in Fig. 1. Data can also be fit to the proper kinetic equations.[49] The rate equations for inhibition of a single substrate or a pseudo-single substrate system is as follows:

Competitive inhibition: $v = V_{max}S/[S + K_m(1 + I/K_i)]$

Noncompetitive inhibition: $v = V_{max}S/[(S + K_m)(1 + I/K_i)]$

Uncompetitive inhibition: $v = V_{max}S/[K_m + S(1 + I/K_i)]$

The authors recommend Segal's "Biochemical Calculations" for a more detailed treatment of analysis of enzyme inhibition.[50]

Substrate Protection. The quickest and highest through-put method for characterizing how an inhibitor interacts with JNK relative to its substrates is through substrate competition experiments. An ATP-competitive inhibitor will have attenuated potency at saturating ATP concentrations compared with a K_m-level ATP (Fig. 2). IC_{50} values are obtained at high and low concentrations of substrate. Six-point dose–response curves (DRCs) at 5 and 50 μM ATP are performed at half-log increments of inhibitor concentration. If the IC_{50} shifts >5-fold at the high ATP concentration, the inhibitor may be ATP competitive. For example, replotting the double-reciprocal data for SB203580 inhibition of JNK2 as a function of ATP (Fig. 1) in the form of dose–response plots show that the potency of an ATP-competitive inhibitor will decrease as the ATP concentration increases (Fig. 2). Evaluating the phosphoacceptor is more problematic because 10-fold the K_m value would be 3600 μg/ml. Usable c-Jun concentrations are as follows: 0.55 μM (0.020 mg/ml) and 11 μM (0.40 mg/ml).

Substrate protection is a simple and quick experiment for assessing if an inhibitor is competitive with one of the substrates. Allosteric inhibition can be implied only if the inhibitor is not competitive with either substrate. Also, irreversible inhibitors may be affected by the amount of protein (e.g., c-Jun) in the reaction and appear to be substrate competitive. Thus, knowing the mechanism of an inhibitor is critical to meaningful analysis.

Mechanism of Action

Inhibitor potency may be due to multiple mechanisms including reversible inhibition, tight binding reversible inhibition, and irreversible inactivation. For meaningful comparison of inhibitors, the mode of inhibition should

[49] I. H. Segal, "Enzyme Kinetics: Behavior and Analysis of Rapid Equilibrium and Steady-State Enzyme Systems." John Wiley & Sons, New York, 1975.

[50] I. H. Segal, "Biochemical Calculations," 2nd Ed. John Wiley & Sons, New York, 1976.

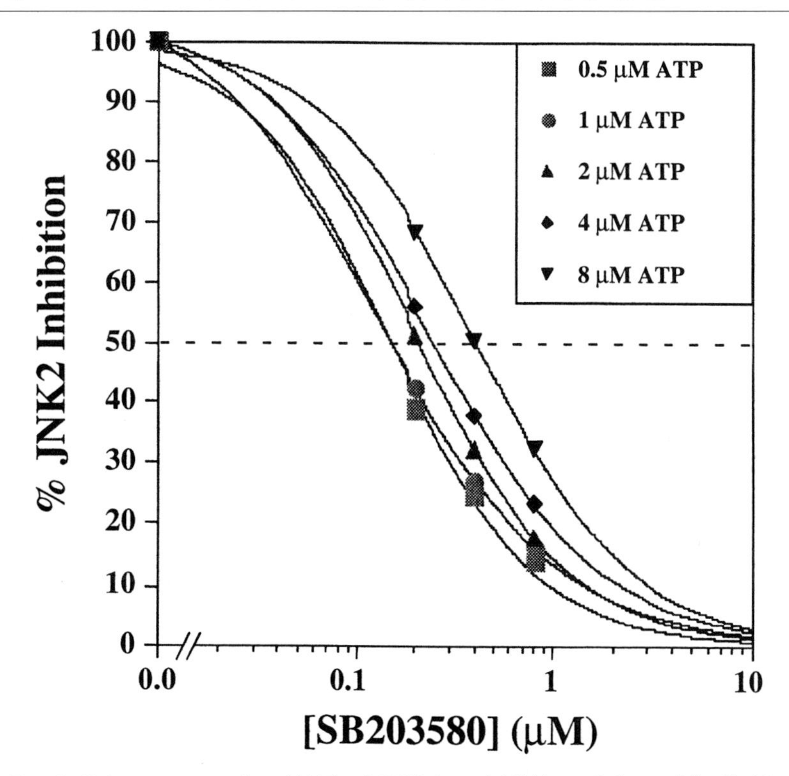

FIG. 2. Substrate protection (ATP) of JNK from inhibition of the pyridinylimidazole SB203580. Shown is the replot of the double-reciprocal data (Fig. 1A) in the format of a dose–response plot. Higher ATP concentrations decrease the observed potency of the ATP-competitive inhibitor SB203580. The effect would be substantially larger if the comparison were of dose–response curves at two concentrations of ATP: the K_m level and greater than 10 times the K_m level.

be determined: simple reversible inhibition or time-dependent inhibition (irreversible or tight binding).

Time Dependence. The potency of reversible inhibitors is governed by the enzyme–inhibitor equilibrium while irreversible or tight-binding inhibitors are dependent on both equilibrium and kinetic constraints. As such, preincubation of an inhibitor with the enzyme prior to the kinase reaction will have little to no effect on the observed potency of a reversible inhibitor whereas a longer preincubation time will enhance the observed potency of irreversible or tight-binding inhibitors. This method is a variation of the standard *in vitro* kinase assay described in an earlier section. The inhibitor concentration is varied (at six different concentrations, in triplicate) with and without a 30-min room temperature JNK preincubation.

The inhibitor preincubation with JNK contains all the assay reagents except c-Jun, ATP, and [γ-^{32}P]ATP. The kinase reaction is subsequently initiated with the addition of K_m levels of the substrate and [γ-^{32}P]ATP. Reversible inhibitors such as SB203580 will exhibit less than a 5-fold change in potency whereas irreversible or tight inhibitors will inhibit JNK with a >5-fold enhanced apparent potency.

Washout Experiments. On dilution, simple reversible inhibitors should dissociate from JNK. Tight-binding inhibitors or irreversible inactivators have little to no off-rate and as such will not be subject to dilution effects. JNK (20–30 nM) is preincubated with a saturating level of inhibitor in a 0.1-ml volume for 30 min. Two controls are included: no inhibitor and 6 μM SB203580 (10 times the IC_{50} concentration). Two sets of reactions are prepared in duplicate: one set to be analyzed before the dilution to ensure full initial inhibition and one set to be analyzed after the dilution to detect loss of inhibition. Inhibited JNK is subjected to sequential filtration with a Microcon 10 (Millipore, Bedford, MA). The JNK solution is diluted to 0.5 ml with a dilution buffer [50 mM HEPES (pH 7.4), 50 mM NaCl, 2 mM DTT] and filtered to a final volume of 5–10 μl. The concentrated JNK is diluted to 0.5 ml with the dilution buffer and filtered. This procedure is repeated until a >10,000 dilution is achieved. The activity of the recovered JNK is determined with a standard JNK kinase activity (see previous section). For a reversible inhibitor such as SB203580, the samples analyzed before the dilution should show total inhibition whereas the samples analyzed after the dilution should show <20% inhibition relative to the no-inhibitor control.

Specificity of Inhibition

Evaluating a JNK inhibitor against a limited panel of kinases will indicate gross specificity problems. This might also help identify chemically reactive compounds. A useful panel of protein kinases is as follows: protein kinase A (PKA), p38, ERK, JNKKs, and p56^{lck}. Thus an inhibitor is tested against the most well-known Ser/Thr protein kinase (PKA), MAP kinase family members (p38 and ERK), upstream activators (JNKK1/JNKK2), and a tyrosine kinase (p56^{lck}). Currently many kinases are commercially available, some in kit form, for quick testing of compounds: Upstate Biotechnology (Lake Placid, NY), Cerep (l'Evescault, France), Biomol (Plymouth Meeting, PA), and others. Many other kinases could potentially be utilized.

Evaluation of JNK Inhibitors in Cellular Context

Once the biochemical JNK–inhibitor interactions are characterized, then an inhibitor can be evaluated in the more complex cellular context.

Among the challenges an effective inhibitor must overcome are membrane permeability, quenching caused by nonspecific binding interactions, metabolism-induced chemical modification, nonspecific toxicity, and targeting JNK in the context of a multiprotein complex(es). In this section we describe assays for measuring both inhibition of endogenous JNK directly, and the inhibition of molecular events that are dependent on JNK activity.

Direct Analysis of JNK Enzyme Activity

Evaluating JNK Inhibitors on Endogenous JNK. It is useful to confirm the effects of an inhibitor on the endogenous enzyme or enzyme complex as compared with the baculovirally expressed, His_6-tagged, recombinant JNK used in the biochemical assays. Two assays are described below for measuring activity of endogenous JNK: GST–c-Jun capture assay and immune complex assay. If the compound being used is known to act as an irreversible inhibitor of JNK, as determined by biochemical assays (previous section), the compound may be added to the cells and the JNK then isolated and assayed. However, if the compound is a reversible inhibitor, it will be washed out during the isolation of JNK. In this situation, it is necessary to add the compound just prior to, or concurrently with, the [γ-^{32}P]ATP in the kinase reaction.

The c-Jun capture assay uses a c-Jun bait substrate to bind JNK in a cell lysate and has been described in detail in an earlier volume in this series[42] and elsewhere.[8] This assay has been the method of choice for several reasons: First, the assay will measure all JNK isoforms. Second, there is no kinetic interference due to bound antibody. After the first catalytic cycle, JNK will be released from the c-Jun beads. Third, the interaction of JNK with c-Jun is robust, which makes this assay simple and versatile. Stringent washes can be used to remove most other proteins including contaminating kinases.

In brief, GST–c-Jun (1–79) (7 μg per sample) is complexed to glutathione–Sepharose (20 μl per sample; Amersham Pharmacia Biotech, Piscataway, NJ) to produce c-Jun affinity beads. Cell lysate is mixed with the affinity beads and allowed to mix at 4° for 3 hr. The sample is centrifuged to pellet the bead complex and washed three times with lysis buffer containing 300 m*M* NaCl and once with kinase buffer (see *In Vitro* Kinase Assay, above). The bead pellet is then mixed with 30 μl of kinase buffer containing [γ-^{32}P]ATP (5 μM, 10 μCi) (and JNK inhibitor if required) and incubated for 30 min at 30°. The sample is denatured with gel loading buffer (10 μl) prior to fractionation by SDS–PAGE and autoradiography. Studies have shown that this c-Jun capture method does not pull down either p38 kinase or ERK kinase (B. Bennett, unpublished observation, 1999).

The IP–JNK kinase assay utilizes a JNK antibody to immunoprecipitate-activated JNK from a cell lysate and has been described in an earlier volume of this series[42] and elsewhere.[51,52] Limitations of the IP–kinase approach are that (1) with the antibodies now available, all isoforms do not immunoprecipitate to equal levels and (2) the kinase is complexed to an antibody that may affect reaction kinetics. As highly selective nonneutralizing antibodies to individual JNK isoforms become available, the IP-kinase assay will be valuable for analyzing isoform-selective JNK inhibitors. Phospho-specific MAP kinase antibodies are discussed in depth in [24] and [25] in this volume.[52a,b] Recommended commercial JNK antibodies for immunoprecipitation studies are available from PharMingen (Carlsbad, CA) and Santa Cruz Biotechnology (Santa Cruz, CA). These antibodies show preferences but not complete selectivity for specific isoforms.

Immunoblotting of Phospho-c-Jun. Endogenous JNK activity can also be monitored by measuring the phosphorylation status of cellular c-Jun. Kits that contain three c-Jun antibodies (phospho-Ser-63 specific, phospho-Ser-73 specific, and phospho-independent) are commercially available (e.g., New England BioLabs, Beverly, MA). We have observed that the phospho-specific antibodies detecting c-Jun phospho-Ser-73 typically provide more robust results.

Jurkat T cells of human origin (ATCC TIB-152) are routinely cultured at a density of 0.2–1.0 $\times$ 10^6 cells/ml in RPMI 1640 medium with 10% (v/v) heat-inactivated fetal bovine serum at 37° in 95% air, 5% CO_2. Before assay, cells are pelleted and resuspended in fresh warm medium (3 $\times$ 10^6 cells per sample at 1 $\times$ 10^6 cells/ml). Inhibitors should be added 10–20 min prior to stimulation to allow time for the compound to penetrate the cell and bind to JNK. The stimulus, for example, TNF-α at 20 ng/ml (Peprotech, Rocky Hill, NJ) or phorbol myristate acetate (PMA, and 50 ng/ml) phytohemagglutinin (PHA, 1 μg/ml) (Sigma), or anisomycin (50 ng/ml; Sigma), is added as a bolus representing 20% of the final volume and the cells are incubated for 15 min. After incubation, equal volume of ice-cold PBS containing 4$\times$ phosphatase inhibitors (20$\times$; 10 m*M* EGTA, 6 m*M* Na_3VO_4, 0.4 *M* β-glycerophosphate, 0.2 *M* NaF, 20 m*M* benzamidine, 0.2 *M* *p*-nitrophenyl phosphate) is added to each sample and the cells are pelleted at 4°. The cell pellet is lysed in 100 μl of RIPA buffer [1$\times$ PBS, 1% (v/v) Igepal CA-630 (Sigma), 0.5% (w/v) sodium deoxycholate, 1% (w/v) SDS,

[51] Y. Kawakami, S. E. Hartman, P. M. Holland, J. A. Cooper, and T. Kawakami, *J. Immunol.* **161,** 1795 (1998).

[52] T. Herdegen, F. X. Claret, T. Kallunki, A. Martin-Villalba, C. Winter, T. Hunter, and M. Karin, *J. Neurosci.* **18,** 5124 (1998).

[52a] A. J. Whitmarsh and R. J. Davis, *Methods Enzymol.* **332,** [24], 2001 (this volume).

[52b] S. Goneli and B. W. Jarvis, *Methods Enzymol.* **332,** [25], 2001 (this volume).

and protease inhibitor cocktail (Boehringer Mannheim, Indianapolis, IN)] and the protein concentration is determined. Protein (80–100 μg) is fractionated by SDS–PAGE and blotted to nitrocellulose membrane according to the manufacturer recommendations. Time course experiments will highlight the temporal kinetics of JNK/c-Jun activation in the cell, while compound dose-curve experiments will provide an IC_{50} for inhibition of c-Jun phosphorylation. Quantitation of c-Jun phosphorylation is accomplished by densitometry. Inhibitor specificity can be determined by using phosphospecific antibodies against other kinases or transcription factors that are activated within the same cell system, for example, ERK, p38, Elk-1, and ATF2 (New England BioLabs).

Analysis of JNK-Dependent Cell Function by Interleukin 2 Expression

Analysis of JNK1 and JNK2 knockout animals has identified distinct defects in T cell proliferation and differentiation.[53,54] Furthermore, validation experiments with these genetic mutants,[55] dominant negative JNK pathway mutants,[56,57] and JNK-dependent IL-2 mRNA stability[58] have shown that JNK activity is essential for the expression of IL-2, an autocrine cytokine required for T cell activation and proliferation. IL-2 expression may be monitored either as the mRNA or as the secreted protein. Although analysis of secreted protein typically provides higher throughput than mRNA analysis, protein levels may be affected by nonspecific compound effects on posttranscriptional mechanisms.

Interleukin 2 mRNA Quantitation by TaqMan Reverse Transcriptase-Polymerase Chain Reaction. The Perkin Elmer (Norwalk, CT) TaqMan detector for monitoring polymerase chain reaction (PCR) amplification, will provide a moderate throughput analysis JNK inhibitors by measuring steady state IL-2 mRNA levels. In this assay, Jurkat T cells (1 ml at 1 × 10^6 cells/ml) are plated in a 24-well plate as required for the experiment. Compounds are added as a 20-min pretreatment and the cells are then stimulated [PMA (50 ng/ml) and PHA (1 μg/ml); or anti-CD3 and anti-CD28; see Evaluating JNK Inhibitor Effects, below.] and allowed to incubate for 3–5 hr. Temporal kinetics using this assay show that IL-2 levels

[53] C. Dong, D. D. Yang, M. Wysk, A. J. Whitmarsh, R. J. Davis, and R. A. Flavell, *Science* **282,** 2092 (1998).

[54] K. Sabapathy, Y. Hu, T. Kallunki, M. Schreiber, J. P. David, W. Jochum, E. F. Wagner, and M. Karin, *Curr. Biol.* **9,** 116 (1999).

[55] J. Jain, V. E. Valge-Archer, and A. Rao, *J. Immunol.* **148,** 1240 (1992).

[56] M. Faris, N. Kokot, L. Lee, and A. E. Nel, *J. Biol. Chem.* **271,** 27366 (1996).

[57] A. Khoshnan, S. J. Kempiak, B. L. Bennett, D. Bae, W. Xu, A. M. Manning, C. H. June, and A. E. Nel, *J. Immunol.* **163,** 5444 (1999).

[58] C. Y. Chen, F. Del Gatto-Konczak, Z. Wu, and M. Karin, *Science* **280,** 1945 (1998).

increase after 1 hr and peak at 6 hr. Cells are harvested and total RNA isolated with TriReagent (Sigma) or RNeasy (Qiagen, Valencia, CA). RNA is resuspended in RNase-free water and quantified by absorbance at 260 nm (1.0 AU = 40 μg/ml nucleic acid). Real-time PCR is performed according to the manufacturer protocols. For a loading/cytotoxicity control it is possible to multiplex the PCR amplification and analyze a constitutively expressed gene such as glyceraldehyde-3-phosphate dehydrogenase (GAPDH) levels simultaneously with IL-2. Primer pairs and TaqMan probes that have proved effective in this assay are as follows.

IL-2 forward:	5′-CAG ATG ATT TTG AAT GGA ATT AAT AAT TAC AA-3′
IL-2 reverse:	5′-GCC TTC TTG GGC ATG TAA AAC T-3′
GAPDH forward:	5′-GAA GGT GAA GGT CGG AGT C-3′
IL-GAPDH reverse:	5′-GAA GAT GGT GAT GGG ATT TC-3′
IL-2 (FAM) probe:	5′-TCC CAA ACT CAC CAG GAT GCT CAC ATT-3′
GAPDH (Joe) probe:	5′-CAA GCT TCC CGT TCT CAG CC-3′

The amount of RNA should be optimized, but we typically use close to 5 ng per sample. Final reagent concentrations are 1× TaqMan buffer (Perkin-Elmer), 5 m*M* MgCl, 200 μM each of dATP, dCTP, dGTP, and dUTP, 100 n*M* each of IL-2 probe and GAPDH probe, 900 n*M* IL-2 primers, 200 n*M* GAPDH primers, 20 U of RNasin, 12.5 U of murine leukemia virus (MuLV), and 1.25 U of Amplitaq Gold.

Interleukin 2 Protein Quantitation by Enzyme-Linked Immunosorbent Assay. The assay is typically performed in a U-bottomed 96-well plate with a final volume of 250 μl. Because of compound pretreatment, a 25-μl volume of 10× inhibitor is first added to each well. A 200-μl volume of 1×10^6 Jurkat cells in growth medium is then added (0.2×10^6 cells per well) and mixed by pipetting or shaking. After 20 min to allow for compound pretreatment, a 25-μl volume of 10× stimulus (PMA at 500 ng/ml and PHA at 10 μg/ml) is added to each well and mixed. For IL-2 production, cells should be cultured for at least 10 hr. More conveniently, the experiment is set up late in the day and allowed to incubate overnight, for example, 16 hr. The culture plate is then centrifuged to pellet the cells and approximately 160 μl of medium is removed and transferred to a clean plate. IL-2 can typically be measured without dilution with enzyme-linked immunosorbent assays (ELISAs) available from Biosource (Camarillo, CA), Endogen (Woburn, MA), and R&D Systems (Minneapolis, MN). They include all stimulation and vehicle controls. When performing compound titrations for dose curves, it is important to maintain the concentration of the vehicle constant in all samples. It is also important to determine the maximum concentration of vehicle allowable in the assay. For instance, a vehicle such as dimethyl sulfoxide (DMSO) should not be used at concentrations greater than 0.5% (v/v) with Jurkat T cells.

JNK-dependent inhibition of IL-2 production should be due to specific inhibition of JNK, and not to nonselective compound toxicity. Compound toxicity can be simultaneously monitored using the MTT/Alomar Blue cytotoxicity assays available from Promega (Madison, WI); GIBCO-BRL (Gaithersburg, MD), and Biosource. In these assays, a substrate dye is added to the culture and incubated for 1–2 hr. Substrate conversion by metabolic enzymes (evidence of cell viability) changes the absorbance of the medium and can be read in a standard spectrophotometric plate reader. The substrate should be added for the final 1–2 hr of the culture time course. We have verified that the substrate dye and conversion product do not affect the IL-2 ELISA.

Because IL-2 is selectively expressed by T cells, this assay can be readily formatted for primary peripheral blood mononuclear cells (PBMCs: monocytes and lymphocytes) or whole blood diluted 1:5 with PBS. These more complex cell systems provide increased stringency for evaluating inhibitors prior to animal testing.

Evaluating JNK Inhibitor Effects on Primary T Cells by Flow Cytometry

Cell Preparation, Isolation, and Cell Culture

To test candidate, small molecule JNK drugs that inhibit or regulate the immune response of human T cell subsets, an *in vitro* human assay system was developed.[59] Effector T cell subsets used in these studies are obtained from either flow sorting or immunomagnetic separation, using adult peripheral blood or neonatal cord blood. The purified naive T cell subsets are subjected to a specific cytokine polarization environment to transform them into cells of the helper T type 1 (Th1) subtype or the Th2 subtype. Classic identification of the polarized Th1 or Th2 subsets can be done primarily through intracellular staining for their specific cytokine production.[60] Specific cell surface markers have been associated with Th1 and Th2 cells[61–63] and are used in this method to further characterize the cells generated and correlate these with known cell surface activation markers in response to the candidate JNK inhibitor.

[59] E. M. Palmer and G. A. van Seventer, *J. Immunol.* **158,** 2654 (1997).

[60] T. R. Mosmann and R. L. Coffman, *Adv. Immunol.* **46,** 111 (1989).

[61] F. Annunziato, G. Galli, L. Cosmi, P. Romagnani, R. Manetti, E. Maggi, and S. Romagnani, *Eur. Cytokine Netw.* **9,** 12 (1998).

[62] M. L. Alegre, H. Shiels, C. B. Thompson, and T. F. Gajewski, *J. Immunol.* **161,** 3347 (1998).

[63] H. Kanegane, Y. Kasahara, Y. Niida, A. Yachie, S. Sughii, K. Takatsu, N. Taniguchi, and T. Miyawaki, *Immunology* **87,** 186 (1996).

Tissue. Peripheral blood lymphocytes are obtained from normal, random donor buffy coats from the San Diego Blood Bank (San Diego, CA). Neonatal cord blood is obtained from Advanced Bioscience Resources (Alameda, CA) or as selected $CD4^+$ cells (AllCells, Foster City, CA).

Peripheral Blood Mononuclear Cell Isolation. A processed buffy coat (50 ml) is diluted 1:10 in 500 ml of sterile Dulbecco's phosphate-buffered saline (DPBS) containing sodium heparin (8 U/ml; Life Technologies, Rockville, MD) in a sterile tissue culture hood, using universal blood handling precautions, and separated with Lymphoprep (Life Technologies, Rockville, MD) as per the manufacturer instructions.[64] Check the PBMCs for viability and final numbers, using 0.02% (w/v) trypan blue (Life Science Technologies) and a hemacytometer.

$CD4^+$ Subset Isolation by Flow Sorting. Phenotypically naive $CD4^+CD45RA^+$ cells can be isolated with a high degree of purity by flow sorting of PBMCs (Fig. 3). PBMCs are stained with anti-CD45 RA–fluorescein isothiocyanate (FITC), CD4–CyChrome, and CD8–phycoerythrin (PE) (PharMingen). For every 1×10^6 cells, 100 μl of sterile-filtered, staining buffer [DPBS with 0.5% (v/v) FBS] and 20 μl of each antibody are added. The cells are stained on ice for 30 min in the dark and then washed twice by centrifugation in 3 ml of cold staining buffer. The sort sample is suspended at a final cell concentration of $2–5 \times 10^6$ cells/ml in DPBS with 0.5% (v/v) FBS and kept on ice until the initiation of the sort procedure. The sample is then gated for the $CD4^+CD8^-$ population with sort gates set for the $CD45RA^-$ and $CD45RA^+$ fraction. The $CD4^+CD45RA^+CD8^-$ cells are sorted with a FACStarPlus flow cytometer (Becton Dickinson Immunocytometry Systems, San Jose, CA).

$CD4^+$ Cell Selection by Immunomagnetic Separation. $CD4^+$ cells can alternatively be selected by positive immunomagnetic selection with the MultiSort MidiMACS system (Miltenyi Biotech, Auburn, CA). The staining buffer used in the selection process must be cold and degassed. The manufacturer instructions should be followed. After collecting the positive fraction, count the cells and determine viability, using 0.02% (w/v) trypan blue and a hemacytometer. As a quality control step, an aliquot of the cells can be stained with a fluoresceinated anti-CD4 monoclonal antibody and analyzed on the flow cytometer to determine purity. We routinely obtain >97% purity by this method.

Cell Activation

Cells are cultured in all these procedures with RPMI 1640 (HyClone, Logan, UT) supplemented with 10% (v/v) heat-inactivated fetal bovine

[64] A. Boyum, *Scand. J. Clin. Lab. Invest. Suppl.* **97,** 77 (1968).

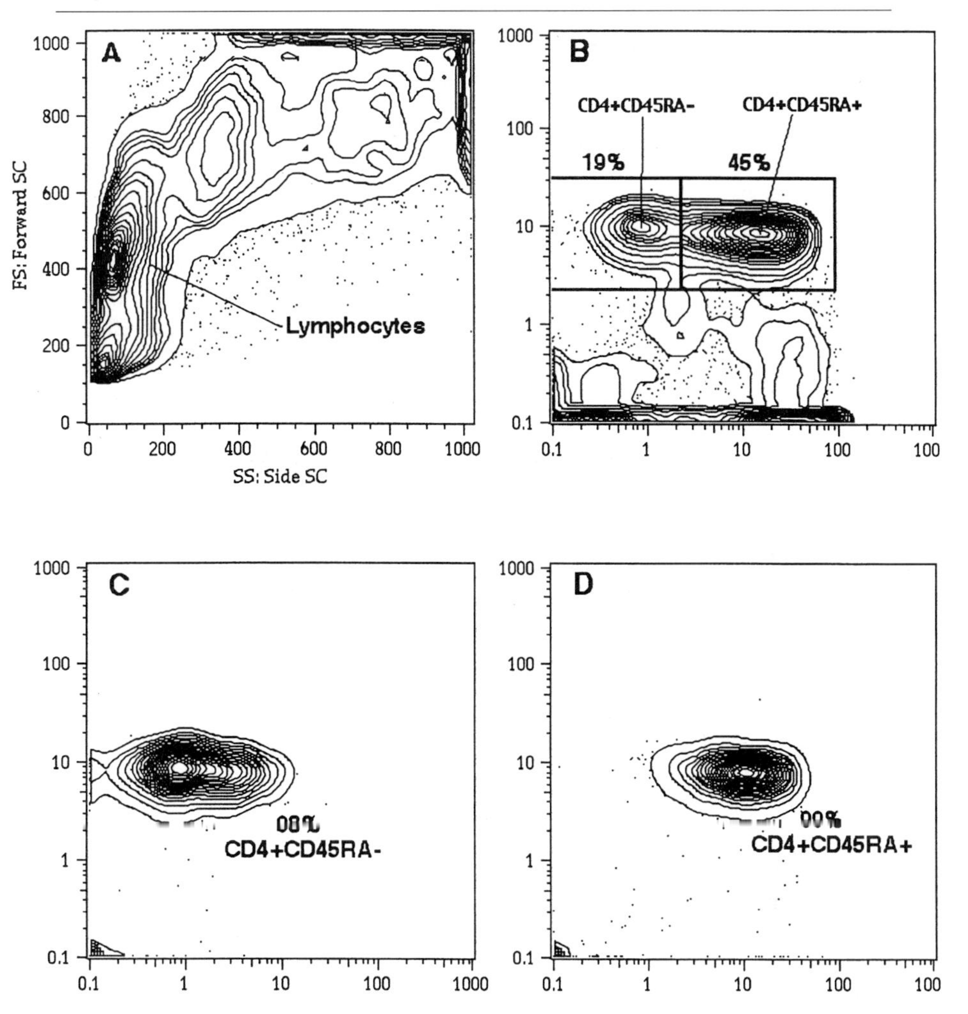

Fig. 3. (A) Correlated forward (y) and side (x) scatter measurements of the PBMC sample on a flow cytometer to define lymphocytes. (B) Correlated CD4–CyChrome (y) and CD45RA–FITC (x) measurements to define the $CD4^+$ subpopulations of interest from the gated lymphoid cells. (C and D) Ungated analysis of the sorted cells to determine purity. Contours are shown as 5% probability plots.

serum (FBS) (HyClone) with 2 mM L-glutamine (HyClone) and 0.1 mg/ml of both penicillin–streptomycin (Pen/Strep) (Hyclone).

Generating Activated $CD4^+$ Cells by CD3 and CD28 Costimulation. Purified $CD4^+CD45RA^+$ cells can be induced to activation using CD3 and CD28 monoclonal antibodies. The CD3 antibody is known to react with the ε chain of the CD3/T cell antigen receptor (TCR) complex and plays a key role in signal transduction during antigen recognition.[65] CD28 is a 44-kDa homodimeric transmembrane glycoprotein on the surface of the $CD4^+$ cell and plays an important costimulatory role in the activation process.[66] Evidence suggests that CD28 initiates and regulates a separate and unique signal transduction pathway from those stimulated only through the TCR complex, although the exact mechanism is being defined.[67] By evaluating new compounds using the CD4 activation sequence, information regarding its effect on key signaling pathways can be elucidated.

PREPARING CD3 ACTIVATION MATRIX. Either 24-well trays or 25-cm^2 flasks are used in these methods and prepared by binding the mouse anit-human monoclonal antibody CD3 to the tissue culture surface of the culture vessel as the primary stimulus. Prepare a 2-μg/ml solution of azide-free/low-endotoxin anti-CD3 antibody (PharMingen) in Ca–Mg-free Dulbecco's PBS. Dispense 1 ml of the solution to each well of the 24-well tray or 3 ml into the 25-cm^2 flask. Incubate at 4° overnight. Just prior to plating the cells, aspirate the antibody solution and then add 2 ml of cold DPBS to wash each well and aspirate again. For 25-cm^2 flasks, add 3 ml of cold DPBS to wash the flask and aspirate.

ACTIVATING $CD4^+$ CELLS. The purified $CD4^+$ naive T cells from the cell sorter or the magnetic column are suspended in RPMI 1640 with 10% (v/v) FBS supplemented with 1% (w/v) L-glutamine and 1% (v/v) Pen/Strep and washed. The cells are adjusted to a density of $0.5–1.0 \times 10^6$ cells/ml. Compounds to be tested are used at concentrations determined by cell testing (see above). The cells are incubated with compound for 1 hr at 37°. The costimulus, anti-CD28 (PharMingen), is then added at a final concentration of 2 μg/ml. If needed, the cross-linker protein G (Sigma/Aldrich) may be added at a final concentration of 2 μg/ml to achieve maximum activation signaling. Cultures are incubated for 72 hr at 37° in humidified 5% CO_2.

The cells are suspended in the well or flask by gentle pipetting then transferred to sterile 15-ml polypropylene centrifuge tubes. The suspensions are spun down at 300*g* for 10 min at 4°. The supernatants are placed in cryovials and frozen at −80°, and saved for later studies of cytokine levels. The cell pellet can now be stained for cell surface markers or cell cycle analysis, or cultured again in a secondary assay.

[65] F. W. Fitch, *Microbiol. Rev.* **50,** 50 (1986).

[66] C. H. June, J. A. Ledbetter, P. S. Linsley, and C. B. Thompson, *Immunol. Today* **11,** 211 (1990).

[67] J. Nunes, S. Klasen, M. D. Franco, C. Lipcey, C. Mawas, M. Bagnasco, and D. Olive, *Biochem. J.* **293,** 835 (1993).

Polarizing Naive CD4$^+$ T Cells to Th1 and Th2 Subtypes. The use of specific effector Th1 and Th2 subsets of CD4$^+$ cells in testing compounds requires that the naive CD4$^+$ cells be cultured with specific cytokines in order to polarize the cells to the correct phenotype.

Th1 Cell Generation. Naive CD4$^+$45RA$^+$ cells from the sorter or immunomagnetic column are first washed in complete medium and then cultured at 0.5×10^6 cells/ml with recombinant IL-12 (3 ng/ml; R&D Systems) and neutralizing mouse anti-human IL-4 (PharMingen) antibody (1 μg/ml). Cells are plated into 24-well trays with bound CD3 antibody (2 μg/ml) and soluble CD28 (2 μg/ml) added 1 hr later. Cells are cultured for 72 hr at 37° in humidified 5% CO_2.

Th2 Cell Generation. Th2 phenotypes are generated from CD4$^+$ cells in medium containing recombinant IL-4 (10 ng/ml; R&D Systems) with neutralizing mouse anti-human IL-12 (PharMingen) antibody (1 μg/ml). This is supplemented with soluble anti-CD28 (2 μg/ml) 1 hr later and plated into wells with bound CD3. The cultures are then incubated for 72 hr at 37° in humidified 5% CO_2.

After 72 hr, the cells are gently suspended in the well (or flask) by gentle pipetting and transferred to a culture tube, where the cells are centrifuged at 300*g* for 10 min at 4°. An aliquot of supernatant is removed and stored at −80° for ELISA testing and the cells are washed in 3 volumes of culture medium warmed to 37°. The cells are then suspended in fresh culture medium to a plating density of 1×10^6 cells/ml. An aliquot of the cells can be taken for immunophenotyping or cell cycle analysis.

At this point, candidate compounds can be added for coculture with the polarized Th1 or Th2 cells. The cells are then plated onto fresh culture trays with bound CD3 and soluble CD28 as previously described, supplemented with the correct complement of growth factors to maintain their polarized state. The cultures are then incubated for 48 to 72 hr at 37° in humidified 5% CO_2. Cell phenotype, cell cycle, and supernatant ELISA of the treatment groups follow this final incubation.

Cell Analysis Methods

The isolated cells from the sorter or immunomagnetic column can be stained or analyzed for cell cycle status to form the baseline profile. Cells that have been cultured in the various treatment groups are also prepared by these procedures in order to ascertain their polarization state and level of activation. In addition, supernatants taken from actively growing cultures can be analyzed for secreted cytokines such as IL-2 (see above).

Immunophenotyping and Sample Analysis. Direct, three-color immunofluorescence staining is performed according to standard methods. Mouse anti-human CD3, CD4, CD8, and CD19 are used to determine the level of T cell purity; CD25, CD69, CD71, CD152 for the degree of activation; CD26 and CCR5 for identifying the Th1 effector cells; and CD30, 62L, and CXCR4 for the identification of the Th2 cells. The reagents used are directly conjugated to FITC, PE, peridinin-chlorophyll protein (PerCP), or Cy-Chrome (CyC) and obtained from PharMingen or Becton Dickinson Biosciences. Immunoglobulin-matched control antibodies with appropriate fluorescent label are also included.

Cells (1×10^6) in 100 μl of cold staining buffer with 0.1% (w/v) sodium azide are added to a 12×75 polystyrene culture tube. Twenty microliters of each appropriate monoclonal antibody is added to the cell suspension and mixed. (*Note:* Both the Becton Dickinson and PharMingen reagents are optimized for use at 20 μl for 10^6 cells. Reagents from other sources may be different.) Incubate the staining cells on ice for 30 min. Add 2 ml of cold staining buffer to wash each tube and spin at 300*g* for 10 min at 4° and then remove the supernatant. Perform the wash step twice. Suspend the cells in 0.5 ml of cold buffer with propidium iodide (0.5 μg/ml) and analyze the samples with a flow cytometer.[68] If the cells cannot be analyzed immediately, the cell pellet is suspended in 0.25 ml of cold DPBS and then added into 0.25 ml of a cold 4% electron microscopy (EM)-grade formaldehyde stock solution (Polysciences, Warrington, PA), using a transfer pipette for fixation.[69] Store the samples at 4° in the dark and then analyze within 3 days.

Samples are analyzed in our laboratory with a single-laser Coulter XL/MLS (Beckman-Coulter, Hialeah, FL) flow cytometer equipped with a 15-mW air-cooled, argon ion laser. The samples are acquired with System II software. A list mode file of 15,000 events is created for each sample. Data analysis and presentation are performed with FlowJo version 2.7.7 (TreeStar, San Carlos, CA). Cell cycle analysis utilizes the Dean and Jett algorithm.[70] The effects of the candidate JNK inhibitors on T cell activation are measured by performing cell cycle measurements on a flow cytometer.

Cell Cycle Status. The cell cycle profile of the experimental sample is determined by a modified digitonin permeabilization and propidium iodide staining protocol.[71] In our hands, this procedure provides and maintains over a longer period of time a low coefficient of variation (CV) and better

[68] D. T. Sasaki, S. E. Dumas, and E. G. Engleman, *Cytometry* **8,** 413 (1987).
[69] J. D. Lifson, D. T. Sasaki, and E. G. Engleman, *J. Immunol. Methods* **86,** 143 (1986).
[70] P. N. Dean and J. H. Jett, *J. Cell Biol.* **60,** 523 (1974).
[71] D. G. Pestov, M. Polonskaia, and L. F. Lau, *BioTechniques* **26,** 102 (1999).

estimations of the cells in the different compartments of the cell cycle. The method also includes RNase to enzymatically eliminate any double-stranded RNA from the analysis.

Stock solutions are prepared as follows: 0.8% (w/v) digitonin (CalBiochem): Dissolve 0.8 g into 100 ml of 100% ethanol by warming to 70°, using a water bath; HEPES buffer: In 500 ml of distilled water add 20 mM HEPES (2.6 g; Sigma), 0.16 M NaCl (4.7 g; Sigma), and 1 mM EGTA (0.19 g; CalBiochem); filter (0.22-μm pore size) and store at room temperature; Propidium iodide (1 mg/ml) (Molecular Probes, Eugene,

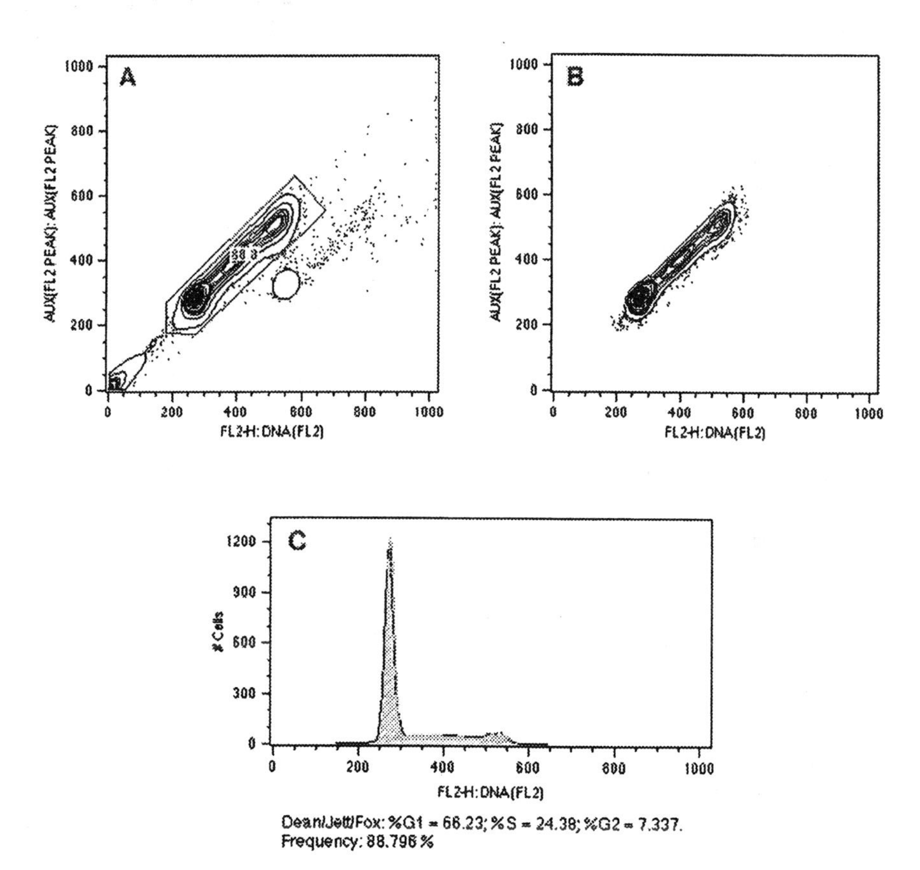

FIG. 4. CD4+ cells simulated for 48 hr with bound anti-CD3 and soluble CD28 were permeabilized with digitonin and stained with propidium iodide. A region was defined in the correlated DNA fluorescence pulse area versus pulse peak (A) to discriminate against doublets and debris. (B) Discriminated data used in the cell cycle calculation; (C) fitted histogram with the areas calculated using FlowJo software.

OR), and RNase (10 mg/ml distilled water) (Sigma): Store frozen as 1-ml aliquots.

Working propidium iodide staining solution should be prepared on the day of staining by combining the stock solutions prepared above as follows: 2.5 ml of stock 0.8% (w/v) digitonin, 100 ml of HEPES buffer, 2 ml of stock RNase, and 1 ml of stock propidium iodide. Filter through a 0.22-μm pore size filter and store refrigerated (dark bottle) at 4°.

Wash 5×10^5 to 1×10^6 cells from each sample in 2 ml of cold DPBS, centrifuge, and remove the supernatant. Suspend the cell pellet in 1 ml of the working stain solution and incubate for 1 hr at room temperature. Keep the samples in the dark during this step. Run the cells on the flow cytometer or place the samples in the refrigerator for up to 5 days.

Establish propidium iodide fluorescence in linear signal amplification and display as propidium iodide fluorescence area versus propidium iodide fluorescence peak. Set a gate around the single-stained cells by excluding the doublets, clumps, and cellular debris (Fig. 4) and then acquire at least 10,000 gated list mode events.

Acknowledgments

The authors thank Weiming Xu for technical expertise with TaqMan RT-PCR, and Dennis Young (Cancer Center, University of California San Diego) for flow cell sorting. We also thank Anthony Manning and John K. Westwick for critical reading of the chapter.

Author Index

Numbers in parentheses are footnote reference numbers and indicate that an author's work is referred to although the name is not cited in the text.

A

Abo, A., 262, 280, 319
Abraham, J. A., 236, 409, 410(15)
Abram, C. L., 89(40), 102
Adami, G. R., 370
Adams, J. L., 335, 336, 433
Adams, L., 66
Adler, V., 432
Aebersold, R., 301, 314
Afar, D., 140, 141(6), 143(6)
Ahern, S. M., 89(41), 102
Ahmad, M. F., 432, 433(2)
Ahmad, T. B., 433
Ahn, N. G., 319, 326(3), 333(3), 337, 343, 387, 388(1), 399, 419, 426, 427, 427(10), 428(15), 429(15), 430, 431
Akedo, H., 259
Al-Alawi, N., 261
Alberol, I., 412
Alberts, A. S., 276
Albright, C. F., 128, 138(1)
Alder, H., 152
Alegre, M. L., 445
Alessandrini, A., 354
Alessi, D. R., 181, 377, 419, 421(8), 422(9), 426(8), 429(8), 430(8)
Alexandrov, K., 204
Allen, J. B., 260
Altman, A., 140
Alton, G., 300
Alvarez, U., 36, 38, 47(6, 20), 49(20)
Ambroziak, P., 103
Ammerer, G., 369
Anderson, D., 433
Anderson, M. J., 232
Anderson, N. G., 344
Anderson, R. G., 302
Andersson, L., 176, 307
Andrade, J., 89(39), 102
Andreev, J., 262
Andreotti, A., 88
Andres, D. A., 10, 195, 196, 197, 198(15), 199, 200(15, 16), 202(15), 203, 204, 205, 205(10), 206(10), 302, 303
Angrand, P. O., 414
Annunziato, F., 445
Aoki, M., 86
Apell, G., 417
Appel, R. D., 308
Arboleda, M. J., 353
Armstrong, S. A., 199
Arnqvist, A., 306
Aronheim, A., 260, 261, 262, 267(13)
Aronsohn, A. I., 87
Aruffo, A., 213
Arvand, A., 222
Aryee, D. N., 297
Ashar, H., 123
Ashby, M. N., 103, 112(2)
Ashworth, A., 280, 319, 377, 419, 422(9)
Aspenstrom, P., 276
Atkins, C. M., 388
Auger, K. R., 184, 193(11)
Ausubel, F. M., 292
Averboukh, L., 233
Aviv, H., 226
Avruch, J., 131, 144, 271, 274(9), 279, 387, 412, 432, 433, 433(2)
Ayrton, A., 336
Aziz, N., 401, 412(5), 413(5)

B

Bach, J. F., 279
Badger, A., 335

Bae, D., 443
Baer, R., 262
Bagnasco, M., 448
Bagrodia, S., 280, 281, 319
Baier, G., 65
Bajaj, V., 414
Baker, T. L., 68, 69(24), 73(24)
Bakin, A. V., 222
Baldwin, A. S., 234
Baltimore, D., 4, 32(4), 35(4), 88, 217, 401, 405(7)
Band, H., 194
Banerjee, M., 262, 432, 433(2)
Bansal, A., 187
Baranes, D., 96, 97(25)
Barbacid, M., 184, 185(7), 191(7), 221
Bardeesy, N., 5
Barr, M., 154, 271, 368
Barrett, S. D., 343, 419
Barrett, T., 320, 321, 330, 333, 333(19), 334(9, 19), 432, 433, 433(1), 434
Bar-Sagi, D., 88, 276, 280
Barsoum, J., 37
Bartel, P. L., 142, 153, 277, 294, 370
Barthel, A., 417
Bartleson, C., 174
Bartlett, B. A., 341
Bartram, C., 353
Bastians, H., 222
Basu, T. N., 354
Batalao, A., 86
Batista, O., 86
Batzer, A., 88
Bauer, B., 65, 128
Baumgartner, W. K., 4
Beach, D., 166, 211, 271
Beccelaere, K. V., 343
Becherer, K., 277
Beck, T. W., 140
Becker-Hapak, M., 37, 38, 38(15), 39(15), 41(15), 49(15, 17)
Beg, A. A., 234
Belisle, B., 262
Bellacosa, A., 86
Belshaw, P. J., 412
Bemis, G. W., 435
Bender, M. A., 213
Benjamin, D. R., 152
Benjamini, E., 177
Bennett, B. L., 432, 441, 443
Beranger, F., 133
Berg, D. E., 306
Bergeron, J. J., 80
Berks, B. C., 83
Berman, M., 299
Bernard, H. U., 213
Berndt, P., 315
Betscholtz, C., 432
Bhagwat, S., 433
Bhat, R. V., 341
Bhatt, R. R., 337
Biermann, B. J., 302
Binder, M. D., 354
Binetruy, B., 432
Birchmeier, C., 139
Birnbaum, M. J., 417
Birnboim, H. C., 217, 220(12)
Birrer, M. J., 434
Bishop, J. M., 401, 407(4), 409(4), 413
Bishop, W. R., 115, 122(6), 123, 123(6), 124, 127(6)
Bjarnegard, M., 432
Black, S., 123
Blaikie, P., 185
Blake, D. J., 80(21), 82
Blank, J. L., 433
Blechner, S., 171, 174
Blenis, J., 174
Block, C., 159, 161(16), 162(16), 163
Blumer, K. J., 433
Boehm, J. C., 336
Boeke, J., 260
Boettner, B., 151, 166, 168(24)
Boge, A., 417
Boguski, M. S., 255, 279
Bohmann, D., 433
Bokoch, G. M., 319
Bollag, G., 5, 103, 245
Bonatti, S., 80
Bond, R. W., 115, 124
Bonnefoy-Bérard, N., 140
Booden, M. A., 64, 68, 69(24), 73, 73(24), 75(30), 82
Boren, T., 306
Borg, J. P., 187, 193(21), 194, 194(21)
Börsch-Haubold, A. G., 336
Bortner, D., 234
Bos, J. L., 128, 152, 221, 274, 353
Bosch, E., 259, 410, 412, 412(17), 413(20), 415(20), 417(20)

Botfield, M. C., 100
Botstein, D., 255
Boucheron, C., 427
Boulain, J. C., 89
Boulton, T. G., 341, 344
Bourne, H. R., 66, 78, 203, 210(1, 5)
Bouvier, M., 66
Bowcock, A. M., 262
Boyartchuk, V. L., 103, 112(2)
Boyum, A., 446
Brachmann, R. K., 260
Brady, H., 434, 435(43)
Brand, L., 159
Brann, M. R., 4
Brenner, D. A., 432, 434, 441(42), 442(42)
Brennwald, P., 78
Brent, R., 84, 143, 153, 260, 277, 281, 283, 289(4), 292, 294, 294(4), 298(25)
Brickey, D. A., 174
Bridges, A., 343, 366, 418, 419, 430(5)
Bridges, T., 276
Brini, M., 80(23), 82
Brink, S., 83
Broder, Y. C., 262, 267(13)
Brodin, P., 37
Brondello, J.-M., 343
Brown, J. H., 333, 334, 334(25)
Brown, M. L., 435
Brown, M. S., 115, 117, 187, 196, 199, 302, 306
Brown, M. T., 89(39), 102
Brown, P. H., 434
Brown, P. O., 255
Brown, T. W., 417
Bruder, J. T., 131, 144, 279
Bruenger, E., 303
Brugge, J. S., 100
Brunet, A., 338, 343, 343(5), 354
Bruret, A., 368
Bryant, M., 67
Bucci, C., 133
Buckholz, R., 299
Buckle, R. S., 386
Buckman, S. Y., 334
Buday, L., 261
Burakoff, S. J., 194
Burbelo, P. D., 276
Burgering, B. M. T., 128
Burgess, R. R., 308
Burke, B., 184, 193(11)
Burns, K., 109
Burridge, K., 83
Burrier, A., 434
Buss, J. E., 11, 12(25), 13, 14(29), 15(29), 26, 30(26), 64, 65, 68, 69, 69(24), 71, 73, 73(24, 26), 75(30), 80(15), 81, 82, 82(15), 119, 123(12), 195, 303
Bustelo, X. R., 184, 185(7), 191(7)
Butler, G., 262
Butler, M., 89(43), 102

C

Cadwallader, K., 65, 106, 112, 354
Calonge, M. J., 5
Camonis, J. H., 271, 276(8)
Campbell, D., 330, 377
Campbell, K. S., 184, 193(11)
Campbell, S. L., 78, 79, 273
Campbell-Burk, S. L., 11, 12(24), 306
Canaani, E., 152
Canada, F. J., 112
Canagarajah, B., 389
Candia, J. M., 399, 419, 427(10)
Cano, E., 386, 432
Cantley, L. C., 171, 174, 184, 185, 186, 187, 188, 193(18), 194, 194(22)
Cantor, S. B., 128
Capua, M., 97(43), 102
Carbone, R., 432
Cardone, M. H., 181
Carlile, G. W., 433
Carlson, M., 277
Carnero, A., 166, 211
Caron, P. R., 435
Carpenter, C., 184
Carpenter, J. W., 306
Carr, D., 115, 124
Carraway, C. A., 88
Carraway, K. L., 88
Carrier, A., 89
Carson-Jurica, M. A., 410
Cartwright, P., 405
Casey, P. J., 13, 14(29), 15(29), 67, 103, 119, 123(12), 195, 301
Castagnoli, L., 96, 97(24)
Catino, J. J., 115, 122(6), 123(6), 124, 127(6)
Catling, A. D., 344, 348(8), 368, 369, 370, 370(14), 371(20), 373(14, 20), 379, 384(14, 20), 430

Cattolico, L., 89
Cavanagh, J., 301, 321, 333, 335, 369, 433
Cavigelli, M., 386
Cecillon, M., 279
Celis, J. E., 204, 205(8)
Cepko, C. L., 4, 32(6), 35(6)
Cerione, R. A., 280, 319
Cesareni, G., 96, 97(24)
Chaddock, A. M., 83
Chalfie, M., 84
Chambon, P., 405
Chang, P. F., 142
Chang, T. Y., 5
Chant, J., 276
Chantry, D., 274
Chardin, P., 128, 203, 210(4), 279
Chassaings, G., 37
Chaudhuri, M., 171, 184, 185, 187(6), 189, 189(6), 191(6)
Chellaiah, M. A., 36, 38, 47(6, 20), 49(20)
Chelsky, D., 112
Chen, C. Y., 443
Chen, D., 409
Chen, H., 97(43, 44), 102
Chen, J., 88, 280, 354
Chen, L. L., 37
Chen, L. M., 210
Chen, P. L., 277
Chen, S., 262
Chen, S. Y., 11, 12(23)
Chen, T. K., 434
Chen, W.-J., 187, 196
Chen, X., 211
Chen, X. R., 409
Chen, Y.-W., 128, 131(10), 133(10), 186
Cheng, P. F., 414
Chern, Y., 210
Chernoff, J., 277, 281, 319
Chertkov, H., 84, 277, 289(4), 294(4)
Cherwinski, H., 259, 401, 409, 410, 412, 412(5), 413(5, 20), 415(20), 417(20)
Chevrier, D., 89
Chew, C. E., 10
Chiappinelli, V. A., 26
Chiariello, M., 319
Chien, C. T., 142, 153, 277, 370
Chien, K. R., 333, 334, 334(25)
Childs, J. E., 50, 195, 262
Chin, L., 5
Chirgwin, J. J., 226
Chiu, V. K., 10, 51, 56(6), 59(6), 61(6), 62(6), 69, 105, 109(6), 112
Cho, Y.-J., 233
Choi, K. Y., 368, 369, 369(3)
Choi, W. E., 186
Choi, Y.-J., 5, 103, 245
Choy, E., 10, 50, 51, 56(6), 59(6), 61(6), 62(6), 69, 105, 109(6), 112
Christensen, A., 87
Christopherson, K. S., 414
Chrzanowska-Wodnicka, M., 83
Chu, Y., 371, 384(24)
Cirillo, D., 36, 47(4)
Claret, F. X., 319, 386, 433, 442
Clark, G. J., 7, 15, 28(31), 30(31), 79, 211, 218(5), 222
Clark, M. J., 89(41), 102
Clark, R., 279
Clarke, P. R., 433
Clarke, S., 50, 105, 112(5), 195
Clerk, A., 435
Cleveland, J. L., 140
Clift, R. E., 431
Coats, S. G., 26, 69, 73, 73(26), 75(30), 80(15), 81, 82, 82(15)
Cobb, M. H., 334, 337, 341, 344, 370, 373(21), 387, 388, 389, 419, 420, 430, 432, 433, 433(2)
Cochet, C., 174
Coda, L., 96
Coffey, R. J., Jr., 222
Coffman, R. L., 445
Cohen, A., 262, 419, 421(8), 422(9), 426(8), 429(8), 430(8)
Cohen, F. E., 88
Cohen, G. B., 88
Cohen, P., 321, 335(12), 336, 341, 377, 431, 435
Colicelli, J., 139, 140, 141, 141(5, 6), 143, 143(5, 6), 146(5), 147
Collicelli, L., 271, 274(6)
Collier, L. S., 369, 370(14), 373(14), 384(14), 430
Collins, F. S., 354
Colombo, L. L., 258
Comer, A. R., 97(41), 102
Confalonieri, S., 96
Cong, F., 334
Conklin, D. S., 166, 211
Conti, C. J., 258
Conze, D., 434

Cool, R. H., 128, 160, 161(22)
Cooper, G. M., 11, 12(22), 402
Cooper, J. A., 89(39), 102, 145, 153, 154(7), 187, 193(23), 194(23), 271, 274, 274(3), 277, 442
Cooper, J. B., 83
Copeland, R. A., 343, 366, 419, 421(7), 430(7)
Copeland, T. D., 184, 193(12)
Cordon-Cardo, C., 5
Cormier, J. J., 51, 61(7)
Corsini, A., 303
Cosmi, L., 445
Coso, O. A., 319
Courtneidge, S. A., 89(40), 102
Covacci, A., 306
Cowley, S., 354, 377, 419, 422(9)
Cox, A. D., 3, 10, 11, 12(26), 15, 28(30, 31), 30(27, 31), 65, 103, 115, 195, 211, 218(5), 300, 301, 302(11), 303, 313, 314(41)
Crane, K. M., 435
Craxton, M., 336
Creasy, C. L., 281
Crespo, P., 319
Crews, C. M., 354
Crick, D. C., 196, 197, 198(15), 199, 200(15), 202(15), 205, 302, 303
Critchley, D. R., 80(27), 83
Croce, C., 152
Crowell, D. N., 302
Cuenda, A., 341, 419, 421(8), 426(8), 429(8), 430(8), 435
Curran, T., 222
Cywin, C. L., 435

D

Daar, I. O., 412
Dahan, S., 80
Dai, Q., 105, 109(6)
Dai, T., 432, 433, 433(2)
Daikh, Y., 37
Dalesio, O., 363
Daly, R. J., 354
Danesi, R., 303
Dang, A., 370, 373(21), 419
Dang, C. V., 222
Danielian, P. S., 405
Danishefsky, S., 37, 49(14)
Das, P., 276
Datta, S. R., 181
Daulerio, A. J., 343, 366, 419, 421(7), 430(7)
David, J. P., 443
Davies, K. E., 80(21), 82
Davis, L. J., 116, 369
Davis, R. J., 301, 319, 320, 321, 326(1), 330, 331, 333, 333(19), 334, 334(9, 19, 26), 335, 336(35), 341, 368, 369, 386, 432, 433, 433(1, 2), 434, 435, 442, 443
Day, R., 354
Deak, J. C., 433
Dean, A. D., 71
Dean, P. N., 450
De Camilli, P., 89(43, 44), 102
Decker, S. J., 366, 418, 430(5)
Degetyarev, M. Y., 66
de Gunzburg, J., 133
de Hoop, M. J., 204
Dejgaard, K., 80
Del Gatto-Konczak, F., 443
Della, N. G., 10
del Rosario, M., 184
Del Vecchio, R., 379, 381(27)
DeMaggio, A. J., 174
Demetrick, D., 271
Demo, S. D., 128, 131(10), 133(10)
Deng, T., 320, 334(9), 432, 433(1)
Denny, C. T., 222
Dent, P., 344
DePinho, R. A., 5, 344
Deprez, J., 181
Der, C. J., 3, 4, 7(8), 10, 11, 12(21–26), 13, 14(29), 15, 15(29), 26, 28(30, 31), 30(27, 31), 65, 68, 69, 69(24), 73, 73(24, 26), 78, 79, 80(15), 81, 82(15), 96, 97(25), 99, 103, 115, 119, 123(12), 139, 186, 193(18), 195, 211, 218(5), 221, 222, 232, 234, 273, 301, 302(11, 12), 313, 314(41), 371, 384(24)
Derijard, B., 319, 320, 321, 330, 333, 333(19), 334, 334(9, 19), 432, 433, 433(1, 2)
DeRisi, J. L., 255
Derossi, D., 37
Deschenes, R. J., 68
Desjardins, M., 133
Dhaka, A., 140, 141(6), 143(6)
Dhand, R., 144
Dhe-Paganon, S., 194
Dhillon, N., 277
Dickens, M., 320, 335, 341, 369
Diefenthal, T., 210

Di Fiore, P. P., 96, 97(43–45), 102, 128, 432
Dingwall, C., 80(22), 82
DiRocco, R., 341
Distel, B., 78, 80(4), 81(4)
Dobrowolska, G., 176
Dohi, K., 133, 134(22)
Dokken, C. G., 279
Dolan, L., 432
Dolfi, F., 386
Doll, R., 124
Doly, J., 217, 220(12)
Dominguez, M., 80
Donaldson, J. G., 89(39), 102
Dong, C., 443
Dong, L., 102
Donoghue, D. J., 65, 72(4)
Doria, M., 96
Dowd, S., 338, 343(5)
Dowdy, S. F., 36, 37, 38, 38(15), 39(15), 41(15), 46(21), 47(20), 49(15–22)
Downward, J., 131, 144, 181, 261, 276, 354
Doza, Y. N., 435
Drachman, J. G., 428
Drevet, P., 89
Drugan, J. K., 273
Druker, B. J., 184, 193(11)
D'Souza-Schorey, C., 65, 78
Ducancel, F., 89
Duckworth, B., 184
Dudek, H., 181
Dudley, D. T., 343, 366, 418, 419, 421(8), 426(8), 429(8), 430(5, 8), 431
Dumas, S. E., 450
Dunphy, J. T., 66, 68(14)
Dupont, D. R., 312
Dupree, P., 204
Durfee, T., 277
Duronio, V., 431
Dusanter-Fourt, I., 427
Duttweiler, H. M., 292
Dzudzor, B., 143

E

Earl, R. A., 343, 366, 419, 421(7), 430(7)
Ebert, D., 420
Eberwein, D., 353
Eblen, S. T., 344, 348(8), 368, 369, 370(14), 373(14), 384(14), 430
Ebright, R. H., 272
Eck, M. J., 194
Eckenrode, V. K., 51, 61(7)
Ecker, J. R., 210
Edman, C. F., 222
Edwards, M. C., 260
Eilers, M., 401, 405
Eisen, M. B., 255
Ekman, P., 176
Elion, E. A., 368, 369, 369(3)
Elledge, S. J., 131, 144, 260, 262, 277, 279, 370
Ellenberg, J., 64
Eller, N., 210
Elliott, E. M., 433
Elloit, G., 37
Elly, C., 140
Endo, M., 134
Engelberg, D., 261
Engelhardt, M., 159, 162(20), 163(20)
Engleman, E. G., 450
English, J., 341, 387, 389
Engstrand, L., 306
Enslen, H., 333, 334(26)
Epstein, W. W., 303
Erickson, A. K., 344, 368
Erikson, R. L., 354
Errede, B., 369, 433
Escobedo, J. A., 184, 417
Espenshade, P. J., 299
Estojak, J., 143, 279, 299
Euskirchen, G., 84
Evan, G. I., 405
Evangelista, C., 260
Evans, R. M., 245
Eyers, P. A., 336
Ezhevsky, S. A., 37, 38, 38(15), 39(15), 41(15), 49(15, 16)

F

Fabbro, D., 366
Fabian, J. R., 145, 412
Fanger, G. R., 332
Fantl, W. J., 140, 184, 369
Farh, L., 68
Faris, M., 443
Farnsworth, C. C., 195, 303
Farrar, M. A., 412
Faust, J., 196

Favata, M. F., 343, 366, 419, 421(7), 430(7)
Fawell, S., 37
Fazal, A., 80
Feeser, W. S., 343, 366, 419, 421(7), 430(7)
Feig, L. A., 128, 354
Felder, S., 189
Feldmann, K. A., 210
Feng, S., 88
Feoktistov, M., 10, 51, 56(6), 59(6), 61(6), 62(6), 69, 112
Feramisco, J. R., 206, 222
Ferrell, J. E., 337
Fesik, S. W., 194
Fetter, C. H., 26
Fiddes, J. C., 206
Fields, S., 86, 141, 142, 153, 260, 277, 294, 370
Filhol, O., 174
Filmus, J., 258
Finco, T. S., 234
Fink, G. R., 368, 369
Finley, R., 294
Finley, R. L., Jr., 153
Finlin, B. S., 10, 197, 198(15), 200(15, 16), 202(15), 203, 205, 302, 303
Finney, F., 151
Fiordalisi, J. J., 3, 26, 301, 302(12)
Fitch, F. W., 448
Fitzgerald, C., 36, 47(6)
Fitzgibbon, M. J., 435
Flanagan, J., 89
Flavell, R. A., 333, 334(26), 434, 443
Fletcher, J. A., 354
Fluerdelys, B., 116
Forman-Kay, J., 187, 194(22)
Forsburg, S. L., 281
Fowlkes, D., 92
Fox, R. O., 88, 96
Franco, M. D., 448
Frank, R., 187, 193(23), 194(23)
Franke, T. F., 181
Frankel, A. D., 36
Frantz, B., 431
Franza, B. R., 301
Fre, S., 89(43), 96, 102
Frech, M., 279
Freed, E., 140
Freeman, J. L., 280
Frick, I. M., 306
Friedman, E., 66
Frierson, H. F., Jr., 344
Frisch, S., 181
Fritsch, A., 262
Fritsch, E. F., 19, 197, 201(14), 204, 213, 217(11)
Fritsch, E. M., 116
Frost, J. A., 370, 373(21), 419, 430
Fry, M. J., 144
Fu, H., 181
Fujii, M., 82
Fujimura-Kamada, K., 109, 114
Fujisawa, H., 321
Fujita-Yoshigaki, J., 159
Fukuda, Y., 155, 373
Fullekrug, J., 80
Furth, M. E., 116
Furuse, M., 124, 232, 245

G

Gaarde, W. A., 434, 435(43)
Gabay, L., 344, 345(9)
Gairdina, G., 432
Gajewski, T. F., 445
Galanis, A., 330
Galasinski, S. C., 427, 428(15), 429(15)
Gale, R. P., 152
Gallagher, T. F., 336, 435
Galli, G., 445
Gallione, C. J., 279
Galullo, V., 435
Gangarosa, L. M., 222
Garcia, A. M., 115
Garcia Ramirez, J. J., 409
Garcia-Ranea, J. A., 78
Gardner, A. M., 433
Gartner, A., 210, 369
Gebhardt, D. K., 434
Gehrke, L., 254, 258(7)
Geiger, T., 366
Gelb, M. H., 301, 303, 314
Gelinas, R. E., 213
Geneste, O., 276
Geppert, T. D., 420
Gerber, S. A., 314
Gertler, F. B., 89(41, 42), 102, 187, 193(23), 194(23)
Gerwins, P., 332, 433
Geyer, M., 159, 161(16), 162(16), 163
Geysen, H. M., 96, 102

Gibbs, J. B., 115, 234, 354
Gibson, T., 89(40), 102
Giddings, B. W., 128, 138(1)
Giedlin, M. A., 417
Gierasch, L. M., 187
Gietz, R. D., 294
Gigowski, R., 26
Gilbert, B. A., 107, 112
Gilman, A. G., 66
Gil-Nagel, A., 279
Gimeno, R. E., 299
Gioeli, D., 343, 344
Giseli, D. G., 423
Gish, G., 171, 184, 187(6), 189(6), 191(6)
Gisselbrecht, S., 427
Gius, D. R., 38, 49(16)
Glaser, S. M., 96
Glaser, V., 313
Glomset, J. A., 195, 196, 303
Glover, T. W., 354
Godowski, P. J., 414
Goedert, M., 336
Goff, S. P., 334
Goi, T., 128
Goldman, M. E., 434, 435(43)
Goldsmith, E. J., 334, 389
Goldstein, J. L., 115, 117, 187, 196, 302, 306
Golemis, E. A., 77, 84, 85, 143, 277, 279, 279(8), 280(8), 281, 283, 283(24), 287(8), 289(4, 36), 293, 294(4), 298(24, 25), 299
Golstein, J. L., 199
Goneli, S., 442
Goody, R. S., 159, 160, 162(20), 163(20)
Gorbsky, G. J., 344, 348(8)
Gordon, A. S., 80(29), 83
Gordon, J., 67
Gorg, A., 308
Gorlich, D., 78
Gorman, C., 130
Gorman, J. J., 312
Gorycki, P. D., 336
Gotoh, Y., 181, 321, 334, 373, 433
Gottesman, S., 37, 49(13)
Goud, B., 133
Goueli, S. A., 337, 343
Gould, S. J., 81
Gout, I., 144
Govek, E., 166, 168(24)
Gowan, R. C., 343, 419
Graham, S. M., 7, 15, 28(31), 30(31), 195, 211, 218(5)
Grant, G. A., 26
Graves, D. J., 174
Gray, P. W., 274
Graziani, A., 184
Green, E. D., 279
Green, M., 36
Green, O. M., 100
Green, S., 405
Greenberg, M. E., 181, 335, 341, 369
Greene, L. A., 184, 193(12)
Greenwald, J. E., 71
Gregory, J. S., 341
Griswold, D. E., 336
Gromov, P. S., 204, 205(8)
Groner, B., 89
Groninga, L., 186, 193(17)
Gross, R. W., 140
Grover-Bardwick, A., 432
Grunicke, H. H., 65
Grussenmyer, T., 357
Gu, Y., 152
Guan, K. L., 369
Guan, Z., 334
Guarente, L., 143
Guesdon, J. L., 89
Gullick, W. J., 184, 193(10)
Gunther, S., 308
Gupta, S., 320, 321, 330, 331, 433
Gutierrez, L., 66
Gutkind, J. S., 319
Gutkind, S., 432
Gutman, O., 10
Gutmann, D. H., 354
Guy, A. M., 96, 97(25)
Guzman, R. C., 211
Gygi, S. P., 301, 314
Gyuris, J., 84, 277, 281, 289(4), 294(4), 298(25)

H

Habets, G. G., 5, 105, 109(6), 245
Hai, T., 434
Hall, A., 36, 47(1, 2), 66, 276, 280, 319
Hall, R., 336
Hall Jackson, C. A., 336
Halpern, J. R., 335, 369
Han, J., 319, 320, 333, 334, 334(25), 433

Han, L., 140, 141(5, 6), 143(5, 6), 146(5), 147, 271, 274(6)
Han, M., 369
Hanafusa, H., 184, 185(7), 191(7), 334, 433
Hanahan, D., 234, 258
Hancock, D. C., 405
Hancock, J. F., 50, 65, 66, 106, 112, 195, 262, 303, 313, 314(20), 354
Hannah, V. C., 199
Hannon, G. J., 166, 211, 271
Harada, N., 152
Hardison, N. L., 96, 97(24)
Hardman, N., 89
Harlan, J. E., 194
Harlow, E., 119, 320, 339
Harper, J. W., 370
Harrowe, G., 369
Hart, K. C., 65, 72(4)
Hart, L. P., 89
Hartl, F. U., 37, 49(14)
Hartman, S. E., 442
Harwerth, I. M., 89
Hasegawa, T., 131
Haser, W. G., 171, 184, 187(6), 189(6), 191(6)
Hateboer, G., 405
Hattori, S., 159
Hauser, C. A., 13, 409
Hazzalin, C. A., 386, 432
Heaney, F., 4
Heberle-Bors, E., 210
Hedge, P., 336
Heidecker, G., 432
Helfman, D. M., 206
Helin, K., 405
Hellstrom, I., 96
Hellstrom, K. E., 96
Hemmings, B. A., 80(19), 82
Hemmings, H. C. J., 433
Henis, Y. I., 10
Hennemann, H., 262
Her, J. H., 344, 368, 379, 381(27)
Herdegen, T., 442
Herrera, D., 151
Herrera, R., 343, 419
Herrmann, C., 133, 151, 159, 160, 161(16, 19, 23), 162(16, 20), 163, 163(20), 279
Herrmann, R. G., 83
Herschman, H., 140, 141(6), 143(6)
Herskowitz, I., 87
Hespenheide, B. M., 89
Hettema, E. H., 78, 80(4), 81(4)
Hibi, M., 320, 321, 324(13), 330(13), 331(13), 334(9), 432, 433(1, 8), 441(8)
Hibner, B., 353
Hildebrand, J. D., 80(24), 83
Hill, C. S., 245
Himmler, G., 89
Hinoi, T., 128, 131, 132, 134(18), 136(18)
Hirata, H., 82
Hirota, Y., 128
Hirschberg, K., 64
Hirschl, S., 89
Hisaka, M. M., 4, 7(8), 11, 12(24), 30(26), 195
Hittle, J. C., 344, 348(8)
Ho, A., 37, 38, 38(15), 39(15), 41(15), 46(21), 49(15, 19, 21)
Hoagland, N., 171, 174
Hobbs, F., 343, 366, 419, 421(7), 430(7)
Hobohm, U., 315
Hochholdinger, F., 65
Hockenberry, T. N., 115, 122(6), 123(6), 127(6)
Hoekstra, M. F., 171, 174, 274, 433
Hofer, F., 141, 159
Hoffman, N. G., 92, 96, 97(24)
Hoffmann, B., 102
Hoffmuller, U., 102
Holash, J., 5
Holland, P. M., 442
Hollenberg, S. M., 142, 145, 153, 154(7), 271, 274(3), 277, 414
Holtzmann, D., 274
Hooper, S., 401, 413(6)
Hopkin, K., 260
Horiuchi, K. Y., 343, 366, 419, 421(7), 430(7)
Horn, G., 133, 159, 161(19), 279
Horner, J. W. II, 5
Hoshi, M., 321
Houtteville, J. P., 279
Howell, B. W., 187, 193(23), 194(23)
Hruska, K., 36, 38, 47(6, 20), 49(20)
Hrycyna, C. A., 105, 112(5)
Hsi, K. L., 312
Hsiao, K., 435
Hu, P., 189
Hu, Y., 443
Huang, L., 141, 159
Huang, S., 333, 334(25)
Huang, T., 211
Hubank, M., 223

Hubbert, N. L., 87
Huber, L. A., 203, 204, 205(2)
Hude-DeRuyscher, R., 102
Hudson, J., Jr., 255
Hue, L., 181
Huebner, K., 152
Huff, S. Y., 11, 12(23, 24)
Hughes, J. A., 87
Hughes, R. T., 433
Hughes, S. H., 206
Hunt, D. F., 344, 368
Hunt, J., 431
Hunter, T., 171, 174, 176, 233, 245, 305, 334, 337, 434, 435(43), 442
Huse, W. D., 96
Hussain, N. K., 96, 97(25)
Hussaini, I. M., 344, 350(10)
Hutchinson, M. A., 357
Hwang, Y.-W., 273
Hynes, N. E., 89

I

Ibl, M., 89
Ikeda, M., 128, 132, 133, 134(18, 22), 136, 136(18)
Ilver, D., 306
Imai, K., 312
Imamura, F., 259
Immanuel, D., 185
Incecik, E. T., 306
Inouye, C., 277
Ip, N. Y., 344, 433
Ip, Y. T., 319, 326(1)
Ishida, O., 128, 136
Ishino, M., 89(38), 102
Ishizaka, K., 140
Ito, H., 155
Ito, Y., 159
Ivanov, I. E., 10, 51, 56(6), 59(6), 61(6), 62(6), 69, 112
Iwamatsu, A., 128, 152
Iwasaki, T., 259
Iyer, V. R., 255

J

Jacinto, E., 433
Jackson, P., 80(27), 83, 401, 405(7)
Jain, J., 443
Jaitner, B. K., 159, 161(16), 162(16), 163
James, G. L., 115, 302
James, L., 115, 122(6), 123(6), 124, 127(6)
Janssen, J. W. G., 353
Jarpe, M. B., 332
Jarvik, J. W., 117
Jarvis, B. W., 337, 341, 442
Jayaraj, P., 334
Jeffrey, S. S., 255
Jendrisak, J. J., 308
Jett, J. H., 450
Jiang, M. S., 335
Jiang, R., 277
Jiang, Y., 333
Jin, M. H., 128
Jin, W. H., 123
Jo, H., 233
Jochum, W., 443
John, J., 279
Johnson, D., 67
Johnson, D. I., 281
Johnson, E. W., 279
Johnson, G. L., 332, 433
Johnson, K. S., 100, 323
Johnson, M. L., 159
Johnson, R., 258
Johnson, R. L. II, 3, 234
Johnston, J. F., 366
Joliot, A. H., 37
Jones, P., 80(27), 83
Jones, T. L. Z., 66, 279
Joneson, T., 276, 280
Jordan, J. D., 11, 30(27), 313, 314(41)
Joutel, A., 279
Juang, J. L., 89(41), 102
Jumar, S., 335
June, C. H., 443, 448
Jung, H. H., 279

K

Kadono-Okuda, K., 10, 197, 200(16), 204, 205(10), 206(10)
Kaibuchi, K., 133, 152
Kaiser, C. A., 299
Kalbitzer, H. R., 159, 161(16), 162(16), 163
Kallunki, T., 433, 442, 443
Kam, J., 262

Kamakura, S., 431
Kamata, H., 82
Kameshita, I., 321
Kanegane, H., 445
Kapeller, R., 184
Kaplan, D. R., 184, 193(12)
Kapoor, T., 88
Kardalinou, E., 386
Karin, M., 4, 131, 245, 261, 262, 276, 319, 320, 321, 324(13), 330(13), 331(13), 334(9), 386, 387, 432, 433, 433(1, 8), 441(8), 442, 443
Karnauchov, I., 83
Kasahara, Y., 445
Kashishian, A., 274
Kaspar, R., 254, 258(7)
Kassis, S., 335, 336
Kataoka, T., 131
Katayama, H., 132
Kato, K., 11, 12(24), 195
Katz, M. E., 139, 270
Katz, S., 262, 267(13)
Kaur, K. J., 270
Kaushansky, K., 428
Kavanaugh, W. M., 185, 417
Kawai, G., 159
Kawakami, M., 321
Kawakami, T., 442
Kawakami, Y., 442
Kawata, M., 133
Kay, B. K., 88, 89(45), 92, 94, 96, 97(21, 24, 25), 99, 100, 102
Kay, L. E., 187, 194(22)
Kay, R., 4, 35(5), 211, 212(2), 218(2)
Kay, S. A., 50, 59(5), 62(5)
Kaziro, Y., 134
Kellendonk, C., 414
Keller, G. A., 81
Kelley, C., 433
Kelly, K., 371, 384(24)
Kemp, B. E., 177
Kemp, P., 354
Kempiak, S. J., 443
Kerschbaumer, R., 89
Kersulyte, D., 306
Keyomarsi, K., 370
Keyse, S., 338, 343(5)
Khazak, V., 277, 279(8), 280(8), 287(8)
Khokhlatchev, A., 341, 389
Khoshnan, A., 443
Khosravi-Far, R., 78, 79, 273, 371, 384(24)
Khwaja, A., 276
Kibbelaar, R. E., 363
Kieber, J. J., 210
Kikuchi, A., 127, 128, 131, 131(10), 132, 132(15), 133, 133(10), 134, 134(18, 22), 135(27), 136, 136(18), 138(27), 140
Kilburn, A. E., 277
Kim, C. M., 117
Kim, E., 89(37), 102, 103
Kim, J., 432
Kim, S.-H., 141, 159
Kimura, A., 155
King, F., 171, 184, 187(6), 189(6), 191(6)
Kingston, R., 292
Kinoshita, S., 211
Kinsella, T. M., 32
Kinzler, K. W., 234
Kirk, H., 4, 35(5), 211, 212(2), 218(2)
Kirn, D., 280, 354
Kirschmeier, P., 115, 122(6), 123, 123(6), 124, 127(6)
Kishida, S., 128, 131, 132, 134, 134(18), 135(27), 136, 136(18), 138(27)
Kitamura, T., 211
Klagsbrun, M., 409
Klasen, S., 448
Klatt, T., 431
Kleuss, C., 66
Klippel, A., 417
Kloog, Y., 10
Klosgen, R. B., 83
Kluthe, C. M., 199
Knaus, U. G., 319
Knepper, M., 5, 245
Knoll, L., 67
Kohl, N. E., 115, 234
Kohn, A. D., 417
Koide, H., 159
Koivunen, E., 87
Kokot, N., 443
Komagome, R., 259
Koo, D., 143
Kooistra, A., 363
Koshland, D. E., Jr.., 112
Kosofsky, B., 279
Kossman, K. L., 78
Koury, E., 341
Kovacina, K. S., 417
Kovar, H., 297

Koyama, S., 127, 128, 132, 133, 134, 134(18, 22), 135(27), 136(18), 138(27)
Kozak, M., 5
Kramer, A., 102
Krammer, G., 213
Kranz, J. E., 369
Krauss, A., 369, 370(14), 373(14), 384(14), 430
Krebs, E. G., 177, 433
Krieger, M., 196
Kuan, C. Y., 434
Kuehl, P. M., 279
Kuhn, L. A., 89
Kumar, S., 335
Kumar, V., 405
Kurakin, A., 92, 102
Kuriyama, K., 152
Kuriyama, M., 152
Kuriyan, J., 194
Kuroda, S., 152
Kurth, J. H., 279
Kwon, Y. T., 174
Kyriakis, J. M., 131, 144, 279, 387, 432, 433, 433(2)

L

Labauge, P., 279
Laberge-le Couteulx, S., 279
Lacal, J. C., 65
Lacal, P. M., 65
Laemmli, U. K., 308
Lai, C. C., 11, 12(23)
Lakhani, S., 344
L'Allemain, G., 344, 368
Lamarche, N., 276
Lambeth, J. D., 280
Land, H., 4, 5(3), 166, 259, 405, 414
Lane, D., 119, 320, 339
Lange-Carter, C., 419, 433
Langen, H., 315
Langer, S., 234
Lanier, L. M., 187, 193(23), 194(23)
Lapetina, E. G., 204, 205(7)
Lashkari, D., 255
Laskey, R. A., 80(22), 82
Latham, D. G., 37, 38(15), 39(15), 41(15), 49(15)
Lau, L. F., 450
Laudano, A. P., 402
Laurance, M. E., 384
Lauro, I., 96
Lautwein, A., 160
Law, S. F., 85, 293
Lawrence, J. C., Jr., 417
Layton, M., 353
Lear, A. L., 80(27), 83
Leblus, B., 37
Lechleider, R. J., 171, 184, 187(6), 189(6), 191(6)
Ledbetter, J. A., 448
Leder, P., 89, 226
Lee, C. F., 5
Lee, C. H., 5, 194
Lee, C. H. J., 10
Lee, J. C., 186, 193(17), 255, 335, 336, 369, 433, 435
Lee, J. E., 140
Lee, J. Y., 431
Lee, L., 443
Lee, L. A., 222
Lee, W. H., 277
Lee-Lin, S. Q., 279
Lees, E., 259, 412, 413(20), 415(20), 417(20)
Leevers, S. J., 65, 87, 419, 422(9)
Legrain, P., 153
Leining, L. M., 303
Lenormand, P., 338, 343, 343(5), 368
Lenzen, C., 160, 161(22)
Leonetti, M., 89
Leopold, W. R., 343, 419
Lepp, S., 433
Lescoat, C., 279
Levy, E., 187, 193(21), 194(21)
Lewis, A., 433
Lewis, B. C., 222
Lewis, M. D., 115
Lewis, T. S., 319, 326(3), 333(3), 337, 387, 388(1), 430, 431
Ley, S. C., 433
Li, A., 271, 274(9)
Li, N., 185, 261
Li, Q., 222
Li, S. C., 187, 194(22)
Li, Z., 333
Liang, C., 433
Liang, P., 233, 234(1), 235, 240
Liebl, E. C., 83, 89(41), 102
Lifson, J. D., 450
Lim, W. A., 88, 96

Lim, Y. M., 141
Lin, A., 319, 321, 324(13), 330(13), 331(13), 334, 432, 433, 433(2, 8), 441(8)
Lin, R., 280
Lin, S., 333
Linder, M. E., 66, 68(14)
Linnemann, T., 159, 161(16), 162(16), 163
Linsley, P. S., 448
Linz, J. E., 89
Lipcey, C., 448
Lippincott-Schwartz, J., 64
Lisitsyn, N., 223
Lissy, N. A., 37, 38, 38(15), 39(15), 41(15), 49(15, 17, 18)
Littlewood, T. D., 405
Liu, D., 183
Liu, H., 369
Liu, H. S., 5
Liu, J., 128, 138(1)
Liu, Y. C., 140
Lloyd, A. C., 259, 414
Lock, P., 89(40), 102
Lockshon, D., 260
Loeb, D. M., 184, 193(12)
Loewenstein, P. M., 36
Lotti, L. V., 80
Louie, D. F., 431
Louis, D. N., 279
Louvion, J. F., 401, 414(1)
Lowy, D. R., 87, 133
Lu, K. P., 174
Luo, Y., 86
Luo, Z., 412
Lupher, M. L., Jr., 194
Lutcke, A., 204
Lynch, A., 186, 193(17)
Lynch, M., 103
Lyons, D. M., 368, 369, 369(3)
Lyons, J., 353

M

Ma, Y. T., 107, 112
Ma, Z., 115
Macara, I. G., 4
MacDonald, M. A., 65
MacDonald, M. J., 11, 12(21, 25)
MacDonald, R. J., 226
MacDonald, S. G., 65, 106, 140, 354
MacLean, M., 96
MacNicol, A. M., 140
Madhani, H. D., 368
Madsen, P., 204, 205(8)
Maertienssen, R., 80(32), 83
Maestro, R., 166, 211
Magee, A. I., 50, 66, 75, 195, 262
Maggi, E., 445
Magolda, R. L., 343, 366, 419, 421(7), 430(7)
Mahadevan, L. C., 386, 432
Mahanty, S. K., 369
Mahon, G. M., 211
Maillere, B., 89
Maity, A., 222
Maller, J. L., 344
Manandhar, M., 369
Mandell, J. W., 344, 350(10)
Mandiyan, V., 88, 194
Manetti, R., 445
Maniatis, T., 19, 116, 197, 201(14), 204, 213, 217(11)
Mann, M., 301, 312
Manning, A. M., 433, 434, 443
Manning, M. C., 399, 419, 427(10)
Manor, D., 280
Manos, E. J., 343, 366, 419, 421(7), 430(7)
Mansour, S. J., 399, 419, 427(10)
Mant, A., 83
Marchisio, P. C., 36, 47(4)
Marchuk, D. A., 279
Marcus, S., 154, 271, 368
Marcy, V. R., 341
Marechal, E., 279
Margolis, B., 185, 187, 193(21), 194, 194(21)
Mark, M. R., 414
Markworth, C., 102
Maroun, M., 262
Marshall, C. J., 50, 65, 66, 87, 112, 195, 245, 259, 262, 270, 354, 377, 401, 413(6), 419, 422(9)
Marshall, M. S., 131, 144, 279, 412
Marshall, T. K., 68
Marte, B. M., 276
Martenson, C. H., 100
Martin, C. B., 186, 193(18)
Martin, G. A., 160, 161(23), 184
Martin, G. S., 83, 141, 159
Martin, J. A., 140
Martinetto, H., 433
Martino, P., 344

Martin-Villalba, A., 442
Massague, J., 5
Masters, S., 181
Mastino, P., 368
Masuyama, N., 334, 433
Mathias, P., 333
Matsubara, K., 132, 134, 135(27), 138(27)
Matsuno, R., 204
Matsuura, J. E., 399, 419, 427(10)
Matsuura, Y., 132, 134, 134(18), 135(27), 136, 136(18), 138(27)
Mattingly, R. R., 4
Mattioni, T., 401, 414(1)
Maurer, R. A., 384
Maurizi, M. R., 37, 49(13)
Mawas, C., 448
Mayer, B. J., 96, 433
Mayhew, M. W., 83
McAllister, S., 38, 47(20), 49(20)
McCaffrey, P. G., 435
McCarthy, S. A., 236, 405, 409, 410(15)
McComick, F., 139, 203, 210(1), 270
McCormick, F., 78, 140, 159, 184, 203, 210(5), 279, 280, 354
McGeady, P., 303
McGlade, J., 184, 185(7), 191(7)
McKay, I. A., 66
McLaughlin, J., 140
McLellan, C. A., 303
McMahon, G., 433
McMahon, M., 236, 259, 401, 405, 407(4), 409, 409(4), 410, 410(15), 412, 412(5, 17), 413, 413(5, 20), 415(20), 417(20)
McPherson, P. S., 88, 96, 97(24, 25)
Meadows, R. P., 194
Medin, A., 176
Mehul, B., 75
Meier, R., 341, 435
Meijer, C. J., 363
Meisenhelder, J., 305
Melemed, A. S., 427
Mendelsohn, A. R., 260, 277
Mendler, J. H., 38, 49(17)
Menez, A., 89
Mercola, D., 432
Mercurio, F., 433, 434
Mermelstein, S. J., 38, 49(19)
Mettler, G., 279
Metzger, D., 405
Meyer, T., 10, 100
Michaelis, S., 105, 109, 109(6), 112, 112(5), 114
Michaelson, D., 10, 51, 56(6), 59(6), 61(6), 62(6), 69
Michalak, M., 109
Miki, T., 319
Miller, A. D., 213
Miller, C. D., 64
Miller, P. J., 281
Milner, R., 109
Minden, A., 4, 131, 245, 276, 319, 321, 324(13), 330(13), 331(13), 432, 433, 433(8), 441(8)
Minden, J. S., 313
Minenkova, O., 96
Misra-Press, A., 384
Mitchell, D. A., 68
Mitina, O. V., 277
Miyajima, A., 211
Miyao, S., 128
Miyawaki, T., 445
Miyazaki, M., 186, 193(17)
Miyazawa, T., 159
Mochly-Rosen, D., 80(29), 83
Monaghan, A. P., 414
Monia, B. P., 366
Moodie, S. A., 234
Mooi, W. J., 363
Moore, C., 37
Moore, D., 292
Moore, G., 433
Moore, I., 210
Moore, T., 255
Morehead, T. A., 302
Morgan, M. E., 313
Morgenbesser, S. D., 344
Morgenstern, J. P., 4, 5(3), 166
Moriguchi, T., 334, 431, 433
Morii, N., 36, 47(5)
Morimoto, T., 10, 51, 56(6), 59(6), 61(6), 62(6), 69, 112
Morinaka, K., 128
Moroni, M. C., 405
Morrice, N., 336
Morrison, A. R., 334
Morrison, D. K., 145, 412
Morrison, L., 279
Mosmann, T. R., 445
Mosser, S. D., 354
Mossier, B. M., 297
Mottola, G., 80
Mufti, G. J., 353

Mukai, M., 259
Muller, B. H., 89
Muller, H., 405
Muller, M., 366
Mullick, A., 405
Mulligan, R. C., 4, 32(6), 35(6)
Mumby, S. M., 66, 68(15), 71, 73(15), 76(28)
Murai, H., 136
Murakami, H., 36, 47(5)
Murata, K., 155
Murcko, M. A., 435
Murray, B. W., 432, 433, 434, 435(43)
Muslin, A. J., 140
Mustelin, T., 140
Muszynska, G., 176
Muto, Y., 159
Myers, C. E., 303
Myrand, S. P., 431

N

Nagahara, H., 37, 38, 38(15), 39(15), 41(15), 49(15, 16)
Nagano, Y., 204
Nagasu, T., 115
Nahreini, T. S., 427, 428(15), 429(15)
Nakafuku, M., 134, 152
Nakamura, I., 36, 47(5)
Nakamura, T., 152
Nakashima, S., 128
Naldini, L., 36, 47(4)
Nandi, S., 211
Narshimhan, R. P., 433
Narumiya, S., 36, 47(5)
Nasmyth, K., 369
Nassar, N., 159
Nebreda, A. R., 341
Nel, A. E., 443
Nemerow, G. R., 333
Nguyen, J. T., 88
Nichols, D. W., 402
Nicolette, C., 4, 131, 139, 245, 276
Niebuhr, K., 89(42), 102
Niedbala, M., 103
Niida, Y., 445
Nikolakaki, E., 432, 433(2)
Nilsson, T., 80
Nimmesgern, E., 37, 49(14)
Nishida, E., 321, 333, 334, 373, 431, 433
Nishimura, S., 159
Niv, H., 10
Njoroge, G., 124
No, D., 245
Nobes, C. D., 36, 47(2)
Noda, M., 132, 133
Nogalo, A., 65
Nolan, G. P., 4, 32, 32(4), 35(4), 211, 217
Nolte, R. T., 194
Nordstrom, J., 432
Norris, J. L., 234
Norris, K., 87
North, A. K., 103
Nouvet, F. J., 114
Novick, P., 78
Nowell, P. C., 152
Nunes, J., 448
Nunez-Oliva, I., 115, 122(6), 123(6), 127(6)
Nye, S. H., 344

O

O'Brien, P. J., 314
O'Bryan, J. P., 96, 97(25), 186, 193(18)
Ogawa, T., 82
Ogren, J., 306
Ogris, E., 184, 193(11)
O'Hagan, R., 5
Ohanion, V., 80(27), 83
O'Hare, P., 37
Ohba, T., 89(38), 102
Ohtsubo, M., 82
Okano, H., 128
Okawa, K., 128
Okazaki, M., 131, 133, 134(22)
Okazaki-Kishida, M., 136
O'Keefe, S. J., 431
Oldham, S. M., 222
Olejniczak, E. F., 194
Oliff, A., 115, 234, 301, 302(13)
Olive, D., 448
Olivier, P., 184, 185(7), 191(7)
Olkkonen, V., 204
Olson, M. F., 259, 280, 319
O'Malley, B. W., 410
O'Neill, S. A., 312
Ong, S. J., 210
Onishi, M., 211
Ooi, J., 187, 193(21), 194(21)

Opas, M., 109
Orfanoudakis, G., 89
Ostrowski, M. C., 7, 234, 409
Ottilie, S., 281
Ottinger, E., 186, 193(17)
Otto, J. C., 103
Ouerfelli, O., 37, 49(14)
Owaki, H., 420

P

Pabo, C. O., 36
Paccaud, J. P., 80
Pages, G., 354, 368
Pai, J.-K., 115, 122(6), 123, 123(6), 127(6)
Pallas, D. C., 184, 193(11)
Palme, K., 210
Palmer, E. M., 445
Palmer, T. D., 213
Pan, B. T., 11, 12(22)
Panayotatos, N., 344
Pang, L., 236, 244(16), 366, 418, 430(5)
Pang, M., 431
Pantginis, J., 5
Paoluzi, S., 96
Papageorge, A. G., 87
Pappin, D., 276, 433
Pardee, A. B., 233, 234(1), 235
Pargellis, C. A., 435
Park, I., 405
Parker, M. G., 405
Parmr, K., 140
Parrini, M. C., 96
Parry, D., 259, 412, 413(20), 415(20), 417(20)
Parsons, J. T., 80(24), 83
Parton, R. G., 133
Pasching, C. L., 232
Pascual, R. V., 211
Pasqualini, R., 87
Pasquet, S., 336
Pastan, I., 50
Patel, A., 140
Patel, B., 80(27), 83
Paterson, H. F., 50, 65, 66, 87, 259, 354, 401, 413(6)
Pathak, R. K., 302
Patton, R., 115, 124
Pav, S., 435
Pawson, T., 88, 171, 183, 184, 185(7), 187, 187(6), 189(6), 191(6, 7), 193(12), 194(22)
Payne, D. M., 344, 368
Pazhanisamy, S., 435
Pear, W. S., 4, 32(4), 35(4), 217
Pearson, G., 387
Pelham, H. R., 78, 80(1)
Pelicci, P. G., 96, 128
Penix, L. A., 333, 334(26)
Pennington, C. Y., 65
Pentland, A. P., 334
Pepinsky, B., 37
Pérez-Sala, D., 112
Pergamenschikov, A., 255
Perisic, O., 159
Perlmutter, R. M., 412
Perou, C. M., 255
Pestka, J. J., 89
Pestov, D. G., 450
Peter, M., 204
Petermann, R., 297
Peterson, D., 410
Petraitis, J. J., 388
Petrin, J., 124
Petroni, G. R., 344
Petros, A. M., 194
Philipp, A., 405
Philips, M. R., 10, 50, 105, 107, 109(6), 112, 112(10)
Phillips, C., 80(21), 82
Picard, D., 401, 405(7), 414(1)
Pillinger, M. H., 107, 112(10)
Pitts, W. J., 343, 366, 419, 421(7), 430(7)
Piwnica-Worms, H., 171, 174
Pizon, V., 133, 279
Platko, J. D., 4, 7(8)
Plattner, R., 232
Pleiman, C. M., 433
Plowman, G. D., 171, 262
Polonskaia, M., 450
Polverino, A., 4, 131, 154, 245, 271, 276, 368
Pomerantz, J., 5
Ponglikitmongkol, M., 405
Ponting, C. P., 80(21), 82, 152
Porath, J. O., 176, 307
Porteu, F., 427
Posada, J., 399
Posas, F., 369
Posewitz, M. C., 307
Postel, W., 308

Pouyssegur, J., 338, 343, 343(5), 354, 368
Powers, S., 11, 12(23), 221
Pozzan, T., 80(23), 82
Prasad, R., 152
Prasher, D. C., 51, 61(7), 84
Prekeris, R., 83
Prendergast, F. G., 51, 61(7)
Presley, J. F., 64
Price, G. J., 80(27), 83
Price, J. R., 302
Prigent, S. A., 184, 193(10), 222
Primavera, M. V., 36, 47(4)
Pritchard, C. A., 236, 409, 410, 410(15), 412(17)
Prochiantz, A., 37
Prosperini, E., 405
Przbyla, A. E., 226
Przybranowski, S., 343, 419
Punke, S. G., 68, 69(24), 73(24)
Putz, H., 262

Q

Qiu, R. G., 280, 354
Quilliam, L. A., 11, 12(24), 13, 26, 69, 73(26), 80(15), 81, 82(15), 92, 99
Quinlan, R. A., 336

R

Rabilloud, T., 311
Rabun, K. M., 11, 12(24)
Radhakrishna, H., 89(39), 102
Radziejewska, E., 344
Radzio-Andzelm, E., 433
Raingeaud, J., 320, 330, 333, 333(19), 334(19, 26), 341, 433
Raitano, A., 335, 369
Rakic, P., 434
Ramakrishnan, M., 143
Ramjaun, A. R., 96, 97(25)
Randall, R. E., 4
Randall, S. K., 302
Randazzo, P. A., 89(39), 102, 262
Rando, R. R., 107, 112
Rands, E., 115
Rao, A., 443
Rapp, U. R., 131, 140, 144, 279, 377, 432
Ratner, L., 38, 49(18)
Ratnofsky, S., 171, 184, 187(6), 189(6), 191(6)
Ravichandran, K. S., 194
Ray, L. B., 344
Recht, M., 333, 334(26)
Reddy, G. R., 370, 371(20), 373(20), 384(20)
Reed, J. C., 181
Reep, B. R., 204, 205(7)
Rees, C. A., 255
Reineke, U., 102
Reiners, J. J., 431
Reinhard, M., 89(42), 102
Reiss, Y., 306
Ren, R., 88
Rensland, H., 160
Renzi, L., 344, 348(8)
Resh, M. D., 66, 75, 78, 80(16), 81, 81(3), 314
Resing, K. A., 417, 431
Reuter, C. W., 370, 371(20), 373(20), 379, 384(20)
Reuter, G., 73
Reuther, G. W., 26, 301
Rewerts, C., 102
Rich, S. S., 279
Richards, F. M., 88, 96
Rickles, R. J., 100
Rider, J. E., 92, 100
Rider, M. H., 181
Ridley, A. J., 36, 47(1, 3), 276
Riezman, H., 80
Rigaut, G., 301
Rigel, D. F., 434
Riggs, M., 139
Rilling, H. C., 303
Rincón, M., 333, 334, 334(26), 434
Rine, J., 103, 112(2)
Rizzuto, R., 80(23), 82
Robbins, D. J., 344, 419, 420
Roberson, M. S., 384
Roberts, B., 4, 32(6), 35(6), 184, 187(6), 189(6), 191(6)
Roberts, R. L., 369
Roberts, T. M., 171, 184, 193(11)
Robinson, C., 83
Robinson, L. J., 66
Robinson, M. J., 334, 433
Robles, A. I., 258
Rochon, Y., 301
Rodgers, L., 139

Rodriguez-Viciana, P., 144
Rodriguez-Vicinia, P., 276
Roecklein, J. A., 294
Rogers, J. S., 320, 335, 369
Rogers, S., 435
Rogers-Graham, K., 301, 302(12)
Rojnuckarin, P., 428
Rolando, A., 431
Romagnani, P., 445
Romagnani, S., 445
Roman, G., 210
Romano, A., 311
Romano, J., 105, 109(6)
Ron, D., 330
Ronai, Z., 432
Rosen, N., 37, 49(14), 115
Rosenberg, N., 140
Rosenwald, I., 254, 258(7)
Ross, D. T., 255
Ross, J., Jr., 333, 334(25)
Rossman, K. L., 79
Rossomando, A. J., 344, 368, 379, 381(27)
Roth, R. A., 417
Rothenberg, M., 210
Rotin, D., 88
Roudebush, M., 354
Rouse, J., 341
Rousseau, D., 254, 258(7)
Roussel, M., 234
Roux, D., 338, 343(5)
Rouyez, M. C., 427
Rowekamp, W. G., 213
Rowell, C., 115
Roy, N., 181
Rubie, E. A., 432, 433(2)
Rubin, G. M., 369
Ruderman, J. V., 222
Ruggieri, R., 140
Ruoslahti, E., 87
Ruskin, B., 112
Russell, D. W., 196
Rutter, W. J., 226
Rutz, B., 301
Ryder, J. W., 427

S

Sabapathy, K., 443
Sabatini, D. D., 262
Sabe, H., 184, 185(7), 191(7)
Sager, R., 233
Sah, V. P., 333, 334, 334(25)
Sahoo, T., 279
Saito, H., 152, 369
Saito, S., 36, 47(5)
Saito, Y., 377
Sakai, H., 321
Salcini, A. E., 89(45), 96, 102
Salmeron, A., 433
Salser, S. J., 401
Saltiel, A. R., 343, 366, 418, 419, 421(8), 426(8), 429(8), 430(5, 8)
Salvesen, G. S., 181
Sambrook, J., 19, 116, 197, 201(14), 204, 213, 217(11)
Samuels, M. L., 236, 401, 407(4), 409, 409(4), 410, 410(15), 412(17)
Sanchez, I., 433
Sander, C., 203, 210(4)
Sanders, D. A., 78, 203, 210(1, 5)
Sandhu, G., 434
Santolini, E., 89(45), 96, 102
Sapperstein, S. K., 105, 112(5)
Sardet, C., 368
Sasaki, D. T., 432, 450
Sasaki, H., 89(38), 102
Sasaki, T., 89(38), 102
Sasaki, Y., 204
Sasazuki, T., 5, 124, 232, 245
Sather, S., 433
Sato, K. Y., 232
Satoh, T., 134
Satoh, Y., 433
Satterberg, B., 368, 369, 369(3)
Sattler, M., 194
Sawamoto, K., 128
Sawyers, C. L., 335, 369
Schaefer, E., 271, 276(8), 341, 389
Schaeffer, H. J., 368, 369, 370, 370(14), 371(20), 373(14, 20), 384(14, 20), 430
Schaeffer, J. P., 11, 12(25), 65
Schaffer, A., 432
Schaffhausen, B. S., 184, 193(11)
Schaller, M. D., 80(24), 83
Schatz, D. G., 223
Scheid, M. P., 431
Scheidtmann, K. H., 357
Schell, J., 210
Scherer, A., 159

Scherle, P. A., 343, 366, 419, 421(7), 430(7)
Schiestl, R. H., 294
Schimmoller, F., 80
Schipper, A., 222
Schlessinger, J., 88, 184, 189, 194, 261, 262
Schmidt, R. A., 196
Schmidt, W. K., 109
Schneider, C., 37, 49(14), 196
Schneider-Mergener, J., 102
Schoenen, F., 102
Schrader, W. T., 410
Schreiber, S., 88, 443
Schroder, S., 80
Schuler, G., 255
Schutz, G., 414
Schwager, C., 89
Schwartz, J., 124
Schwarze, S., 38, 46(21), 49(19, 21, 22)
Scita, G., 432
Scolnick, E. M., 50, 116
Scott, J. D., 88, 183
Scott, M. L., 4, 32(4), 35(4), 217
Scott, P. H., 417
Seabra, M. C., 195, 306
Sebolt-Leopold, J. S., 343, 419
Seed, B., 213, 214(9)
Seery, J., 37
Sefton, B. M., 176
Segal, I. H., 438
Seger, R., 344, 345(9), 399, 433
Seidman, J., 292
Selcher, J. C., 388
Sells, M.-A., 281, 319
SenGupta, D., 294
Sepp-Lorenzino, L., 37, 49(14), 115
Seraphin, B., 301
Serebriiskii, I., 277, 279, 279(8), 280(8), 281, 283(24), 287(8), 289(36), 293, 298(24, 25), 299
Seth, A., 331
Settiawan, N. B., 89
Settleman, J., 131, 144, 279
Settles, A. M., 80(32), 83
Sewing, A., 259, 414
Shabanowitz, J., 344, 368
Shahinian, S., 66
Shalon, D., 255
Shannon, K., 103
Shao, H., 10, 197, 198(15), 200(15, 16), 202(15), 205, 302
Shapiro, P. S., 319, 326(3), 333(3), 337, 343, 387, 388(1), 426, 427, 428(15), 429(15), 430
Sharrocks, A. D., 330, 334, 336(35), 386, 435
Shaw, P. E., 430
Shaywitz, D. A., 299
Shen, Q., 5
Sherman, F., 271
Sherr, C., 234
Shevchenko, A., 301, 312
Shi, W., 258
Shibasaki, Y., 36, 47(5)
Shibuya, A., 211
Shields, J. M., 221
Shiels, H., 445
Shih, T. Y., 50
Shilo, B. Z., 344, 345(9)
Shim, H., 222
Shinkai, K., 259
Shirasawa, S., 5, 124, 232, 245
Shirouzu, M., 159
Shivakumar, L., 276
Shoelson, S. E., 171, 184, 185, 185(7), 186, 187(6), 189, 189(6), 191(6, 7), 193(17), 194
Shore, P., 386
Silletti, J., 10, 51, 56(6), 59(6), 61(6), 62(6), 69
Silvius, J. R., 66
Simmon, J., 88
Simon, J. P., 262
Simpson, A. W., 80(23), 82
Singer-Kruger, B., 80
Sisodia, S. S., 114
Sithanandam, G., 377
Slebos, R. J., 363
Slepnev, V., 89(43, 44), 102
Slivka, S. R., 105, 109(6)
Slunt, H., 114
Sluss, H. K., 321, 433
Smeal, T., 321, 324(13), 330(13), 331(13), 432, 433(2, 8), 441(8)
Smith, D. B., 100
Smith, J. A., 292
Smith, L. M., 434
Smith, S. B., 314, 323
Smith-McCune, K., 415
Snyder, E. L., 37, 38(15), 39(15), 41(15), 49(15)
Soderling, T. R., 174
Soga, N., 38, 47(20), 49(20)

Solomon, D. L., 405
Solski, P. A., 11, 12(25), 13, 14(29), 15(29), 26, 30(27), 65, 68, 69, 69(24), 73(24, 26), 80(15), 81, 82(15), 119, 123(12), 313, 314(41), 371, 384(24)
Soltoff, S., 184
Sonenberg, N., 254, 258(7)
Song, O., 86, 141, 260, 277
Songyang, Z., 171, 174, 183, 184, 185, 185(7), 186, 187, 187(6), 188, 189(6), 191(6, 7), 193(18), 194, 194(22)
Sonoda, G., 279
Sorenson, M., 336
Soriano, P., 89(42), 102
Sorisky, A., 4
Sorkin, A., 128
Southern, J. A., 4
Spaargaren, M., 128, 133, 159, 161(19), 279
Spain, B. H., 143
Sparks, A. B., 92, 99
Spiegel, A. M., 66
Spiegelman, B., 234, 258
St. Jean, A., 294
St. Jules, R. S., 314
Stacy, D. W., 354
Stahl, P. D., 65
Stam, J., 363
Stambrook, P. J., 5
Stanbridge, E. J., 181, 232
Stanton, V. P., Jr., 402
Starr, L., 96
Staudt, L. M., 255
Stauffer, T. P., 100
Steen, H., 430
Stein, B., 433
Steitz, S. A., 105, 109(6)
Stenglenz, R., 370
Stennicke, H. R., 181
Stephens, R. M., 184, 193(12)
Sternglanz, R., 142, 153, 277
Stewart, F., 414
Stewart, S., 369
Stickney, J. T., 64, 69
Stinchcombe, T. J., 96
Stines Nahreini, T., 417
Stippec, S. A., 334
Stock, J. B., 107, 108, 112(10)
Stokoe, D., 65, 106, 354
Stork, P. J., 384
Storm, S. M., 140
Stowers, L., 276
Stradley, D. A., 343, 366, 419, 421(7), 430(7)
Struhl, K., 292
Stuab, A., 405
Sturgill, T. W., 344, 368, 379, 381(27)
Styles, C. A., 369
Su, B., 320, 334, 334(9), 432, 433, 433(1)
Su, I. J., 5
Su, M. S.-S., 333, 334, 334(26), 336(35), 435
Su, W., 184, 193(11)
Subramani, S., 81
Suders, J., 199
Sudol, M., 88
Sugden, P. H., 435
Sullivan, K. F., 50, 59(5), 62(5)
Summers, S. A., 417
Sumortin, J., 5, 245
Sun, L., 433
Sun, P., 166, 211
Sundaram, M., 369
Sung, A., 140
Sung, Y.-S., 273
Sutherland, R. L., 354
Swanson, S., 38, 47(20), 49(20)
Sweatt, J. D., 388
Sydor, J. R., 159, 162(20), 163(20)
Symons, M., 65, 103, 106, 140, 280, 354
Syto, R., 115, 124

T

Tabak, H. F., 78, 80(4), 81(4)
Tai, J. H., 210
Takahashi, N., 36, 47(5)
Takai, Y., 133, 204
Takatsu, K., 445
Takei, K., 89(43), 102
Takeuchi-Suzuki, E., 131, 144, 279
Tam, A., 5, 109, 114
Tamada, M., 131
Tan, E. W., 112
Tang, C., 433
Taniguchi, N., 445
Tao, X., 181
Tapon, N., 276
Tassi, E., 96
Tate, S. E., 302
Tavitian, A., 128, 133, 279
Taylor, B., 103

Taylor, J. A., 100
Taylor, S. S., 433
Tecle, H., 343, 419
Telmer, C. A., 117
Templeton, D. J., 334, 433
Tempst, P., 307
Tenca, P., 432
Ten Harmsel, A., 26
Teramoto, H., 319
Terrian, D. M., 83
Testa, J. R., 279
Teti, A., 36, 47(4)
Therrien, M., 369
Thomas, D. Y., 80
Thomas, G. P., 206
Thomas, J. W., 279
Thompson, A. D., 222
Thompson, C. B., 445, 448
Thorn, J. M., 99
Thorner, J., 277
Tibbles, L. A., 319, 326(2)
Toby, G. G., 77, 85
Tocci, M. J., 431
Tognon, C., 4, 35(5)
Tolwinski, N. S., 343, 417
Tomerup, N., 204, 205(8)
Tong, L., 435
Tonks, N. K., 344
Tora, L., 405
Torrisi, M. R., 80
Touchman, J. W., 279
Tournier, C., 301, 333, 335, 369, 433
Tournier-Lasserve, E., 279
Touse, J., 435
Toussaint, L. G. III, 300, 301, 302(12)
Toyooka, T., 312
Toyoshima, F., 334, 433
Travis, J., 262
Treisman, R., 221, 245, 276
Trent, J. M., 255
Trienies, I., 401, 413(6)
Trigo-Gonzalez, G., 4, 35(5)
Tronche, F., 414
Trub, T., 186, 193(17), 194
Trzaskos, J. M., 343, 366, 388, 419, 421(7), 430(7)
Tsai, L. H., 174
Tsichlis, P., 86
Tsigelny, I., 433
Tsukamoto, T., 211
Tu, Y., 84
Turck, C. W., 88, 184
Turecek, F., 314
Tzivion, G., 412

U

Uberall, F., 65
Udagawa, N., 36, 47(5)
Ulevitch, R. J., 319, 320, 333, 433
Ülkü, A. S., 3
Ullrich, A., 189
Ullrich, O., 204
Unlu, M., 313
Urano, T., 128
Urena, J., 189

V

Vale, T., 271, 276(8)
Valencia, A., 78, 203, 210(4)
Valge-Archer, V. E., 443
Van Aelst, L., 4, 78, 131, 151, 152, 153, 154, 154(10), 155(10), 166, 168(24), 245, 271, 274(7), 276
Van Becelaere, K., 419
VandenBerg, S. R., 344, 350(10)
van den Ijssel, P., 336
Vanderbilt, C. A., 420
Vander Heyden, N., 38, 49(18)
van de Rijn, M., 255
Vanderschueren, R. G., 363
Van Dyk, D. E., 343, 366, 419, 421(7), 430(7)
Van Dyk, L., 38, 49(17)
Vanhaesebroeck, B., 144
van Seventer, G. A., 445
van't Hof, W., 66, 81, 314
van Zandwijk, N., 363
Vavvas, D., 271, 274(9), 412
Veer, L. J. V., 128
Vertommen, D., 181
Vicente, O., 210
Viciana, P. R., 131
Vidal, M., 153
Vigo, E., 405
Vik, T. A., 427
Vila, J., 344

Vincent, S. J., 187, 194(22)
Virta, H., 204
Vito, M., 128, 138(1)
Vives, E., 37
Vocero-Akbani, A., 36, 37, 38, 38(15), 39(15), 41(15), 46(21), 49(15–19, 21)
Vogelstein, B., 234
Vogt, P. K., 86
Vojtek, A. B., 139, 145, 153, 154(7), 271, 274, 274(3), 277, 301
Volker, C., 107, 108, 112(10)
Volkmer-Engert, R., 102
von Willebrand, M., 140
Vorm, O., 312
Vries-Smith, A. M. M., 128

W

Wachowicz, M. S., 417
Wade, W. S., 194
Waechter, C. J., 196, 199, 303
Wagenaar, S. S., 363
Wager, C., 102
Wagner, D., 102
Wagner, E. F., 443
Waksman, G., 38, 49(19)
Walberg, M. W., 260
Walker, E. H., 159
Wallach, B., 417
Walter, G., 357
Walton, K. M., 341
Wang, D., 38, 47(20), 49(20)
Wang, H. Y., 66
Wang, L., 115, 124
Wang, X., 330
Wang, Y., 139, 333, 334, 334(25)
Wani, M. A., 5
Ward, W. W., 51, 61(7), 84
Warne, P. H., 131, 144, 276
Wartmann, M., 301
Waterfield, M. D., 144, 276
Watson, M. A., 299
Watson, S. P., 336
Webb, E. F., 336
Webb, R. L., 434
Weber, M. J., 343, 344, 348(8), 350(10), 368, 369, 370, 370(14), 371(20), 373(14, 20), 379, 381(27), 384(14, 20), 401, 407(4), 409(4), 423, 430
Weberand, M. J., 234
Wedegaertner, P. B., 66
Wehland, J., 89(42), 102
Wei, M., 38, 49(16)
Wei, N., 370
Weinberg, R. A., 128, 138(1)
Weiner, M. P., 299
Weintraub, H., 142, 414
Weiss, E., 89
Weiss, L., 334
Weiss, M., 186
Weissman, B. E., 11, 12(21)
Weitzel, C., 432
Welford, S. M., 222
Wels, W., 89
Weng, Z., 100
West, R. W., 262
Westwick, J. K., 13, 234, 300, 432, 434, 441(42), 442(42)
Whalen, A. M., 427, 428(15), 429(15)
Whelan, J., 434
White, B. A., 22
White, D. M., 435
White, M. A., 4, 131, 140, 141(6), 143(6), 152, 189, 245, 270, 271, 273, 274(7), 276, 276(8), 334
White, M. F., 186, 193(17)
Whitehead, I., 4, 35(5), 211, 212(2), 218(2)
Whitmarsh, A. J., 301, 319, 321, 330, 333, 333(19), 334, 334(19), 335, 336(35), 368, 369, 386, 433, 434, 435, 442, 443
Whyte, D. B., 115, 122(6), 123(6), 127(6)
Wickmer, S., 37, 49(13)
Widmann, C., 332
Wigler, M. H., 4, 131, 139, 152, 154, 223, 245, 271, 274(7), 276, 276(8), 368
Wiland, A., 343, 419
Wiley, S., 187, 194(22)
Wilkins, M. R., 308
Williams, C. F., 255
Williams, H., 431
Williams, K. L., 308
Williams, L. T., 128, 131, 131(10), 132(15), 133(10), 140, 184, 185, 369, 417
Williams, M., 4, 7(8)
Williamson, M., 88
Willingham, M. C., 50
Willumsen, B. M., 65, 87, 133, 234
Wilm, M., 301, 312
Wilsbacher, J., 387, 389

Wilson, C., 210
Wilson, K. P., 435
Wilson, O., 115, 123
Wilson, R., 401, 413(6)
Windsor, W. T., 115, 124
Winge, P., 128
Winkler, D., 102
Winter, C., 442
Wirth, P. J., 311
Wisdom, R., 234, 258
Wiseman, B., 259, 414
Witte, O., 140, 141(6), 143(6)
Wittinghofer, A., 128, 133, 139, 159, 160, 161(16, 19, 22, 23), 162(16, 20), 163, 163(20), 203, 210(4), 279
Wojnowski, L., 140
Wolf, G., 186, 193(17)
Wolfgang, C. D., 434
Wolfman, A., 234
Wolthuis, R. M. F., 128, 274
Wolven, A., 314
Wong, A. M., 369
Wong, D., 140, 141(6), 143(6)
Wong, M., 5
Wong, M. J., 258
Wong, S., 140
Wood, C., 274
Woodgett, J. R., 319, 326(2), 432, 433, 433(2)
Woods, D., 259, 412, 413, 413(20), 415(20), 417(20)
Woods, R. A., 294
Wu, C. S., 222
Wu, I.-H., 320, 333, 334(9), 432, 433, 433(1)
Wu, J., 185, 379, 381(27), 433
Wu, L. C., 262
Wu, M. F., 280
Wu, P., 341, 389
Wu, Z., 433, 443
Wyckoff, H. W., 89(37), 102
Wysk, M., 443

X

Xia, Z., 335, 369
Xiang, J., 334
Xie, L. Y., 166, 211
Xie, W., 140, 141(6), 143(6)
Xu, C. W., 277
Xu, N., 319
Xu, S., 341, 389, 419
Xu, W., 443
Xu, X., 5

Y

Yachie, A., 445
Yagisawa, H., 82
Yajnik, V., 185
Yamabhai, M., 88, 89(45), 94, 96, 97(21, 24, 25), 102
Yamada, C., 128
Yamaguchi, A., 128
Yamamoto, D., 152
Yamamoto, K. R., 401
Yamamoto, T., 152
Yamasaki, K., 36, 47(5), 159
Yamashita, T., 321
Yan, M., 433
Yancopoulos, G. D., 5, 344
Yang, B. S., 409
Yang, D. D., 334, 434, 443
Yang, H.-H., 435
Yang, S.-H., 330, 334, 336(35)
Yang, Y., 277
Yao, T. P., 245
Yashar, B. M., 433
Yasuda, J., 301, 335, 369
Ye, Z.-H., 128, 131(10), 133(10)
Yee, K., 433
Yeh, J., 431
Yeh, S. H., 277
Yelton, D., 96
Yen, A., 4, 7(8)
Yen, T. J., 344, 348(8)
Yi, T., 184, 185(7), 191(7)
Yin, Y., 434
Yokoe, H., 10
Yokoyama, N., 124, 159, 232, 245
Yoshida, H., 140
Yoshikawa, S., 128
Yoshimata, K., 115
Yoshioka, K., 259
Young, D. F., 4
Young, P. R., 435
Young, S. G., 103
Yu, H., 88
Yu, J.-C., 319
Yu, W., 369

Yu, X., 262
Yuan, P. M., 312
Yuan, Q., 89

Z

Zaal, K. J. M., 64
Zabramski, J. M., 279
Zack, D. J., 10
Zambonin-Zallone, A., 36, 47(4)
Zambryski, P. C., 277
Zandi, E., 262
Zapton, T., 333, 334(26)
Zarsky, V., 210
Zeceure, M., 423
Zecevic, M., 343, 344, 348(8), 350(10)
Zerial, M., 133, 203, 205(2)
Zhang, A., 272
Zhang, D., 36, 47(5)
Zhang, F. L., 67, 115, 195
Zhang, H., 233, 240
Zhang, R., 115, 233, 240
Zhang, S., 319
Zhang, X. F., 131, 144, 271, 274(9), 279
Zhang, Y., 369
Zhang, Z., 194
Zhen, E., 420
Zheng, C., 334
Zheng, T. S., 434
Zhengui, X., 341
Zhon, M., 189
Zhou, H., 5, 86
Zhou, M. M., 194
Zhou, Y., 272
Zhou, Z., 369
Zhu, J., 413
Zhu, L., 86
Zhu, S. X., 255
Zhu, W., 233
Zlatkine, P., 75
Zohn, I. M., 78
Zoller, M. J., 100
Zon, L. I., 433
Zu, K., 344
Zuckermann, R. N., 88
Zwahlen, C., 187, 194(22)
Zwickl, M., 89

Subject Index

A

AF-6
ALL1 fusion, 152
domains, 152
Ras/Rap1 interaction assays
Ras-binding domain interaction studies
fluorescence titration, 159–160
guanine nucleotide dissociation inhibition assay, 160–162
rationale, 158–159
stopped-flow assay of intrinsic fluorescence changes, 161–163
retroviral system for stable transfection
advantages, 165
effects on cell function, 168
media and solutions, 166
packaging cell line, 166
rationale, 163, 165
transfection and infection, 166–168
yeast two-hybrid system
filter assay, 158
histidine prototrophy assay, 157
liquid assay for β-galactosidase, 155–157
media, 153
plasmids and constructs, 154–155
principle, 152–153
solutions, 154
transformation, 155–156
yeast strain, 153
Akt, substrate prediction using degenerate peptide libraries, 180–182
Alkaline phosphatase fusion proteins
applications in protein interactions, 101–102
generation
cell growth and induction, 91
secretion, 91
transformation, 91
troubleshooting, 91–92
vector, 90–91, 101
SH3 domain–ligand interaction assays
binding strength assay, 92–94
membrane assay format, 94–96
microtiter plate assay format, 100
plastic pin assay format, 96–97, 99–100
specificity of binding, 92–94
structure of bacterial protein, 88–90

B

Bcr/Abl, RIN1 interactions, 140

C

Calcium phosphate, Ras vector transfection, 28–29
Cassette mutagenesis, modification of multiple cloning sites in Ras vectors
cassette generation and ligation, 25
posttranslational modification site introduction, 26
primer design, 25
rationale, 24–25
CDC42, cytoskeleton regulation, 36
Chloroplast, targeting sequences, 83
Confocal microscopy
Ras–green fluorescent protein fusion protein, 62, 64
Tat–ATPase fusion protein transduction analysis, 46
Cytoskeleton, targeting sequences, 83

D

Differential display
applications, 234–235
Ras-regulated genes
advantages, 244
confirmation of differentially expressed complementary DNAs, 240, 243

denaturing polyacrylamide gel electrophoresis of DNA, 239
design considerations, 235–236
DNase I treatment of RNA, 237
expression kinetics of target genes, 243
materials, 236–237
mitogen-activated protein kinase inhibitor studies, 244
polymerase chain reaction, 238–239
reamplification of complementary DNA bands, 239–240
reproducibility, 243
reverse transcription, 237–238
DNA microarray analysis, K-Ras-induced genes
applications, 259–260
BioMind database and data analysis, 255–256
complementary DNA microarray hybridization and data normalization, 254–255
ecdysone receptor-expressing cells
characterization, 247–249
generation, 246–247
ecdysone-inducible K-Ras plasmid generation, 249–250
normalization of data, 256–257
poly(A) RNA isolation, 253–254
Ras-inducible cell clone generation, 250–251, 253
Ras knockout cell clones, 245
reverse transcriptase–polymerase chain reaction data comparison, 256, 258
specific gene results, 257–259
Dual bait system, *see* Yeast two-hybrid system

E

ELISA *see* Enzyme-linked immunosorbent assay
Endoplasmic reticulum
protein processing, 77–78
targeting signals
dilysine signal, 80
KDEL retention signal, 78–80
Enzyme-linked immunosorbent assay, interleukin-2 and Jun N-terminal kinase inhibitor analysis, 444–445
Epitope tag, vectors, 9–10
ER, *see* Endoplasmic reticulum; Estrogen receptor
ERK, *see* Extracellular signal-regulated kinase
Estrogen receptor, hormone-binding domain fusion proteins for Raf–MEK–MAPK pathway conditional activation
advantages, 401–402
cell culture considerations, 415–416
controls, 415
detection of fusion proteins, 416
fusion terminus selection, 414–415
ΔMEK1:ER construction, 413
principle, 401
prospects, 416–417
Raf-1 regulation by hormone-binding domain of estrogen receptor, 412
ΔRaf-1:ER
conditional transformation, 406–407
cycloheximide independence of activation, 407, 409
design and construction of conditionally active forms, 402, 405–406
green fluorescent protein fusion protein, 412–413
inactivation following hormone withdrawal, 409–410
kinase assay, 407
mechanism of activation, 410
ΔA-Raf conditional transformation, 410, 412
ΔB-Raf conditional transformation, 410, 412
retroviral vectors, 416
Expression cloning
farnesylated proteins
colony protein filter preparation, 198–199
colony purification, 200–202
library plating and screening, 197–198
prenyl labeling of filters, 199–200
prospects, 202
small GTPase identification
colony protein filter preparation, 205–207
colony purification, 207–210
GTP overlay probing, 207
library plating and screening, 204–205
principle, 204

prospects, 210
Ras-related proteins in retina and mouse embryo, 209–210
Extracellular signal-regulated kinase
activation assay using phosphorylation-specific antibodies
antibody production
antigen selection, 338–339
characterization of specificity, 345–347
peptide immunogen conjugation and immunization, 339, 345
purification, 339–340, 345
applications, 344, 353
immunofluorescence
fixed cells, 347–348
subcellular localization, 348–349
PC12 cell extract studies, 341–343
tissue immunostaining
controls, 351–352
imunohistochemistry, 350–352
prostate tissue preparation, 349–350
scoring, 352
Western blotting, 340–341
kinase kinases, *see* Mitogen-activated protein kinase/extracellular signal-regulated kinase kinase; MKK1/2
purification of activated kinase from recombinant *Escherichia coli*
affinity tagging, 390–391
anion-exchange chromatography, 396–397
cell lysis, 395
ERK1 expression with kinases on two plasmids, 393–394
gel filtration, 397
histidine-tagged ERK2 expression with kinases on single plasmid, 392
kinase assay, 394–395, 398
nickel affinity chromatography, 395–396
single-plasmid versus two-plasmid approach, 390
solutions, 399–400
upstream kinase coexpression and plasmids, 389–390
Western blot analysis, 397–398
reconstitution of Raf–MEK–MAPK pathway
advantages and limitations of system, 383–384
ERK purification, 379
MEK–glutathione *S*-transferase fusion protein purification from recombinant *Escherichia coli*
affinity chromatography, 378
cell growth, induction, and lysis, 377–378
storage and yield, 378–379
MEK phosphorylation conditions, 380–381
MP1 effects
ERK activation, 381, 383
MEK activation, 381
MP1 purification of histidine-tagged protein, 379–380
Raf purification from baculovirus–insect cell system, 379
signaling cascade, 337, 368
specificity of signaling, 368–369
types, 387

F

Farnesylation, *see* Prenylation
Farnesyltransferase inhibitors
clinical trials, 301
Ras inhibition in cells, 115
TC10 inhibition, 123
Flow cytometry
Tat–ATPase fusion protein transduction analysis, 46
T cell analysis of Jun N-terminal kinase inhibitors
activation of $CD4^+$ cells, 446–448
$CD4^+$ subset isolation, 446
cell cycle analysis, 450–452
immunophenotyping, 449–450
peripheral blood lymphocyte isolation from blood, 445–446
polarizing naive $CD4^+$ cells to Th1 and Th2 subtypes, 449
Fluorescein isothiocyanate, labeling of Tat–ATPase fusion proteins
confocal microscopy of transduction, 46
flow cytometry analysis of transduction, 46
labeling reaction, 46

rationale, 45–46
troubleshooting, 46

G

Geranylgeranylation, *see* Prenylation
GFP, *see* Green fluorescent protein
Golgi, retention signal, 80
Green fluorescent protein
advantages for cell biology studies, 50
ΔRaf-1:ER fusion effects on conditional activation, 412–413
Ras fusion proteins, *see* Ras
Guanine nucleotide dissociation stimulator, *see* RalGDS

H

High-performance liquid chromatography, prenylation analysis of small GTPases, 120, 122–123, 127

I

Immune-complex protein kinase assay
antibody binding to protein G–Sepharose, 324
cell lysis, 323–324
gel electrophoresis, 324
incubation conditions, 324
principle, 320–321
In gel protein kinase assay
autoradiography, 326
cell lysis, 325
denaturation/renaturation of gels, 325–326
electrophoresis, 325
incubation conditions, 326
principle, 321–322
Interleukin-2, Jun N-terminal kinase inhibitor analysis
enzyme-linked immunosorbent assay, 444–445
rationale, 443
reverse transcriptase–polymerase chain reaction, 443–444

J

JNK, *see* Jun N-terminal kinase
Jun N-terminal kinase
activation assay using phosphorylation-specific antibodies
antibody production
antigen selection, 338–339
peptide immunogen conjugation and immunization, 339
purification, 339–340
caveats, 326–327
immunocytochemistry, 329–330
materials, 327–328
overview, 326
PC12 cell extract studies, 341–343
Western blotting, 328–329, 340–341
activity assays
immune-complex protein kinase assay
antibody binding to protein G–Sepharose, 324
cell lysis, 323–324
gel electrophoresis, 324
incubation conditions, 324
principle, 320–321
in gel assay
autoradiography, 326
cell lysis, 325
denaturation/renaturation of gels, 325–326
electrophoresis, 325
incubation conditions, 326
principle, 321–322
materials, 322–323
solid-phase protein kinase assay
cell lysis, 325
gel electrophoresis, 325
incubation conditions, 325
principle, 321
substrate binding to glutathione–Sepharose, 324–325
AP-1 transactivation, 432
ATP affinity for types, 436–437
inhibitors
direct assay in cells
capture assays, 441–442
Western blot analysis, 442–443
enzymatic analysis
double-reciprocal analysis, 436–438

kinase assay, 434–436
kinase specificity determination, 440
substrate protection studies, 438
time-dependence of inhibition, 439–440
washout experiments for reversibility of inhibition, 440
interleukin-2 assays
enzyme-linked immunosorbent assay, 444–445
rationale, 443
reverse transcriptase–polymerase chain reaction, 443–444
JNK-interacting protein-1, 335
SB203580, 437–438
T cell flow cytometry analysis
activation of CD4$^+$ cells, 446–448
CD4$^+$ subset isolation, 446
cell cycle analysis, 450–452
immunophenotyping, 449–450
peripheral blood lymphocyte isolation from blood, 445–446
polarizing naive CD4$^+$ cells to Th1 and Th2 subtypes, 449
kinase kinases in activation, 332–333, 432–433
mutant alleles of kinases in activation studies
activated mutant alleles, 333–334
dominant-interfering mutant alleles, 334–335
pathogenesis role, 433–434
reporter gene assays
β-galactosidase assay, 332
luciferase assay, 332
materials, 331
principles, 330–331
signaling cascade, 319, 337, 432–433
stress activation, 387–388

L

Lipidation signals, *see* Myristoylation; Palmitoylation; Prenylation
Lipofection, Ras vector transfection, 29–30

M

MAPK, *see* Mitogen-activated protein kinase
Mass spectrometry, proteomic analysis of Ras signaling
databases, 316
isotope-coded affinity tags, 314
peptide mass fingerprinting, 314–316
MEK, *see* Mitogen-activated protein kinase/extracellular signal-regulated kinase kinase
Microarray, *see* DNA microarray analysis, K-Ras-induced genes
Mitochondria
targeting sequences, 82
targeting vectors, 86
Mitogen-activated protein kinase, *see also* Extracellular signal-regulated kinase; Jun N-terminal kinase; p38 mitogen-activated protein kinase
mutant alleles of kinases in activation studies
activated mutant alleles, 333–334
dominant-interfering mutant alleles, 334–335
dominant negative MEK mutant studies, *see* Raf–MEK–MAPK pathway
Raf activation, *see* Raf–MEK–MAPK pathway
signaling cascade, 319, 332–333, 337, 388
Mitogen-activated protein kinase/extracellular signal-regulated kinase kinase, *see also* MKK1/2
activation of mitogen-activated protein kinase, 319, 332–333, 337
dominant negative mutant studies, *see* Raf–MEK–MAPK pathway
dual specificity, 388–389
kinases, 388
purification of activated kinase from recombinant *Escherichia coli*
glutathione *S*-transferase–MEK2 fusion protein expression and purification, 398–399
single-plasmid versus two-plasmid approach, 390
solutions, 399–400
upstream kinase coexpression and plasmids, 389–390
reconstitution of Raf–MEK–MAPK pathway

advantages and limitations of system, 383–384
ERK purification, 379
MEK–glutathione *S*-transferase fusion protein purification from recombinant *Escherichia coli*
affinity chromatography, 378
cell growth, induction, and lysis, 377–378
storage and yield, 378–379
MEK phosphorylation conditions, 380–381
MP1 effects
ERK activation, 381, 383
MEK activation, 381
MP1 purification of histidine-tagged protein, 379–380
Raf purification from baculovirus–insect cell system, 379
Mitogen-activated protein kinase/extracellular signal-regulated kinase kinase-partner 1
protein–protein interactions
coimmunoprecipitation with MEK1 or ERK1
advantages and limitations of assay, 376–377
extraction, 371, 374
normalization of protein levels, 375–376
precipitation, 371, 374
serum-starved cultured cell studies, 373–376
transfection, 370–371, 373–374
Western blot analysis, 371, 373, 375
specificity, 369–370
purification of histidine-tagged protein, 379–380
Raf–MEK–MAPK reconstituted pathway effects
ERK activation, 381, 383
MEK activation, 381
reporter assay for overexpression effects on extracellular signal-regulated kinase signaling
advantages and limitations of assay, 386
enhancement of signaling, 385–386
luciferase assay, 384
overview, 384
transfection, 384
signaling specificity role of scaffolding proteins, 368–370
MKK1/2
activity assays
ERK2 substrate preparation, 420
inhibitor effects, 421
MEKK-C activation, 420–421
inhibitors
assays
MKK1 activity inhibition, 421
phosphorylation assays, 422–423
phospho-specific antibodies, 423–424
commercial sources, 419
kinase specificity evaluation, 430–431
PD98059
ERK suppression in response to serum stimulated cells, 424, 426
mechanism of action, 429–430
megakaryocyte differentiation inhibition, 427–429
stability, 426
structure, 418
PD184352 structure, 418
U0126
ERK suppression in response to serum stimulated cells, 424, 426
mechanism of action, 429–430
MKK1 inhibiton following transient transfection, 427
structure, 418
kinases, 417–418
stimulators, 417
substrate specificity, 418
MP1, *see* Mitogen-activated protein kinase/extracellular signal-regulated kinase kinase-partner 1
Myristoylation, lipidation signal introduction
Akt fusions, 86
functional activation, 65
membrane binding effects, 65–66, 81
signals, 66–67, 81
targeting vectors, 84–86

N

NLS, *see* Nuclear localization sequence
Nuclear localization sequence

applications, 87
targeting vectors, 84–86
types, 82
vector design for protein targeting, 84

O

Oncogene, *see also* Ras
abundance, 233
retroviral complementary DNA library screening
advantages, 211
complementary DNA
ligation, 215
linkering, 214–215
preparation, 213–213
size fractionation, 215
plasmid library
conversion to viral libraries, 217–218
generation, 215–216
recovery from soft agar, 216–217
poly(A) RNA isolation, 213
screening
cloning of amplification products, 220
infection, 218
polymerase chain reaction amplification of proviral complementary DNA inserts, 218–220
troubleshooting, 221
vector construction, 212–213

P

p38 mitogen-activated protein kinase
activation assay using phosphorylation-specific antibodies
antibody production
antigen selection, 338–339
peptide immunogen conjugation and immunization, 339
purification, 339–340
caveats, 326–327
immunocytochemistry, 329–330
materials, 327–328
overview, 326
PC12 cell extract studies, 341–343
Western blotting, 328–329, 340–341
activity assays
immune-complex protein kinase assay
antibody binding to protein G–Sepharose, 324
cell lysis, 323–324
gel electrophoresis, 324
incubation conditions, 324
principle, 320–321
in gel assay
autoradiography, 326
cell lysis, 325
denaturation/renaturation of gels, 325–326
electrophoresis, 325
incubation conditions, 326
principle, 321–322
materials, 322–323
solid-phase protein kinase assay
cell lysis, 325
gel electrophoresis, 325
incubation conditions, 325
principle, 321
substrate binding to glutathione–Sepharose, 324–325
inhibitors
Ras-regulated gene differential display studies, 244
SB203580, 335–336
kinase kinases in activation, 332–333
mutant alleles of kinases in activation studies
activated mutant alleles, 333–334
dominant-interfering mutant alleles, 334–335
reporter gene assays
β-galactosidase assay, 332
luciferase assay, 332
materials, 331
principles, 330–331
signaling cascade, 319, 337
stress activation, 387–388
types, 388
Palmitoylation
lipidation signal introduction
lack of signal sequence, 68
membrane binding effects, 66
metabolic labeling in verification, 76
polymerase chain reaction constructs, 76

strategies
C-terminal palmitoylation, 75
N-terminal dual acylation, 75
N-terminal palmitoylation, 73, 75
subcellular fractionation combination with hydrophobic partitioning, 76–77
prior lipidation requirements, 68, 73
PCR, *see* Polymerase chain reaction
PD98059, MKK1/2 inhibition
assays
MKK1 activity inhibition, 421
phosphorylation assays, 422–423
phospho-specific antibodies, 423–424
ERK suppression in response to serum stimulated cells, 424, 426
mechanism of action, 429–430
megakaryocyte differentiation inhibition, 427–429
stability, 426
structure, 418
PD184352, structure, 418
PDZ domain, membrane binding, 82
Peptide library
protein kinase screening with degenerate libraries
Akt substrate prediction, 180–182
design of library
degenerate positions, 174–175
leading sequence, 175
orientation, 174
inhibitor development, 183
kinase assays
anion-exchange chromatography of phosphopeptides, 176–177
ferric chelation column chromatography of phosphopeptides, 176–179
incubation conditions, 176
kinase preparation, 176
pool screening, 173
principle, 171–172
substrate specificity
data analysis, 179–180
prediction, 181, 183
synthesis of library, 175
three-step screening, 172–173
SH2 or phosphotyrosine-binding domain, screening for library interactions
data analysis, 189–190
design of library
double-degenerate phosphopeptide library, 185
secondary structure-constrained phosphopeptide library, 185–187
tyrosine-degenerate peptide library, 187
glutathione *S*-transferase–binding domain fusion proteins
affinity purification and sequencing, 188–189
preparation, 188
phosphotyrosine-binding domain binding preferences
nonphosphotyrosine-containing sequences, 193–194
turn–loop structure recognition, 194–195
tyrosine N-terminal sequences, 191, 193
SH2 domain binding preferences for phosphotyrosine C-terminal sequences, 190–191
synthesis of library, 187–188
Peroxisome, targeting sequences, 78, 81
PH domain, *see* Pleckstrin homology domain
Phosphatidylinositol 3-kinase γ, Ras/Rap1 interaction assay with yeast two-hybrid system, 157–158
Phosphotyrosine-binding domain
binding preferences
nonphosphotyrosine-containing sequences, 193–194
turn–loop structure recognition, 194–195
tyrosine N-terminal sequences, 191, 193
peptide library screening for interactions, *see* Peptide library
phosphotyrosine interactions, 185
protein distribution, 185
Pleckstrin homology domain, membrane binding, 82
Podosome, formation induction in osteoclasts by Tat–Rho, 47–49
Polymerase chain reaction, *see also* Differential display; Representational difference analysis
lipidation signal constructs

palmitoylation, 76
prenylation
ligation, 70–71
one-step polymerase chain reaction, 69–70
sequencing, 71
Ras–green fluorescent protein fusion protein construction, 52–54
Ras vectors, generation of new restriction sites
amplification conditions, 22
cloning products into vector, 22–23
coding sequence reading frame alignment with epitope tag, 21–22
confirmation of insert sequence, 23–24
primer design, 21
reverse transcriptase–polymerase chain reaction
interleukin-2, Jun N-terminal kinase inhibitor analysis, 443–444
Ras-regulated gene analysis, 256, 258
Posttranslational modification, *see* Myristoylation; Palmitoylation; Prenylation
Prenylation
assay of small GTPases in COS cells
cell culture, 116
cell lysis, 119
high-performance liquid chromatography
reversed-phase chromatography, 122
standards, 122
TC10 analysis, 120, 122–123
immunoprecipitation, 119
methyl iodide cleavage
acetone extraction, 119–120
cleavage reaction, 120
prenyl alcohol extraction, 120
trypsin digestion, 120
mevalonate labeling, 117–119
overview, 115–116
vectors
Ras, 116
Rho, 117
transfection, 118
assay of small GTPases in human colon tumor cell lines
cell lysis, 125
high-performance liquid chromatography analysis, 127
immunoprecipitation, 125, 127
mevalonate labeling, 125
overview, 116, 123
stable transformant characterization, 124
transfection, 123–124
enzymes, overview, 195–196
expression cloning of farnesylated proteins
colony protein filter preparation, 198–199
colony purification, 200–202
library plating and screening, 197–198
prenyl labeling of filters, 199–200
prospects, 202
function, 195
inhibitors, *see* Farnesyltransferase inhibitors
lipidation signal introduction
functional activation, 65
membrane association verification using subcellular fractionation, 72
membrane binding effects, 65–66
metabolic labeling in verification, 71–72
polymerase chain reaction constructs
ligation, 70–71
one-step polymerase chain reaction, 69–70
sequencing, 71
selection of C-terminal CaaX-containing sequence, 68–69, 81–82
signals, 67–68
mevalonate labeling in protein discovery, 196–197
proteomic analysis of Ras signaling, metabolic labeling with isoprenoids
breast carcinoma cells, 303, 305–306
inhibitor utilization, 302–303
rationale, 302
Ras, 103
substrate specificity of prenylation enzymes, 115
Prenylcysteine carboxymethyltransferase
human enzyme
assay
membrane preparation, 107
scintillation proximity assay, 107
complementary DNA isolation, 105–106
function, 105, 114

immunofluorescence studies of subcellular localization, 106, 109
inhibitors, 107–108
kinetics of postprenylation processing, 109, 111–112, 114
metabolic labeling and membrane partitioning of K-RasD12, 108
Ras membrane localization role, 112, 114
sequence homology analysis, 103–104, 109
STE14 of yeast, 105
synthetic substrates, 106
Promoters
Ras mammalian expression vectors, 4–5
targeted protein vectors, 84
Protein kinase, *see also specific kinases*
classification, 171
number in organisms, 171
peptide library screening for substrates, *see* Peptide library
phosphotyrosine interacting domains, 183–184
PTB, *see* Phosphotyrosine-binding domain

R

Rac, cytoskeleton regulation, 36
Raf–MEK–MAPK pathway
antisense knockdown of Raf-1, 366
carcinogenesis role, 365–367
dominant negative MEK mutant studies
anchorage-independent growth assays, 356, 360–361, 366
cell culture, 355
cell proliferation assays, 355–356, 359–360
endogenous MEK activity reduction, 358–359
implications, 367
kinase assays, 357–358
mutation, 354–355
plasmids, 355
statistical analysis, 358
transfection, 355
tumorigenicity inhibition *in vivo*
animal model, 356–357
human tumor cell line results, 361–363
metastasis, 362–363
survival as end point, 363–365
vascularization, 366–367
Western blot analysis of pathway members, 357–358
overview, 354, 402
Raf genes, 402
Raf–Ras/Rap1 interaction assay with yeast two-hybrid system, 157–158
reconstitution of extracellular signal-regulated kinase pathway
advantages and limitations of system, 383–384
ERK purification, 379
MEK–glutathione *S*-transferase fusion protein purification from recombinant *Escherichia coli*
affinity chromatography, 378
cell growth, induction, and lysis, 377–378
storage and yield, 378–379
MEK phosphorylation conditions, 380–381
MP1 effects
ERK activation, 381, 383
MEK activation, 381
MP1 purification of histidine-tagged protein, 379–380
Raf purification from baculovirus–insect cell system, 379
steroid receptor hormone-binding domain fusion proteins for conditional activation
advantages, 401–402
cell culture considerations, 415–416
controls, 415
detection of fusion proteins, 416
fusion terminus selection, 414–415
ΔMEK1:ER construction, 413
principle, 401
prospects, 416–417
ΔA-Raf conditional transformation, 410, 412
ΔB-Raf conditional transformation, 410, 412
ΔRaf-1:ER
conditional transformation, 406–407
cycloheximide independence of activation, 407, 409

design and construction of conditionally active forms, 402, 405–406
green fluorescent protein fusion protein, 412–413
inactivation following hormone withdrawal, 409–410
kinase assay, 407
mechanism of activation, 410
Raf-1 regulation by hormone-binding domain of estrogen receptor, 412
retroviral vectors, 416
steroid receptor domain selection, 413–414
Ral
binding proteins, 128
functions, 128
GDP/GTP exchange protein, *see* RalGDS
RalGDS
family members, 128
Ral specificity for GDP/GTP exchange, 127–128
Ras interactions
COS cell studies
extracellular signal-dependent complex formation, 131–132
immunocytochemistry, 132–133
immunoprecipitation, 131
subcellular localization regulation by Rap1, 133
subcellular localization regulation by Ras, 132
transient expression, 130–131
materials for analysis, 129–130
overview, 128–129
Ral activation assays
GDP/GTP exchange of Ral in intact cells, 134–135
Rap1 activation, 138
reconstituted liposome assays, 135–138
in vitro interactions
full-length protein preparation and immunoprecipitation, 133
Ras-interacting domain preparation, 134
Western blot analysis, 130
Ras/Rap1 interaction assay with yeast two-hybrid system, 157–158
Rap1
AF-6 interactions, *see* AF-6
RalGDS interactions, *see* RalGDS
Ras
activation, 354
AF-6 interactions, *see* AF-6
differential analysis of regulated genes, *see* Differential display; DNA microarray analysis, K-Ras-induced genes; Representational difference analysis
effector-selective mutant isolation, *see* Yeast two-hybrid system
expression cloning in identification, *see* Expression cloning
expression vectors, *see* Vector, mammalian expression of Ras
family members, 78–79, 203
green fluorescent protein fusion proteins
expression construct
C-terminal constructs, 56–57
fluorescent mutants, 51–52
ligation, 54–55
polymerase chain reaction primers and cloning, 52–54
restriction digestion, 54
transformation of bacteria, 54, 56
vector, 51–52
fluorescence microscopy
cameras, 63
cell culture substrates, 60
confocal microscopy, 62, 64
filter sets, 62
Golgi, 63–64
light source, 63
objectives, 63
transfection
cell culture, 59–60
cell type selection, 57–59
kinetics of fluorescence appearance, 60–61
stable transformants, 61–62
techniques for transfection, 60
GTP probing of blots in identification, 203–204
mitogen-activated protein kinase activation, *see* Mitogen-activated protein kinase
mutation in cancer, 221–222, 233–234, 353–354

posttranslational modification, *see* Myristoylation; Palmitoylation; Prenylcysteine carboxymethyltransferase; Prenylation; Ras-converting enzyme
proteomics analysis of signaling
affinity purification of Ras-associated proteins
immobilized metal affinity chromatography for phosphopeptide recovery, 307
receptor activity-directed affinity tagging, 306–307
applications, 301–302
mass spectrometry
databases, 316
isotope-coded affinity tags, 314
peptide mass fingerprinting, 314–316
metabolic labeling with isoprenoids
breast carcinoma cells, 303, 305–306
inhibitor utilization, 302–303
rationale, 302
two-dimensional polyacrylamide gel electrophoresis
detection technique sensitivity, 311–312
extraction of proteins, 308–309
first dimension, 310
fluorescence detection, 312–313
immobilized pH gradient strips, 308–310
principle, 308
radioactive labeling and autoradiography, 313–314
reduction and alkylation, 310
second dimension, 310–311
silver staining, 309, 311–312
RalGDS interactions, *see* RalGDS
RIN1 interactions, *see* RIN1
sequence homology within subfamilies, 203
signaling, 139, 221–222, 245, 300–301
subcellular localization, 50, 57–59
trafficking, 50–51
Ras-converting enzyme
human enzyme
assay
membrane preparation, 107
scintillation proximity assay, 107
complementary DNA isolation, 105–106
function, 103, 114
immunofluorescence studies of subcellular localization, 106, 109
inhibitors, 107–108
isoforms, 114
kinetics of postprenylation processing, 109, 111–112, 114
metabolic labeling and membrane partitioning of K-RasD12, 108
Ras membrane localization, 112, 114
sequence homology analysis, 103–104, 109
substrate specificity, 103
synthetic substrates, 106
yeast enzyme, 103
Ras recruitment system
advantages over other systems
Sos recruitment system, 262–263
yeast two-hybrid system, 270
bait test, 265, 267
library screening, 267–268
materials and solutions, 263–264
plasmid isolation from yeast, 268
principle of protein interaction detection, 262
reporter assays
chloramphenicol acetyltransferase, 269
luciferase, 269
transfection of human embryonic kidney cells, 269
transformation of yeast, 264–265
RCE1, *see* Ras-converting enzyme
RDA, *see* Representational difference analysis
Representational difference analysis
principle, 222–224
Ras-regulated genes
biological significance assays, 232
cell culture, 225–226
cloning
downregulated genes, 225
upregulated genes, 225
comparison of difference products, 229
complementary DNA generation, 226
differential expression verification, 231
final difference product cloning, 230
first difference product generation, 228–229
identification of genes, 230–231
messenger RNA isolation, 226

primers, 226–227
screening of cloned genes, 231–232
second difference product generation, 229
subtractive hybridization, 228
tester and driver complementary DNA preparation, 227–228
third difference product generation, 229
Retroviral vector
oncogenes, complementary DNA library screening
advantages, 211
complementary DNA
ligation, 215
linkering, 214–215
preparation, 213–213
size fractionation, 215
plasmid library
conversion to viral libraries, 217–218
generation, 215–216
recovery from soft agar, 216–217
poly(A) RNA isolation, 213
screening
cloning of amplification products, 220
infection, 218
polymerase chain reaction amplification of proviral complementary DNA inserts, 218–220
troubleshooting, 221
vector construction, 212–213
Ras expression
advantages, 31–32
infection, 33–34
packaging lines, 32
safety, 32–33
titering, 34–35
types, 7, 9, 32
virus production, 33
targeted proteins, 84
Rho
cytoskeleton regulation, 36
prenylation, *see* Prenylation
transduction using Tat fusion protein delivery
fluorescein isothiocyanate labeling of fusion proteins
confocal microscopy of transduction, 46
flow cytometry analysis of transduction, 46
labeling reaction, 46
rationale, 45–46
troubleshooting, 46
materials, 38–39
mechanism, 37, 49
podosome formation induction in osteoclasts by Tat–Rho, 47–49
principle, 36–38
purification of fusion protein
cell growth and induction, 39, 41
nickel affinity chromatography, 41–42
troubleshooting, 42
vector, 39
size limitations for fusion, 49
solubilization of fusion protein into aqueous buffer
desalting column, 45
fast protein liquid chromatography, 42–43
gravity columns and batch preparations, 43–44
ion-exchange chromatography, 42–44
troubleshooting, 43, 45
Western blot analysis, 39, 41
RIN1
Bcr/Abl interactions, 140
14-3-3 interactions, 140, 143–144
Ras interactions
Ras-binding domain, 139–140
in vitro binding assays
baculovirus expression systems, 146–148
buffers, 144–145
coimmunoprecipitation from mammalian cells, 151
Ras-binding domain pull down of Ras from cell extracts, 149–150
RIN1–Ras-binding domain interactions, 146–149
tagging of proteins, 144
in vitro translation of RIN1 and pull down assay, 145–146
yeast two-hybrid analysis
plasmids, 141–142
signal detection, 142–143
yeast strains, 142
splicing variants, 140–141

tissue expression pattern, 140
RRS, *see* Ras recruitment system

S

SB203580, mitogen-activated protein kinase inhibition, 335–336, 437–438
SH2 domain
  peptide library screening for interactions, *see* Peptide library
  phosphotyrosine interactions, 184
  preferences for phosphotyrosine C-terminal sequences, 190–191
SH3 domain
  molecular recognition property determination, *see* Alkaline phosphatase fusion proteins
  protein interactions, 88
Solid-phase protein kinase assay
  cell lysis, 325
  gel electrophoresis, 325
  incubation conditions, 325
  principle, 321
  substrate binding to glutathione–Sepharose, 324–325
Sos recruitment system
  advantages over conventional two-hybrid system, 270
  bait test, 265, 267
  Cdc25 mutant utilization, 261
  library screening, 267–268
  materials and solutions, 263–264
  plasmid isolation from yeast, 268
  principle of protein interaction detection, 261–262
  reporter assays
    chloramphenicol acetyltransferase, 269
    luciferase, 269
  transfection of human embryonic kidney cells, 269
  transformation of yeast, 264–265
SRS, *see* Sos recruitment system
STE14, *see* Prenylcysteine carboxymethyltransferase
Stopped-flow fluorescence, Ras/Rap1–AF-6 interaction assay, 161–163
Stress-activated protein kinases, *see* Jun N-terminal kinase; p38 mitogen-activated protein kinase

T

Tat–ATPase fusion protein transduction
  fluorescein isothiocyanate labeling of fusion proteins
    confocal microscopy of transduction, 46
    flow cytometry analysis of transduction, 46
    labeling reaction, 46
    rationale, 45–46
    troubleshooting, 46
  materials, 38–39
  mechanism, 37, 49
  podosome formation induction in osteoclasts by Tat–Rho, 47–49
  principle, 36–38
  purification of fusion protein
    cell growth and induction, 39, 41
    nickel affinity chromatography, 41–42
    troubleshooting, 42
    vector, 39
  size limitations for fusion, 49
  solubilization of fusion protein into aqueous buffer
    desalting column, 45
    fast protein liquid chromatography, 42–43
    gravity columns and batch preparations, 43–44
    ion-exchange chromatography, 42–44
    troubleshooting, 43, 45
  Western blot analysis, 39, 41
T cell, flow cytometry analysis of Jun N-terminal kinase inhibitors
  activation of $CD4^+$ cells, 446–448
  $CD4^+$ subset isolation, 446
  cell cycle analysis, 450–452
  immunophenotyping, 449–450
  peripheral blood lymphocyte isolation from blood, 445–446
  polarizing naive $CD4^+$ cells to Th1 and Th2 subtypes, 449
Transfection, Ras vectors
  calcium phosphate precipitation, 28–29
  lipofection, 29–30
  overview of techniques, 27–28
  stable transfection, 30–31
  technique effect on biological activity, 30
  transient transfection, 30

Two-dimensional polyacrylamide gel electrophoresis, proteomic analysis of Ras signaling
detection technique sensitivity, 311–312
extraction of proteins, 308–309
first dimension, 310
fluorescence detection, 312–313
immobilized pH gradient strips, 308–310
principle, 308
radioactive labeling and autoradiography, 313–314
reduction and alkylation, 310
second dimension, 310–311
silver staining, 309, 311–312

U

U0126, MKK1/2 inhibition
assays
MKK1 activity inhibition, 421
phosphorylation assays, 422–423
phospho-specific antibodies, 423–424
ERK suppression in response to serum stimulated cells, 424, 426
mechanism of action, 429–430
MKK1 inhibiton following transient transfection, 427
structure, 418

V

Vector, mammalian expression of Ras
cassette mutagenesis modification of multiple cloning sites
cassette generation and ligation, 25
posttranslational modification site introduction, 26
primer design, 25
rationale, 24–25
constitutive versus inducible expression, 5
construction
insert orientation determination, 26–27
overview, 18
polymerase chain reaction generation of new restriction sites
amplification conditions, 22
cloning products into vector, 22–23
coding sequence reading frame alignment with epitope tag, 21–22
confirmation of insert sequence, 23–24
primer design, 21
record keeping and sequence maps, 27
shuttle vectors, 24
subcloning
dephosphorylation of digested vector, 20
digested DNA purification, 19–20
ligation, 20–21
vector and insert preparation, 18–19
epitope tagging vectors, 9–10
expression levels of different vectors, 11, 13
functional activity in different vectors, 11, 13
H-Ras mutants expressed in pZIP-NeoSV(X)1, 11–12
promoters, 4–5
properties, 6–7
restriction maps, 8
retroviral vectors
advantages, 31–32
infection, 33–34
packaging lines, 32
safety, 32–33
titering, 34–35
types, 7, 9, 32
virus production, 33
selection factors for assays, 3–4, 17, 35
transfection
calcium phosphate precipitation, 28–29
lipofection, 29–30
overview of techniques, 27–28
stable transfection, 30–31
technique effect on biological activity, 30
transient transfection, 30
transformation assays, 15–17
transient expression signaling assays, 13–15

W

Western blot
Jun N-terminal kinase inhibitor analysis, 442–443

mitogen-activated protein kinase activation assay, 328–329, 340–341, 357–358
RIN1–Ras interaction analysis, 146, 148–149
Tat–ATPase fusion protein, 39, 41

Y

Yeast two-hybrid system
dual bait system for small GTPase signaling protein interactions
activation domain fusion plasmid, 285
applications, 280–281
bait transcriptional activation profile and expression
activation testing of all reporters, 292
bait 2 introduction and colorimetric reporters, 292
chloroform overlay assay for test activation, 292–293
expression assay, 289, 291–292
initial activation assay for bait 1, 289
plating, 287
transformation of yeast, 287
Western blot analysis, 289, 291–293
Cdc42 baits, 280
cI fusion plasmids, 285
cloning into bait vectors, 281, 283
commercial libraries, 294
confirmation of positive interactions
approaches, 297
polymerase chain reaction, 298
sequencing, 299
transformation with amplification product, 298–299
LexA fusion plasmids, 284
physiological validation, 299–300
principles, 277, 279, 281
reporter plasmids, 286
screening for interacting proteins, 295–297
transforming library in bait strain, 294–295
validation, 279–280
yeast strains, 286
limitations, 260–261
Ras effector-selective mutant isolation
interaction strengths with specific targets, 276
library of randomly mutated Ras genes
display in yeast, 272–273
production, 272
overview, 270–271, 276–277
plasmid rescue, 275–276
plasmids, 271
Ras target display, 273–274
replica-plate mating assay, 274–275
yeast strains and media, 271
Ras/Rap1–AF-6 interaction assay
filter assay, 158
histidine prototrophy assay, 157
liquid assay for β-galactosidase, 155–157
media, 153
plasmids and constructs, 154–155
principle, 152–153
solutions, 154
transformation, 155–156
yeast strain, 153
RIN1–Ras interaction analysis
plasmids, 141–142
signal detection, 142–143
yeast strains, 142

ISBN 0-12-182233-8